# HYPERSTATIC STRUCTURES

Volume I

# HYPERSTATIC STRUCTURES

AN INTRODUCTION TO THE THEORY OF
STATICALLY INDETERMINATE STRUCTURES

VOLUME I

J. A. L. MATHESON
M.B.E., M.Sc., Ph.D., M.C.E., M.I.C.E., M.I.Struct.E.
*Vice-Chancellor
Monash University,
Australia*

*With Chapters by*
N. W. MURRAY
B.E., Ph.D.
*and*
R. K. LIVESLEY
M.A., Ph.D.

LONDON
BUTTERWORTHS

THE BUTTERWORTH GROUP

ENGLAND: BUTTERWORTH & CO. (PUBLISHERS) LTD.
LONDON: 88 Kingsway, WC2B 6AB

AUSTRALIA: BUTTERWORTH & CO. (AUSTRALIA) LTD.
SYDNEY: 586 Pacific Highway Chatswood, NSW 2067
MELBOURNE: 343 Little Collins Street, 3000
BRISBANE: 240 Queen Street, 4000

CANADA: BUTTERWORTH & CO. (CANADA) LTD.
TORONTO: 14 Curity Avenue, 374

NEW ZEALAND: BUTTERWORTH & CO. (NEW ZEALAND) LTD.
WELLINGTON: 26-28 Waring Taylor Street, 1
AUCKLAND: 35 High Street, 1

SOUTH AFRICA: BUTTERWORTH & CO. (SOUTH AFRICA) (PTY) LTD.
DURBAN: 152-154 Gale Street

*First published in 1959*
*Second impression 1964*
*Third impression 1966*
*E.L.B.S. edition published 1966*
*Second edition 1971*

© Butterworth & Co. (Publishers) Ltd., 1971

ISBN 0 408 70174 9

*Printed photo-litho by J. W. Arrowsmith Ltd.
Bristol 3, England*

*To My Wife*

# PREFACE TO THE SECOND EDITION

ALTHOUGH there has been an explosion in the use of computers since the first edition appeared there remains an important place for the classical techniques of structural analysis. They are needed for preliminary and simpler design work, for checking computer results and they help the embryo structural designer understand how structures respond to load systems.

The second edition does not depart greatly from the first. In view of the change to S.I. units in some parts of the English speaking world and the use of other units in other parts of the world the book has been rewritten without reference to any particular system of units. Initially it was thought that this would complicate matters but it turns out that some simplifications are realised. An appendix on energy methods has been added. It explains an alternative approach to energy methods and brings out a number of important ideas such as that of duality in structural theory, the range of validity of certain of the energy methods and the relationship of energy methods to the concepts of equilibrium, compatibility and the stress-strain relationship.

J. A. L. MATHESON
N. W. MURRAY

# TO THE FIRST EDITION

THE THEORY of statically indeterminate structures has developed rather spasmodically. After the early work of such men as Müller-Breslau, Mohr, Maxwell and Castigliano there was a long period when few new ideas appeared. Indeed some of the older ideas seem to have been forgotten, for some textbooks published in the early years of this century dealt very sketchily with topics which had been treated more adequately long before.

More recently there has been a tremendous revival of interest in the subject. The development of reinforced concrete and welding in civil engineering construction and the increasing size and complexity of aircraft structures confronted designers with formidable theoretical problems which, until 1930, could not be satisfactorily dealt with. In that year Hardy Cross published an account of his method of Moment Distribution which gave designers for the first time a simple and convenient method for analysing rigid frames. The idea of successive approximations, upon which Hardy Cross's method depends, was at once recognized as being of very great utility; many variations and developments of the basic idea have now been described in the periodical literature, and most recent textbooks include accounts of the analysis of rigid frames by these methods.

At about the same time Southwell described a method of dealing with over-stiff braced frameworks by a successive correction method which he called the 'Relaxation of Constraints'. This has proved to be of very great generality, for it was found possible to replace a solid body such as a flat slab or an irregular bar under torsion by an analogous network which could be solved by relaxation.

The work of Hardy Cross and Southwell and their followers, although exceedingly valuable and important, is essentially concerned with the technique of dealing with equations which would otherwise be fantastically laborious or even completely intractable. The equations themselves spring from the classical theory of structures.

The student of the subject now finds himself faced with a variety of theorems and methods, each useful in its own field, but apparently bearing little relation to the others. The available textbooks emphasize mainly the classical theories or the modern methods, according to the taste of the author, but the subject has yet to be treated as an organic and coherent whole.

This is the aim of the present work. It attempts to demonstrate that these different techniques and theorems are in fact related to one another and that they fit into a pattern which can be likened to a family tree. As each theorem is reached its relationship to its neighbour is emphasized and where an alternative proof is available the one which brings out that relationship most clearly is emphasized. The diagram overleaf is intended to demonstrate the relationships between some of the theorems and methods of the theory of hyperstatic structures. The genealogy is not really so definite as the diagram perhaps implies for many of the theorems can be thought of as evolving in different ways from that shown. The broad distinction between

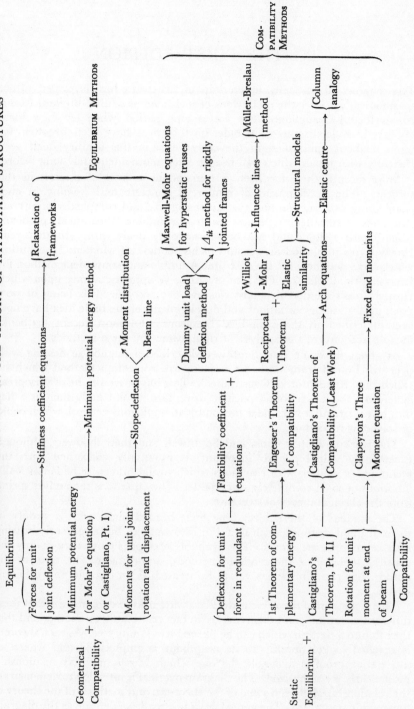

# PREFACE

'equilibrium' and 'compatibility' methods, according to the order in which the two sets of conditions are established, is of major importance, however, and it enables a clear division to be made in the diagram.

A complication which is difficult to show on this diagram is one which is emphasized by contemporary interest in plasticity. The Principle of Minimum Potential Energy is quite general and can be used to establish the conditions of equilibrium whether the deflexion of the structure is proportional to the applied load or not. Many of the methods to be described, however, are restricted to linear elastic problems where deflexion is proportional to load. As most real structures satisfy this requirement, at least to an adequate extent, the restriction is more important from a theoretical standpoint than in practice. The diagram will probably be of most value and interest to the student when he has progressed some way beyond the opening chapters and can recognize the cross-connexions which exist between the methods and can devise alternative arrangements of the diagram.

It is interesting to look for unsuspected relationships; one such is Müller-Breslau's graphical treatment of structures with two or three redundant reactions in which the redundants are specially chosen so as to be independent instead of interconnected. This turns out to be essentially the same as the elastic centre method for analysing arches, which in turn can be shown to be identical with the column analogy.

At the present time the accepted philosophy of structural design is under attack. Until quite recently the behaviour of structures was always analysed on the assumptions of elastic theory, although engineers were well aware that the ductility of mild steel had occasionally to come to their rescue. Van den Broek, following his old master Kist, proposed his method of 'Limit Design' as a means of resolving this paradoxical situation. The actual working stresses in the components of a structure were considered to be of little importance if it were certain that the frame as a whole would not collapse under working loads by excessive plastic deformation. Baker and his group at Cambridge have extensively developed these ideas and demonstrated experimentally that they provide a rational and economical method of design for single storey structures with welded joints and with unencased members. Their long series of experiments has enormously increased our knowledge of the real, as opposed to the assumed, behaviour of steel building frames and has profoundly affected the thinking of all designers, even those working in the rival medium of reinforced concrete. The appealing simplicity of van den Broek's proposals, however, is becoming less convincing now that investigation is revealing the difficulties that may arise with more complex structures.

It might be thought that the contemporary concentration of research on the theory of plasticity and its consequences would make it unnecessary to reconsider the methods deriving from elastic theory. But the new, as we are now beginning to see, has not supplanted the old; it has, in an odd way, strengthened it by showing where its importance and significance really lie. Thus the old problem of elastic stability, for example, is attracting increasing attention as a direct consequence of the new insistence on the importance of studying the mechanics of structual collapse. Recent work, too, on composite action can be appreciated best by those who understand both the principles underlying the elastic analysis of hyperstatic structures and the importance of post-elastic behaviour.

These considerations led to the decision to include an account of contemporary ideas on elastic stability in Chapter 8, which has been written by Dr N. W. Murray. In this Chapter emphasis is laid on the behaviour of complete structures rather than on individual struts and it will be seen that much use is made of methods that are direct developments of the simpler procedures described earlier. This recent work has not yet found its way into engineering practice but research is proceeding actively in several laboratories and it is to be expected that before long a stage will be reached when designers will be able to make use of the new approach.

One of the most striking developments of recent years has been in the field of electronic digital computers. Many different models are now commercially available and are being employed to perform a wide and rapidly growing range of calculations. Dr R. K. Livesley, who contributes Chapter 9, was one of the first to develop methods for analysing complicated structures on computers and his programmes for the Ferranti machine at Manchester University have now been used on a number of occasions. The mathematics used in these programmes is not very difficult but it is rather unfamiliar and for this reason the Chapter opens with an account of such parts of matrix theory as are relevant. The remainder of the Chapter describes the principles used in developing programmes for making structural calculations on computers. Since the importance of this development lies in the facility with which exceedingly elaborate calculations can now be made it has naturally not been practicable to demonstrate the full power of these methods but it is hoped that the Chapter will prove to be a useful introduction to the subject.

Throughout the book rather simple worked examples have been used to illustrate the various theorems and methods but to have included more complicated examples and problems for solution by the student himself would have extended the text unduly. A second volume, consisting mainly of examples, is therefore in preparation by the author in collaboration with Professor A. J. Francis.

The author records with gratitude the help he has received in discussions and correspondence with many colleagues and friends including particularly Mr T. M. Charlton and Professor A. J. Francis. Special thanks are due to his former colleagues Dr N. W. Murray who contributed Chapter 8 and Dr R. K. Livesley who contributed Chapter 9, to Mr J. Nutt who has checked the calculations and to Mrs C. W. Williams and Mrs Edith Roberts who typed the manuscript.

Many of the ideas now incorporated in the book were originally suggested by the late Professor Cyril Batho who combined a remarkable knowledge of structural theory with great gifts of exposition. This indeed is an attempt to produce the treatise which his friends hoped that Batho would write, a project which was frustrated by failing health and his death soon after retirement.

J. A. L. MATHESON

# CONTENTS

1. **HYPERSTATIC OR STATICALLY INDETERMINATE STRUCTURES**
   - 1.1. Introduction ... 1
   - 1.2. Statically Determinate and Indeterminate Structures ... 2
     - 1.2.1. Definition ... 2
     - 1.2.2. External Redundancy ... 3
     - 1.2.3. Internal Redundancy of Pin-Jointed Trusses ... 8
     - 1.2.4. Combination of Internal and External Conditions ... 13
     - 1.2.5. Three-Dimensional or Space Structures ... 15
     - 1.2.6. Rigidly Jointed Frames in Two Dimensions ... 15
     - 1.2.7. Combination of Pin- and Rigidly Jointed Frames ... 18
   - 1.3. A General Method for Evaluating the Redundants ... 21
   - 1.4. The Principle of Superposition and Hooke's Law ... 23
   - 1.5. Intuitive Methods for Approximate Solutions ... 27
     - 1.5.1. Structures in which Bending Stress Predominates ... 27
     - 1.5.2. Structures in which Direct Stress Predominates ... 30
   - 1.6. Conclusion ... 32

2. **THE ENERGY THEOREMS OF STRUCTURAL ANALYSIS**
   - 2.1. Introduction ... 33
   - 2.2. Work and Energy ... 33
   - 2.3. Strain and Complementary Energy ... 34
   - 2.4. Slowly and Suddenly Applied Forces ... 37
   - 2.5. The Principle of Superposition ... 38
   - 2.6. The Energy Theorems ... 39
     - 2.6.1. The Theorem of Minimum Potential Energy ... 40
     - 2.6.2. The Principle of Virtual Work ... 49
       - 2.6.2.1. Application to Determinate Frameworks ... 50
       - 2.6.2.2. Mohr's Equation of Virtual Work ... 55
       - 2.6.2.3. Identity of Mohr's Equation and Principle of Minimum Potential Energy ... 57
       - 2.6.2.4. Examples of application of Mohr's Equation ... 58
       - 2.6.2.5. The Calculation of Deflexions ... 60
       - 2.6.2.6. The Versatility of the Virtual Work Method ... 68
     - 2.6.3. Castigliano's Theorem, Part I ... 69
       - 2.6.3.1. Identity of Castigliano's Theorem, Part I and Mohr's Equation of Virtual Work ... 70
     - 2.6.4. First Theorem of Minimum Strain Energy ... 70
     - 2.6.5. First Theorem of Complementary Energy ... 71
       - 2.6.5.1. Deflexions by Complementary Energy ... 73
       - 2.6.5.2. Identity of the first Theorem of Complementary Energy and Mohr's Equation of Virtual Work as used in Deflexion Calculations ... 74
     - 2.6.6. Castigliano's Theorem, Part II ... 75
       - 2.6.6.1. Deflexions by Strain Energy ... 75
     - 2.6.7. Engesser's Theorem of Compatibility ... 76
     - 2.6.8. Castigliano's Theorem of Compatibility ... 79
     - 2.6.9. Nomenclature ... 82
   - 2.7. Relations between the Theorems ... 84

3. **GENERAL THEOREMS FOR LINEAR ELASTIC STRUCTURES**
   - 3.1. Introduction ... 86
   - 3.2. Development of the Principle of Virtual Work ... 86
     - 3.2.1. The Equation of Virtual Work in Terms of Stresses ... 86
     - 3.2.2. The Virtual Work Equation in Terms of Axial Force, Bending Moment and Transverse Shear ... 90

3.3. General Expressions for Strain Energy ... 92
   3.3.1. Strain Energy due to Axial Forces ... 93
   3.3.2. Strain Energy due to Bending and Shearing ... 94
   3.3.3. Strain Energy under Combined Stress ... 94
   3.3.4. Alternative Derivations of Strain Energy Expressions ... 95
3.4. General Expressions for the Deflexion of a Structure ... 97
   3.4.1. Derivation from the Virtual Work Equation ... 97
   3.4.2. Derivation from the Strain Energy Equation ... 98
   3.4.3. Example of General Deflexion Calculation ... 98
3.5. Reciprocal Relationships ... 100
   3.5.1. Maxwell's Reciprocal Theorem ... 100
   3.5.2. Betti's Reciprocal Theorem ... 102
3.6. The Deflexion of Pin-jointed Frameworks ... 103
   3.6.1. The Dummy Unit Load Method ... 104
   3.6.2. Williot-Mohr Construction ... 109
3.7. The Deflexion of Beams ... 115
   3.7.1. Sign Convention ... 116
   3.7.2. Deflexions by Integration ... 117
   3.7.3. Deflexions by the Dummy Unit Load Method ... 124
   3.7.4. The Moment-Area Method ... 126
   3.7.5. Hoadley's Method ... 133

## 4. GENERAL METHODS FOR LINEAR HYPERSTATIC STRUCTURES

4.1. Introduction ... 138
4.2. The Stiffness Coefficient Equations (Equilibrium Method) ... 138
   4.2.1. Derivation of the Equations ... 138
   4.2.2. Details of Application ... 140
4.3. Relaxation ... 142
   4.3.1. Introduction ... 142
   4.3.2. Linear Simultaneous Equations ... 145
   4.3.3. Direct Relaxation of Frameworks ... 150
   4.3.4. Conclusion ... 152
4.4. The Flexibility Coefficient Equations (Compatibility Method) ... 153
   4.4.1. Introduction ... 153
   4.4.2. Redundant Reactive Restraints ... 154
   4.4.3. Redundant Internal Forces and Moments ... 157
4.5. The Maxwell-Mohr Equations ... 158
   4.5.1. Derivation ... 158
   4.5.2. Completion of Analysis of Figure 4.9. ... 160
   4.5.3. Comments on the Maxwell-Mohr Equations ... 162
   4.5.4. Prestressing ... 163
   4.5.5. Temperature Stresses ... 165
   4.5.6. Hyperstatic Space Frames ... 166
4.6. The $\Delta_{ik}$ Method ... 170
   4.6.1. Derivation ... 170
   4.6.2. Application ... 173
4.7. Applications of Castigliano's Theorem of Compatibility ... 176
   4.7.1. Derivation of the Maxwell-Mohr Equations ... 176
   4.7.2. Derivation of the $\Delta_{ik}$ method ... 177
   4.7.3. Examples ... 178
4.8. Deflexion Calculations for Hyperstatic Structures ... 184

## 5. MOVING LOADS ON STRUCTURES

5.1. Introduction ... 189
5.2. Influence Lines ... 189
   5.2.1. Influence Lines for Beams ... 189
   5.2.2. Indirect Loading ... 190
   5.2.3. Influence Lines for Trusses ... 193
   5.2.4. Müller-Breslau's Principle ... 197

5.3. Influence Lines for Hyperstatic Structures ... 199
   5.3.1. Structures with a Single Redundant Reaction ... 199
   5.3.2. Trusses with a Single Redundant Member ... 209
   5.3.3. Structures with Two Redundant Reactions ... 212
   5.3.4. Special Selection of Redundants ... 214
5.4. Structural Models ... 217
   5.4.1. Models of Statically Determinate Structures ... 218
   5.4.2. Models of Hyperstatic Structures ... 219
      5.4.2.1. Laws of Elastic Similarity ... 219
      5.4.2.2. The Indirect Model Method ... 221
      5.4.2.3. Application to Structures with Several Redundants ... 224
      5.4.2.4. Practical Details ... 228
      5.4.2.5. Direct Model Methods ... 229

## 6. FRAMES WITH RIGID JOINTS

6.1. Introduction ... 233
6.2. Hyperstatic Beams ... 233
   6.2.1. The Propped Cantilever ... 233
   6.2.2. The Beam with Fixed Ends ... 235
   6.2.3. Continuous Beams ... 238
6.3. The Slope-Deflexion Equations ... 249
   6.3.1. Derivation of the Equations ... 249
   6.3.2. Application to Frames in which Sway cannot occur ... 251
   6.3.3. Application to Frames in which Sway can occur ... 255
   6.3.4. Comments on the Slope-Deflexion Procedures ... 264
6.4. The Moment Distribution Method ... 264
   6.4.1. Derivation of the Basic Equations ... 265
   6.4.2. Analysis of Frames in which Sway cannot occur ... 269
   6.4.3. Analysis of Frames in which Sway can occur ... 277
   6.4.4. Developments of the Method ... 290
      6.4.4.1. Naylor's Method ... 290
      6.4.4.2. Bolton's Method ... 293
   6.4.5. Comments on the Moment Distribution Method ... 299
6.5. Further Methods for the Analysis of Rigid Frames ... 300
   6.5.1. The Geometry of Bending Moment Diagrams ... 300
   6.5.2. The Concept of 'Degree of Fixity' ... 306
   6.5.3. The Beam Line Method ... 309
6.6. The Analysis of Collapse Loads ... 318
   6.6.1. Introduction ... 318
   6.6.2. Basic Ideas of Analysis of Collapse Loads ... 321
   6.6.3. General Methods for Finding Collapse Loads ... 325

## 7. ARCHES

7.1. Introduction ... 332
   7.1.1. The Principles of Arch Action ... 333
   7.1.2. Types of Arch ... 335
   7.1.3. The Linear Arch ... 337
7.2. The Three-Hinged Arch ... 339
   7.2.1. Determination of the Horizontal Thrust ... 339
   7.2.2. Determination of Thrust and Shear ... 341
   7.2.3. Determination of Stress ... 341
   7.2.4. Influence Lines ... 343
7.3. The Two-Hinged Arch ... 345
   7.3.1. Derivation of the Expression for Horizontal Thrust ... 345
   7.3.2. The Tied Arch ... 350
   7.3.3. The Effect of Temperature ... 351
   7.3.4. The Effect of Axial Thrust ... 351
7.4. The Fixed Arch ... 353
   7.4.1. The Elastic Centre ... 353
   7.4.2. Formulae for Members of Polygonal Arch ... 360
   7.4.3. Example of Fixed Arch Analysis ... 361
   7.4.4. Influence Lines ... 364

- 7.5. The Column Analogy ... 367
  - 7.5.1. Principle ... 367
  - 7.5.2. Examples ... 369
  - 7.5.3. Comments on the Column Analogy ... 374
- 7.6. Interconnected Arches ... 375
  - 7.6.1. Introduction ... 375
  - 7.6.2. Generalized Slope-Deflexion Procedure ... 376

# 8. STABILITY OF STRUTS AND FRAMEWORKS

- 8.1. Stability of Isolated Struts ... 382
- 8.2. Effect of Axial Load on Simple Cantilever ... 386
  - 8.2.1. s- and c- Functions ... 386
  - 8.2.2. s″-Functions ... 391
  - 8.2.3. Sway Functions ... 391
  - 8.2.4. Other Tabulated Functions ... 392
- 8.3. Elastic Behaviour of Ideal Frameworks ... 392
  - 8.3.1. Rectangular Type Framework ... 392
  - 8.3.2. Braced Framework ... 396
- 8.4. Methods of Evaluating the Critical Loads of Frameworks ... 399
  - 8.4.1. Moment Distribution Method ... 399
  - 8.4.2. Bolton's Approximate Method ... 406
  - 8.4.3. Experimental Methods ... 409
- 8.5. Actual Behaviour of Frameworks ... 411
  - 8.5.1. Elastic Analysis ... 412
  - 8.5.2. Plastic Analysis ... 414
  - 8.5.3. Combined Elastic and Plastic Behaviour ... 416
- 8.6. Discussion and Conclusion ... 418

# 9. MATRIX METHODS

- 9.1. Introduction ... 421
- 9.2. Matrix Algebra ... 422
  - 9.2.1. General Ideas and Notation ... 422
  - 9.2.2. Some Properties of Square Matrices ... 425
  - 9.2.3. Some Notes on Numerical Techniques ... 428
- 9.3. The Matrix Analysis of Continuous Beams ... 432
- 9.4. Problems Involving Coordinate Transformations ... 436
  - 9.4.1. Plane Pin-jointed Frames ... 436
  - 9.4.2. Plane Rigidly Jointed Frames ... 440
  - 9.4.3. Space Frames ... 441
- 9.5. Non-Linear Effects ... 442
- 9.6. Partitioning Techniques ... 444
- 9.7. Dynamic Problems ... 446
- 9.8. Conclusion ... 447

## APPENDIX A: NON-UNIFORM BEAMS

1. Deflexion ... 449
2. Beams with Fixed Ends ... 450
3. The Slope-Deflexion Equations ... 451
4. Moment Distribution Factors ... 453
5. Evaluation of Coefficients ... 456

## APPENDIX B: NOTES ON THE DESIGN OF HYPERSTATIC STRUCTURES ... 461

## APPENDIX C: ENERGY THEOREMS BY THE ARGYRIS AND KELSEY APPROACH ... 468

NAME INDEX ... 495

SUBJECT INDEX ... 497

# SYMBOLS

Symbols are defined when they first appear and are listed here for convenience

| | |
|---|---|
| $A$ | Cross-sectional area of member. Elastic weight of arch. General matrix |
| $A^*$ | Area of part of cross-section (shear calculations) |
| $A_m$ | Area of bending moment diagram |
| $B$ | Specific breadth of beam. General matrix |
| $C$ | Complementary energy. Constant of integration. General matrix |
| $D$ | Column vector representing angular and linear displacement |
| $E$ | Young's modulus |
| $F$ | Axial force in member. Column vector representing forces and moments acting as end loads on beam |
| $G$ | Property of non-uniform beam. Constant |
| $H$ | Horizontal component of reaction |
| $I$ | Second moment of area. Unit matrix |
| $I_p$ | Polar moment of inertia |
| $J$ | Second polar moment of area |
| $K$ | $I/L$ |
| $L$ | Length of member |
| $M$ | Bending moment |
| $M_p$ | Plastic moment of resistance |
| $N$ | Modulus of rigidity |
| $P$ | External load |
| $P_E$ | Euler load |
| $Q$ | Shearing force |
| $R$ | Reaction. Radius of curvature. Resultant |
| $S$ | Axial force (in matrix analysis) |
| $T$ | Thrust. Coordinate transformation matrix |
| $U$ | Strain energy |
| $U_i$ | Strain energy ($i$ emphasises internal nature of $U_i$) |
| $U_i^*$ | Complementary Strain energy ($i$ emphasises internal nature of $U_i^*$) |
| $V$ | Potential energy. Vertical component of reaction. Volume |
| $W$ | Work. Elastic 'load' on arch. External load system |
| $W_c$ | Critical load |
| $W_e$ | Work done on structure by external forces |
| $W_e^*$ | Complementary work of external forces |
| $X$ | Redundant force. Vector |
| $Y$ | Redundant reaction (arch). Vector. Stiffness matrix (of structure) |
| $Z$ | Redundant couple (arch). Section modulus. Vector. Stiffness matrix (of supports) |
| $a$ | Dimension of length |
| $a_{ij}$ | General matrix coefficient |
| $a_o$ | Lack of straightness of strut |
| $b$ | Dimension of length |

## SYMBOLS

| | |
|---|---|
| $c$ | Dimension of length. Carry-over function |
| $d$ | Increment (as a prefix). Dimension of length |
| $e$ | Change of length of member |
| $f$ | Direct stress. Function |
| $f°$ | Degree of fixity |
| $h$ | Vertical dimension. Rise of arch |
| $i$ | Slope |
| $j$ | Number of joints in frame. Coefficient in fixed point method |
| $k$ | Coefficient. Axial stiffness. $EI/L$. Stiffness of joint support |
| $l$ | Direction cosine |
| $m$ | Number of members in frame. Coefficient in beam-line methods. Number of independent collapse mechanisms. Sway function. Direction cosine |
| $n$ | Degree of redundancy. Coefficient in fixed point and beam-line methods. Direction cosine |
| $p$ | Stiffness coefficient. Number of possible hinge locations. Internal load in partitioned structure |
| $q$ | Shearing stress |
| $r$ | Number of reactive restraints. Radius of gyration |
| $s$ | Dimension of length, especially along arch rib. Model scale. Stiffness function |
| $t$ | Change of temperature |
| $w$ | Uniformly distributed load per unit length |
| $x$ | Coordinate |
| $y$ | Coordinate. Distance from central axis of beam. Ordinate to arch axis. Increase of lateral deflexion of strut |
| $y_c$ | Increase of lateral deflexion at centre of strut |
| $z$ | Coordinate |
| | |
| $\alpha$ | Angle. Coefficient, especially of variation of flexural rigidity |
| $\alpha_t$ | Coefficient of linear thermal expansion |
| $\beta$ | Numerical deflexion coefficient |
| $\gamma$ | Strut deflexion coefficient |
| $\delta$ | Displacement. Increment |
| $\Delta$ | Deflexion. Displacement vector |
| $\epsilon$ | Unit change of length. Strain. Eccentricity |
| $\zeta$ | Eigenvalue |
| $\theta$ | Angle. Angular rotation |
| $\kappa$ | Coefficient of shearing deflexion |
| $\lambda$ | Lack of fit |
| $\mu$ | Moment of elastic 'load'. $\dfrac{\pi}{2}\sqrt{\dfrac{F}{P_E}}$ |
| $\nu$ | Poisson's ratio. Shape factor |
| $\rho$ | $L/AE$ |
| $\sigma$ | Component of stress |

## SYMBOLS

| | |
|---|---|
| $\phi$ | Angular distortion. $\delta/L$. Effective load |
| $\psi$ | Angle |
| $\omega$ | Natural frequency |

### Suffixes and Indices

| | |
|---|---|
| $\delta_A^B$ | Displacement of $A$ from the tangent $B$ |
| $\Delta_{ab}$ | Displacement of point $A$ caused by unit load at point $B$ |
| $F_a, F_b$ | Force in member of primary structure caused by force $X_A = 1, X_B = 1$ |
| $F_o$ | Force in member of primary structure caused by actual loading |
| $F'$ | Force in member of framework caused by specified unit load |
| $F'_c$ | Force in member of framework caused by specified unit couple |
| $M_{AB}$ | Bending moment at end $A$ of member $AB$ |
| $M_{FAB}$ | Fixing moment at end $A$ of fixed ended beam $AB$ |
| $M_o$ | Bending moment produced in primary structure by actual loading |
| $M_{SAB}$ | Sway correction moment |
| $M'$ | Bending moment produced by specified unit load |
| $M'_c$ | Bending moment produced by specified unit couple |
| $P_i$ | Imaginary external load |
| $\bar{P}, \bar{F}$ | External loads and corresponding bar forces of the force system used in Mohr's equation of virtual work |
| $X_a, X_b$ | Quasi-external forces or moments acting in place of redundants |

CHAPTER 1

# HYPERSTATIC OR STATICALLY INDETERMINATE STRUCTURES

## 1.1. INTRODUCTION

IN the course of his first introduction to the subject of the Theory of Structures, the student usually studies beams and frames of a kind which can be analysed by means of a direct application of the laws of statics. Although problems of this sort can arise which require some ingenuity to solve, no great difficulty of principle is usually encountered. The equations of statics and their use in practice, by the processes of resolving in two directions at right angles and of taking moments about some convenient point[†], are normally emphasized in school courses on mechanics. The extension of the ideas to cover the basic engineering structures is an easy step.

When the student advances to the study of statically indeterminate structures he quite often finds himself seriously confused first by the clumsy title, with its suggestion that in this field the familiar laws of statics are invalid, and later by the bewildering variety of apparently unrelated theorems and methods.

The purpose of this book is to show how all these theorems fit together into a coherent pattern and consequently methods and proofs which best illustrate the family relationships between the various theorems are emphasized. These are not always the most familiar or even the most direct proofs but they provide a solid and consistent basis for the understanding of the subject.

Although the qualification 'statically indeterminate' is almost universally applied to the structures with which we are concerned the adjective 'hyperstatic'—meaning beyond statics—is both shorter and more exact. There seems to be every reason for following Southwell's[1] lead in using this title. Some authors write of 'redundant' structures but in this book we reserve that word for the superfluous members, restraints or stress components[‡] which make the structure hyperstatic.

It is assumed that the reader is familiar with the ideas and methods of the elementary theory of structures. The main topics which will be relied upon are:

The statics of coplanar force systems treated graphically and analytically.
The analysis of pin-jointed trusses: Bow's notation; the method of joints; the method of sections.

---

[†] Except where space frames are specifically mentioned, the discussion is restricted to plane frames.
[‡] The convenient but inaccurate term 'stress components' stands for 'bending moment, shearing or axial force in a member'.
[1] *Loc. cit.* §§ 45 (see References, p. 32).

The analysis of beams: Shearing force and bending moment diagrams. Moving loads on beams and trusses: Influence lines.

These matters are treated in many textbooks; those by Timoshenko and Young[2], Wilbur and Norris[3], and Merchant and Bolton[4] will be found to be reliable and informative. The student who refers to more than one book will soon find that there is no agreed convention of signs or symbols but this is a difficulty he must learn to contend with. In this book we have adopted the most widely used symbols wherever possible, but care has been taken to define each symbol as it appears and to give a complete list on page xiii. This is a subject in which symbols must often be particularized by superscripts and subscripts whose exact meaning must be fully understood.

Although the basic theorems of statics and some of the special methods derived from them apply to all types of structure this book is mainly concerned with the analysis of hyperstatic structures. It is essential that the engineer should be able to recognize and classify structures of this type and to distinguish them from statically determinate structures which can be dealt with by simple methods.

In the following Sections we discuss the rules which enable a structure to be diagnosed as statically determinate or otherwise. These rules are a little uncertain in operation, however, and the student should not rely exclusively upon them; rather should he study the action of his structures from first principles and, by adding and taking away members and restraints, satisfy himself that he fully understands how the applied forces are transferred to the foundations.

## 1.2. STATICALLY DETERMINATE AND INDETERMINATE STRUCTURES

### 1.2.1. Definition

A structure is said to be 'statically determinate' when the forces and bending moments acting can everywhere be determined by means of the equations of static equilibrium.

It is 'statically indeterminate' or 'hyperstatic' when it possesses more members or is supported by more reactive restraints than are strictly necessary for stability: the excess members or restraints are described as 'redundant'. In such cases the equations of statics, by themselves, are insufficient to provide a solution and, as we shall see, must be supplemented by equations of compatibility of displacement for it is the relative flexibility of the different parts of the structure that determines the partition of load between them.

It is convenient to introduce this topic by distinguishing between 'external redundancy', when a structure has redundant reactive restraints and 'internal redundancy', which occurs when trusses possess redundant members and are overstiff. We shall see later, however, that these two types of redundancy are not independent.

The redundancy involved in continuous membranes such as slabs, domes and shells is outside the scope of this book but it is analogous to the internal redundancy mentioned above.

## 1.2.2. EXTERNAL REDUNDANCY

We begin by considering the external redundancy of beams and trusses. Problems involving the former will be regarded as solved when the bending moment and shearing force have been determined at all points; it is interesting to note, however, that the calculation of the internal stress in a beam is essentially a statically indeterminate problem since it involves considerations of internal deformation or strain. It was only when this was recognized that Navier[5] was able to correct the work of his predecessors, from Galileo onwards, and give the first completely satisfactory account of the stress distribution in beams subjected to bending.

It can hardly be accidental that Navier was also the first to evolve a general method for dealing with such statically indeterminate problems as the propped cantilever (FIGURE 1.1(b)).

*Figure 1.1.*(a)  This simple cantilever, carrying a uniformly-distributed load, is statically determinate; (b) The additional support at $B$ is redundant and makes the cantilever hyperstatic

We begin by considering the simple cantilever shown in FIGURE 1.1(a); the shearing force and bending moment can be computed at once for any section and the arrangement is therefore statically determinate.

If, however, there is an additional support, as in FIGURE 1.1(b), the problem will be found to be insoluble by the methods of statics. It is only after we have found the reaction at $B$ by some other method that we can, by resolving vertically and taking moments, find the shearing force and bending moment diagrams.

Thus, by a simple trial, we have discovered that this propped cantilever is statically indeterminate. It is also clear that there is only one redundant quantity, and the reaction at $B$ seems to be singled out for this distinction.

*Figure 1.2.*(a)  This simply-supported beam is statically determinate; (b) Fixing the end $A$ makes the beam hyperstatic (Compare *Figure 1.1*(b))

The same problem can be stated in slightly different terms. Consider the beam $AB$ of FIGURE 1.2(a). This can be analysed quite as easily as the cantilever by means of the laws of statics. If, however, the end $A$ were fixed to a wall the same dilemma would arise as before; the end $A$ would no longer be free to rotate as the beam deflected, and a restraining moment $M_A$ would be called into play as in FIGURE 1.2(b). If $M_A$ can be found the problem again becomes soluble and, moreover, it is seen to be the same problem as before but on this occasion it seems as though the restraining moment $M_A$ is the redundant quantity.

It seems, therefore, that when a structure turns out to be statically indeterminate there is some freedom of choice in selecting the member or reaction to be regarded as redundant.

This is in fact the case, but it will often be found that one choice involves less labour to provide a solution than another. Unfortunately there is no way of predicting which will prove to be the happiest choice but this often becomes apparent soon after the calculations have been begun.

These ideas can be extended to cover the case of a cantilever with two props, as shown in FIGURE 1.3. Here the value of two redundant quantities must be found by some means before the equations of statics can be successfully applied.

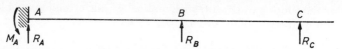

*Figure 1.3.* The cantilever with two props has two redundant reactions

These redundants can be taken as the bending moment at some point ($A$, for example) and either $R_B$ or $R_C$, or they can be $R_B$ and $R_C$.

After a little practice it becomes a simple matter to recognize this type of problem and its extensions to more complex beam arrangements. FIGURE 1.4, for instance, will be recognized as a cantilever with three props, but a

*Figure 1.4.* This can be regarded as a cantilever with three props or as a beam continuous over four supports and with one end fixed

more usual and more convenient technique is to treat it as a beam continuous over four supports, with one end fixed. Either approach involves the evaluation of three redundants as the first step. FIGURE 1.5 illustrates a beam with fixed ends. Extending the previous discussion we see that there must be two redundant restraints and these are usually taken as the 'fixing moments' $M_{FAB}$ and $M_{FBA}$ at the ends.

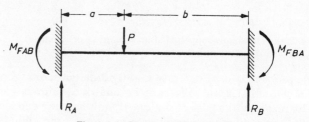

*Figure 1.5.* Beam with fixed ends

In order to draw general conclusions from these examples we recognize that a body which is stable in itself, such as a beam or an appropriately constructed truss, can be attached to its foundation in a variety of ways. At each point of contact with the ground there can be a roller bearing, or a

## 1.2. DETERMINATE AND INDETERMINATE STRUCTURES

hinge or a fixed joint which allows no rotation; these are illustrated in FIGURE 1.6. The first method, shown at (a), involves a single reactive force or restraint which, because its direction is known, can be found from a single equation. The hinge shown at (b) can offer both horizontal and vertical restraint so that two equations are called for. Three equations are necessary to find the reactive couple and the horizontal and vertical restraining forces which can act at a fixed support such as that shown at (c).

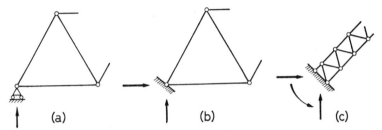

*Figure 1.6.*(a) One restraint only acts at a simple support; (b) Two restraints act at a hinge; (c) Three restraints, one of which is a couple, act at a fixed end

The laws of statical equilibrium, when applied to a plane structure, provide three equations, which can be written

$$\Sigma P_x = 0; \quad \Sigma P_y = 0; \quad \Sigma M = 0 \qquad \ldots (1.1)$$

Here $\Sigma P_x$ refers to the algebraic sum of the horizontal components of the acting forces (i.e. in the $x$-direction);
$\Sigma P_y$ refers to the vertical components ($y$-direction);
$\Sigma M$ is the sum of their moments about some point.

These three equations serve to find the values of three reactive restraints in terms of the applied loads. If a beam or truss is supported by three restraints then it is statically determinate; any restraints in excess of three are redundant. The beam $ABCD$ shown in FIGURE 1.7 is supported by seven

*Figure 1.7.* This beam has seven restraints, four of which are redundant

restraints—three at the fixed end $A$, one each at the roller supports $B$ and $C$ and two at the hinge $D$; four of these are redundant. If the examples illustrated in FIGURES 1.1–1.5 are tested in this manner the results will, at first sight, seem to be inconsistent with those obtained from first principles. In those examples, however, we were concerned with vertical loads only so that the three equations of statics (1.1) are effectively reduced to two, namely

$$\Sigma P_y = 0; \quad \Sigma M = 0 \qquad \ldots (1.1(a))$$

In the absence of any horizontal loads there is only one restraint at a hinged support and two at a fixed end.

Thus, in the fixed beam of FIGURE 1.5 there are two unknown restraints at each of the ends; the two equations of statics suffice to determine two of these, leaving two redundants as before†.

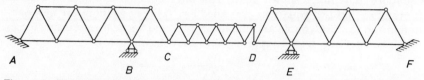

*Figure 1.8.* This bridge has six restraints, but the hinge and rocker carrying the suspended span make the whole structure statically determinate

Multiple-span bridges are often constructed by interconnecting adjacent beams or trusses and this can be done in such a way as to make the structure determinate. FIGURE 1.8, for example, shows a bridge whose supports exert six restraints. At $C$, however, a hinge has been introduced and at this point there can be no bending moment. At $D$ there is a rocker arrangement which can neither transmit bending moment nor horizontal thrust. We can therefore write three extra equations

$$\Sigma M_C = \Sigma M_D = \Sigma P x_D = 0$$

which, together with the three equations of equilibrium, just suffice to find the six unknown restraints in terms of any applied loads; the bridge is therefore statically determinate.

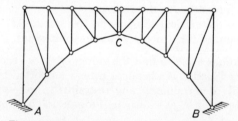

*Figure 1.9.* The crown hinge is introduced to make this spandrel-braced arch statically determinate

FIGURE 1.9 illustrates a similar point. Here an arch bridge is hinged to its two abutments $A$ and $B$, so that there are four restraints. There can be no bending moment at the crown hinge $C$, however, so we can write four equations for the four restraints.

It is also possible to support a structure by means of a suitable arrangement of hinged bars. FIGURE 1.10(c) shows a body being carried by three bars; comparison with the more usual arrangements at (a) and (b) show that it is just as effectively restrained and, as there are just three unknown bar forces which can be found from the three equations of statics, the problem is determinate.

FIGURE 1.10(d) and (e) illustrates special cases of some interest; if the three bars are parallel they will clearly offer no resistance to horizontal

---

† The axial force induced in the beam by the bending deflexion is usually regarded as negligible.

## 1.2. DETERMINATE AND INDETERMINATE STRUCTURES

motion and the arrangement is unstable. If the bars are concurrent, as at (e), then their point of intersection becomes an instantaneous centre of rotation and a small but significant movement can take place until the supporting

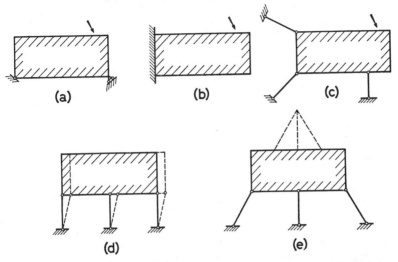

*Figure 1.10.*(a), (b) and (c) Three ways of supporting a rigid body in a stable and determinate manner; (d) If the supporting bars are parallel horizontal loads cannot be resisted; (e) Slight movement is permitted when the supporting bars intersect at a single point

bars have moved so that their directions no longer intersect. This movement can only be accommodated by small changes in the length of the members; such distortions are not usually important in statically determinate problems and are normally neglected. In the present instance they cannot be ignored since they determine the partition of load between the three bars; the *critical configuration* under discussion is therefore essentially hyperstatic. The designer is well advised to avoid such a sensitive arrangement.

Similar remarks apply to the three-hinged arch of FIGURE 1.9. The essential elements of such an arch are shown in FIGURE 1.11(a) which is analysed as follows:

Resolving vertically,
$$V_A + V_B - P = 0 \quad \text{(i)}$$

Resolving horizontally,
$$H_A - H_B = 0 \quad \text{(ii)}$$

Taking moments about $B$,
$$V_A(a + b) + H_A(h_B - h_A) - Pb = 0 \quad \text{(iii)}$$

Taking moments about $C$,
$$V_A a - H_A h_A = 0 \quad \text{(iv)}$$

Hence
$$V_A = H_A \frac{h_A}{a}$$

$$H_A(a+b) \frac{h_A}{a} + H_A(h_B - h_A) = Pb$$

i.e.
$$H_A = \frac{Pab}{bh_A + ah_B}$$

The result shows that $H_A$, and therefore $H_B$, will be zero if either $a$ or $b$ is zero, as in FIGURE 1.11(b).

If
$$bh_A + ah_B = 0$$
that is if $C$ lies in the straight line $AB$, then $H$ becomes infinite.

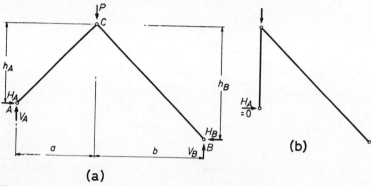

Figure 1.11. The three-hinged arch has critical forms when $a$ or $b$ is zero, as shown at (b), or when $A$, $B$ and $C$ are collinear

This is a simple example of the more general treatment given by Fife and Wilbur[6] who explain how the equations of equilibrium of a structure, such as (i) to (iv) above, can be solved by means of determinants. It turns out that the same determinant appears in the denominator of the expression for each of the restraints and, as above, a critical arrangement corresponds to zero value of this determinant.

### 1.2.3. INTERNAL REDUNDANCY OF PIN-JOINTED TRUSSES

Many engineering structures, especially bridges and roof trusses, are built up of individual members connected to form a framework or truss and we must now consider the ways in which this can be done. For the time being we assume that the members are joined by simple hinges or 'pinned joints' which offer no resistance to bending.

FIGURE 1.12(a) shows a simple pin-jointed bridge truss; it will soon be found by trial that the reactions and the forces in the members can be obtained by any of the familiar methods. If, however, the central panel of the truss possessed a second diagonal, as in FIGURE 1.12(b), the reactions could be found but the forces in the members of the centre panel could not. Prior knowledge of the force in $CF$, however, would enable the remaining member forces to be found in the usual way.

It will now be clear that a truss is statically indeterminate internally if it possesses more members than are strictly essential for stability. The non-essential or redundant members are those which can safely be removed;

## 1.2. DETERMINATE AND INDETERMINATE STRUCTURES

they can be recognized by visualizing what will happen if the frame is deprived of one or other of its members.

Referring again to FIGURE 1.12(b) it is obvious that member *CF* is redundant since FIGURE 1.12(a) shows the truss in a perfectly stable condition

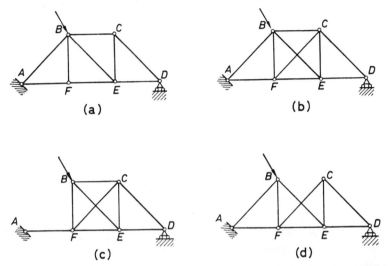

*Figure 1.12.* (a) This pin-jointed truss is stable and determinate both internally and externally; (b) The truss is now indeterminate so far as member forces are concerned; (c) The panel *BCEF* is overstiff while *ABF* is unstable; (d) The truss is again determinate

without it. If *AB* is removed, as in FIGURE 1.12(c), the panel *ABF* is no longer stable so this member can evidently not be selected as the redundant.

FIGURE 1.12(d) illustrates the frame with *BC* removed, and a little consideration will show that it is still stable.

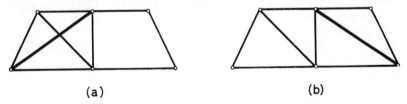

*Figure 1.13.* Although the count of members and joints indicates that there are no redundant members the arrangement (b) is satisfactory while (a) is not

The essential idea emerging from this discussion is that a stable and determinate frame can be built up as an assemblage of triangles. The first triangle, having 3 members and 3 joints, is just stable; each additional joint must be supported by two additional members and we therefore conclude that

$$m = 2j - 3 \qquad \ldots (1.2)$$

where $m$ is the number of members and $j$ is the number of joints.

If there are more members than are indicated by equation 1.2 then the truss has so many redundant members; if it has fewer members then it is unstable.

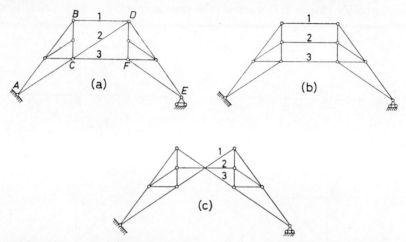

*Figure 1.14.* (a) A compound truss formed by interconnecting two simple trusses; Critical forms occur if the connecting members (b) are parallel, or (c) intersecting

FIGURE 1.13(a) shows a truss which satisfies equation 1.2 but it is seen to be overstiff in one part and not rigid in another. Changing the position of one of the bars, as at (b) converts the truss into a stable one and we conclude that the equation must be applied to the parts of a truss as well as to the whole in order to provide a complete test.

It is also possible to build up trusses by interconnexion. A simple example of such a *compound truss* is shown in FIGURE 1.14(a), where the two trusses *ABC* and *DEF* are shown connected by the three bars which comparison with FIGURE 1.10(c) suggests will be necessary. The combined truss has 17 members and 10 joints and so it satisfies equation 1.2. Critical forms can of course arise if the junction bars are parallel or concurrent, as indicated at (b) and (c).

FIGURE 1.15 illustrates another case of a critical form; the arrangements shown at (a) and (c) are stable and determinate but when the three hinges $A$, $B$ and $C$ lie in a straight line, as at (b), a slight movement of $B$ at right angles to this line would be permitted by small changes of length of the bars or by such play in the joints as is inevitable in practice. This is essentially the same problem, of course, as was discussed in connexion with FIGURE 1.11. It underlines the point made previously, that critical configurations of bars are essentially hyperstatic, since (b) is just a simple overstiff frame.

This suggests the 'pre-stressing test' for critical forms. It will be shown later (Chapter 4) that hyperstatic structures can be pre-stressed by arranging that the lengths of the members are such that they have to be forced into place when being assembled. If this is done some members will be in tension and others in compression before any external loads are applied, and advantage is sometimes taken of this characteristic in order to induce a favourable initial stress distribution.

## 1.2. DETERMINATE AND INDETERMINATE STRUCTURES   1.2.3

It follows that if it is found that a plane truss can be pre-stressed then it must be hyperstatic; if it also satisfies equation 1.2 then it has no redundant members and so its configuration must be critical. The central bar of FIGURE 1.10(e), for example, could be given an initial tension while the outer inclined bars were compressed; this could not be done with the non-critical arrangement of FIGURE 1.10(c).

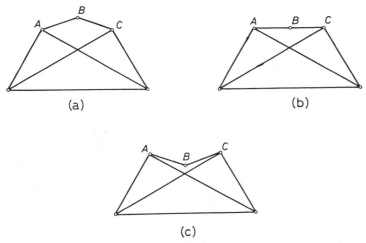

*Figure 1.15.* A critical form, obviously hyperstatic, occurs if the joints $A$, $B$ and $C$ are collinear

Rather obscure examples of critical forms arise in certain cases of *complex trusses*. These are frameworks whose members and joints satisfy equation 1.2 but whose configuration is such that the ordinary methods of analysis break down. Such a truss is shown in FIGURE 1.16; it will soon be found by trial that, since three members meet at every joint (the members cross at $G$ but are not joined there), the method of joints† cannot be employed. Nor can a section be found which will permit the use of the method of sections†.

General methods for the solution of such trusses are given by Timoshenko and Young[2] while a very convenient technique for analysing this particular truss is described by Wilbur and Norris[3]. The latter point out that if the horizontal component of the force in a suitable member, say $FE$, is assumed to be $H$ then a solution can be obtained as follows:

Joint $F$:

Resolving horizontally: Horizontal component of force in $FA = H$

Resolving vertically: Force in $FC = 2\dfrac{x}{20}H = \dfrac{xH}{10}$

---

† The 'method of joints' is used to find the forces in the members of a truss by considering the equilibrium of each of the joints, in turn, under the action of the member forces together with any external loads. The 'method of sections' is used to find the forces in certain members by cutting those members and considering the equilibrium of the part of the truss so isolated.

11

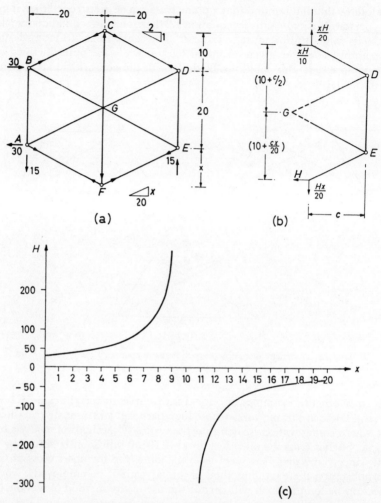

*Figure 1.16.* This compound truss has to be analysed by special methods. A critical form occurs when $x = 10$, and the force in the members is very high when $x$ lies between about 6 and 14.

Joint $C$:

Resolving vertically:   Vertical component of force in $BC$ and $CD$
$$= \frac{1}{2}\frac{xH}{10} = \frac{xH}{20}$$

Resolving horizontally:   Horizontal component of force in $BC$ and $CD$
$$= 2\frac{xH}{20} = \frac{xH}{10}$$

A section is now used to isolate the element of the truss shown at (b). Considering the equilibrium of this element and taking moments about $G$,

## 1.2. DETERMINATE AND INDETERMINATE STRUCTURES

we have
$$15 \times 20 + \frac{xH}{10}(10 + C/2) - H\left(10 + \frac{cx}{2}\right) = 0$$

i.e.
$$H = \frac{300}{10 - x}$$

This solution shows that when $x = 10$ and the truss is symmetrical the configuration is critical. The graph at (c) depicts the variation of $H$ with $x$ and demonstrates very clearly the unfavourable proportions of the truss when $x$ approaches the critical value of 10.

It will be noticed, too, that when the truss is symmetrical it can be pre-stressed by making a diagonal member too short so that it has to be forced into position at assembly. This observation also serves to identify the critical form.

### 1.2.4. COMBINATION OF INTERNAL AND EXTERNAL CONDITIONS

The simple truss shown in FIGURE 1.17(a) is stable and determinate both externally and internally. The three reactive restraints can just be found from the three equations of equilibrium; the 29 members and 16 joints exactly satisfy equation 1.2.

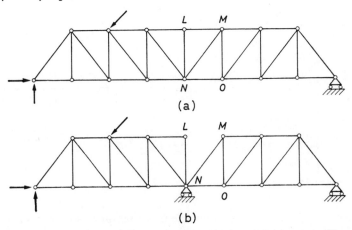

*Figure 1.17.*(a) If the member $LM$ is removed this truss becomes unstable; (b) Stability can be restored by introducing an appropriate restraint

If the bar $LM$, for example, is removed the truss becomes unstable but stability can be restored either by replacing $LM$ or by inserting another bar ($LO$ would serve) or by introducing an additional restraint at an appropriate joint such as $N$. Evidently a restraint is equivalent to a member from the point of view of stability so that the conditions for internal and external determinacy can be obtained from equation 1.2 by writing

$$m + (r - 3) = 2j - 3$$

i.e.
$$m + r = 2j \qquad \ldots (1.3)$$

where $r$ is the number of reactive restraints. It is assumed, of course, that the members and restraints are suitably disposed so as to avoid critical forms and so that the component parts of the structure, as well as the complete structure, are stable and determinate.

Another approach to this problem is to recall that at each hinged joint there is a concurrent coplanar force system in equilibrium, giving $2j$ equations of the type

$$\Sigma P_x = 0; \quad \Sigma P_y = 0$$

The number of unknown forces and restraints is $m + r$ so that a pin-jointed truss will be stable and determinate if

$$m + r = 2j \qquad \ldots (1.3)$$

As a final illustration consider the plane frame shown in FIGURE 1.18 which has 12 members, 8 restraints and 10 joints in all, including those pinned to the wall. Equation 1.3 is therefore satisfied. The arrangement of the bars, however, is rather different from that contemplated during the derivation of equation 1.3 but the following discussion shows that the equation is valid

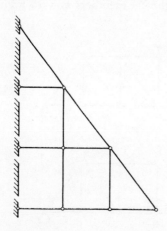

*Figure 1.18.* The stability of this truss, which appears at first sight to have too few members, depends on the eight restraints at the supports

for a plane frame of this type. Suppose that the frame is pinned to the wall at $w$ of its $j$ joints. The wall braces $w$ joints together and, from equation 1.2, is therefore equivalent to $2w - 3$ members.

Similarly $2j - 3$ members would be required to brace the joints if there were no wall so that the number of members required in addition to the wall is

$$(2j - 3) - (2w - 3) = 2(j - w)$$

Hence $$m = 2(j - w) \qquad \ldots (i)$$

Each joint pinned to the wall introduces two reactive restraints

i.e. $$r = 2w \qquad \ldots (ii)$$

Combining (i) and (ii), we have

$$m + r = 2j \qquad \text{as before.}$$

## 1.2.5. Three-Dimensional or Space Structures†

The discussion of plane structures just concluded can be extended to cover the case of space structures by means of appropriate amendments, as follows:

External redundancy

A rigid body is in equilibrium under a system of forces in space provided

$$\Sigma P_x = 0 \quad \Sigma P_y = 0 \quad \Sigma P_z = 0$$
$$\Sigma M_x = 0 \quad \Sigma M_y = 0 \quad \Sigma M_z = 0$$

where the first three equations refer to the algebraic sums of the components of the acting forces along three co-ordinate axes, and the second three equations express the condition that the algebraic sums of the moments of the forces about the three axes also vanish.

Accordingly we conclude that the reactions are statically determinate if a body is supported in such a way that there are six reactive restraints, and that any restraints in excess of six are redundant.

Critical cases can also arise. For instance if the reactive restraints are parallel, or lie in parallel planes, or if they intersect a line, then slight movement cannot be prevented.

Internal redundancy

A stable and determinate space frame can be built up by starting with a tetrahedron (6 members, 4 joints) and then adding joints one at a time, each joint being attached to the parent frame by three members. Proceeding in this way, we see that

$$m = 3j - 6 \qquad \ldots\ (1.4)$$

Critical cases will again arise, as for instance when one of the joints so added lies in the same plane as the three joints to which it is attached. The pre-stressing test can now be used to advantage, as it is often not at all obvious that a space frame has a critical form.

Combination of internal and external conditions

Arguing by analogy with the two-dimensional case, we conclude that the internal and external cases can be combined by writing

$$m + r = 3j \qquad \ldots\ (1.5)$$

The above is only a brief introduction to one aspect of the geometry of space frames, a subject which lies outside the scope of the present work. The formulae given are included by Henderson and Bickley[7], in a concise but general treatment of statical indeterminacy. An interesting account of the synthesis of determinate space frames is given by Roxbee Cox[8].

## 1.2.6. Rigidly Jointed Frames in Two Dimensions

In the preceding Sections we have been concerned essentially with pin-jointed trusses whose stability depended upon a suitable configuration of the members which were carrying axial forces only. In actual practice the joints

---

† A comprehensive account of space structures is given by Timoshenko and Young[2].

of such trusses are usually rivetted or welded so that angular adjustments, which would occur at pin-joints when the members changed length under load, are prevented and bending moments are induced in the members. The *primary stresses* produced in the members are first calculated on the assumption that the joints are pinned; the *secondary bending moments* arising from the rigidity of the joints can then be assessed, if thought necessary.

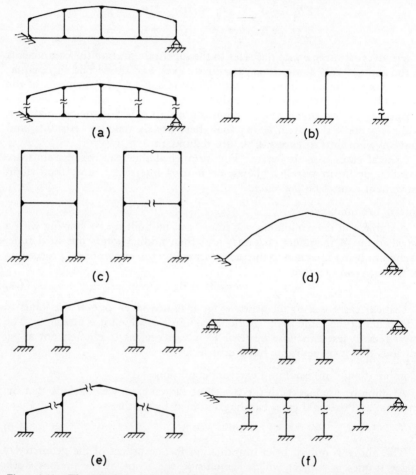

*Figure 1.19.* The members of frames with rigid joints are usually subject to bending moment and shearing force as well as to axial force. The degree of redundancy can be found by cutting members until the frame is determinate; each member so cut contributes three redundant stress components to the total

The assumption here is that the important forces in triangulated structures are the axial forces which are not affected appreciably by the effects of bending moments, although these can and do appreciably alter the distribution of stress in individual members. We return to this subject in Chapter 8.

These trusses are to be distinguished from rigidly jointed frames which depend on the fixity of the joints for their stability. Several examples of such

## 1.2. DETERMINATE AND INDETERMINATE STRUCTURES

frames are given in FIGURE 1.19 and it can be seen at once that they would be quite useless if the joints were pinned. The application of external loads produces important bending moments in the members and, indeed, it is now the axial forces which play the minor part.

Frames of this kind are often highly redundant and special methods must be employed to analyse them. The exact degree of redundancy is not usually required when these methods are used but it can be found as follows.

The degree of external redundancy of structures which are independently supported, such as the Vierendeel truss of FIGURE 1.19(a), can be found in the usual way. More frequently the frame and its foundations must be considered together, the total number of unknowns being the number of restraints acting at the supports together with the number of independent stress components acting in the members. In general each member will be under the action of bending moments together with shearing and axial forces so that the number of unknown stress components is $3m$.

At each joint three equations of equilibrium (1.1) can be written, giving $3j$ equations in all. The number of redundants is therefore

$$n = 3m + r - 3j \qquad \ldots\ldots (1.6)$$

Any extra equations arising from the presence of internal hinges or rockers reduces the degree of redundancy, by one or two respectively, as in Section 2.2.

A more instructive approach, however, is to make such cuts in the structure as are necessary to make it determinate, recalling that three redundants have been neutralized for every member cut. The examples of FIGURE 19 are now examined from the above standpoint.

(a) The Vierendeel truss is carried on a hinge and a roller; there are three reactive restraints and the support system is determinate. There are 13 members and 10 joints, and equation 1.6 gives $n = 12$ which agrees with the consideration that four members must be cut to make the structure completely determinate.

(b) Simple portal     $m = 3$    $r = 6$    $j = 4$    $n = 3$

(c) Two-storey frame  $m = 6$    $r = 4$    $j = 6$    $n = 4$

Note that this frame is made determinate by cutting the lower beam and inserting a hinge in the upper; this eliminates four redundants.

(d) Arch rib          $m = 6$    $r = 3$    $j = 7$    $n = 0$

The arrangement is determinate. A single curved arch rib replacing the polygonal rib would also be determinate, for we should have

$m = 1$    $r = 3$    $j = 2$    $n = 0$

(e) Workshop shed     $m = 10$   $r = 12$   $j = 11$   $n = 9$

(f) Viaduct           $m = 9$    $r = 14$   $j = 10$   $n = 11$

Here the structure can most easily be made determinate by cutting the four columns and changing one roller into a hinge; this eliminates twelve redundants and introduces a new one.

## 1.2.7. Combination of Pin- and Rigidly Jointed Frames

It occasionally happens that pin- and rigidly jointed frames are used in conjunction and this provides the occasion to combine equations 1.3 and 1.6 in a single comprehensive test for the degree of redundancy. The method was originally described in a paper by Matheson[9].

We begin by noticing that the number of unknown stress components acting in a member depends on the nature of its end conditions. (We assume, for the present, that members are loaded only through the joints to which they are attached.) Thus, as we have seen, a pin-jointed member is under axial load only (one stress component) while a rigidly jointed member can be subject to bending moment as well as to axial and shearing force (three stress components). The complete classification is as follows:

| One end | Other end | Number of stress components | Number of members of this type |
|---|---|---|---|
| Fixed | Fixed | 3 | $m_3$ |
| Fixed | Pinned | 2 | $m_2$ |
| Fixed | Sliding | 1 | $m_1$ |
| Pinned | Pinned | 1 | $m_1$ |
| Pinned | Sliding | 0 | $m_0$ |

The total number of unknown stress components is therefore

$$m_1 + 2m_2 + 3m_3 \qquad \ldots \text{(i)}$$

Note that a member which is pinned at one end and free to slide at the other cannot transmit any load from one end to the other and so its contribution to the number of stress components is zero.

We have already observed (Section 1.2.2 and FIGURE 1.6) that the number of reactive restraints depends on the type of support. Let there be

$r_1$ roller supports (one restraint)
$r_2$ hinged supports (two restraints)
and $r_3$ fixed supports (three restraints)

The total number of restraints is therefore

$$r_1 + 2r_2 + 3r_3 \qquad \ldots \text{(ii)}$$

The three equations of statical equilibrium 1.1 can be written at a rigid joint but at a pinned joint the moment equation disappears and at a sliding joint there is only one effective equation. Let there be

$j_1$ sliding joints (one equation)
$j_2$ pinned joints (two equations)
$j_3$ rigid joints (three equations)

The total number of equations is therefore

$$j_1 + 2j_2 + 3j_3 \qquad \ldots \text{(iii)}$$

The number of redundants, $n$, is the excess of unknown stress components over available equations,

i.e. $\quad n = m_1 + 2m_2 + 3m_3 + r_1 + 2r_2 + 3r_3 - j_1 - 2j_2 - 3j_3 \ldots (1.7)$

## 1.2. DETERMINATE AND INDETERMINATE STRUCTURES

The only possible ambiguity arises when some members are rigidly connected and others pinned at the same joint. The derivation of (iii) makes it clear, however, that if two or more members are rigidly connected at such a joint then the third equation of (1.1) is relevant and the joint counts as a rigid one.

If all the joints are pinned equation 1.7 reduces to

$$n = m_1 + (r_1 + 2r_2 + 3r_3) - 2j_2$$

which is essentially identical with equation 1.3. If all the joints are rigid, we have

$$n = 3m_3 + (r_1 + 2r_2 + 3r_3) - 3j_3$$

which coincides with equation 1.6.

Several examples are given in FIGURE 1.20, to which TABLE 1.1 and the notes below apply.

TABLE 1.1.
CALCULATION OF DEGREES OF REDUNDANCY OF FRAMES OF FIGURE 1.20

| Fig. | $m_1$ | $m_2$ | $m_3$ | $r_1$ | $r_2$ | $r_3$ | $j_1$ | $j_2$ | $j_3$ | $n$ |
|---|---|---|---|---|---|---|---|---|---|---|
| a   |   | 2 |    | 1 | 1 |   |   | 2 | 1  | 0  |
| (a)*| 1 | 1 |    | 1 | 1 |   | 1 | 1 | 1  | 0  |
| b   | 1 |   |    | 1 | 1 |   |   | 2 |    | 0  |
| (b)*|   |   |    | 1 | 1 |   | 1 | 1 |    | 0  |
| c   |   |   | 13 | 1 | 1 |   |   |   | 10 | 12 |
| d   |   | 2 | 8  |   |   | 4 |   |   | 11 | 7  |
| e   |   | 4 | 6  |   |   | 4 |   | 2 | 9  | 7  |
| f   | 2 | 2 | 1  | 2 | 3 |   | 2 | 3 | 3  | 0  |
| g   | 36| 2 | 6  | 1 | 3 |   |   | 18| 7  | 8  |
| h   | 31|   |    | 3 | 3 |   |   | 20|    | 0  |

(a) and (b) We began by excluding the possibility that members could be loaded between joints but this example, which will be recognized as a single span beam supported in a statically determinate manner, shows the restriction to be unnecessary. At (a) an imaginary rigid joint has been introduced at the load point; at (b) the load is applied at one end in accordance with the original assumption. In each case the calculations, which are shown in the accompanying table, give the correct result that the number of redundants is zero.

It may be thought that the right hand end of the beam should be counted with the $j_1$ sliding joints since the roller support can offer no horizontal restraint. If this is done then the classification of the member must also be adjusted but the final conclusion, that the arrangement is determinate, is unaffected. These alternative designations are shown in brackets on FIGURES 1.20(a) and (b) and by an asterisk in TABLE 1.1.

On the whole the designation given first is to be preferred.

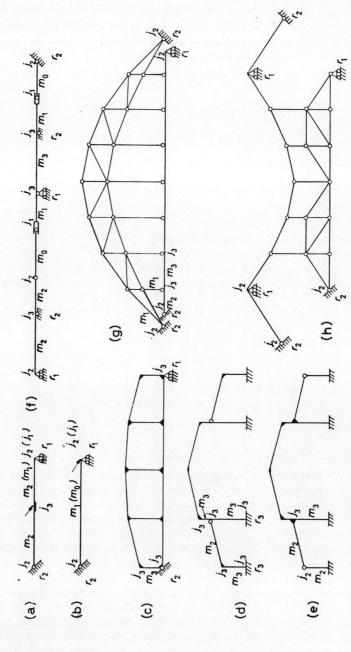

*Figure 1.20.* Another method of assessing the degree of redundancy of frames which have some rigid joints is to classify the joints members and supports, as in Section 1.2.7 and then to use Equation 1.7

(c) The Vierendeel truss of FIGURE 1.19(a) is recalculated according to the present method.
(d) and (e) Hinges have been introduced into the workshop shed of FIGURE 1.19(e) in two different positions. The frames are statically equivalent, however, and the two calculations give the same result.
(f) In this example a cantilever plate-girder bridge is shown schematically. Note that two of the members are pinned at one end and sliding at the other. They are labelled $m_o$ and are ignored in the count of members.
(g) This bowstring arch has a plate girder, for stiffening the roadway, to which the hangers are pinned; these joints are classed as rigid.
(h) It is instructive to compare the arrangement shown, which is determinate, with that which would follow if the stiffening truss were replaced by a plate girder, as in (g) above.

When the number of redundants has been determined the next step is to identify the *primary structure* ; this is the statically determinate structure remaining when the redundants have been temporarily removed.

Attention has already been drawn to the fact that there is often some choice in the selection of the redundants and this is especially true if the structure is redundant both externally and internally.

The main difficulty, that of evaluating the redundants, has so far been evaded. This indeed forms the main theme of this book, but in the next paragraph the underlying principle is discussed.

### 1.3. A GENERAL METHOD FOR EVALUATING THE REDUNDANTS

The discussion of Section 1.2 can be summarized by saying that the structure is statically determinate when the laws of statics provide as many equations as there are unknowns. Any more unknowns must be the redundants which can only be evaluated when appropriate additional equations have been found.

These new equations can always be found from considerations of geometrical compatibility—in other words from a knowledge that the component parts of a structure fit together both before and after they have been strained by the applied forces.

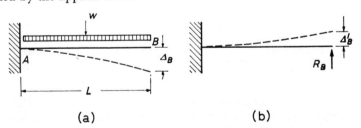

*Figure 1.21.* (a) $\varDelta_B$ is the deflexion of the cantilever due to the uniformly distributed load $w$ acting alone; (b) $\varDelta'_B$ is the deflexion due to $R_B$ acting alone

A simple example is provided by the propped cantilever of FIGURE 1.1(b), (page 3). If the reaction at $B$ is regarded as the redundant the relevant compatibility equation emerges from the fact that there is no deflexion at $B$.

# HYPERSTATIC STRUCTURES

Referring to FIGURE 1.21(a), we have†

$$\Delta_B = \frac{wL^4}{8EI}$$

and in FIGURE 1.21(b)

$$\Delta'_B = \frac{R_B L^3}{3EI}$$

Since the nett deflexion of $B$ is zero,

$$\Delta_B = \Delta'_B$$

hence
$$R_B = \frac{3wL}{8}$$

In such simple cases, where the deflexions required to express compatibility can be so easily written down, there is no advantage in using the more sophisticated methods described in later chapters. However in many problems the labour needed for this method would be unduly great or the complications such that errors would be likely.

Another example is illustrated in FIGURE 1.22(a). Clearly the load $P$ could be supported by any two of the three members and one, say $AC$, is therefore redundant. The necessary additional equation must express the

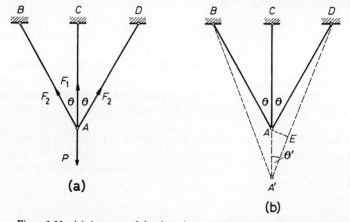

*Figure 1.22.* (a) Any one of the three bars can be taken as the redundant, but symmetry suggests that it will be best to take $AC$; (b) The distortion of the frame is exaggerated here so as to emphasize the geometrical relationship between the extensions of the members

fact that the three members are connected at $A$ both before and after the load is applied. The extension of $AC$ is therefore equal to the vertical component of the extension of $AB$ (or $AD$).

Resolving vertically at $A$, we have

$$F_1 + 2F_2 \cos \theta = P$$

---

† These are standard cantilever deflexion formulae which will be found in Chapter 3, Section 3.7.2.1.

To obtain the deflexions in terms of $F_1$ and $F_2$ consider the diagram 1.22(b) which shows the frame in its distorted position with sufficient exaggeration to make the procedure clear.

$$\frac{\text{Extension of } AD}{\text{Extension of } AC} = \frac{A'E}{A'A} = \cos \theta' \simeq \cos \theta$$

i.e.
$$\frac{F_2 L_{AD}/(AE)_{AD}}{F_1 L_{AC}/(AE)_{AC}} = \cos \theta$$

and
$$\frac{L_{AD}}{L_{AC}} = \frac{1}{\cos \theta}$$

If $AE$ is the same for all three members

$$\frac{F_2}{F_1} = \cos^2 \theta$$

i.e.
$$F_1 = \frac{P}{1 + 2\cos^3 \theta}$$

This problem has been worked in full since it illustrates an important point. An essential feature of the solution was the assumption that $\theta = \theta'$. In other words the distortion of the frame, although finite, is assumed to be small. The *ratio* of the distortion of the component members of the frame is not small, and indeed it determines the distribution of load between them. This consideration underlies nearly all the methods of solving statically indeterminate structures to be discussed, but it is often not obvious that this assumption is being made.

## 1.4. THE PRINCIPLE OF SUPERPOSITION AND HOOKE'S LAW

In solving the propped cantilever problem, illustrated in FIGURES 1.21(a) and (b), we used a method which made direct use of the Principle of Superposition. The deflexion of the point $B$ under two different loadings was calculated, and the ratio of these deflexions was used to evaluate the unknown reaction $R_B$. In so doing we assumed that the value of $\Delta'_B$, for instance, depended only on $R_B$, and would be the same whatever other loading happened to be acting. This assumption is nearly always made in engineering calculations, but it is only true when there is a linear relationship between load and deflexion, that is when Hooke's Law is obeyed[†].

Suppose FIGURE 1.23 represents the load-extension diagram for a bar of light alloy under increasing axial load. An increment of load $\delta F$ produces a change of length $\delta e$; so long as the diagram is linear the ratio $\delta F/\delta e$ is constant. In the non-linear range, however, this ratio is clearly not constant and the extension produced by $\delta F$ depends on the load acting before it is applied. Structures which include such members cannot be treated in the simple manner described above since the Principle of Superposition cannot be invoked. Nor can the Principle be used unless the same straight-line

---

[†] A formal proof of the Principle of Superposition is given by Southwell[1], p. 4.
Further reference to this question is made in Chapter 2, especially Section 2.5.

## HYPERSTATIC STRUCTURES

law connects load and deflexion both for increasing and for decreasing loads. Fortunately it is usually permissible to assume that engineering structures obey Hooke's Law, but the limitations of this assumption should be discussed.

Structural steel obeys Hooke's Law up to a fairly well defined elastic limit, and well-cured concrete has an almost linear stress-straining relationship after some months. These materials, if uniformly stressed, can be considered to satisfy the assumption. If they are used as beams, however, the imposed deflexion must be small or the expression for the curvature

$$\frac{1}{R} = \frac{d^2y/dx^2}{\{1 + (dy/dx)^2\}^{3/2}}$$

cannot be simplified in the usual way to

$$\frac{1}{R} = \frac{d^2y}{dx^2}$$

In trusses the deflexion must be small for two reasons. Referring to FIGURE 1.22(b), we assumed that $\theta \simeq \theta'$ first when we assumed that the resolution of the forces at joint $A$ was unaffected by the deflexion, and again when we set up an equation to describe the relative elongation of the

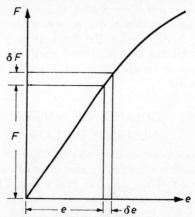

Figure 1.23. Non-linear load-extension relationship

members. Clearly if the members had been helical springs which stretched appreciably under load we could not have equated $\theta$ and $\theta'$ and the working would at once have become much more complicated, even in such a simple frame.

Timoshenko[10] has pointed out that in this frame the usual assumption is not valid when $\theta$ is close to a right angle†. In a similar way the three-hinged arch analysis of FIGURE 1.11 breaks down if the hinges are almost collinear‡.

---

† *Loc. cit.* p. 347. See also Chapter 2. FIGURES 2.8 and 2.9(b).

‡ The arch then acts like a toggle-joint; it will be remembered that the mechanical advantage of this lever system theoretically approaches infinity as the joint crosses the central position.

## 1.4. PRINCIPLE OF SUPERPOSITION

The theory of these rather exceptional cases, which do not often occur in the practice of structural engineering†, is dealt with at some length in Chapter 2. Their importance is there seen to lie in the light they shed on the differences between the various energy methods for analysing hyperstatic structures.

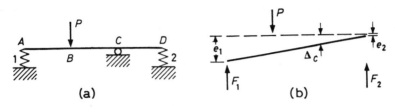

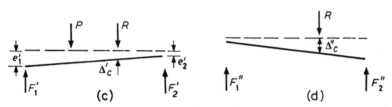

*Figure 1.24.* (a) The rigid beam $ABCD$ is carried on a roller and two springs; spring 1 is non-linear; (b) When the roller at $C$ is removed the load $P$ produces a deflexion $\Delta_C$ at $C$; (c) When $R$ is added the deflexion at $C$ becomes $\Delta'_C$; (d) The deflexion at $C$ produced by $R$ acting alone is $\Delta''_C$. Because of the non-linearity of spring 1, $\Delta'_C - \Delta_C \neq \Delta''_C$

A problem which illustrates the difficulty of dealing with non-linear structures is illustrated in FIGURE 1.24(a). The rigid beam $ABCD$ is carried on a roller at $C$, while the ends $A$ and $D$ are attached to springs which are initially unstressed. A load $P$ is applied at $B$; $AB = BC = CD$. The spring 1 is assumed to be non-linear, so that its extension or contraction $e_1$ is connected with the force $F_1$ in it by the relationship

$$e_1 = \alpha_1 F_1^2$$

The load-extension relationship for the second spring is

$$e_2 = \alpha_2 F_2$$

The reaction at $C$ can conveniently be taken as the redundant quantity. It is clear that the deflexion at $C$ is zero and appropriate use of this fact should enable a solution to be obtained.

In FIGURE 1.24(b), the beam is shown with the hinge at $C$ removed.

Then $\qquad F_1 = 2P/3$ and $F_2 = P/3$

---

† The long-span suspension bridge undergoes deflexions which are too large to be neglected and a special and rather complicated method of analysis is required.

## HYPERSTATIC STRUCTURES

Hence $\quad e_1 = \alpha_1 F_1{}^2 = \dfrac{4\alpha_1 P^2}{9} \quad$ and $\quad e_2 = \alpha_2 F_2 = \dfrac{\alpha_2 P}{3}$

From which $\quad \varDelta_C = \dfrac{e_1}{3} + \dfrac{2e_2}{3} = \dfrac{4\alpha_1 P^2}{27} + \dfrac{2\alpha_2 P}{9} \quad \ldots$ (i)

Let an unknown load $R$ be added at $C$, as in FIGURE 1.24(c).

We now have $\quad F_1' = \dfrac{1}{3}(2P + R) \quad$ and $\quad F_2' = \dfrac{1}{3}(P + 2R)$

From which the deflexion of $C$ is found to be

$$\varDelta_{C}' = \dfrac{\alpha_1}{27}(2P + R)^2 + \dfrac{2\alpha_2}{9}(P + 2R) \quad \ldots \text{(ii)}$$

Subtracting the value of $\varDelta_C$ given by (i) we have

$$\varDelta_{C}'' = \dfrac{\alpha_1}{27}(4PR + R^2) + \dfrac{4\alpha_2 R}{9} \quad \ldots \text{(iii)}$$

where $\varDelta_C''$ is the contribution of $R$ to the deflexion of $C$. If $R$ had been applied independently, as in FIGURE 1.24(d), instead of in addition to $P$, the deflexion of $C$ would have been given by

$$\varDelta_{C}'' = \dfrac{\alpha_1 R^2}{27} + \dfrac{4\alpha_2 R}{9} \quad \ldots \text{(iv)}$$

These two expressions (iii) and (iv) for the deflexion of $C$ differ by the cross-product term $\dfrac{4\alpha_1 PR}{27}$ and this difference remains to be explained.

The discrepancy between the two results is due, of course, to the non-linearity of the load-deflexion relationship of this structure. The deflexion produced by $R$ is therefore not constant but depends upon the load and deflexion existing before its application. Now the value of $R$ required for the solution of this problem is that which will just bring the deflexion of $C$ to zero with $P$ also in position; the first approach, in which the deflexion of $C$ was calculated under the action of $P$ and $R$ together, is therefore the correct one.

Hence, equating the expression (ii) to zero

$$\dfrac{4\alpha_1}{27}(2P + R)^2 + \dfrac{2\alpha_2}{9}(P + 2R) = 0$$

from which $R$ can be calculated.

The second approach, in which $R$ was applied by itself, is thus seen to be invalid.

This example shows that the Principle of Superposition does not apply to non-linear structures. In such cases it is not permissible to calculate the different effects independently and then combine them; the structure and its loads must be treated as a whole.

Non-linear structures are examined in some detail in Chapter 2 where the various energy theorems are described.

## 1.5. INTUITIVE METHODS FOR THE APPROXIMATE SOLUTION OF STATICALLY INDETERMINATE PROBLEMS

We have seen that when the equations of statics are insufficient to solve a problem further equations must be sought and that these equations can be found by considering the deflexion of the structure.

### 1.5.1. Structures in which Bending Stress Predominates

Sufficient additional equations can sometimes be found approximately by other means, as in the following example. The propped cantilever of previous examples is shown again in FIGURE 1.25(a). FIGURE 1.25(b) gives an indication to an enlarged vertical scale of the shape of the cantilever in its deflected position. There must be a point of contraflexure somewhere along the beam and it seems reasonable to estimate the position of this as being in the neighbourhood of the point marked $C$. The beam is convex upwards

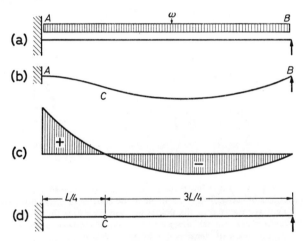

*Figure 1.25.* (a) The cantilever $AB$ is propped at $B$; (b) The deflected shape is sketched to an enlarged vertical scale. There must be a point of contraflexure in the neighbourhood of $C$; (c) The bending moment diagram must be of this form; (d) An imaginary hinge, at $C$, converts the hyperstatic structure into a determinate one

between the points $A$ and $C$ and concave upwards between $C$ and $B$, so the bending moment diagram must be of the general shape indicated by FIGURE 1.25(c). Since the bending moment is zero at a point of contraflexure, an imaginary hinge can be inserted at $C$, FIGURE 1.25(d), and if this be done the problem can at once be solved by statics.

In analytical terms the equation

$$\Sigma M = 0$$

has been obtained for the point $C$, whose position has been guessed, and this enables the single redundant quantity to be evaluated.

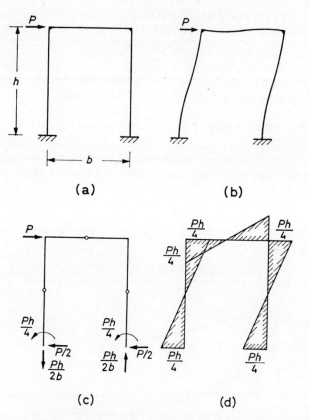

*Figure 1.26.* (a) The portal frame has three redundants; (b) The sketch of the deflected form shows that points of contraflexure occur near the centres of the beam and columns; (c) Imaginary hinges are inserted at the points of contraflexure; the restraints can then be found; (d) The bending moment diagram can then be drawn; moments are plotted on the tension side of the members

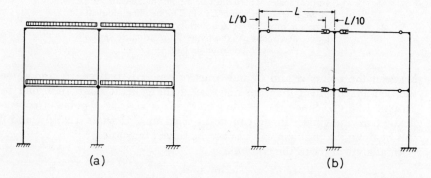

*Figure 1.27.* (a) The rigidly-jointed frame has 12 redundants; (b) Introducing 8 hinges and 4 sliding joints makes the frame determinate

## 1.5. INTUITIVE APPROXIMATE SOLUTIONS 1.5.1

Inspection shows that a reasonable guess would put $C$ at the quarter point, and with this estimate as a basis, $R_B$ is found by taking moments about $C$.

$$R_B \times \frac{3L}{4} = w \times \frac{3L}{4} \times \frac{3L}{8} \qquad R_B = \frac{3wL}{8}$$

Taking moments about $A$,

$$M_A = \frac{wL^2}{2} - R_B \times L \qquad M_A = +\frac{wL^2}{8}$$

These are in fact the correct values, and we conclude that we have guessed the position of $C$ accurately. It is found that the assumption made about the position of $C$ directly affects the value of $M_A$.

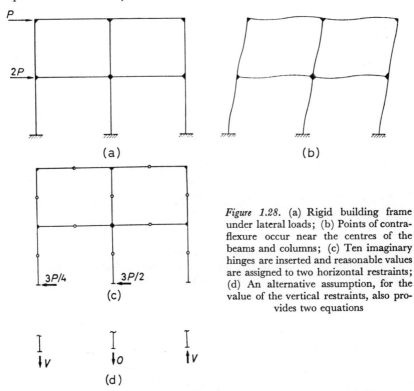

Figure 1.28. (a) Rigid building frame under lateral loads; (b) Points of contraflexure occur near the centres of the beams and columns; (c) Ten imaginary hinges are inserted and reasonable values are assigned to two horizontal restraints; (d) An alternative assumption, for the value of the vertical restraints, also provides two equations

Evidently it is necessary to make one assumption of this kind for every redundant. In the simple example just given it was quite easy to make a good estimate of the position of a single point of contraflexure but the task is more difficult in highly redundant frames. Nevertheless an order of accuracy can often be achieved by this means, even in elaborate building frames of the skyscraper type, that is quite acceptable because the superimposed loading can never be precisely known. Further developments of this approach are illustrated in the following examples.

FIGURE 1.26(a) shows a portal frame under a side load; there are three redundants so that three approximate equations are required. Consideration of the sketch of the deflected shape at (b) indicates that there are points of contraflexure near the middle of the three members and if imaginary hinges are introduced at those points, as at (c), the frame becomes determinate. The reactions and the bending moment diagram are then as shown. The horizontal components of the reactions are seen to be equal, which is reasonable in view of the symmetry of the frame, and this could have been used to provide an equation instead of assuming the point of contraflexure in the beam. If the columns had been unequal it would have been easier to proportion the horizontal restraints in a rational manner than to guess the shape of the beam.

Vertical loads on multi-storey rigid frames can be treated similarly. FIGURE 1.27(a) shows such a frame under symmetrical distributed loading. There are 12 redundant stress components, 8 of which can be obtained by assuming suitable positions for the points of contraflexure; in this case the beams will act as if their ends were partially fixed and hinges can be inserted at about one-tenth of the span. The remaining equations come from the consideration that the axial force in the beams is negligible so that four of the hinges can be treated as sliding joints, as at (b).

Lateral loads on such frames produce the sort of distortion indicated in FIGURE 1.28(b). Ten positions for imaginary hinges immediately suggest themselves; two further equations follow from assumptions about the reactions. The horizontal restraints, for instance, can be proportioned suitably, a common procedure being to assign equal forces to all the interior columns and half that force to each of the external columns, as shown at (c). Alternatively the vertical forces in the columns can be assumed to be proportional to the distance of the columns from the centre of the frame as at (d).

### 1.5.2. STRUCTURES IN WHICH DIRECT STRESS PREDOMINATES

Although simplifying assumptions are often made to enable overstiff braced structures to be treated as though they were determinate, it is not always easy to justify the procedure by postulating reasonable proportions for the deflected shape. The truss of FIGURE 1.29(a), for example, which has two

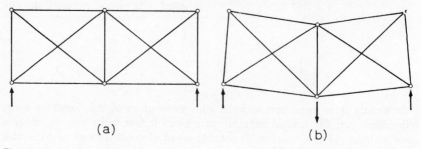

*Figure 1.29.* (a) This truss has two redundant members which can be taken to be one diagonal in each panel; (b) In the deflected position the contraction of the compression diagonals is about equal to the extension of the tension diagonals

redundant members would sometimes be analysed by assuming that the diagonal members in each panel were equally loaded. The justification for this is that when the truss has deflected into the shape indicated at (b) the extension of one diagonal is about the same as the compression of the other.

The same stress will be induced in each and, if the cross-sectional areas are equal, the forces in each will be numerically equal but of opposite sign.

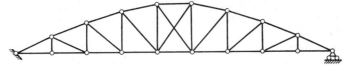

*Figure 1.30.* The counterbraced centre panel can be designed on the assumption that the slender compression diagonal carries negligible load

While such a procedure might serve for a preliminary analysis it would hardly be adequate for the final design of a bridge, for example, where the loadings are usually known with fair accuracy.

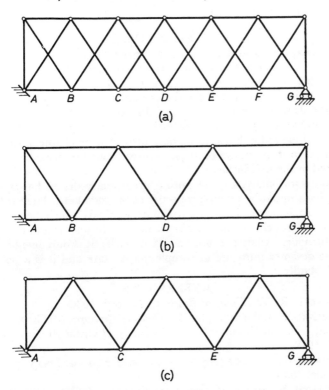

*Figure 1.31.* Multiple system trusses, as at (a) are sometimes split up into component trusses (b) and (c). The deflexions of these trusses are not independent, however, and any applied load must be shared between them

In light lattice structures, such as transmission towers, the alternative assumption that the long slender diagonals cannot carry compressive stress seems to be reasonable. The same assumption is often made in dealing with the counterbraced central panels of otherwise determinate bridge trusses, such as that shown in FIGURE 1.30. The designer now has the opportunity of detailing the members to encourage conformity with the assumption by proportioning the cross-section of these members so that the second moment of area, about one principal axis, is rather small; the buckling load, which effectively defines the maximum value of the compressive stress, is correspondingly small.

On the other hand the practice of dividing a multiple-system truss into two or more component trusses may be wrong in principle. The truss shown in FIGURE 1.31(a) would be split up, according to this procedure, into the two trusses (b) and (c); a load at joint $C$ would then be assigned exclusively to truss (c), while truss (b) is idle. In the complete truss, however, such a load is inevitably shared between the two trusses since truss (b) cannot deflect without carrying truss (c) with it. This method, then, leads to over-design and should only be adopted in minor structures where a more elaborate analysis would not be justified.

## 1.6. CONCLUSION

In this chapter an attempt has been made to encourage the reader to visualize the mode of action of the various structures which have been considered rather than to rely exclusively upon formulae. We saw that such a rational attitude was essential for the proper interpretation of the rules for finding the degree of redundancy and that the selection of the redundant members and restraints must inevitably be a subjective process. The methods for finding these redundants, to be described in later chapters, can only be effectively used if the statics of the structures concerned have been properly interpreted in the first instance.

Nor can a full analysis begin until a preliminary design has been made for it depends upon the elastic properties of the members. Here is the most important application of the approximate methods which have just been described and which call for an almost instinctive understanding of hyperstatic behaviour. Analysis is not an end in itself; it is only one of the tools which the designer must use in completing his task and it is a poor one if it is used blindly.

## REFERENCES

[1] SOUTHWELL, R. V. *Theory of Elasticity*. Oxford. 1936
[2] TIMOSHENKO, S. and YOUNG, D. H. *Theory of Structures*. McGraw-Hill. 1945
[3] WILBUR, J. B. and NORRIS, C. H. *Elementary Structural Analysis*. McGraw-Hill. 1948
[4] MERCHANT, W. and BOLTON, A. *An Introduction to the Theory of Structures*. Blackie. 1956
[5] TIMOSHENKO, S. *History of Strength of Materials*. McGraw-Hill. 1945
[6] FIFE, W. M. and WILBUR, J. B. *Theory of Statically Indeterminate Structures*. McGraw-Hill. 1937
[7] HENDERSON, J. C. de C. and BICKLEY, W. G. 'Statical Indeterminacy of a Structure'. *Aircraft Engineering*. Dec. 1955
[8] ROXBEE COX, H. 'Theorems for the synthesis of simply-stiff framework'. *J. Roy. Aer. Soc.* Vol. 49. 1945. p. 21
[9] MATHESON, J. A. L. 'Degree of redundancy of plane frameworks'. *Civ. Eng. and P.W. Review*. Vol. 52. 1957. p. 655
[10] TIMOSHENKO, S. *Strength of Materials*. Part I. Macmillan. 1930

CHAPTER 2

# THE ENERGY THEOREMS OF STRUCTURAL ANALYSIS

## 2.1. INTRODUCTION

In Chapter 1 we saw that the equations of equilibrium are not in themselves sufficient for the analysis of hyperstatic structures and that the deformation of the structure must be taken into account to give the equations of compatibility. In simple cases, a direct comparison of displacements gave a solution, but more powerful methods are usually required in engineering practice.

These methods can nearly all be traced back to one or other of the group of fundamental energy theorems to be described in this chapter. We shall see too that although each of these theorems has its own special field of application and its own limitations, which must be clearly recognized by the student, there are interesting family relationships between the members of the group.

Although these theorems have been known individually for many years their interconnection has not been at all widely recognized until quite recently. The first systematic approach seems to have been by Williams[1]; and Brown[2] has given a more complete account paying particular attention to the validity of the theorems in cases when Hooke's Law does not apply.

A further important study of energy theorems by Argyris and Kelsey[19] has been included as an appendix at the end of this book.

## 2.2. WORK AND ENERGY

It is perhaps only necessary here to remind the reader of the fundamental ideas which spring from Newton's Laws of Motion.

A *force*, it will be recalled, is that which can change the motion of a mass and its magnitude is measured in terms of the acceleration which it gives to unit mass. When a force acting on a mass moves its point of application then *work* is said to be done, equal to the product of the force and the distance moved, which manifests itself as an increase of the *kinetic energy* of the mass if it is accelerated to a higher velocity, or *potential energy* if it is elevated in the gravitational field.

Forces distort the structures to which they are applied and, as a result, work is done which is stored up in the structure as a particular form of potential energy known as *strain energy*. Some mechanical devices, such as clocks, employ springs as reservoirs of strain energy from which useful work can be drawn as required. The two forms of potential energy used in weight- and spring-actuated clocks respectively should be noted.

# THE ENERGY THEOREMS OF STRUCTURAL ANALYSIS

We are accustomed, then, to the idea that work and energy can be interchanged and that energy can be converted from one form to another.

It is instructive to trace the sequence of events depicted in FIGURE 2.1, where the original potential energy of the sphere is converted successively to kinetic, strain and kinetic energy until the sphere comes to rest again at the end of the cycle with its original potential energy restored. In this, as

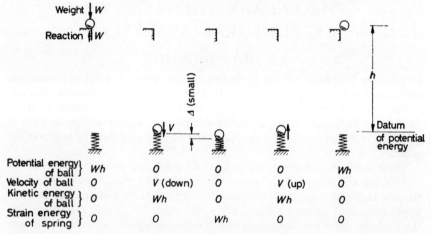

*Figure 2.1.* When the weight $W$ is allowed to fall its potential energy, which is due to its elevated position in the gravitational field, is converted in turn to kinetic energy, to strain energy of the compressed spring, and to kinetic energy again. It finally comes to rest at the original elevation, with its original potential energy restored, if no energy is lost during these transformations

in many engineering problems, we are assuming that the Principle of Conservation of Energy applies in a rigorous way to the system under consideration. Mechanical energy is assumed to be convertible from one *useful* form to another without loss; no energy is dissipated as heat. Although this convenient postulate is never realized in practice, the divergence between the ideal and the real does not often lead to serious error in structural problems, at least when steady as opposed to fluctuating loads are involved.

The most interesting stage in the cycle, for our present purpose, is when the spring is compressing; in the following section we examine that stage in more detail.

## 2.3. STRAIN AND COMPLEMENTARY ENERGY

In considering the energy balance of a system consisting of a series of loads acting on a structure, it is essential to distinguish between the work done and the energy lost by the loads on the one hand and the strain energy gained by the structure on the other. The latter is governed entirely by the load-deflexion characteristics of the structure and these may or may not be linear; if the structure is elastic, however, so that no energy is wasted during its distortion, the same relationship will connect load and deflexion both for increasing and for decreasing loads. It is most instructive to consider the non-linear case first and then to introduce such modifications as are permissible because deflexion is proportional to load.

## 2.3. STRAIN AND COMPLEMENTARY ENERGY

Thus the curve shown in FIGURE 2.2(a) is supposed to represent the displacement characteristic of a spring which can now be regarded as typifying a more general structure. This curve has been obtained, we are to imagine, by compressing the spring slowly with a hydraulic jack and recording simultaneously the thrust of the jack and the compression of

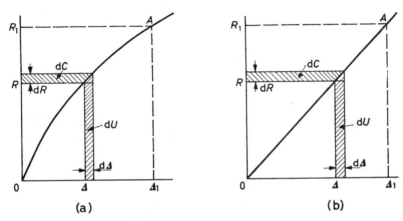

Figure 2.2. The reactive thrust $R$ exerted by an elastic body, a spring for example, whose deflexion is $\Delta$. (a) Non-linear; (b) Linear

the spring. This technique ensures that at all times the reaction of the spring is equal to the action of the jack, and from now on we can be certain that when the spring is compressed an amount $\Delta$ it will exert a resistance $R$, as indicated by the line $OA$, both when the load is increasing and when it is decreasing.

From the definition of work given above we see that as the thrust of the spring increases from $R$ to $R + dR$ and the compression increases from $\Delta$ to $\Delta + d\Delta$ the work done will be approximately

$$\left(R + \frac{dR}{2}\right) d\Delta \simeq R \cdot d\Delta$$

which is the small shaded area marked $dU$ in FIGURE 2.2.
The total work done or strain energy stored as the spring is compressed an amount $\Delta_1$ from the relaxed state is evidently the area under the curve $OA$.

Hence, Strain Energy $\qquad U = \int_0^{\Delta_1} R \cdot d\Delta \qquad \ldots$ (2.1)

It follows at once from this definition that $\dfrac{\partial U}{\partial \Delta} = R$; this equation will later (page 70) take its place, as equation 2.7, as one of the basic energy theorems under the title 'Castigliano's Theorem, Part I'.

In the linear case shown at (b) in FIGURE 2.2 the area under the line $OA$ is known at once and we have

$$U = \tfrac{1}{2} R_1 \Delta_1 \qquad \ldots \text{(2.2)}$$

## THE ENERGY THEOREMS OF STRUCTURAL ANALYSIS

If the spring is now replaced by a single bar then $R$, the reaction to the load, is equal to $F$, the force in the bar; the expression 2.1 for the strain energy then becomes

$$U = \int_0^{e_1} F \cdot de \qquad \ldots \ldots (2.1(a))$$

where $e$ is the change of length of the bar, and for a structure composed of an assemblage of $n$ bars we can write

$$U = \sum_n \int_0^{e_n} F_n \, de \qquad \ldots \ldots (2.1(b))$$

If each bar obeys Hooke's Law and has a modulus of elasticity $E$, a length $L$ and a cross-sectional area $A$, then the change of length $e$ and the stress $f$ are given by

$$e = \frac{FL}{AE} \text{ and } f = \frac{F}{A}$$

The expression 2.2 can then be written

$$U = \frac{F^2 L}{2AE} = \frac{AEe^2}{2L} = \frac{f^2 AL}{2E} \qquad \ldots \ldots (2.2(a))$$

and for an assemblage of such bars

$$U = \sum \frac{F^2 L}{2AE} = \sum \frac{AEe^2}{2L} = \sum \frac{f^2 AL}{2E} \qquad \ldots \ldots (2.2(b))$$

The area to the left of the line $OA$ in FIGURE 2.2, made up of such elements as that marked $dC$, is evidently

$$\int dC = C = \int_0^{R_1} \Delta \cdot dR \qquad \ldots \ldots (2.3)$$

This quantity is called the 'complementary energy' and, as will appear, important conclusions can be drawn about it.

For example, we see at once that by definition $\dfrac{\partial C}{\partial R} = \Delta$. This is given later (page 73) in a more general form as equation 2.10, under the title 'First Theorem of Complementary Energy'.

The complementary energy of a structure composed of an assemblage of bars is

$$C = \sum_n \int_0^{F_n} \Delta_n dF_n \qquad \ldots \ldots (2.3(a))$$

In the linear case (b) it appears that the strain and complementary energies are equal. For

$$U = \tfrac{1}{2} R_1 \Delta_1 = C \qquad \ldots \ldots (2.3(b))$$

## 2.4. SLOWLY AND SUDDENLY APPLIED FORCES

In describing the calibration of the spring we specified that at all times the applied force $P$ was to be kept equal to the thrust $R$ of the spring by slowly increasing the pressure in the jack. If, however, the applied force is increased in an arbitrary manner then equilibrium is only possible when the deflexion is such that $P$ and $R$ are equal; in addition the system will only be at rest if the kinetic energy is zero.

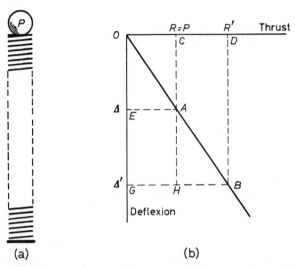

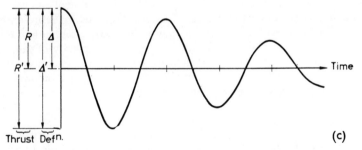

*Figure 2.3.* Sudden application of load. (a) With the spring relaxed the load is suddenly released; (b) The thrust $R$ in the spring and its deflexion $\Delta$ are related by the line $OAB$

(c) When the load is suddenly applied the reactive thrust in the spring temporarily reaches $R'$ but oscillation takes place about the equilibrium position where $P=R$

Suppose that the spring shown in FIGURE 2.3(a) has the thrust-deflexion relationship shown at (b). A load $P$ is brought into contact by some external agency and then suddenly released. As the gravitational attraction on the load is constant the force $P$ has the constant value indicated by the line $CAH$. The reaction $R$ of the spring grows, according to the line $OAB$, until it is equal to $P$ when the deflexion is $\Delta$.

37

At this stage the potential energy lost by the load $P$ is $P\Delta$, represented by the area $OCAE$, while the strain energy gained by the spring is only the area $OAE$. The difference between these two, the area $OCA$, measures the kinetic energy of the load at this stage. Further movement must therefore occur until the energies are balanced when the deflexion is $\Delta'$. The area $OCHG$, representing the potential energy lost by the load, is then equal to the area $OBG$ which represents the strain energy of the spring. The spring reaction is then $R'$, which is obviously much greater than $P$, and the spring will rebound.

It can be seen that oscillation will occur about the position where the deflexion is $\Delta$, as indicated by FIGURE 2.3(c) until the energy represented by the area $OCA$ has been dissipated. It is apparent that if $OAB$ is a straight line then

$$\Delta' = 2\Delta \text{ and } R' = 2P$$

so that the maximum thrust in the spring is twice that in the equilibrium position†.

On the other hand if we imagine the weight $P$ to be slowly lowered on to the spring by the external supporting agency then there will be a gradual reduction of the force in the support and a corresponding growth of the reaction of the spring; at any stage of the lowering operation these two forces just balance the weight. When the deflexion reaches the value $\Delta$ the whole of the weight is carried by the spring and the external agency is completely unloaded.

The loss of potential energy of the weight, as it is lowered, is at every stage equal to the work done on the spring (its strain energy) together with the work done against the resistance of the support. In the final position these energies are represented by the areas $OCAE$, $OAE$ and $OCA$ respectively.

Thus whether the load is applied slowly or suddenly the strain energy of the spring is eventually equal to the area $OAE$ and is controlled entirely by the force-deflexion law of the spring. In both cases energy represented by the area $OCA$ has been lost, in the one case as heat and in the other as work done on the support. This area is the complementary energy of the spring.

It thus appears that when a load deforms a structure the complementary energy of the structure is that part of the potential energy lost by the load which cannot be transformed into strain energy of the structure.

## 2.5. THE PRINCIPLE OF SUPERPOSITION

The assumption underlying the method of the direct comparison of displacements, which was used in Chapter 1 for analysing simple hyperstatic structures, is that the deflexion of a point on a structure is independent of the order of application of the loads producing it. This is equivalent to the statement that the deflexion produced by several forces is the sum of the deflexions which would have been produced by the individual forces acting separately.

---

† An additional reservation should really be made in the case of rapidly applied loads, for the same stress will only occur simultaneously in all parts of the structure if its physical dimensions are not too great. The acceleration stresses in long winding ropes, for example, are affected by the fact that a stress wave takes an appreciable time to travel from one end to the other[3].

This statement, which is known as the Principle of Superposition, can be shown to be valid only when Hooke's Law, that deflexions are proportional to the forces which produce them, is obeyed. Southwell's[4] proof consists of applying two forces to a body, one after the other, and then removing them in reverse order. The deflexion is evaluated at each stage and it is eventually shown that the final deflexion has the expected value of zero only when Hooke's Law is obeyed. In the discussion attention is concentrated on the deflexion of a structure as a whole rather than on the contributions of its individual members.

There are, however, two aspects of this matter of superposition which should be considered.

*The superposition of forces* is achieved if the same increment of external load always produces the same increment of force in the members whatever total load is acting.

Statically determinate structures satisfy this requirement provided the loads are not such as to produce gross distortion; for in analysing these structures no account need be taken of the elastic properties of the members. The force in them depends only on the geometry of the frame.

Hyperstatic structures only satisfy this requirement if their individual members obey Hooke's Law because the distribution of load between the members, as we have seen, depends on relative deflexions; if these do not bear a constant ratio to one another then the load distribution cannot be constant.

*The superposition of deflexions* is achieved when the same increment of strain in a member always makes the same contribution to the deflexion of the structure as a whole whatever the magnitude of the deflexion existing when the strain is imposed.

This means that gross distortion of the structure must be excluded, for if the angles between the members change to an appreciable extent then their sines and cosines also change and the essential geometry of the structure is altered. This can also happen if some of the angles concerned are close to 0° or 90°, for then quite small changes of angle involve rather large changes in sines or cosines.

In seeking to apply one or other method of analysis we must be sure that the assumption of superposition, whether of forces or of deflexions, is only made when it is justifiable.

## 2.6. THE ENERGY THEOREMS

We are now in a position to use the conceptions of work, potential and strain energy in the derivation of the basic theorems of structural analysis. In the following sections these theorems are presented in what seems to be the most logical order and they are given titles by which they can be readily identified. The strain energy theorems introduced by Castigliano are given his original titles rather than those, sometimes used by later writers, which are given in footnotes.

We begin by introducing those theorems which concern the total potential energy of the structure and its superimposed loads: then follow the theorems which involve only the strain energy of the structure itself and finally the theorems involving its complementary energy are given.

## 2.6.1. The Theorem of Minimum Potential Energy

The general theorem in mechanics that 'the total potential energy of a system has a stationary value when that system is in equilibrium and a minimum value when the equilibrium is stable' is often illustrated by reference to a body on various surfaces as shown in Figure 2.4. The only forces involved here are the weight of the body and the reaction of the surface and the only energy is the potential energy associated with the body's

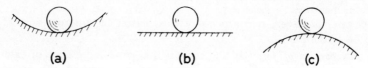

Figure 2.4. Stationary Potential Energy. (a) Potential energy minimum: equilibrium stable; (b) Potential energy constant: equilibrium neutral; (c) Potential energy maximum: equilibrium unstable

position in the gravitational field. In each case the body is in equilibrium and the potential energy has a stationary value, for slight lateral movements of the body involve no corresponding vertical movements. Moreover the equilibrium is stable, neutral and unstable in Figure 2.4(a), (b) and (c) when the potential energy is respectively least, constant and greatest.

In order to use this theorem in structural problems it is necessary to combine the potential energy $V$ of the loads with the strain energy $U$ of the structure to give the total potential energy of the system in terms of the relevant independent displacements.

The results of small changes of position are obtained by differentiating the total potential energy with regard to these displacements. The theorem states that such small displacements from the equilibrium position involve insignificant changes of total potential energy so that the equilibrium position is given by

$$\frac{\partial}{\partial \Delta_1}(U+V) = \frac{\partial}{\partial \Delta_2}(U+V) = 0 \qquad \ldots (2.4)$$

In hyperstatic problems of the kind with which we are concerned this procedure amounts to a preliminary satisfaction of the compatibility conditions followed by expression of the equilibrium condition of minimum potential energy. No question of linearity, or otherwise has so far arisen so that the procedure is applicable whatever the law relating deformation and load, but large deformations involving gross distortion may be difficult to handle because of the mathematical complications.

### 2.6.1.1. *Examples of the use of the Principle of Minimum Potential Energy*

(a) The framework shown in Figure 2.5(a) consists of uniform bars which have the same cross-section and modulus of elasticity, i.e. they obey Hooke's Law. Find the force in the bars. This will be recognized as the same problem as in Figure 1.10, which was solved by taking $AC$ as a redundant member and equating the vertical components of the deflexions of the three bars on the assumption that the angle $\theta$ did not change appreciably.

## 2.6.1. THEOREM OF MINIMUM POTENTIAL ENERGY 2.6.1.1

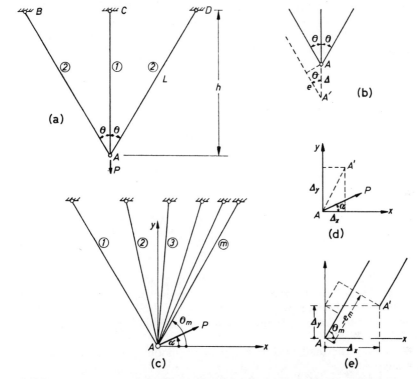

*Figure 2.5.* Analysis of a hyperstatic structure by means of the Principle of Minimum Potential Energy. (a) Simple symmetrical frame; (b) Compatibility relationship at $A$ between $\Delta$ and $e$; (c) Unsymmetrical frame; (d) The joint $A$ is supposed to be displaced to $A'$; (e) Compatibility relationships

We now make the same assumption about $\theta$ in writing the compatibility relationship, and then use the Principle of Minimum Potential Energy to give the equilibrium condition. The diagram at (b) shows the connexion between $e$ and $\Delta$ after a small vertical deflexion has occurred. Then the strain energy of the bars, from equation 2.2(b) is

$$U = \sum \frac{AEe^2}{2L} \qquad \ldots \text{(i)}$$

and writing, for the inclined bars $AB$ and $AD$

$$e = \Delta \cos\theta \text{ and } L \cos\theta = h$$

$$U = 2U_{AB} + U_{AC}$$

$$= \frac{AE(\Delta \cos\theta)^2 \cos\theta}{h} + \frac{AE\Delta^2}{2h} \qquad \ldots \text{(ii)}$$

The potential energy of the load is given by

$$V = V_0 - P\Delta \qquad \ldots \text{(iii)}$$

where $V_0$ is the arbitrary potential energy at the start of the proceedings and $P.\varDelta$ is the loss incurred during the displacement. We now contemplate a small displacement $d\varDelta$ about the equilibrium position involving zero change in the potential energy.

Thus
$$\frac{d}{d\varDelta}(U+V) = \frac{2AE\varDelta \cos^3 \theta}{h} + \frac{AE\varDelta}{h} - P \quad \ldots \text{(iv)}$$

$$= 0 \text{ if}$$

$$\varDelta = \frac{Ph}{AE(1 + 2\cos^3 \theta)}$$

The force in the bar $AC$, for example, is given by

$$F_{AC} = \frac{AE\varDelta}{h}$$

$$= \frac{P}{1 + 2\cos^3 \theta} \quad, \text{ as before.}$$

(b) The method shows to greater advantage if there are many bars as in FIGURE 2.5(c). In the absence of symmetry we can no longer assume that the deflexion $\varDelta$ occurs in the line of action of the load $P$, and we begin by considering the loss of potential energy involved in the displacement shown at (d).

Resolving the load into its components in the $x-$ and $y-$ directions, we have
$$V = V_0 - P\cos\alpha \cdot \varDelta_x - P\sin\alpha \cdot \varDelta_y$$

and
$$\frac{\partial V}{\partial \varDelta_x} = -P\cos\alpha \qquad \frac{\partial V}{\partial \varDelta_y} = -P\sin\alpha$$

The diagram at (e) shows the connexion between the extension of the $m$th bar and the component deflexions $\varDelta_x$ and $\varDelta_y$.

$$e_m = -\varDelta_y \sin\theta_m - \varDelta_x \cos\theta_m \quad \ldots \text{(i)}$$

Hence, generally
$$\frac{\partial e}{\partial \varDelta_x} = -\cos\theta \qquad \frac{\partial e}{\partial \varDelta_y} = -\sin\theta$$

As
$$U = \sum_m \frac{AEe^2}{2L}$$

we have
$$\frac{\partial}{\partial \varDelta_x}(U+V) = \sum_m \frac{AEe}{L}\frac{\partial e}{\partial \varDelta_x} + \frac{\partial V}{\partial \varDelta_x}$$

$$= \sum_m \frac{AE}{L}(\varDelta_y \sin\theta + \varDelta_x \cos\theta)\cos\theta - P\cos\alpha = 0$$

## 2.6.1. THEOREM OF MINIMUM POTENTIAL ENERGY   2.6.1.1

Hence

$$\Delta_x \sum_m \frac{AE}{L} \cos^2 \theta + \Delta_y \sum_m \frac{AE}{L} \sin \theta \cos \theta = P \cos \alpha$$

Similarly

$$\frac{\partial}{\partial \Delta_x}(U+V) = \Delta_x \sum_m \frac{AE}{L} \sin \theta \cos \theta + \Delta_y \sum_m \frac{AE}{L} \sin^2 \theta = P \sin \alpha$$

.... (ii)

Solution of these two equations for $\Delta_x$ and $\Delta_y$ enables the extension, and hence the stress, of each bar to be obtained from (1).

It is to be remarked that if equation 1.3 is employed to determine the number of redundant bars in the framework, we find that

$$m = m; \quad r = 2m; \quad j = m + 1$$

Hence

$$m + r - 2j = m - 2$$

The method of solution given in Chapter 1, and indeed any method in which the bar forces are treated as the redundant quantities and evaluated directly, would require the solution of $m - 2$ simultaneous equations of compatibility.

In the present method two equations only, corresponding to the two possible components of deflexion of the joint $A$, have been required and this is clearly a great improvement†. In general we can say that the Minimum Potential Energy method usually shows to good advantage when the number of degrees of freedom of the joints is less than the number of redundant bars.

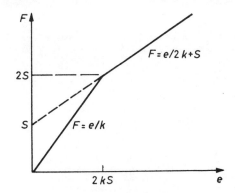

*Figure 2.6.* Assumed load-extension diagram for material with proportional limit of $2S$

(c) Returning to Figure 2.5(a) we now examine the case when the bars have the non-linear load-extension law shown in Figure 2.6. No generality is lost if $\theta$ is taken equal to $60°$; the solution can then be compared with that given by Westergaard[5].

---

† It may be thought that there is some inconsistency here; but if the conditions of equ - librium and compatibility are written down formally it will be found that both methods of approach use $m$ equations to find the $m$ unknown bar forces.

Let suffix 1 refer to $AC$ and suffix 2 to $AB$ and $AD$.
When $F_1 < 2S$; $F_2 < 2S$

The strain energy $U$ of a bar is

$$U = \int_0^e F.de = \int e/k \, de \ldots e^2/2k$$

As $\quad e_1 = \Delta$ and $e_2 = \Delta \cos \theta = \Delta/2$ for $\theta = 60°$

$$U + V = \frac{\Delta^2}{2k} + 2\left(\frac{\Delta^2}{8k}\right) + V_0 - P\Delta$$

$$\frac{d}{d\Delta}(U + V) = \frac{\Delta}{k} + \frac{\Delta}{2k} - P$$

$$= 0 \text{ if } \Delta = \frac{2Pk}{3}$$

i.e. $\quad \left. \begin{array}{l} F_1 = 2P/3 \\ F_2 = P/3 \end{array} \right\}$

This solution holds so long as $P < 3S$; above that load the bar $AC$ will have passed its proportional limit and its force will follow the law

$$F = e/2k + S$$

When $F_1 > 2S$; $F_2 < 2S$

i.e. $\quad F_1 = e_1/2k + S$; $\quad F_2 = e_2/k$

$$U_1 = \int F.de = \int_0^{2kS} e/k \, de + \int_{2kS}^{e_1}(e/2k + S)de = \frac{e_1^2}{4k} + Se_1 - kS^2$$

$$U_2 = \int_0^{e_2} F.de = e_2^2/2k$$

$$U + V = \frac{\Delta^2}{4k} + S\Delta - kS^2 + 2\left(\frac{\Delta^2}{8k}\right) + V_0 - P\Delta$$

$$\frac{d}{d\Delta}(U + V) = \frac{\Delta}{2k} + S + \frac{\Delta}{2k} - P$$

$$= 0 \text{ if } \Delta = (P - S)k$$

i.e. $\quad \left. \begin{array}{l} F_1 = \dfrac{P + S}{2} \\ F_2 = \dfrac{P - S}{2} \end{array} \right\}$

This solution holds so long as $P < 5S$; above that load all the bars will have passed the proportional limit.

## 2.6.1. THEOREM OF MINIMUM POTENTIAL ENERGY

When $F_1 > 2S$; $F_2 > 2S$

i.e. $F = e/2k + S$ for both bars.

$$U_1 = e_1^2/4k + Se_1 - kS^2 = \frac{\Delta^2}{4k} + S\Delta - kS^2$$

$$U_2 = e_2^2/4k + Se_2 - kS^2 = \frac{\Delta^2}{16k} + \frac{S\Delta}{2} - kS^20$$

$$U + V = \frac{\Delta^2}{4k} + S\Delta - kS^2 + 2\left(\frac{\Delta^2}{16k} + \frac{S\Delta}{2} - kS^2\right) + V_0 - P\Delta$$

$$\frac{d}{d\Delta}(U + V) = \frac{\Delta}{2k} + S + \frac{\Delta}{4k} + S - P$$

$$= 0 \text{ if } \Delta = \frac{(P - 2S)\,4k}{3}$$

i.e.
$$F_1 = \frac{2P - S}{3}$$
$$F_2 = \frac{P + S}{3}$$

Finally it should be noted that in this example the signs of the member forces are self-evident. In more general cases, especially where the bars have different load-extension and load-compression laws, alternative solutions may need to be explored.

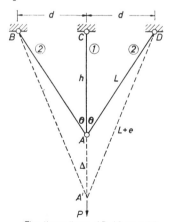

The three lines $AB, AC$ and $AD$ represent flexible helical springs

*Figure 2.7.* Frame constructed of flexible springs so that the deflexion can no longer be considered small in relation to the lengths of the members

(d) In the previous examples we have imposed the restriction that the displacement $\Delta$ is so small that $\theta$ can be regarded as constant. We now relax this requirement, and consider the case shown in FIGURE 2.7, when

$\Delta$ is of the same order as the other dimensions. The load-extension relationship is taken to be linear so that

$$F_1 = a_1 \Delta \quad \text{and} \quad F_2 = a_2 e$$

Hence $\quad U_1 = \tfrac{1}{2} a_1 \Delta^2 \quad \text{and} \quad U_2 = 2 \cdot \tfrac{1}{2} \cdot a_2 e^2 = a_2 e^2$

The geometry of the diagram gives

$$2eL + e^2 = 2\Delta h + \Delta^2 \qquad \ldots \text{(i)}$$

i.e. $\qquad 2L.de + 2e.de = 2h.d\Delta + 2\Delta.d\Delta$

$$\frac{de}{d\Delta} = \frac{\Delta + h}{e + L} \qquad \ldots \text{(ii)}$$

Solving (i) for $e$ gives $\quad e = -L + \sqrt{L^2 + \Delta^2 + 2\Delta h} \qquad \ldots \text{(iii)}$

and hence (ii) becomes $\quad \dfrac{de}{d\Delta} = \dfrac{\Delta + h}{\sqrt{L^2 + \Delta^2 + 2\Delta h}} \qquad \ldots \text{(iv)}$

Now $\qquad \dfrac{d}{d\Delta}(U + V) = \dfrac{dU_1}{d\Delta} + \dfrac{dU_2}{d\Delta} - P$

$$= a_1 \Delta + 2 a_2 e \frac{de}{d\Delta} - P \qquad \ldots \text{(v)}$$

i.e.

$$0 = a_1 \Delta + 2 a_2 \{ -L + \sqrt{L^2 + \Delta^2 + 2\Delta h} \} \frac{\Delta + h}{\sqrt{L^2 + \Delta^2 + 2\Delta h}} - P \quad \ldots \text{(vi)}$$

There is no ready means of solving this equation explicitly for $\Delta$ but it can be solved by trial and error.

It is to be noticed that if $\Delta$ is small compared with the other dimensions and if $a_1 = AE/h$ and $a_2 = \dfrac{AE \cos \theta}{h}$

then $\qquad \sqrt{L^2 + \Delta^2 + 2\Delta h} \simeq \sqrt{L^2 + 2\Delta h}$

$$\simeq L \left( 1 + \frac{2\Delta h}{L^2} \right)^{\frac{1}{2}}$$

$$\simeq L \left( 1 + \frac{\Delta h}{L^2} \right) = L + \Delta \cos \theta$$

then $\qquad \dfrac{\Delta + h}{\sqrt{L^2 + \Delta^2 + 2\Delta h}} \simeq \dfrac{\Delta + h}{L + \Delta \cos \theta}$

$$\simeq \frac{h}{L} = \cos \theta$$

## 2.6.1. THEOREM OF MINIMUM POTENTIAL ENERGY

Equation (v) then becomes

$$0 = \frac{AE\Delta}{h} + \frac{2AE \cos \theta}{h} (\Delta \cos \theta) \cos \theta - P$$

giving 
$$P = \frac{AE\Delta}{h} (1 + 2 \cos^3 \theta) \qquad \ldots\ (vii)$$

as in example (a).

FIGURE 2.9(a) shows the results obtained for a frame, similar to that of FIGURE 2.7 having d = 30, h = 40 and L = 50; the bars AB, BC and CD are made of steel and have a cross sectional area of 1. The full line gives the load-deflexion relationship as obtained from equation (vi) above, while the chain line is obtained from equation (vii). It will be seen that there is no appreciable departure from linearity until $\Delta/h$ is about 3 per cent, corresponding to $P/AE \approx 0 \cdot 06$; this is far in excess of $P/AE \approx 0 \cdot 003$ that induces yield in the central bar. It therefore seems that non-linear behaviour caused by gross distortion is very unlikely to occur in normal steel structures.

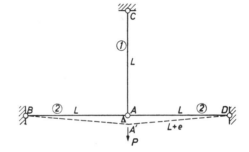

*Figure 2.8.* In this case the junction A initially lies on the straight line BD so that the angle $\theta$ of *Figure 2.7* is now a right angle

(e) Example (d) can be simplified in a different way if the two side bars are initially horizontal, as in FIGURE 2.8. In this case the load taken by the bars AB and AD depends essentially on the occurrence of appreciable deflexion for if $\Delta$ is zero then the forces in AB and AD have no vertical components and AC takes all the load.

Putting $h = 0$ in the equation (iii) of example (d) we obtain

$$e = -L + \sqrt{L^2 + \Delta^2}$$
$$= -L + L\left(1 + \frac{\Delta^2}{L^2}\right)^{\frac{1}{2}}$$
$$\simeq \Delta^2/2L \text{ if } \frac{\Delta}{L} \text{ is small} \qquad \ldots\ (iii(a))$$

and hence 
$$\frac{de}{d\Delta} = \frac{\Delta}{L} \qquad \ldots\ (ii(a))$$

Now 
$$U + V = \tfrac{1}{2}a_1\Delta^2 + a_2 e^2 + V - P\Delta$$

as in Example (d), and hence

47

$$\frac{d}{d\varDelta}(U+V) = a_1\varDelta + 2a_2 e \frac{de}{d\varDelta} - P$$

$$= a_1\varDelta + 2a_2 \frac{\varDelta^2}{2L} \cdot \frac{\varDelta}{L} - P = 0$$

i.e. $$P = a_1\varDelta + \frac{a_2 \varDelta^3}{L^2}$$

from which $\varDelta$ can be found.

It is to be observed that if $\varDelta/L > 10$ per cent (say) the approximation used in the derivation of this formula may no longer be sufficiently accurate; equation (vi) of the previous example must then be employed.

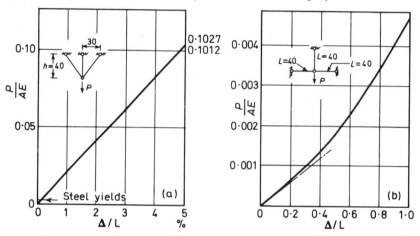

Figure 2.9. (a) Relation between $P$ and $\varDelta/h$ for the frame of Figure 2.7 if the members are supposed to be mild steel and to have a cross-sectional area of 1. The relationship only becomes appreciably non-linear when the load is many times that required to induce yield in the most highly-stressed member. The full line shows the result obtained from equation (vi) of Section 2.6.1.1. Example (d), while the chain line is obtained from equation (vii); (b) Relation between $P$ and $\varDelta/L$ for the frame of Figure 2.8, if the identical members have stiffness $a$ equal to 1. The full line shows the deflexion calculated from equation (vi) of Section 2.6.1.1. Example (d) while the chain line is the deflexion on the assumption that the central bar $AC$ takes all the load

FIGURE 2.9(b) shows the load-deflexion relationship for the frame of FIGURE 2.8, taking $L = 40$ and $a_1 = a_2 = 1$. When the load (and consequently $\varDelta$) is very small it is carried mainly by the central bar. As the deflexion increases the side bars carry a growing share of the load and the stiffness of the frame increases.

2.6.1.2. *Comments on the use of the method of Minimum Potential Energy*

The solutions to the problems given above are sufficient to show that the method under discussion is valid for all frameworks, linear or otherwise, and that even gross distortions present no obstacle in principle.

In actual practice, however, considerable computational difficulties may arise in the following circumstances.

(a) If the numbers of degrees of freedom of the joints exceeds two or three then the formal solution of twice as many simultaneous equations becomes tedious and inexact.

(b) In non-linear problems with more than one degree of freedom the resulting simultaneous equations may be quite intractable by formal means. For instance, if in Example (d) the frame had been unsymmetrical then two equations like (v) would have resulted for $\Delta_x$ and $\Delta_y$.

It should also be noted that no account has been taken of the new situation that arises in practice when a member is first stressed beyond the elastic limit and then unloaded. It is well known that in such circumstances the load-deflexion diagram for decreasing loads is not the same as that for increasing loads, so that the principles described above cease to apply. We must therefore restrict the use of these methods to the first application of the loads unless it is certain that the structure obeys the same load deflexion law for increasing and for decreasing loads; this is the case when the non-linearity arises from gross distortion of the structure while the members remain within the elastic limit as, for example, when slack cables are involved.

2.6.2. THE PRINCIPLE OF VIRTUAL WORK

In FIGURE 2.10 a particle $A$ is shown under a system of concurrent coplanar† forces $P_1 P_2 \ldots$ having a resultant $R$.

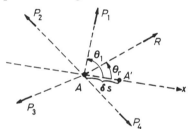

*Figure 2.10.* The particle $A$, in equilibrium under the forces $P_1$, $P_2$, is displaced a distance $\delta s$ in the direction $Ax$

Suppose that an arbitrary imaginary or 'virtual' displacement $\delta s$, such that the direction of the forces is unchanged, is given to the particle in a specified direction, $Ax$. The work done by the forces is then given by

$$P_1 \, \delta s \, \cos \theta_1 + P_2 \, \delta s \, \cos \theta_2 + \ldots$$

The total work done is the sum of these terms, which can be written

$$\Sigma F . \, \delta s . \, \cos \theta = \delta s \, \Sigma F . \cos \theta = \delta s . \, R . \cos \theta_r$$

where $\theta_r$ is the angle between the line of action of the resultant of the forces and the direction of the displacement.

---

† This restriction is here made only for simplicity; the proof of the principle can easily be extended to cover a system of forces in space.

Hence the nett work done by the forces is equal to the work done by this resultant. If arbitrary displacements $\delta s_x$ and $\delta s_y$ are imposed in two directions at right angles and the total work is found to be zero in both cases, then $R$ must be zero and the particle is in equilibrium.

Alternatively, if the particle is known to be in equilibrium then the work done is zero whatever the direction of the displacement, and we have

$$\Sigma P\, \delta s \cos \theta = 0$$

The same argument will apply if couples are acting on the particle, when the virtual work will be

$$\Sigma M.\, \delta \theta$$

The terms $P\, \delta s \cos \theta$ can be regarded either as the product of $P$ and the projection of the displacement $\delta s$ on the direction of $P$, the 'corresponding displacement', or as the product of the displacement $\delta s$ and the component of $P$ in the direction of $\delta s$.

It must be understood that the expression $\Sigma P\, \delta s \cos \theta$ vanishes because of the ratios which the terms $\delta s \cos \theta$ bear to one another and not because they become infinitesimally small individually. It is thus possible to divide each term in the equation of virtual work by a constant and if this constant be a small interval of time then we have

$$P_1 \frac{\delta s \cos \theta_1}{\delta t} + P_2 \frac{\delta s \cos \theta_1}{\delta t} + \ldots = 0$$

which is known as the Principle of Virtual Velocities[6].

When the Principle of Virtual Work is applied to a framework the equal and opposite interactions between adjacent particles within a single member need not be accounted for. The forces exerted by the members on the joints to which they are attached can be brought within the range of the Principle if the arbitrary displacements imposed on the joints are mutually consistent. Moreover, it is often convenient to select the direction of the imposed displacements so that unwanted forces disappear from the equation. In this way reactive forces can often be disposed of, for if the displacement is consistent with the method of support then reaction components normal to those supports will do no work and can be ignored.

2.6.2.1. *Application of the Virtual Work Method to Statically Determinate Frameworks*

(a) Consider the three-hinged arch shown in Figure 2.11 and suppose that $H_B$, the horizontal component of reaction at $B$, is to be determined. If a small horizontal movement of $B$ to $B'$ is allowed no virtual work will be done by $R_B$ since it is vertical. This movement is accomplished by temporarily removing the pin at $B$, thus giving the frame one degree of freedom; any motion of $A$ and $C$ must be consistent with that of $B$.

Clearly the only motion occurring at $A$ is a rotation of the member $AC$, and since no couple is acting at $A$ no virtual work is done. The rotation of $AC$ about $A$ brings the joint $C$ to $C'$, and the arc $CC'$ can be represented by a straight line normal to $AC$ at $C$.

The virtual work done at the joint $C$ must now be examined; here $P$ has moved its point of application in its own direction through $CC''$ the vertical component of $CC'$. There remains the work which may have been done by the forces in the bars $AC$ and $BC$, and by the thrust exerted by one side of

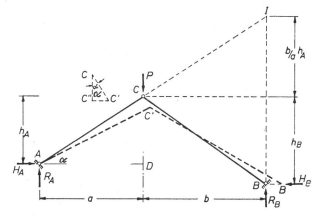

*Figure 2.11.* In order to use the Principle of Virtual work to find the reactive thrust $H$ a virtual horizontal displacement of $B$ is supposed to occur

the hinge upon the other. As these thrusts are equal and opposite, the virtual work done by them will be equal and of opposite sign, and need not

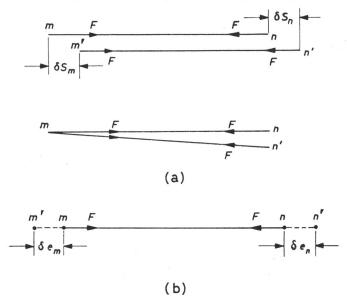

*Figure 2.12.* (a) Translation or rotation of an inextensible member involves no virtual work; (b) Negative virtual work results when a *tension* member is given a virtual *extension*

be considered further. As to the bar forces, consider the virtual work performed if an inextensible bar changes its position. FIGURE 2.12(a) shows such a bar $mn$ isolated from the frame to which it belongs. For convenience it is supposed to be in tension. If *mn* undergoes a small translation, without change of length, to *m'n'*, the total virtual work done is

$$F.\ \delta s_m - F.\ \delta s_n = 0$$

It is also seen that no virtual work is done if the inextensible bar *mn* is rotated.

Returning to the arch of FIGURE 2.11, we have reduced the forces actually doing virtual work to two, namely $P$ and $H_B$, and we must now find the components of motion of $C$ and $B$ in the directions of $P$ and $H_B$ respectively. Since $C$ moves normally to $AC$ while $B$ moves horizontally the instantaneous centre of rotation of $BC$ is at $I$ and we have

$$CC' : BB' = IC : IB$$

Now the triangles $CC'C''$ and $ACD$ are similar,

hence
$$CC'' : CC' = AD : AC$$

i.e.
$$CC'' : BB' = \frac{IC}{AC} \cdot \frac{AD}{IB}$$

$$= \frac{b}{a} \cdot \frac{a}{bh_A/a + h_B} = \frac{ab}{bh_A + ah_B}$$

Now the total virtual work is zero,

i.e.
$$P.CC'' - H_B.BB' = 0$$

$$H_B = P \frac{ab}{bh_A + ah_B}$$

(b) Simple frameworks are usually analysed by considering, in turn, the equilibrium of their joints under the forces in the members. This can be done analytically or graphically, using a force diagram, but in either case only two unknown forces can be found at each joint. The joints $J$ and $O$ of the roof truss shown in FIGURE 2.13(a) support five members each, however, and the analysis comes to a halt when one of these joints is reached. If the force in one of the members, say $JM$, can be found by some other method then the calculations or the force diagram can be successfully completed.

The force in $JM$, for example, can be conveniently found by Virtual Work, if that number is temporarily replaced by the forces $F_{JM}$ which it previously exerted on the joints $J$ and $M$. The panel $LMN$ is thus given one degree of freedom and its instantaneous centre is at $N$ (FIGURE 2.13(b)).

The method of the previous example† can then be used to find the force in $JM$.

(c)‡ In recent years a new philosophy of structural design has appeared, based upon the fact that a steel structure can usually sustain heavier loads

---

† Various geometrical procedures for improving this technique are described by Timoshenko and Young[7].

‡ See also Chapter 6, Section 6.6.

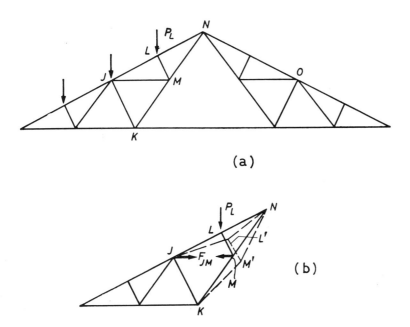

*Figure 2.13.* In order to find the force $F_{JM}$ the member $JM$ is replaced by the equal and opposite forces $F_{JM}$ and a virtual displacement is given to $M$

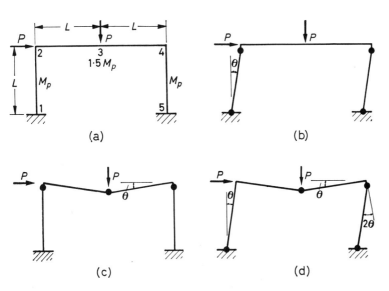

*Figure 2.14.* (a) The plastic moments of resistance of beam and column are $1 \cdot 5\, M_p$ and $M_p$ respectively; (b) Collapse Mode I; (c) Collapse Mode II; (d) Collapse Mode III

than are necessary to initiate yielding at the most highly stressed part. The idea is that the working load should be computed by dividing the collapse load by a suitable load factor; the exact stresses occurring are not regarded as significant, provided that the load factor can be chosen so that such permanent strain as may occur does not build up cumulatively.

Attention is now concentrated on the load which will cause the whole structure, or part of it, to collapse, rather than on the local stresses in individual members. Several methods for identifying the critical collapse mechanism and computing the collapse load have been devised for cases where failure is predominantly due to plastic bending as opposed to elastic or quasi-elastic instability.

When a steel beam is subjected to an increasing bending moment a stage is eventually reached when, at a bending moment known as 'the plastic moment of resistance', angular distortion at the most highly stressed cross-section increases indefinitely. A 'plastic hinge' is said to be formed at this section.

The estimation of the collapse load of a structure requires the preliminary recognition of the possible modes of failure and of the potential plastic hinges. Thus the portal frame shown in FIGURE 2.14(a) could fail in any of the ways indicated at (b), (c) and (d). Plastic hinges are imagined at the points indicated by the black circles† and when sufficient of these are functioning the frame has become a mechanism. It is now possible to invoke virtual work to obtain the relationship between the loads and the plastic moments of resistance.

In FIGURE 2.15 the collapse of a beam with fixed ends, by the formation of three plastic hinges, is envisaged; the plastic moments of resistance, $M_p$, will develop in such a way as to oppose the motion and so will act as shown.

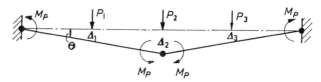

Figure 2.15. The collapse of a beam due to the formation of plastic hinges

The total virtual work done in this displacement, assuming that the forces and moments have constant values is therefore

$$\Sigma P.\varDelta - \Sigma M_p\, \theta = 0$$

i.e.
$$\Sigma P.\varDelta = 4M_p\, \theta$$

The relationship between $\varDelta$ and $\theta$ is a geometrical one which is known, so that the values of $P_1$, $P_2$ and $P_3$ (assumed to bear a constant ratio to one another) for collapse are known in terms of $M_p$.

---

† The hinges at 2 and 4 are shown in the columns rather than in the stronger beam.

## 2.6.2. PRINCIPLE OF VIRTUAL WORK

Applying these ideas to FIGURE 2.14(a), (b) and (c) in turn we have

Collapse Mode I $\quad PL\,\theta = 4M_p\,\theta \quad\quad\quad\quad\quad\quad P = \dfrac{4\,M_p}{L}$

Collapse Mode II $\quad PL\,\theta = (2M_p + 2 \times 1{\cdot}5\,M_p)\theta \quad\quad P = \dfrac{5\,M_p}{L}$

Collapse Mode III $\quad 2PL\,\theta = (2M_p + 2 \times 1{\cdot}5\,M_p)\theta + M_p\,2\theta \quad P = \dfrac{3{\cdot}5\,M_p}{L}$

The lowest value of $P$ for collapse is given by Collapse Mode III which has thus been identified as the correct one, provided all the possible modes have been examined.

In the paper by Neal and Symonds[10], from which this example is taken, the number of possible modes is shown to be connected with the number of equations required to specify the equilibrium conditions of the frame; in this way it is possible to ensure that all possible modes and combinations of modes have been covered.

### 2.6.2.2. *Mohr's Equation of Virtual Work*

In the solution of the statically determinate arch of Section 2.6.2.1.(a) the elastic distortions which must accompany stress were ignored on the usual understanding that they were so small that the geometry of the structure was not altered. This was equivalent to assuming that the truss members were inextensible and made no contribution to the virtual work of the whole system.

These small distortions are responsible for the deflexions of the joints and so play an essential part in the analysis of hyperstatic structures. We must therefore now consider the virtual work performed when a member, in which the force is $F$, changes its length by a small amount $\delta e$ (FIGURE 2.12(b)). This is an arbitrary virtual extension not necessarily produced by the action of the force $F$.

The sign convention is that *tension* and *extension* are positive, but in calculating the virtual work we must be careful to distinguish between the force *in* the member and the forces which the member exerts *on* the joints. These are equal in magnitude but opposite in direction, and we concentrate our attention on the forces acting *on* the joints. The virtual work done when the member extends is

$$-F.\,\delta e_m - F.\,\delta e_n = -F.\,\delta e$$

Suppose that a series of forces $\overline{P}$ is acting on a structure and that the corresponding bar forces are $\overline{F}$ as indicated in FIGURE 2.16(a). Each individual joint is in equilibrium under the external and bar forces acting on it and we can therefore apply the principle of virtual work if a suitable virtual displacement can be imposed. This is achieved by imagining that all the members are inextensible except $mn$ which is given a virtual extension $\delta e$ and that in consequence the joints are caused to move to new positions. We can now proceed to examine the work done by all the forces.

The reactions can be quickly disposed of; $R_A$ and $R_C$ do no work since joints $A$ and $C$ are only capable of rotation, while any movement of $B$ is at right angles to the direction of action of $R_B$.

So far as the bar forces, which occur in pairs, are concerned it is clear from the discussion of FIGURE 2.12, on page 55, that the extensible bar $mn$ is the only one that contributes to the virtual work, the amount being $-\bar{F}.\,\delta e$.

It remains to account for the external loads: at joint 3, for example, the component of the displacement in the direction of $\bar{P}_3$ is $\delta\varDelta_3$ and the virtual work done is therefore equal to $\bar{P}_3.\,\delta\varDelta_3$.

Over the whole structure, therefore, we have

$$\Sigma\bar{P}.\,\delta\varDelta - \bar{F}.\,\delta e = 0$$

If all the bars are considered to be extensible we have

$$\Sigma\bar{P}.\,\delta\varDelta = \Sigma\bar{F}.\,\delta e \qquad \dots\,. \,(2.5)$$

This is the Principle of Virtual Work expressed in infinitesimal form and in a way suitable for application to elastic structures. It was apparently first used for this purpose by Mohr after whom it is sometimes named.

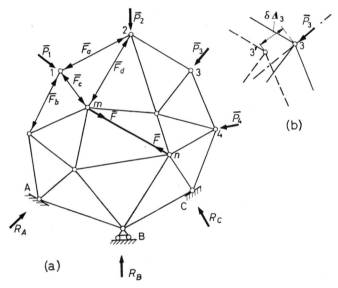

*Figure 2.16.* Derivation of Mohr's Equation of virtual work for a braced structure

It remains only to emphasize that $\delta\varDelta$, while being due to some agency other than $\bar{P}$, corresponds to $\bar{P}$ in the sense that it is measured along the line of action of $\bar{P}$ and in the same direction. (See FIGURE 2.16(b).)

In Mohr's Equation, 2.5, $\bar{P}$ and $\bar{F}$ are consistent external and bar forces, respectively, which can conveniently be described as the 'force system', while $\delta\varDelta$ and $\delta e$ are consistent displacements and bar extensions and comprise the 'displacement system'. These two systems are independent

## 2.6.2. PRINCIPLE OF VIRTUAL WORK 2.6.2.3

and are chosen to suit the circumstances of a particular problem. In order to emphasize the connexion between the terms, Mohr's Equation can be written

$$\underbrace{\Sigma \overline{P} \times \{\delta \Delta\}}_{\text{force system}} = \underbrace{\Sigma \overline{F} \times \{\delta e\}}_{\text{displacement system}}$$

It will be evident that none of the force of the argument will be lost if we substitute couples $\overline{M}$ for the forces $\overline{P}$, provided that we also replace the displacements $\delta \Delta$ of the loaded joints by their rotations $\delta \theta$. We then have

$$\Sigma \overline{M}. \{\delta \theta\} = \Sigma \overline{F}. \{\delta e\} \qquad \ldots (2.5(a))$$

Here $\overline{F}$ represents the bar forces produced by the external couples $\overline{M}$.

Finally we observe that if the elements of the deflexion system bear a constant ratio to one another, which is invariably the case unless there is gross distortion, then we can dispense with the differentials and write equation 2.5 as

$$\Sigma \overline{P}. \{\Delta\} = \Sigma \overline{F}. \{e\} \qquad \ldots (2.6)$$

which is Mohr's Equation in finite form.

### 2.6.2.3. *Identity of Mohr's Equation and the Principle of Minimum Potential Energy applied to frameworks*

It has perhaps already struck the reader that essentially the same conception underlies these two methods. In both cases we imagine that a small 'virtual' displacement from the equilibrium position has taken place; in the former we then go on to make use of the fact that zero 'virtual work' has been done, while in the latter the same fact is employed by equating the change of total potential energy to zero.

If Mohr's Equation be rewritten

$$\Sigma \overline{F}. \delta e - \Sigma \overline{P}. \delta \Delta = 0$$

it is clear that the terms are identical with those of the equation of Minimum Potential Energy

$$d(U + V) = 0$$

for the first term represents the increment of strain energy while the second is the differential of the potential energy of the loads

$$V = V_0 - \Sigma \overline{P} \Delta$$

The solutions of the problems worked in Section 2.6.1.1 could therefore be regarded as exercises in the method of virtual work in which the strain energy terms of Mohr's Equation were converted to functions of the bar extensions, dimensions and elastic properties only. While it is often convenient thus to dispose of $\overline{F}$ this is by no means an essential step as the following calculations show.

### 2.6.2.4. *Examples of the application of Mohr's Equation*

Some of the examples of Section 6.1.1. are worked below by the Virtual Work method; the reader should contrast the solutions with those given previously.

(a) See FIGURE 2.5(a) (page 41). In seeking to apply Mohr's Equation to this problem we first observe that, as gross distortion is excluded, we can use the finite form

$$\Sigma \bar{P} \{\Delta\} = \Sigma \bar{F} \{e\} \qquad \ldots\ldots (2.6)$$

We can then conveniently choose the 'force system' to be the actual bar forces $F_1$ and $F_2$ corresponding to the applied load $P$, while the 'displacement system' can consist of an arbitrarily imposed displacement $\Delta$ of $A$ in the line of action of $P$.

Thus 
$$P.\{\Delta\} = F_1 \{e_1\} + 2F_2 \{e_2\} \qquad \ldots\ldots (i)$$

Compatibility of strain gives

$$\Delta = e_1 \text{ and } e_1 \cos\theta = e_2 \qquad \ldots\ldots (ii)$$

Substituting from (ii) in (i) we have

$$P = F_1 + 2F_2 \cos\theta \qquad \ldots\ldots (iii)$$

which is, of course, the equation of equilibrium.

The real strains of the bars produced by the load $P$ constitute a perfectly legitimate displacement system since, whatever the value of $P$, $e_1$ bears the constant ratio to $e_2$ given by equation (ii). (Gross distortion, sufficient to alter $\theta$ appreciably, is here excluded.) We can therefore say

$$\frac{e_1}{e_2} = \frac{\text{real extension of } AC}{\text{real extension of } AB} = \frac{F_1 h/AE}{F_2 L/AE} \qquad \ldots\ldots (iv)$$

and hence, from (ii), that $F_1 \cos^2\theta = F_2$.

Substitution in (iii) gives $\qquad F_1 = \dfrac{P}{1 + 2\cos^3\theta}$, as before.

It will be noticed that the above is essentially identical with the solution of this problem given in Chapter 1 (page 23), the only difference being that the equilibrium equation was here derived by virtual work instead of by direct resolution.

If, having derived (iii), we had rewritten it in terms of displacements, by using (ii) and (iv), we should have obtained

$$P = \frac{AE\Delta}{h} + \frac{2AE\Delta \cos^3\theta}{h}$$

which appeared in the minimum potential energy solution of Section 2.6.1.1.

It therefore appears—and this turns out to be generally true—that the Virtual Work method leads to the same equations as are obtained by the direct comparison of strains if the work is carried out in terms of forces, and to the same equations as appear in the minimum potential energy method if the work is carried out in terms of displacements.

## 2.6.2. PRINCIPLE OF VIRTUAL WORK 2.6.2.4

(b) The unsymmetrical case of FIGURE 2.5(c) involves, as we have seen, two degrees of freedom and we proceed by choosing, as the displacement system, displacements $\varDelta_x$ and $\varDelta_y$ in the $x$- and $y$-directions. The corresponding extensions of the $m$th bar are

$$- \varDelta_x \cos \theta_m \text{ and } + \varDelta_m \sin \theta_m \qquad \ldots \text{(i)}$$

Since the $x$- and $y$-displacements are independent, two independent virtual work equations can be written, namely

and
$$\begin{aligned} P \cos \alpha \, \{\varDelta_x\} &= \Sigma F_m \{ - \varDelta_x \cos \theta_m \} \\ P \cos \alpha \cdot \varDelta_x &= - \varDelta_x \sum_m F_m \cdot \cos \theta_m \\ P \sin \alpha \cdot \varDelta_y &= - \varDelta_y \sum_m F_m \cdot \sin \theta_m \end{aligned} \Biggr\} \quad \ldots \text{(ii)}$$

As in the previous example it is convenient now to identify the real extensions and displacements with the virtual ones and write, using (i)

$$F_m = \frac{A_m E_m}{L_m} e_m = \frac{A_m E_m}{L_m} (- \varDelta_x \cos \theta_m - \varDelta_y \sin \theta_m)$$

Substitution in (ii) then gives, as before

$$\left. \begin{aligned} P \cos \alpha &= \varDelta_x \sum_m \frac{AE}{L} \cos^2 \theta + \varDelta_y \sum_m \frac{AE}{L} \sin \theta \cos \theta \\ P \sin \alpha &= \varDelta_x \sum_m \frac{AE}{L} \sin \theta \cos \theta + \varDelta_y \sum_m \frac{AE}{L} \sin^2 \theta \end{aligned} \right\}$$

In this case we chose to work in terms of displacements so that only two equations, corresponding to the two degrees of freedom, were eventually obtained for solution. If we had worked in terms of forces $n - 2$ equations would have emerged.

(c) See FIGURES 2.5(a) and 2.6, making $\theta = 60°$.

We can see that Mohr's equation is not restricted to cases of linear elasticity by reference to example (c) of Section 2.6.1.1; it will be recalled that the bars of the framework have the load-extension law of FIGURE 2.6. Three cases were considered and we now repeat the second of these, when $F_1 > 2S$ and $F_2 < 2S$.

We begin by recalling the equilibrium equation (iii) of Example (a) of the present Section, which was obtained by means of Mohr's Equation

$$P = F_1 + 2F_2 \cos \theta$$

As $\theta = 60°$ in this case, we have

$$P = F_1 + F_2 \qquad \ldots \text{(iii)}$$

Equation (iv) now becomes

$$\frac{e_1}{e_2} = \frac{\text{real extension of } AC}{\text{real extension of } AB} = \frac{(F_1 - S) 2k}{F_2 k} \qquad \ldots \text{(iv)}'$$

The compatibility relationship, equation (ii), is now

$$e_1 = 2e_2 \qquad \ldots \text{(ii)}'$$

And so, substituting from (ii)' and (iv)' in (iii)' we have

$$F_1 = \frac{P+S}{2} \quad \text{, as before.}$$

(d) When there is gross distortion, as in Example (d) of Section 2.6.1.1. (FIGURE 2.7) it is essential to use the infinitesimal form of Mohr's Equation.

$$\overline{\Sigma P}\{\delta\Delta\} = \overline{\Sigma F}\{\delta e\}$$

i.e. $\qquad P\{\delta\Delta\} = F_1\{\delta\Delta\} + 2F_2\{\delta e\}$

The load-extension relationships of the bars are

$$F_1 = a_1\Delta \quad \text{and} \quad F_2 = a_2 e$$

Hence, substituting for $F_1$ and $F_2$

$$P - a_1\Delta = 2a_2 e \,\frac{de}{d\Delta}$$

This is equation (v) of the previous solution; hence, after introducing the compatibility relationships (ii) and (iii), we obtain (vi) and the solution proceeds as before.

This example illustrates very well the point that although non-linear problems can be solved, in principle, by the methods under discussion the final solution of the resulting equations may be difficult or even impossible. This is especially true in cases of multiple redundancy.

In this event the only effective procedure is one of successive approximations such as that given in Example (b) of Section 2.6.2.5.

### 2.6.2.5. *The calculation of deflexions by Virtual Work*

No new points arise when the remaining examples of Section 2.6.1.1. are treated by virtual work, and we can therefore turn to a new class of problem where Mohr's Equation shows to good advantage. A convenient example has been worked by Brown[2]; we give a solution here in a rather more concise form.

(a) The pin-jointed truss of FIGURE 2.17 has eccentrically loaded compression members $AB$, $BC$ and $CD$ whose load-extension relation is

$$F = \frac{Be}{L} + De^2 \qquad \ldots \text{(i)}$$

For the remaining members $\qquad F = \dfrac{AEe}{L} \qquad \ldots$ (ii)

It is required to find the vertical deflexion of $E$. We note first that the truss is statically determinate and therefore that the bar forces $F$ can be written down, at once, by resolution at the various joints in turn.

## 2.6.2. PRINCIPLE OF VIRTUAL WORK

The relationships (i) and (ii) enable the extensions $e$ of the various members to be calculated; at this stage, with Mohr's Equation in mind, we can see that a displacement system, consisting of the actual distortions of the truss, has suggested itself. A force system must now be selected which will enable

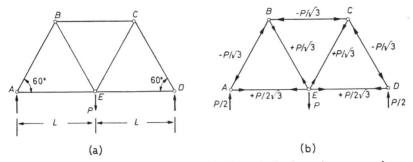

Figure 2.17. The compression members of this statically determinate truss obey a non-linear load-deflexion law

the deflexion of $E$ to be computed; the obvious approach is to let $\overline{P}$ be unit vertical load acting at $E$ while $\overline{F}$ is the corresponding bar force which can now be written $F'$

We thus have
$$1.\{\Delta_E\} = \Sigma F'.\{e\} \quad \ldots \text{(iii)}$$

Equation (iii) can best be evaluated in tabular form.

| Member | $F$ | $e$ | $F'$ |
|---|---|---|---|
| $AB, BC, CD$ | $-P/\sqrt{3}$ | $-B/2DL + (B^2/4D^2L^2 - P/D\sqrt{3})^{\frac{1}{2}}$ | $-1/\sqrt{3}$ |
| $BE, CE$ | $+P/\sqrt{3}$ | $PL/\sqrt{3}AE$ | $+1/\sqrt{3}$ |
| $AE, DE$ | $+P/2\sqrt{3}$ | $PL/2\sqrt{3}AE$ | $+1/2\sqrt{3}$ |

The required deflexion is obtained by multiplying $F'$ by $e$ for each member and summing the results.

Thus
$$\Delta_E = \frac{\sqrt{3}B}{2DL} + \frac{5PL}{6AE} - \left(\frac{3B^2}{4D^2L^2} - \frac{P\sqrt{3}}{D}\right)^{\frac{1}{2}}$$

Charlton[11] has pointed out that this method can be used for determining the deflexion of hyperstatic structures. When the forces in the members have been found, and their extensions calculated, the displacement of the joints and hence the deflexion of the frame can be found by considering a convenient determinate frame whose members connect the joints. Thus in the simple hyperstatic framework of FIGURE 2.18(a) the forces and extensions of the members are supposed to have been determined. The final position of $A$ is controlled by the extensions $e$ of any two members, say $AB$ and $AG$, and can be found, according to the present method, by applying unit load to the determinate frame $ABG$, finding the corresponding bar forces $F'$, and finding the sum of $F'e$ for these two members, as at (b)†.

† See also Chapter 4, Section 8.

Again we can be sure that the displacements of the joints of the hyperstatic frame of FIGURE 2.19(a) are the same as in the determinate frame, shown at (b), whose bars deform by the same amounts as do the corresponding bars in frame (a).

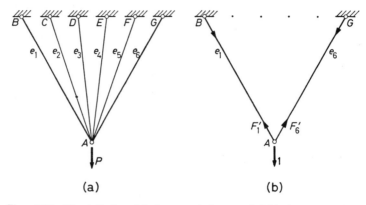

*Figure 2.18.* The deflexion of the hyperstatic framework (a) is the same as that of the simple frame (b)

It is to be noted that the direction in which a joint displaces is generally known only when the frame and its loading are symmetrical. In many cases the vertical component only of the deflexion is required, and this is obtained by applying unit vertical load. If the absolute movement is required then two unit loads, at right angles, must be imposed in turn.

We should perhaps examine the above from the point of view of superposition: we have imposed an imaginary or 'virtual' force system on the frame in addition to the real displacements, and we must be satisfied that

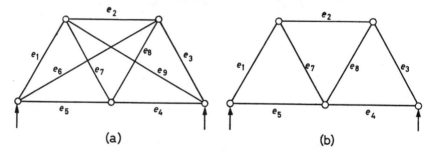

*Figure 2.19.* The two frames (a) and (b) have identical deflexions

this is legitimate. It was pointed out, in Section 5.1, that the same increment of external load always produced the same increment of force in the members of a determinate frame, whatever total load was acting, unless there was gross distortion. As we have seen, the distortion of a hyperstatic frame is the same as that of a corresponding determinate frame, and we can therefore be assured that the method only breaks down when the distortion of a frame is large enough to alter its essential geometry.

## 2.6.2. PRINCIPLE OF VIRTUAL WORK

(b) The method just described has been extended by King[12] to deal with problems which are intractable by other means because of mathematical complications. FIGURE 2.20(a) shows the cantilever frame considered by King and FIGURE 2.21 the load-extension diagrams for the members. The equations to these curves are

for member $AB$: $\quad F = \dfrac{200}{\sqrt{3}} e - \dfrac{1000}{\sqrt{3}} e^2$

for all other members: $\quad F = \dfrac{100}{\sqrt{3}} e - \dfrac{250}{\sqrt{3}} e^2$

The technique used is essentially that of the direct comparison of displacements described in Chapter 1 but, since the non-linear characteristic of the material invalidate the application of the principle of superposition, the solution obtained is only approximate. A second calculation, based on the first solution, gives a closer approximation; this process is continued until a solution is reached which differs by an acceptably small margin from its predecessor.

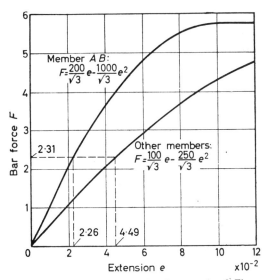

*Figure 2.21.* The bars of the framework of *Figure 2.20*(a) deform according to these load-extension laws

The first step is to remove the hinge at $E$ and, by means of Mohr's Equation of Virtual Work, to calculate the horizontal and vertical components of the displacement of the point $E$ of the primary structure (FIGURE 2.20(b)) under the loads $P = 5$, $R = 1$ and $S = \sqrt{\frac{3}{2}}$; the last values, of $R$ and $S$, are guessed. (The value for $R$ was reached by taking 50 per cent of the value it would have had if the member $AB$ and the horizontal reaction $S$ had been omitted; $S$ was given the value of $\sqrt{\frac{3}{2}}$, which is about the mean value of the

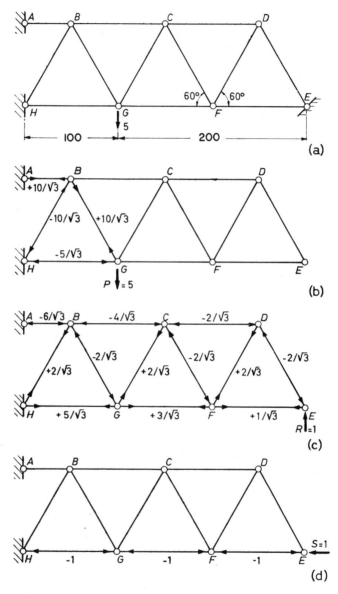

*Figure 2.20.* (a) The members of this hyperstatic cantilever obey the load-extension laws of *Figure 2.21*; (b) The forces in the primary structure caused by the load $P = 5$; (c) The forces caused by the load $R = 1$; (d) The forces caused by the load $S = 1$

## 2.6.2. PRINCIPLE OF VIRTUAL WORK

force in $HE$ with $P = 5$ and $R = 1$.) The deflexions so obtained are adjusted so as to be equal to zero by making corrections $\delta R$ and $\delta S$ to the guessed values of $R$ and $S$.

Thus
$$\left. \begin{array}{l} \Delta_{R(P,R,S)} + \Delta_{R(\delta R)} + \Delta_{R(\delta S)} = 0 \\ \Delta_{S(P,R,S)} + \Delta_{S(\delta R)} + \Delta_{S(\delta S)} = 0 \end{array} \right\} \quad \ldots \text{(i)}$$

Here the first suffix, $R$ or $S$, identifies the direction of the deflexion of $E$ while the second, in brackets, gives the forces acting.

The six terms of equations (i) are given by Mohr's Equation, in the following manner:

For $\Delta_{R(P,R,S)}$: $\Sigma \overline{P} \{\Delta\} \qquad = \Sigma \overline{F} \{e\}$

$$1 \cdot \{\Delta_{R(P,R,S)}\} = \Sigma F_{(R=1)} \{e_{(P,R,S)}\}$$

i.e. $\qquad \Delta_{R(P,R,S)} \qquad = \Sigma F_{(R=1)} \cdot e_{(P,R,S)} \qquad \ldots$ (a)

For $\Delta_{R(\delta R)}$: A small change in $\delta R$ in the value of $R$ alters the force in a member by $\delta R \dfrac{\partial F}{\partial R}$ and its extension by $\delta R \dfrac{\partial F}{\partial R} \cdot \dfrac{de}{dF}$

i.e.  $\qquad 1 \cdot \{\Delta_{R(\delta R)}\} = \displaystyle\sum F_{(R=1)} \left\{ \delta R \dfrac{\partial F}{\partial R} \cdot \dfrac{de}{dF} \right\}$

$$= \delta R \sum F^2_{(R=1)} \dfrac{de}{dF} \quad \text{for } \dfrac{\partial F}{\partial R} = \text{numerical}$$

value of $F_{(R=1)}$

$$= \delta R \cdot \alpha \qquad \ldots \text{(b)}$$

Here $\dfrac{de}{dF}$ is the reciprocal of the slope of the $F - e$ curve of FIGURE 2.21 at the estimated bar force $F_{P,R,S}$ given in column 4 of TABLE 2.1.

For $\Delta_{R(\delta S)}$: $\quad 1 \cdot \{\Delta_{R(\delta S)}\} = \displaystyle\sum F_{(R=1)} \left\{ \delta S \cdot \dfrac{\partial F}{\partial S} \cdot \dfrac{de}{dF} \right\}$

$$= \delta S \sum F_{(R=1)} F_{(S=1)} \dfrac{de}{dF} \quad \text{for } \dfrac{\partial F}{\partial S} = \text{numerical}$$

value of $F_{(S=1)}$

$$= \delta S \cdot \beta \qquad \ldots \text{(c)}$$

Similar reasoning gives

$$\Delta_{S(P,R,S)} = \Sigma F_{(S=1)} e_{(P,R,S)} \qquad \ldots \text{(d)}$$

$$\Delta_{S(\delta R)} = \delta R \sum F_{(S=1)} F_{(R=1)} \dfrac{de}{dF}$$

$$= \delta R \cdot \beta \qquad \ldots \text{(e)}$$

$$\Delta_{S(\delta S)} = \delta S \sum F^2{}_{(S=1)} \frac{de}{dF}$$

$$= \delta S \cdot \gamma \qquad \qquad \ldots \text{(f)}$$

Hence equations (i) become

$$\left. \begin{array}{l} \Delta_{R(P,R,S)} + \delta R \cdot \alpha + \delta S \cdot \beta = 0 \\[6pt] \Delta_{S(P,R,S)} + \delta R \cdot \beta + \delta S \cdot \gamma = 0 \end{array} \right\} \quad \ldots \text{(ii)}$$

Solving these for $\delta R$ and $\delta S$ we have

$$\left. \begin{array}{l} \delta R = \dfrac{\Delta_{R(P,R,S)} \cdot \gamma - \Delta_{S(P,R,S)}\beta}{\beta^2 - \alpha\gamma} \\[10pt] \delta S = \dfrac{\Delta_{S(P,R,S)}\,\alpha - \Delta_{R(P,R,S)}\beta}{\beta^2 - \alpha\gamma} \end{array} \right\} \quad \ldots \text{(iii)}$$

The necessary calculations are set out in TABLE 2.1, to which the following notes apply.

Column 1, 2, 3: Bar forces obtained from FIGURES 2.20(b), (c) and (d).

Column 4:   $F_{(P,R,S)}$ = bar force with external loads $P = 5$, $R = 1$ and $S = 0.87$ acting simultaneously. Sum of Columns 1, 2 and 3.

Column 5:   $e_{(P,R,S)}$ = bar extension under force $F_{(P,R,S)}$ obtained from FIGURE 2.21.

Column 6:   $F_{(R=1)}$ = numerical value of bar force with external load $R = 1$.

Column 7:   $\Delta_{R(P,R,S)}$ = vertical deflexion of $E$ under loads $P = 5$, $R = 1$ and $S = 0.87$. Product of Columns 5 and 6. Equation (a).

Column 8:   $F_{(S=1)}$ = numerical value of bar force with external load $S = 1$.

Column 9:   $\Delta_{S(P,R,S)}$ = horizontal deflexion of $E$ under loads $P = 5$, $R = 1$ and $S = 0.87$. Product of Columns 5 and 8. Equation (d).

Column 10:   $de/dF$ at $e_{(P,R,S)}$. $\Bigg[$ e.g. for $AB$: $F = \dfrac{200}{\sqrt{3}} e - \dfrac{1000}{\sqrt{3}} e^2$

$$\dfrac{dF}{de} = \dfrac{200}{\sqrt{3}} - \dfrac{2000}{\sqrt{3}} e$$

$$= 115 \cdot 4 - 1154 e$$

at $e = 2 \cdot 26 \cdot 10^{-2} = 89 \cdot 3$

$$\therefore \; \dfrac{de}{dF} = \dfrac{1}{89 \cdot 3} = 0 \cdot 0112 \Bigg]$$

TABLE 2.1

ANALYSIS OF FRAME OF FIGURE 2.20

| | 1 | 2 | 3 | 4 | 5 | 6 | 7 | 8 | 9 | 10 | 11 | 12 | 13 | 2a | 3a | 4a | 5a | 7a | 9a |
|---|---|---|---|---|---|---|---|---|---|---|---|---|---|---|---|---|---|---|---|
| | F due to | | | $F(P,R,S)$ | $e'(P,R,S)$ | $F(R=1)$ | $\Delta_R(P,R,S)$ | $F(S=1)$ | $\Delta_S(P,R,S)$ | $\dfrac{dt}{dF}$ | $\alpha$ | $\beta$ | $\gamma$ | F due to | | $F'(P,R,S)$ | $e'(P,R,S)$ | $\Delta'_R$ | $\Delta'_S$ |
| Member | $P=5$ | $R=1$ | $S=.87$ | | $\times 10^{-2}$ | | $\times 10^{-2}$ | | $\times 10^{-2}$ | | | | | $R=1.226$ | $S=1.17$ | | $\times 10^{-2}$ | $\times 10^{-2}$ | $\times 10^{-2}$ |
| AB | +5·78 | −3·47 | | +2·31 | +2·26 | −3·47 | −7·83 | | | 0·0112 | 0·135 | | | −4·26 | | +1·52 | +1·43 | −4·96 | |
| BC | | −2·31 | | −2·31 | −4·49 | −2·31 | +10·36 | | | 0·0223 | 0·119 | | | −2·84 | | −2·84 | −5·71 | +13·19 | |
| CD | | −1·15 | | −1·15 | −2·10 | −1·15 | +2·42 | | | 0·0194 | 0·026 | | | −1·41 | | −1·41 | −2·61 | +3·00 | |
| HG | −2·89 | +2·89 | −0·87 | −0·87 | −1·57 | +2·89 | −4·54 | −1 | +1·57 | 0·0188 | 0·157 | −0·054 | 0·0188 | +3·54 | −1·17 | −0·52 | −0·93 | −2·69 | +0·93 |
| GF | +1·73 | | −0·87 | +0·86 | +1·57 | +1·73 | +2·72 | −1 | −1·57 | 0·0188 | 0·056 | −0·033 | 0·0188 | +2·12 | −1·17 | +0·95 | +1·72 | +2·98 | −1·72 |
| FE | +0·58 | | −0·87 | −0·29 | −0·52 | +0·58 | −0·30 | −1 | +0·52 | 0·0178 | 0·006 | −0·010 | 0·0178 | +0·71 | −1·17 | −0·46 | −0·82 | −0·48 | +0·82 |
| BH | −5·78 | +1·15 | | −4·63 | −11·08 | +1·15 | −12·73 | | | 0·0389 | 0·051 | | | +1·41 | | −4·37 | −10·11 | −11·63 | |
| BG | +5·78 | −1·15 | | +4·63 | +11·08 | −1·15 | −12·73 | | | 0·0389 | 0·052 | | | −1·41 | | +4·37 | +10·11 | −11·63 | |
| CG | | +1·15 | | +1·15 | +2·10 | +1·15 | +2·42 | | | 0·0194 | 0·025 | | | +1·41 | | +1·41 | +2·61 | +3·01 | |
| CF | | −1·15 | | −1·15 | −2·10 | −1·15 | +2·42 | | | 0·0194 | 0·026 | | | −1·41 | | −1·41 | −2·61 | +3·01 | |
| DF | | +1·15 | | +1·15 | +2·10 | +1·15 | +2·42 | | | 0·0194 | 0·025 | | | +1·41 | | +1·41 | +2·61 | +3·01 | |
| DE | | −1·15 | | −1·15 | −2·10 | −1·15 | +2·42 | | | 0·0194 | 0·026 | | | −1·41 | | −1·41 | −2·61 | +3·01 | |
| $\Sigma$ | | | | | | | −12·95 | | +0·52 | | 0·704 | −0·097 | 0·0554 | | | | | −0·18 | +0·03 |

Column 11:  $\alpha = \sum F^2_{(R=1)} \dfrac{de}{dF}$ (Equation (b))

$= \text{Column 6} \times \text{Column 6} \times \text{Column 10}.$

Column 12: $\beta = \sum F_{(R=1)} F_{(S=1)} \dfrac{de}{dF}$ (Equation (c))

$= \text{Column 6} \times \text{Column 8} \times \text{Column 10}.$

Column 13: $\gamma = \sum F^2_{(S=1)} \dfrac{de}{dF}$ (Equation (f))

$= \text{Column 8} \times \text{Column 8} \times \text{Column 10}.$

The values obtained by summing the appropriate columns are inserted in equations (iii) to give the first estimates of $R$ and $S$, as follows:

$$R = \frac{(-0\cdot1295)(0\cdot0554) - (0\cdot0052)(-0\cdot097)}{(-0\cdot097)^2 - (0\cdot704)(0\cdot0554)} = 0\cdot226$$

$$S = \frac{(0\cdot0052)(0\cdot704) - (-0\cdot1295)(-0\cdot097)}{-0\cdot0296} = 0\cdot300$$

The corrected estimates of $R$ and $S$ are therefore

$$R = 1\cdot226 \quad \text{and} \quad S = 1\cdot17$$

and these are tested by calculating the horizontal and vertical deflexions which they produce. These calculations are shown in the remaining columns of TABLE 2.1, from which it will be seen that the deflexions of $E$ have been reduced to such an extent that a further correction by King's method is probably unnecessary. A final correction, by simple proportion, would give

$$R = 1\cdot000 + \frac{12\cdot95}{12\cdot95 - 0\cdot18} \times 0\cdot226 = 1\cdot229$$

$$S = 0\cdot870 + \frac{0\cdot52}{0\cdot52 - 0\cdot03} \times 0\cdot300 = 1\cdot188$$

This procedure has been described at some length because, although laborious, it is at least a practicable method for dealing with multiply-redundant non-linear structures. The exact methods described earlier in this chapter are apt to lead to insoluble equations in any but rather simple cases of the kind used as illustration.

2.6.2.6. *The Versatility of the Virtual Work Method*

The Virtual Work method was introduced, in Section 2.6.2, as a means of describing the equilibrium of a structure; the examples of Section 2.6.2.1 demonstrated by suitable examples its superiority in certain cases over the direct use of the equations of statics.

The development of the basic idea of Virtual Work to include the extensibility of the members led, in Section 2.6.2.2, to Mohr's Equation which was shown in the next section to be essentially identical with the Principle of Minimum Potential Energy and to be equally free from restrictions on its validity. The examples of Section 6.2.4 demonstrated these points when Mohr's Equation was used to find the equilibrium equations for simple hyperstatic frames; the compatibility equations were obtained from considerations of geometry.

A different line of development was introduced in Section 6.2.5 when Mohr's Equation was shown to provide a convenient means of calculating deflexions which was much more powerful than the simple geometrical procedures used previously.

The final step, in example (b), was to use this method of calculating deflexions to find the compatibility equations for a hyperstatic frame with non-linear members, equilibrium having been established by obtaining the member forces by simple resolution at the joints.

It therefore appears that Virtual Work is a method of great versatility capable of yielding, by a suitable approach, both the equilibrium and the compatibility relations essential to the analysis of hyperstatic structures.

## 2.6.3. Castigliano's Theorem, Part I

The title originally given to this theorem by its author[13] was ' Theorem of the Differential Coefficients of the Internal Work, Part I '. Unfortunately, later writers have used other titles for this theorem until, at the present time, the nomenclature of this and other energy theorems associated with Castigliano's name is exceedingly confused.

Southwell[4] uses the title ' Theorem complementary to Castigliano's First Theorem ' and refers to it shortly as the ' Complementary Theorem '.

Brown[2] describes it as ' Castigliano's Theorem of Equilibrium ' a title which, although unfamiliar, has the merit of putting the theorem into its proper place in the scheme of things.

There seems to be every reason, in this confused situation, for returning to the original title which is used throughout this book in the shortened form given at the head of this Section.

Castigliano's description and proof of this theorem is as follows: consider a structure acted upon by a series of external loads $P_1 P_2 \ldots P_j$; and suppose that small increments $dP_1, dP_2 \ldots dP_j$ are given to these loads. The resulting strains will cause the various joints to deflect further amounts $d\Delta_1, d\Delta_2$, etc. and the work done during this displacement, at joint $j$ for example, will be nearly

$$\left(P_j + \frac{dP_j}{2}\right) d\Delta_j \simeq P_j \cdot d\Delta_j \qquad \ldots \text{(i)}$$

The total work done will be equal to the gain of strain energy of the structure, and therefore

$$dU = \Sigma P_j \cdot d\Delta_j \qquad \ldots \text{(ii)}$$

If the strain energy $U$ be expressed as a function of the displacements of the joints then the increase of $U$ produced by an increase of one of the displacements is, by definition

$$\frac{\partial U}{\partial \Delta_j} \cdot d\Delta_j \qquad \ldots \text{(iii)}$$

and hence
$$dU = \sum \frac{\partial U}{\partial \Delta_j} \cdot d\Delta_j \qquad \ldots\ (iv)$$

Equating (ii) and (iv)
$$P_j = \frac{\partial U}{\partial \Delta_j} \qquad \ldots\ (2.7)$$

This is Castigliano's Theorem, Part I.

#### 2.6.3.1. *Identity of Castigliano's Theorem, Part I, and Mohr's Equation of Virtual Work*

Southwell[4] was perhaps the first to indicate that this theorem and the Principle of Stationary Potential Energy—which he called 'a General Theorem in Mechanics'—were essentially one and the same thing. Williams[1] made the point that Mohr's Equation

$$\Sigma \bar{P}\{\delta \Delta\} = \Sigma \bar{F}\{\delta e\}$$

can be written
$$\Sigma P_j \cdot d\Delta_j = dU$$

where $dU$ is the increment of strain energy, equal to $\Sigma F \cdot de$, resulting from the imposition of joint displacements $d\Delta_j$; this is identically the same as (ii) above.

In using this theorem the compatibility relationships are used to enable the strain energy to be expressed as a function of the joint displacements. Differentiation with respect to the joint loads then yields the same equations as would have been given by minimising the total potential energy.

For example, in problem (a) of Section 2.6.1.1 (FIGURE 2.5(a)) the total strain energy is

$$U = \frac{AE(\Delta \cos \theta)^2 \cos \theta}{h} + \frac{AE\Delta^2}{2h}$$

i.e.
$$\frac{dU}{d\Delta} = \frac{2AE\Delta \cos^3 \theta}{h} + \frac{AE\Delta}{h} = P$$

which is identical with equation (iv).

#### 2.6.4. FIRST THEOREM OF MINIMUM STRAIN ENERGY

A special case of some interest arises when Castigliano's Theorem, Part I, is applied to unloaded joints of a structure. At such points we have

$$\frac{\partial U}{\partial \Delta} = 0 \qquad \ldots\ (2.8)$$

This has been called, by Southwell[4] and Brown[2], the 'First Theorem of Minimum Strain Energy', a title which calls attention to the stationary value of the strain energy when unloaded joints are given small displacements about the equilibrium position. (The possibility of a maximum value of the strain energy is excluded by the essentially positive nature of $\partial^2 U/\partial \Delta^2 = \partial P/\partial \Delta$.)

The same conclusion could, of course, have been reached by considering the equilibrium of an unloaded joint from the standpoint of total potential energy.

### 2.6.5. FIRST THEOREM OF COMPLEMENTARY ENERGY

At such a joint
$$\frac{\partial}{\partial \varDelta}(U+V) = 0$$

But, in the absence of applied loads,
$$V = 0$$

i.e
$$\frac{\partial U}{\partial \varDelta} = 0 \quad \text{as before.}$$

In view of the above reasoning there seems to be little advantage in attempting to retain the present theorem in a distinctive position among the energy theorems. Rather do we prefer to regard it simply as a special case of the Principle of Stationary Total Potential Energy.

### 2.6.5. First Theorem of Complementary Energy

We now turn to a theorem which has received little attention since it was first published and whose significance has not been fully appreciated until quite recently. In 1889 Engesser[14] published a paper on statically indeterminate structures which was far in advance of its time; in the course of this paper he enunciated the theorem which can now, in virtue of its place in the pattern of the energy theorems, be given the title quoted above which is that used by Brown[2].

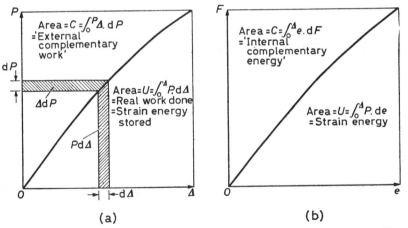

Figure 2.22. (a) Load-deflexion diagram for one of the load points of a non-linear structure: the 'external complementary work' is equal to the area above the line; (b) Force-extension diagram for one of the members of a non-linear structure

Recent interest in the subject of complementary energy was shown first by Westergaard[15] and then by Charlton[16]; in a second paper[17] the latter returned to that comparison of the properties of strain and complementary energy respectively which characterized Engesser's original enquiry.

Complementary energy was briefly defined in Section 3 of this chapter. Extending this definition to the case of a series of forces $P_1, P_2 \ldots P_j$ acting on a structure and producing corresponding deflexions $\varDelta_1, \varDelta_2 \ldots \varDelta_j$, the

# THE ENERGY THEOREMS OF STRUCTURAL ANALYSIS

'complementary work done' or 'external complementary energy' is

$$\sum_j \int_0^{P_j} \varDelta_j \,.\, \mathrm{d}P_j$$

The total complementary energy stored in the $n$ members, the 'internal complementary energy' is

$$\sum_n \int_0^{F_n} e_n \,.\, \mathrm{d}F_n \qquad \text{See Figure 2.22.}$$

Charlton[17] showed that, in the absence of gross distortion, the conservation of complementary energy was assured if real (dynamic) energy was conserved, and we can therefore write

$$C = \sum_j \int_0^{P_j} \varDelta_j \,.\, \mathrm{d}P_j = \sum_n \int_0^{F_n} e_n \,\mathrm{d}F_n \qquad \ldots\ (2.9)$$

The same result was reached by Brown[2], by somewhat different reasoning.

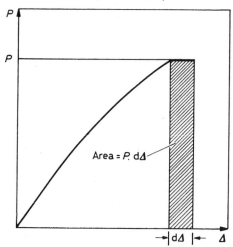

*Figure 2.23.* If one of the load-points of a structure is caused to deflect while the load there remains constant the 'external complementary work' done is zero; the real work is $P.\mathrm{d}\varDelta$

We now contemplate the variation of complementary energy produced by a small increase $\delta P_j$ in one of the applied forces, $P_j$. The gain in complementary energy will, by definition, be

$$\frac{\partial C}{\partial P_j}\,\delta P_j \qquad \ldots\ (\mathrm{i})$$

As a result of this increase in the magnitude of $P_j$ all the loaded joints $1, 2 \ldots j$ will be displaced but, although real work is done when the point of application of a constant force is displaced, complementary work is only

## 2.6.5. FIRST THEOREM OF COMPLEMENTARY ENERGY

done when the force itself increases. (See FIGURE 2.23.) It therefore appears that $\delta P_j$ is the only contributor to the increased complementary work, the amount being

$$\delta P_j \cdot \Delta_j \qquad \ldots \text{(ii)}$$

Equating (i) and (ii), the two alternative expressions for the gain in complementary work, we have

$$\frac{\partial C}{\partial P_j} \cdot \delta P_j = \delta P_j \cdot \Delta_j$$

i.e.
$$\frac{\partial C}{\partial P_j} = \Delta_j \qquad \ldots \text{(2.10)}$$

This is the First Theorem of Complementary Energy.

### 2.6.5.1. *Deflexions by Complementary Energy*

As an example of the use of the First Theorem of Complementary Energy for computing deflexions, we return to Section 6.2.5, example (a) (FIGURE 2.17).

The first step is to work out the complementary energy of the component members.

$AB, BC, CD$: $\quad F = \dfrac{Be}{L} + De^2 \quad$ i.e. $\quad e = -\dfrac{B}{2DL} + \left(\dfrac{B^2}{4D^2L^2} + \dfrac{F}{D}\right)^{\frac{1}{2}}$

$$C = \int_0^F e \cdot dF = -\frac{BF}{2DL} + \frac{2D}{3}\left(\frac{B^2}{4D^2L^2} + \frac{F}{D}\right)^{\frac{3}{2}}$$

but $\quad F = -P/\sqrt{3}$, hence for the three members

$$C = \frac{\sqrt{3}BP}{2DL} + 2D\left(\frac{B^2}{4D^2L^2} - \frac{P}{D\sqrt{3}}\right)^{\frac{3}{2}} \qquad \ldots \text{(i)}$$

For the remaining members

$$F = \frac{AEe}{L} \qquad \text{i.e. } e = \frac{FL}{AE}$$

$$C = \int_0^F e \cdot dF = \frac{F^2L}{2AE}$$

$BE, CE$: $\quad F = +P/\sqrt{3}$

i.e. $\quad C = 2 \cdot \dfrac{P^2L}{2 \cdot 3 \cdot AE} = \dfrac{P^2L}{3AE} \qquad \ldots$ (ii)

$AE, DE$: $\quad F = P/2\sqrt{3}$

i.e. $\quad C = 2 \cdot \dfrac{P^2L}{2 \cdot 4 \cdot 3 \cdot AE} = \dfrac{P^2L}{12AE} \qquad \ldots$ (iii)

The total complementary energy is the sum of (i), (ii) and (iii).

i.e. $$C = \frac{\sqrt{3}BP}{2DL} + \frac{5P^2L}{12AE} + 2D\left(\frac{B^2}{4D^2L^2} - \frac{P}{D\sqrt{3}}\right)^{\frac{3}{2}}$$

$$\Delta_E = \frac{dC}{dP} = \frac{\sqrt{3}B}{2DL} + \frac{5PL}{6AE} - \left(\frac{3B^2}{4D^2L^2} - \frac{P\sqrt{3}}{D}\right)^{\frac{1}{2}} \quad \text{as before.}$$

### 2.6.5.2. *Identity of the First Theorem of Complementary Energy and Mohr's Equation of Virtual Work as used in Deflexion Calculations*

In Section 4 an account of the loading of a spring was given from which the idea emerged that the complementary energy of a structure is that part of the potential energy, lost by the applied loads, which is not available for conversion into strain energy.

Although such a physical interpretation may be helpful at first we are not concerned with complementary energy itself so much as with its derivative; at this point a much more fruitful idea can be seen to emerge, as follows:
From equation 2.10, we have

$$\Delta_j = \frac{\partial C}{\partial P_j} = \frac{\partial C}{\partial F_n} \cdot \frac{\partial F_n}{\partial P_j}$$

hence from equation 2.9 $$\Delta_j = \sum e_n \frac{\partial F_n}{\partial P_j} \quad \quad \ldots \text{(i)}$$

The final step is to write $$\frac{\partial F_n}{\partial P_j} = F_{nj} \quad \quad \ldots \text{(ii)}$$

where $F_{nj}$ is the force in the $n$th member caused by imposing an external force $P_j = 1$ at the joint $j$. We can only take this step if we can be assured that the principle of superposition of forces is valid. (Section 5.1.) Determinate structures, linear or otherwise, present no difficulty but at first sight non-linear hyperstatic structures seem to be excluded. More penetrating consideration†, however, shows that the deflexion calculations in such cases actually refer to the corresponding primary structure, which is statically determinate, loaded with the actual applied loads together with the forces, regarded as constant quasi-external loads, in the redundant reactions and members.

We can therefore confidently combine (i) and (ii) and write

$$\Delta_j = \Sigma e_n \cdot F_{nj}$$

which is the expression for the deflexion of the joint $j$ which would have been given by the direct application of Mohr's Equation to this problem (compare 2.6.2.5(a), equation (iii)).

It now appears that rather than search for the physical meaning of complementary energy we should accept it as a convenient mathematical device for ensuring the correct application of Mohr's Equation to deflexion problems.

---
† See Chapter 4, Section 4.8.

## 2.6.6. Castigliano's Theorem, Part II

This well-known theorem is often called 'Castigliano's First Theorem'.

Its author[13] originally described it as the 'Theorem of the Differential Coefficients of the Internal Work, Part II', and there seems to be no good reason for departing further from his title than the shortened version used at the head of this Section.

The proof follows the general lines of that given for the First Theorem of Complementary Energy in Section 2.6.5 except that it is now the *strain* energy which is considered. Two alternative expressions for the gain in strain energy are obtained by considering a small load $\delta P_j$ to be applied first before and then after the rest of the load system.

When these expressions are equated we have

$$\frac{\partial U}{\partial P_j} = \varDelta_j \qquad \ldots (2.11)$$

which is Castigliano's Theorem, Part II.

There is now no need to give the proof in full since it follows at once from the consideration (see Section 2.3, Equation 2.3b) that for *linear structures* the strain and complementary energies are equal.

Thus if $C = U$

Equation 2.10 becomes $\qquad \dfrac{\partial U}{\partial P_j} = \varDelta_j \qquad \ldots (2.11)$

### 2.6.6.1. *Deflexions by Strain Energy*

Castigliano's Theorem, Part II, must only be used for dealing with **linear structures** but, as these constitute the great majority of actual engineering practice, the theorem is very widely used. It is possible to proceed on very similar lines to those used, in Section 2.6.5.1, to work out deflexions by Complementary Energy. The forces in the members are evaluated, the strain energy computed and summed to give the total strain energy which is finally differentiated with regard to the relevant force in order to obtain the deflexion.

This process can be simplified by carrying out the differentiation at an earlier stage, as follows:

$$\begin{aligned}
\varDelta_j &= \frac{\partial U}{\partial P_j} = \frac{\partial}{\partial P_j} \sum U \\
&= \frac{\partial}{\partial P_j} \sum \frac{F^2 L}{2AE} \quad \text{(see equation 2.2(b))} \\
&= \sum F \frac{\partial F}{\partial P_j} \frac{L}{AE} \\
&= \sum \frac{FF'L}{AE} \qquad \ldots (2.12)
\end{aligned}$$

where $F' = \dfrac{\partial F}{\partial P_j}$ = numerical value of force in a member caused by $P_j = 1$.

THE ENERGY THEOREMS OF STRUCTURAL ANALYSIS

As an example of this we shall work out the vertical deflexion of G and the vertical and horizontal deflexions of E of the frame shown in FIGURE 2.20 but with the pin at E removed. The cross-sectional area of each member is assumed to be 5. Young's modulus E can be taken out as a common factor.

TABLE 2.2
DEFLEXION CALCULATIONS FOR FRAME OF FIGURE 2.20

| Member | 1<br>$L$ | 2<br>$L/A$ | 3<br>$F$ | 4<br>$F'$<br>$(P=1)$ | 5<br>$E\varDelta_G$ | 6<br>$F'$<br>$(R=1)$ | 7<br>$E\varDelta_E$<br>(upward) | 8<br>$F'$<br>$(S=1)$ | 9<br>$E\varDelta_E$<br>(inward) |
|---|---|---|---|---|---|---|---|---|---|
| AB | 50  | 10 | $+10/\sqrt{3}$ | $+2/\sqrt{3}$ | 66·7  | $-6/\sqrt{3}$ | $-200·0$ | 0  |       |
| HG | 100 | 20 | $-5/\sqrt{3}$  | $-1/\sqrt{3}$ | 33·3  | $+5/\sqrt{3}$ | $-166·7$ | $-1$ | 57·9 |
| BH | 100 | 20 | $-10/\sqrt{3}$ | $-2/\sqrt{3}$ | 133·3 | $+2/\sqrt{3}$ | $-133·3$ | 0  |       |
| BG | 100 | 20 | $+10/\sqrt{3}$ | $+2/\sqrt{3}$ | 133·3 | $-2/\sqrt{3}$ | $-133·3$ | 0  |       |
|    |     |    |                |  $\Sigma$      | 366·6 |               | $-633·3$ |    | 57·9 |

The calculations are set out in TABLE 2.2 to which the following notes apply.

Column 3: The force $F$ in the members caused by the actual loading, 5 tons at $G$, acting on the frame (FIGURE 2.20(b)).

Column 4: The numerical value of the force $F'$ in the members caused by unit downward load at $G$ $(= F/5)$.

Column 5: The product of columns 2, 3 and 4 $= E\varDelta_G$.

Column 6: The numerical value of the force $F'$ in the members caused by unit upward load at $E$ (FIGURE 2.20(c)).

Column 7: The product of columns 2, 3 and 6 $= E\varDelta_E$ upwards.

Column 8: The numerical value of the force $F'$ in the members caused by unit inward load at $E$ (FIGURE 2.20(d)).

Column 9: The product of columns 2, 3 and 8 $= E\varDelta_E$ inwards.

The required deflexions are shown against the summation sign at the foot of columns 5, 7 and 9. It is to be noted that only four members enter into the calculation since all other members have zero force under the actual loading of 5 tons at $G$.

The negative value for $E\varDelta_E$ shown at the foot of column 7 shows that $E$ actually deflects downwards, that is in the opposite direction to the unit upward load applied there in FIGURE 2.20(c).

2.6.7. ENGESSER'S THEOREM OF COMPATIBILITY[†]

In Sections 2.6.5 and 2.6.5.1 we saw that the First Theorem of Complementary Energy enabled the deflexions of non-linear structures to be

---

[†] Called by Brown[2] 'The Second Theorem of Complementary Energy'.

## 2.6.7. ENGESSER'S THEOREM OF COMPATIBILITY

calculated. It would therefore appear that it could be used to obtain the compatibility relationships for hyperstatic structures.

The demonstration which follows is not sufficiently general to be described as a proof, but the ideas involved can be developed into a formal proof if desired.

The simple structure of FIGURE 2.24(a) has one redundant bar, numbered 2; we seek to find the force $F_2$ in this member. Moreover, we are to imagine that member 2 is too short, by an amount $\lambda$, to fit into position before the frame is loaded, so that $F_2$ is caused both by the external loads $P$ and by the forces locked up in the frame when member 2 is forced into position. The cross-bar is rigid so that only members 1, 2 and 3 contribute to the deflexion of the frame.

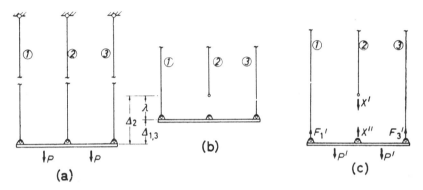

*Figure 2.24.*(a) In this simple hyperstatic framework the middle bar 2 is redundant; (b) The redundant bar 2 is $\lambda$ too short and has to be forced into position; (c) Forces X' and X" have been applied so as to pull the redundant bar 2 and the rest of the framework together

In order to construct the equilibrium and compatibility equations we proceed, in imagination, as follows: the pin holding the redundant bar is removed to convert the hyperstatic structure into the corresponding primary determinate structure shown at (b). With all the members free from stress the lack of fit is $\lambda$. In order to get the frame into its final state (a) the procedure shown at (c) is adopted. Quasi-external equal and opposite forces X' and X" are applied at the pin-holes so as to pull them together, and at the same time the loads P' are gradually increased towards their final values. The equilibrium of the cross-bar at an intermediate stage is given by

$$F'_1 + X'' + F'_3 = 2P' \qquad \ldots (i)$$

when $F'_1$ and $F'_3$ are the temporary values of the forces in the members 1 and 3. Eventually, when X' and X" have been increased to such an extent that the pin-holes coincide and when P' has reached its final value of P, the pin can be replaced and the situation (a) has been achieved.

Equilibrium considerations now tell us that

$$F_1 + X + F_3 = 2P \qquad \ldots (ii)$$

and the physical requirements of the problem that

$$X = F_2 \quad \text{.... (iii)}$$

where $X$ is the final value of $X'$ and $X''$.

The compatibility equation simply expresses the geometrical relationship between the extensions of the members, namely

$$\Delta_2 - \Delta_{1,3} = \lambda \quad \text{.... (iv)}$$

We now use the First Theorem of Complementary Energy to find these deflexions.

$$\left.\begin{array}{l} \Delta_{1,3} = -\left[\dfrac{\partial C}{\partial X''}\right]_{1,3} \\[2ex] \Delta_2 = +\left[\dfrac{\partial C}{\partial X'}\right]_2 \end{array}\right\} \quad \text{.... (v)}$$

The negative sign in the first of these arises because $\Delta_{1,3}$ is opposite in direction to $X''$.

Hence, knowing that $X'$ and $X''$ are numerically equal we can rewrite (iv)

$$\left[\frac{\partial C}{\partial X}\right]_2 + \left[\frac{\partial C}{\partial X}\right]_{1,3} = \lambda$$

i.e.
$$\left[\frac{\partial C}{\partial X}\right]_{1,2,3} = \lambda$$

and, more generally
$$\frac{\partial C}{\partial X_i} = \lambda_i \quad \text{.... (2.13)}$$

The final step indicates that the same procedure can be followed in a more complex structure with several redundant bars in which the forces are $X_a, X_b \ldots X_i$.

Finally we note that if the redundant member had been an exact fit in the unstrained position we should have had

$$\lambda = 0 \text{ and } \frac{\partial C}{\partial X} = 0 \quad \text{.... (2.13(a))}$$

This was the form in which Engesser[14] originally stated the Theorem, from which it appears that the complementary energy, expressed as a function of the external forces $P$ and the quasi-external forces $X$, has a stationary value when compatibility is satisfied. As $\dfrac{\partial^2 C}{\partial X^2}$ is essentially positive the stationary value is evidently a minimum (see Section 2.6.9).

2.6.7.1. *Example of the use of Engesser's Theorem*

The only theoretical limitation which need be placed on this Theorem is that it cannot be used in cases of gross distortion. (See Section 2.6.5.)

## 2.6.8. CASTIGLIANO'S THEOREM OF COMPATIBILITY

Unfortunately, however, mathematical complications prevent an exact solution being reached in any but the simplest of cases. Interest in the Theorem mainly resides in the place it occupies in the logical development of the energy theorems of structural analysis.

For an example we turn once more to the frame of FIGURE 2.5(a) whose members are taken to deform according to FIGURE 2.6. (See Section 2.6.1.1(c) and 2.6.2.4(c).) It is presumed that $F_1 > 2S$ and $F_2 < 2S$; $\theta$ is 60°.

$$F_1 = e_1/2k + S; \quad F_2 = e_2/k$$

$$\left. \begin{array}{l} C_1 = \displaystyle\int e \cdot dF = \int_0^{F_1} (F_1 - S) 2k \cdot dF = kF_1^2 - 2kSF_1 \\[2mm] C_2 = \displaystyle\int_0^{F_2} kF_2 \cdot dF = \dfrac{kF_2^2}{2} \end{array} \right\} \quad \ldots \text{(i)}$$

Taking $AC$ as the redundant bar, and replacing the force $F_1$ in it by quasi-external forces $X$ applies as in FIGURE 2.24(c), the equilibrium of joint $A$ gives

$$X + F_2 = P$$

Hence, from (i)

$$C = C_1 + 2C_2 = kX^2 - 2kSX + \frac{2k(P-X)^2}{2}$$

and, applying equation 2.13(a)

$$\frac{\partial C}{\partial X} = 2kX - 2kS - 2k(P - X)$$

$$= 0 \text{ if } X = \frac{P+S}{2}$$

This gives the required value of $X$ and hence of $F_1$.

It is to be noticed that we have been careful to replace the internal member force $F_1$ by the quasi-external force $X$ before differentiating; this is because $F_1$ is not variable—it is the specific value of $X$ which satisfies the compatibility condition.

### 2.6.8. CASTIGLIANO'S THEOREM OF COMPATIBILITY†

We have already remarked that the complementary and strain energies are equal for linear structures obeying Hooke's Law. It follows, therefore, that for such structures the arguments of Section 2.6.7 can be repeated with Castligiano's Theorem, Part II, replacing the First Theorem of Complementary Energy as the means of finding the deflexions required for the compatibility equation.

Castigliano's Theorem of Compatibility, which applies only to linear structures‡ is thus

---

† Usually known as Castigliano's Theorem of Least Work.

‡ Matheson[18] has showed that in spite of this restriction the theorem can be effectively used in certain special non-linear cases.

$$\frac{\partial U}{\partial X_i} = \lambda_i \qquad \ldots \ (2.14)$$

If there is no initial lack of fit $\lambda_i = 0$ and

$$\frac{\partial U}{\partial X_i} = 0 \qquad \ldots \ (2.14(a))$$

Examples of the use of this theorem follow in Section 2.6.8.1.

Williams[1] and Brown[2] draw a distinction between internal and external redundancy so far as the application of this Theorem is concerned. The former author, drawing an analogy with a theorem in dynamics, shows that the effect of redundant supports is to make the strain energy, under a given load system, greater than in the corresponding simply supported structure. This led him to speak of 'the Theorem of Maximum Strain Energy', a title which was later criticized by Brown who preferred to write of the Second and Third Minimum Strain Energy Theorems for external and internal redundancy respectively.

While there is perhaps some logical justification for the above approach the author doubts whether it contributes greatly either to the understanding or to the use of what can properly be described as a single theorem. The case of a rigid redundant external support can be included within the argument of Section 2.6.7 by specifying that the redundant member is of infinite cross-section.

### 2.6.8.1. *Examples of the Use of Castigliano's Theorem of Compatibility*

(a) *Internal Redundancy*—We return again to the frame FIGURE 2.5(a) which is taken to consist of uniform bars having the same cross-section and Modulus of Elasticity, as in Section 2.6.1.1(a).

With the bar $AC$ temporarily disconnected and the quasi-external forces $X$ acting at the pin-holes, the equation of equilibrium of $A$ is

$$X + 2F_2 \cos \theta = P$$

$$U = \sum \frac{F^2 L}{2AE}$$

$$= \frac{X^2 h}{2AE} + 2 \left(\frac{P-X}{2\cos\theta}\right)^2 \frac{L}{2AE}$$

$$= \frac{X^2 L \cos\theta}{2AE} + \frac{(P-X)^2 L}{4 \cos^2\theta \, AE}$$

Introducing equation 2.14(a)

$$\frac{\partial U}{\partial X} = \frac{XL \cos\theta}{AE} - \frac{(P-X)L}{2 \cos^2\theta \cdot AE}$$

$$= 0 \text{ if } X = \frac{P}{1 + 2\cos^3\theta}$$

This is the required value of $X$ and hence of $F_1$.

## 2.6.8. CASTIGLIANO'S THEOREM OF COMPATIBILITY 2.6.8.1

(b) *External Redundancy*—All the bars of the truss shown in FIGURE 2.25(a) are of equal length, cross-sectional area and Modulus of Elasticity. We define

$F$ = Total force in a member.

$F_0$ = Force in a member of the simply-supported frame under the loads $P$ (FIGURE 2.25(b)).

$F_a$ = Force in a member of the simply-supported frame under unit load replacing the redundant reaction (FIGURE 2.25(c)).

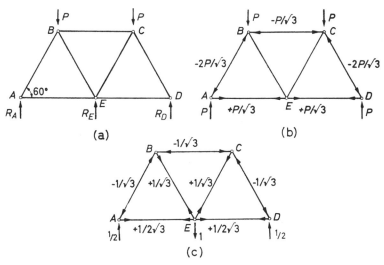

Figure 2.25. (a) This pin-jointed frame has a redundant reaction $R_E$; (b) The forces $F_0$ are developed in the members when the redundant reaction is removed; (c) The forces $F_a$ are developed in the members when unit load is applied at $E$

Then, as the principle of superposition applies to this linear frame,

$$F = F_0 + X \cdot F_a \qquad \ldots \text{(i)}$$

Now
$$U = \sum \frac{F^2 L}{2AE}$$

and, from equation 2.14(a)

$$0 = \frac{\partial U}{\partial X} = \frac{\partial}{\partial X} \sum \frac{F^2 L}{2AE} = \sum F \frac{\partial F}{\partial X} \frac{L}{AE}$$

But, from (i)
$$\frac{\partial F}{\partial X} = F_a$$

i.e.
$$\sum F F_a \frac{L}{AE} = 0$$

81

and, since $\dfrac{L}{AE}$ is constant

$$0 = \Sigma F F_a = \Sigma (F_0 + X F_a) F_a \text{ from (i)}$$
$$= \Sigma F_0 F_a + X \Sigma F_a{}^2$$

i.e.
$$X = -\dfrac{\Sigma F_0 F_a}{\Sigma F_a{}^2}$$

Values of $F_0$ and $F_a$, extracted from FIGURES 2.25(b) and (c), are tabulated in TABLE 2.3 where the required summations are carried out. The value of $X$, the reactive force at $E$, is

$$-\dfrac{\Sigma F_0 F_a}{\Sigma F_a{}^2} = -\dfrac{2P}{11/6} = -\dfrac{12P}{11}$$

From this result we are able, in column 6, to compute the nett force in the members of the frame.

TABLE 2.3
ANALYSIS OF FRAME OF FIGURE 2.25

| Member | 1 | 2 | 3 | 4 | 5 | 6 |
|---|---|---|---|---|---|---|
|  | $F_0$ | $F_a$ | $F_0 F_a$ | $F_a{}^2$ | $XF_a$ | $F_0 + XF_a = F$ |
| AB | $-2P \times 1/\sqrt{3}$ | $-1 \times 1/\sqrt{3}$ | $+2P/3$ | $+1/3$ | $+12P \times 1/11\sqrt{3}$ | $-10P \times 1/11\sqrt{3}$ |
| CD | $-2P$ ,, | $-1$ ,, | $+2P/3$ | $+1/3$ | $+12P$ ,, | $-10P$ ,, |
| BE | $0$ | $+1$ ,, | $0$ | $+1/3$ | $-12P$ ,, | $-12P$ ,, |
| CE | $0$ | $+1$ ,, | $0$ | $+1/3$ | $-12P$ ,, | $-12P$ ,, |
| AE | $+P$ ,, | $+1/2$ ,, | $+P/6$ | $+1/12$ | $-6P$ ,, | $+5P$ ,, |
| ED | $+P$ ,, | $+1/2$ ,, | $+P/6$ | $+1/12$ | $-6P$ ,, | $+5P$ ,, |
| BC | $-P$ ,, | $-1$ ,, | $+P/3$ | $+1/3$ | $+12P$ ,, | $+P$ ,, |
|  |  | $\Sigma$ | $2P$ | $+11/6$ |  |  |

It will be noticed that the calculations in these two examples have been carried out in slightly different ways. In the first the strain energy was fully evaluated before differentiation; in the second, differentiation was carried out at an earlier stage. No difference of principle is implied here, but the second procedure will usually be found to be the more convenient.

### 2.6.9. NOMENCLATURE

The nomenclature used above is perhaps unfamiliar but it has the merit of being logical. So many different titles are now current in the literature that new ones should only be introduced if they can be completely justified.

Although Castigliano's own writing is perfectly clear and logical, later developments have shown that it was unfortunate that he considered pre-stressed structures after, and not before, structures whose members fitted perfectly. This led him to give the title 'Least Work' to his famous Theorem, thus starting a line of false reasoning which has persisted ever since.

Those who dislike short words soon preferred 'Minimum Strain Energy', a title which at once appealed because of the familiar Principle of Minimum Potential Energy. It was an easy step to confuse these; so much so that one

TABLE 2.4

THE ENERGY THEOREMS OF STRUCTURAL ANALYSIS

| | | | |
|---|---|---|---|
| *Theorems expressing conditions of equilibrium* | | | |
| 2.6.1 | Minimum Potential Energy .. .. | $\dfrac{\partial}{\partial \Delta}(U + V) = 0$ | Universal |
| 2.6.2.2 | Mohr's Equation of Virtual Work | | |
| | Infinitesimal Form .. .. .. | $\Sigma \bar{P}\{\delta \Delta\} = \Sigma \bar{F}\{\delta e\}$ | Universal |
| | Finite form .. .. .. .. | $\Sigma \bar{P}\{\Delta\} = \Sigma \bar{F}\{e\}$ | Gross distortion excluded |
| 2.6.3 | Castigliano's Theorem, Part I .. | $\partial U/\partial \Delta_j = P_j$ | Universal | Equivalent to 2.6.1 |
| 2.6.4 | First Theorem of Minimum Strain Energy | $\partial U/\partial \Delta = 0$ | Universal | Special case of 2.6.1 |
| *Theorems used for finding deflexions* | | | |
| 2.6.2.5 | Mohr's Equation of Virtual Work | | |
| | Infinitesimal Form .. .. .. | $\Sigma \bar{P}\{\delta \Delta\} = \Sigma \bar{F}\{\delta e\}$ | Universal |
| | Finite form .. .. .. .. | $\Sigma \bar{P}\{\Delta\} = \Sigma \bar{F}\{e\}$ | Gross distortion excluded |
| 2.6.5 | First Theorem of Complementary Energy | $\partial C/\partial P_j = \Delta_j$ | Gross distortion excluded | Equivalent to finite form of 2.6.2.5 |
| 2.6.6 | Castigliano's Theorem, Part II .. | $\partial U/\partial P_j = \Delta_j$ | Linear frames only | Special case of 2.6.5 |
| *Theorems expressing compatibility in hyperstatic frames* | | | |
| 2.6.7 | Engesser's Theorem of Compatibility .. | $\partial C/\partial X_i = \lambda_i$ | Gross distortion excluded | |
| 2.6.8 | Castigliano's Theorem of Compatibility.. | $\partial U/\partial X_i = \lambda_i$ | Linear frames only | Special case of 2.6.7 |

author has written ' It is often considered sufficient proof of the Theorem (of Least Work) to urge that it is inevitable that the natural expenditure of energy for the establishment of equilibrium in a disturbed body is the least possible '.

Let us therefore use the title ' Castligiano's Theorem of Compatibility ' for this method; in this way we acknowledge the authorship, correctly describe the fundamental idea and avoid confusion with the quite distinct Principle of Minimum Potential Energy.

Identical reasoning suggests that the corresponding Theorem for Complementary Energy should be called ' Engesser's Theorem of Compatibility '.

## 2.7. RELATIONS BETWEEN THE THEOREMS

The rather extensive list of energy theorems which has just been described falls conveniently into three main sections which are indicated in TABLE 2.4; many of the theorems have been found to have close relationships and, in some cases, to be equivalent. These relationships are shown in TABLE 2.4 together with the limitations on the validity on the use of the various theorems.

## REFERENCES

[1] WILLIAMS, D. ' The relations between the energy theorems applicable in Structural Theory '. *Phil. Mag.* Nov. 1938. p. 617
[2] BROWN, E. H. ' The energy theorems of structural analysis '. *Engineering.* March 1955. p. 305
[3] POLLOCK, P. J. and ALEXANDER, G. W. ' Dynamic Stresses in Wire Ropes for use on vertical hoists '. *Proc. Conf. Wire ropes in Mines.* Inst. Min. & Met. London. 1951
[4] SOUTHWELL, R. V. *Theory of Elasticity.* Oxford. 1936. Chapter I.5
[5] WESTERGAARD, H. M. ' On the method of Complementary Energy '. *Trans. Amer. Soc. civ. Engrs.* Vol. 107. 1942. p. 765
[6] LAMB, H. ' The principle of virtual velocities and its application to the theory of elastic structures '. I.C.E. Selected Paper No. 10. 1923
[7] TIMOSHENKO, S. and YOUNG, D. H. *Theory of Structures.* McGraw-Hill. 1945
[8] VAN DEN BROEK, J. A. *Theory of Limit Design.* Wiley. 1948
[9] BAKER, J. F. *The Steel Skeleton.* Vol. II. Cambridge. 1956
[10] NEAL, B. G. and SYMONDS, P. S. ' The rapid calculation of the plastic collapse load for a framed structure '. Proc. I.C.E. Pt. III, Vol. I. 1952. p. 58
[11] CHARLTON, T. M. ' The concepts of real and virtual work '. *Engineering.* July 1955. p. 139
[12] KING, J. W. H. ' Some notes on plane frames not obeying Hooke's Law '. *The Engineer.* July 1953. p. 4
[13] CASTIGLIANO, A. *Théorem de l'Equilibre des Systèmes Elastiques et ses Applications.* Translated by E. S. Andrews as ' Elastic Stresses in Structures '. Scott, Greenwood. 1919
[14] ENGESSER, F. ' Ueber statisch unbestimmte Träger . . . ' Zeitschrift des Architekten — und Ingenieur — Vereins zu Hannover. Vol. 35 (1889)
[15] WESTERGAARD, H. M. ' On the Method of Complementary Energy '. *Trans. Amer. Soc. civ. Engrs.* Vol. 107. 1942. p. 765

## REFERENCES

[16] CHARLTON, T. M. 'Some notes on the analysis of redundant systems by means of the conception of conservation of energy'. *J Franklin Inst.* Vol. 250. 1950. p. 543

[17] CHARLTON, T. M. 'Analysis of Statically Indeterminate Structures by the Complementary Energy Method'. *Engineering*. Sept. 1952. p. 389

[18] MATHESON, J. A. L. 'Castigliano's "Theorem of Compatibility"'. *Engineering*. Dec. 1955. p. 828

[19] ARGYRIS, J. H. and KELSEY, S. *Energy Theorems and Structural Analysis.* Butterworths, 1967.

## ADDITIONAL REFERENCES

HARDY CROSS. 'Virtual Work: A Restatement'. *Trans. Amer. Soc. civ. Engrs.* Vol. 90. 1927. p. 610

NILES, A. S. 'The "Energy Method", which one?' *J. Eng. Education.* Vol. 33. 1943. p. 698

MORICE, P. B. 'Energy approximations applied to problems of structural equilibrium, stability and vibrations'. Cement and Concrete Association, London. Technical Report TRA/157. 1954

MORICE, P. B. 'Further Functional Approximations in Structural Analysis'. Cement and Concrete Association, London. Technical Report TRA/184. 1955

WILLIAMS, D. 'Principle of Minimum Potential Energy in Problems of Static Equilibrium'. Aer. Res. Cttee. R. & M. No. 1827. Vol. II. 1937. p. 1043

CHAPTER 3

# GENERAL THEOREMS FOR LINEAR ELASTIC STRUCTURES

## 3.1. INTRODUCTION

CHAPTER 2 was devoted to a discussion of the energy theorems and their application to structures whose load-deflexion characteristics might, or might not, be linear. The use of these theorems was illustrated by reference to very simple structures but even in these cases non-linear behaviour was found to complicate the analysis enormously.

When the discussion is restricted to structures which obey Hooke's Law, however, the theorems of Chapter 2 can be developed in various ways to yield further theorems and formulae each of which has its own particular range of utility. These are the theorems which are used in the derivation of nearly all the methods of analysing hyperstatic structures to be described in later chapters.

In Chapter 2 we saw that Mohr's Equation of Virtual Work was essentially identical with the fundamental proposition of Stationary Potential Energy. We now use it extensively as a convenient starting point for further work and it might therefore be thought that the resulting formula would be applicable to non-linear structures. This is usually not the case because of restrictions introduced during the course of the argument. The student should note carefully the point at which the analysis becomes dependent on Hooke's Law.

## 3.2. DEVELOPMENT OF THE PRINCIPLE OF VIRTUAL WORK

### 3.2.1. THE EQUATION OF VIRTUAL WORK IN TERMS OF STRESSES

The argument of Chapter 2, Section 2.6.2.2, referring to a pin-jointed structure, can be modified in the following way to apply to a solid body acted on by forces at its boundaries. FIGURE 3.1(a) shows such a body, supported by immovable reactions and acted upon by a series of external forces† $\bar{P}$ at points 1, 2, 3 etc.; we concentrate our attention upon an infinitesimal internal element, conveniently a paralleliped, which is acted upon by the stresses $\bar{f}_x, \bar{f}_y, \bar{f}_z$ etc. caused by the boundary forces $\bar{P}$.

It is supposed that a small change of shape of the body is brought about by some agency, independent of $\bar{P}$, such that the corresponding displacements† at the boundaries are $\varDelta$. This distortion will affect the internal element we are considering, and translation, rotation, distortion and volume change of the element is liable to occur. Suppose the total Virtual Work done

---

† Following Chapter 2, Section 2.6.2.2 we take it that the general term 'force' here includes 'couples'; the displacements 'corresponding to' couples are rotations.

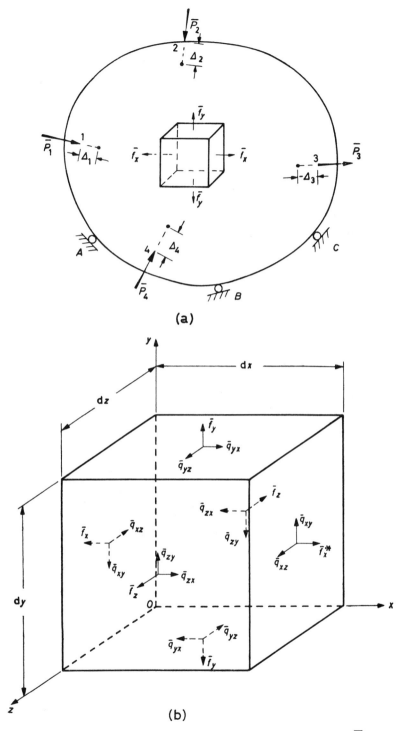

*Figure 3.1.* (a) The body is supported rigidly at $A$, $B$ and $C$; forces $\overline{P}$ act at points 1, 2 ... and produce internal stresses $\overline{f}_x, \overline{f}_y$ etc. A virtual distortion results in corresponding displacements $\Delta$; (b) Stresses on the faces of the element caused by the external loads $\overline{P}$

by all the stresses acting on the element is $dW$ while that part resulting from distortion is $dW_d$. Translation and rotation is thus responsible for $dW - dW_d$ but, since the virtual work associated with the movement of a rigid body in equilibrium under a system of forces is zero, we have

$$dW - dW_d = 0$$

or
$$dW = dW_d$$

Integration over the whole body gives

$$W = W_d$$

where $W$ is the total Virtual Work performed at all the boundaries of the body. At internal boundaries pairs of equal and opposite stresses are displaced by equal amounts and it is only at those points on the external boundaries where forces are applied, that is at points $1, 2, \ldots$, that Virtual Work can be done; at each of these points the Virtual Work done is $\overline{P}\varDelta$ by definition.

Hence $\quad \overline{P}\varDelta = W = W_d \qquad \ldots\ldots (3.1)$

In order to evaluate $W_d$ we examine more closely the stresses $\overline{f}$ etc. which act on the element; these are shown in FIGURE 3.1(b) with the following notation:

$dx, dy, dz$: dimensions parallel to $x$- $y$- $z$- axis.

$\overline{f}_x, \overline{f}_y, \overline{f}_z$: direct stress parallel to axis specified by suffix.

$\overline{q}_{xy}, \overline{q}_{yx} \ldots$: shearing stress acting in the directions specified by the second suffix on the face normal to the axis designated by the first.

It is found that the variation of stress from one side of the element to the other is negligible. For example, the stress marked with an asterisk in FIGURE 3.1(b) should be written in full thus:

$$\overline{f}_x^* = \overline{f}_x + \frac{\partial \overline{f}_x}{\partial x} dx$$

However it transpires that the differential terms result in infinitesimals of a higher order than need be considered and we therefore write $\overline{f}_x^* = \overline{f}_x$.

The equilibrium of the element, from the point of view of rotation about the $z$-axis for instance, is examined by taking moments about that axis. The only stresses concerned are $\overline{q}_{yx}$ and $\overline{q}_{xy}$ and we have

$$\overline{q}_{yx} \cdot dx \cdot dz \times dy = \overline{q}_{xy} \cdot dy \cdot dz \times dx$$

i.e
$$\overline{q}_{yx} = \overline{q}_{xy}$$

Similar results are obtained for the other pairs of complementary shearing stresses, thus

$$\overline{q}_{yz} = \overline{q}_{zy}; \quad \overline{q}_{xz} = \overline{q}_{zx}$$

## 3.2. DEVELOPMENT OF VIRTUAL WORK PRINCIPLE

We can now conveniently use a single suffix to specify the shearing stresses; thus $\bar{q}_z$ stands for the two pairs of stresses $\bar{q}_{xy}$ and $\bar{q}_{yx}$ which act on the faces parallel to the $z$-axis.

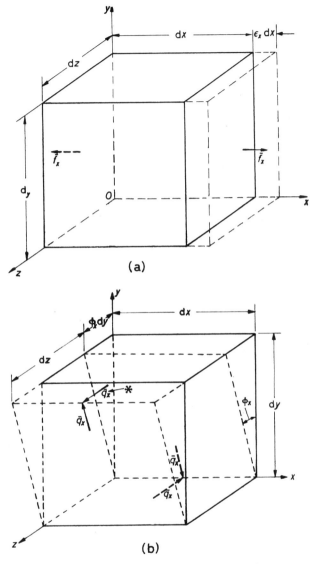

*Figure 3.2.* (a) A virtual extension $\epsilon_x dx$ occurs in the $x$-direction only; (b) A virtual angular distortion $\phi_x$ occurs about the $x$-axis only

It is now possible to compute the work done by all these stresses when a virtual distortion is imposed. We begin by considering separately an extension $\epsilon_x\, dx$ in the $x$-direction as shown in Figure 3.2(a). (If this extension

89

were caused by a stress $\epsilon_x$ would be the strain, but we do not want so to restrict the *cause* of the extension.)

The force associated with the stress $\bar{f}_x$ is $\bar{f}_x.dy.dz$ and the Virtual Work done is therefore

$$\bar{f}_x.dy.dz(\epsilon_x.dx) = \bar{f}_x\,\epsilon_x.dx.dy.dz \qquad \ldots\text{(i)}$$

The next step is to consider an angular distortion of the kind indicated in FIGURE 3.2(b). Of the four shearing stresses $\bar{q}_x$ only that marked with an asterisk contributes to the Virtual Work. The amount is

$$\bar{q}_x\,dx.dz\,(\phi_x.dy) = \bar{q}_x\,\phi_x.dx.dy.dz \qquad \ldots\text{(ii)}$$

Hence the total Virtual Work of the element is the sum of terms such as (i) and (ii),

i.e. $dW_d = [\bar{f}_x\,\epsilon_x + \bar{f}_y\,\epsilon_y + \bar{f}_z\,\epsilon_z + \bar{q}_x\,\phi_x + \bar{q}_y\,\phi_y + \bar{q}_z\,\phi_z]\,dx.dy.dz$

Hence, integrating over the whole body and introducing equation 3.1

$$\Sigma \bar{P}\Delta = \iiint [\bar{f}_x\,\epsilon_x + \bar{f}_y\,\epsilon_y + \bar{f}_z\,\epsilon_z + \bar{q}_x\,\phi_x + \bar{q}_y\,\phi_y + \bar{q}_z\,\phi_z]\,dx.dy.dz$$

$$\ldots\text{(3.2)}$$

In exactly the same way as in equation 2.5 those symbols which are distinguished by the bar belong to the force system while the remainder belong to the displacement system.

In order to employ this important equation in a particular problem we must choose suitable force and displacement systems and then evaluate the stresses $\bar{f}$ and $\bar{q}$ of the force system and the unit distortions $\epsilon$ and $\phi$ of the displacement system. This process cannot conveniently be described in general terms and we therefore proceed to transform equation 3.2 into a more convenient form for dealing with certain classes of problem.

### 3.2.2. The Virtual Work Equation in terms of Axial Force, Bending Moment and Transverse Shear

Most structural problems are treated two-dimensionally and it is therefore convenient to have equation 3.2 in a form suitable for use in cases of beams subjected to direct stress combined with bending in a single plane. As unsymmetrical bending is to be excluded one principal axis of inertia of the cross-section of the beam must lie in the plane of bending.

We take the force system to be a series of loads $\bar{P}$ which produce an axial force $\bar{F}$, a bending moment $\bar{M}$ and a transverse shear $\bar{Q}$ at a section of the beam. If the second moment of area of the cross-section is $I$ and its area is $A$ the stresses at a point on the cross-section are:

$$\left.\begin{array}{l} \bar{f}_x = \dfrac{\bar{M}y}{I} + \dfrac{\bar{F}}{A}\;;\quad \bar{f}_y = 0\;;\quad \bar{f}_z = 0 \\[6pt] \bar{q}_x = 0\;;\quad \bar{q}_y = 0\;;\quad \bar{q}_z = \dfrac{\bar{Q}A^*\bar{y}}{BI} \end{array}\right\} \quad \ldots\text{(i)}$$

## 3.2. DEVELOPMENT OF VIRTUAL WORK PRINCIPLE

The $x$-axis has been taken to coincide with the centroidal axis of the beam and FIGURES 3.3(a) and (b) show the symbols used in the expression† for the shearing stress $\bar{q}_z$ which is obtained by evaluating the difference of the axial bending stresses acting on the area $A^*$. $\bar{q}_z$ is assumed to be uniformly distributed over the breadth $B$.

The displacement system is another set of loads which produce deflexions $\Delta$ at the points of application of $\bar{P}$ and an axial force $F$, a bending moment $M$ and a transverse shear $Q$ at the same cross-section. The unit distortions $\epsilon_x$ etc. are now strains and we have

$$\left. \begin{array}{c} \epsilon_x = \dfrac{My}{EI} + \dfrac{F}{AE} \; ; \quad \epsilon_y = \epsilon_z = 0 \\[1em] \phi_x = \phi_y = 0 \; ; \quad \phi_z = \dfrac{QA^*\bar{y}}{BIN} \end{array} \right\} \quad \ldots \ldots \text{(ii)}$$

Introducing these relations into equation 3.2 we have

$$\sum \bar{P}\Delta = \iiint \left\{ \left( \frac{\bar{M}y}{I} + \frac{\bar{F}}{A} \right) \left( \frac{My}{EI} + \frac{F}{AE} \right) + \frac{\bar{Q}A^*\bar{y}}{BI} \cdot \frac{QA^*\bar{y}}{BIN} \right\} dx.dy.dz$$

$$= \iiint \left\{ \frac{\bar{M}My^2}{EI^2} + \frac{\bar{M}yF}{IAE} + \frac{\bar{F}My}{AEI} + \frac{\bar{F}F}{A^2E} + \frac{\bar{Q}Q}{N}\left(\frac{A^*\bar{y}}{BI}\right)^2 \right\} dx.dy.dz$$

$$= \int \frac{\bar{M}Mdx}{EI^2} \iint y^2.dy.dz \qquad = \int \frac{\bar{M}Mdx}{EI}$$

$$+ \int \frac{\bar{M}Fdx}{IAE} \iint y.dy.dz \qquad + \quad 0$$

$$+ \int \frac{\bar{F}Mdx}{AEI} \iint y.dy.dz \qquad + \quad 0$$

$$+ \int \frac{\bar{F}Fdx}{A^2E} \iint dy.dz \qquad + \int \frac{\bar{F}Fdx}{AE}$$

$$+ \int \frac{\bar{Q}Qdx}{I^2N} \iint \left(\frac{A^*\bar{y}}{B^2}\right)^2 dy.dz \qquad + \kappa \int \frac{\bar{Q}Qdx}{AN}$$

i.e. $$\sum \bar{P}\Delta = \int \frac{\bar{M}Mdx}{EI} + \int \frac{\bar{F}Fdx}{AE} + \kappa \int \frac{\bar{Q}Qdx}{AN} \quad \ldots \ldots \text{(3.3)}$$

We have been able to simplify these integrals by making use of the following:

$$\iint y^2\, dy\, dz = I; \quad \iint y.dy.dz = 0; \quad \iint dy.dz = A$$

The integration of the last term has been carried out in terms of a numerical constant $\kappa$ which depends on the shape of the cross-section; for the rectangular section shown in FIGURE 3.3(c) we have

---

†The proof of this expression will be found in any textbook on the Strength of Materials.

# GENERAL THEOREMS FOR LINEAR ELASTIC STRUCTURES

$$\int \frac{\overline{Q}Q dx}{I^2 N} \int\int \frac{(A^* \bar{y})^2}{B^2} dy.dz = \int \frac{\overline{Q}Q dx}{I^2 N} \int \frac{(A^*\bar{y})^2}{b} \frac{dy}{} \quad \text{since } z = \text{constant} = b$$

$$= \int \frac{\overline{Q}Q dx}{I^2 N} \int_{-d/2}^{+d/2} \frac{\left\{\frac{b}{2}\left(\frac{d^2}{4} - y^2\right)\right\}^2}{b} dy$$

$$= \int \frac{\overline{Q}Q dx}{I^2 N} \cdot \frac{b \; d^5}{120}$$

Now for this section $\quad I^2 = \dfrac{b^2 d^6}{144}$

Hence the integral becomes $\dfrac{144}{120}\int \dfrac{\overline{Q}Q dx}{bd \; N} = \dfrac{6}{5}\int \dfrac{\overline{Q}Q dx}{AN}$

$$= \kappa \int \frac{\overline{Q}Q dx}{AN} \quad \text{say.}$$

Here $A$ is the area of the cross-section. We have seen that $\kappa = 6/5$ for the rectangular section; similar calculations for a circular section give $\kappa = 10/9$ while in the case of an $I$-beam, where the shearing stress is nearly uniformly distributed over the web, it is usually sufficiently accurate to take $\kappa = 1$ and to replace $A$ by the area of cross-section of the web.

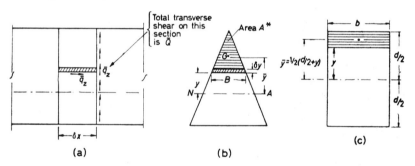

Figure 3.3. At a point distant $y$ from the neutral axis of a beam the horizontal and vertical shearing stresses $q_z$ are given by the equation $\bar{q}_z = \dfrac{QA^*\bar{y}}{BI}$

It must be understood that these calculations have been made on the basis of the elementary theory of bending in which it is assumed that plane sections remain plane. The small error due to the warping of the section caused by the uneven distribution of shear can be allowed for if desired.

The error arising from the assumption that the shearing stress is uniformly distributed across a horizontal interface is usually of negligible consequence[1].

## 3.3. GENERAL EXPRESSIONS FOR STRAIN ENERGY

The Virtual Work equation 3.3 can be used for a variety of purposes but it is especially instructive to use it to derive general expressions for the strain

## 3.3. GENERAL EXPRESSIONS OF STRAIN ENERGY    3.3.1

energy of structures. We begin by recalling that in Chapter 2, Section 2.3, we showed from first principles that the strain energy of a pin-jointed framework was given by

$$U = \sum \frac{F^2 L}{2AE} \qquad \ldots (2.2(b))$$

This result is now obtained by the use of equation 3.3.

### 3.3.1. Strain Energy due to Axial Forces

The force and displacement systems of the Virtual Work equation must be mutually consistent; that is to say that while $\bar{P}$ is an external force acting at a certain point and giving rise to $\bar{F}$, $\bar{M}$ and $\bar{Q}$ at a certain internal section, $\varDelta$ is the 'corresponding displacement' of the same external point produced at the same internal section by an independent loading $F$, $M$ and $Q$. For the time being, however, we exclude bending and shearing forces from consideration and concentrate on the direct forces acting in a pin-jointed framework; we can therefore write

$$\sum \bar{P} \varDelta = \sum_0^m \int_0^L \frac{\bar{F} F \, dx}{AE}$$

$$= \sum_0^m \frac{\bar{F} F L}{AE} \qquad \ldots (3.4)$$

Here we have made use of the fact that the forces $\bar{F}$ and $F$ are constant over the length of the member and, as the joints are pinned, the virtual work of the $m$ members can be added together.

For the purpose of finding the real work done by a single external force $\bar{P}$ we wish to know the deflexion actually produced by $\bar{P}$ itself, and so we choose as the displacement system the deflexion $\delta \bar{\varDelta}$ and the internal force $\delta \bar{F}$ produced by $\delta \bar{P}$, a small increase of $\bar{P}$. There can be no doubt that the force and displacement systems are consistent since one is a small fraction of the other.

Hence
$$\bar{P} . \delta \bar{\varDelta} = \sum \bar{F} . \frac{\delta \bar{F} . L}{AE}$$

or, since the bar over the symbols previously used to distinguish the force system is no longer necessary,

$$P . \delta \varDelta = \sum F . \frac{\delta F . L}{AE}$$

The left-hand side of this equation now represents the real work† done by $P$ as its point of application moves a distance $\delta \varDelta$, and hence it is equal to the increment of strain-energy stored in the framework in consequence.

---

† Since $P$ is no longer constant while an *imaginary* deflexion is effected by some independent agency, but is itself increased by small a amount in order to produce the deflexion, the work done is strictly $P . \delta \varDelta + a . \delta P . \delta \varDelta$, where $a$ is a factor less than unity. The second term, however, is an infinitesimal one degree higher than the first and vanishes on integration. Similar terms are neglected throughout the discussion.

# GENERAL THEOREMS FOR LINEAR ELASTIC STRUCTURES

The total work done, or strain energy stored, as the deflexion increases from 0 to $\Delta$ is obtained by integration between appropriate limits.

i.e.
$$U = \int_0^\Delta P.d\Delta = \sum \int_0^F \frac{F.dF.L}{AE} = \sum \frac{F^2L}{2AE} \quad \ldots (3.5)$$

If $F$ varies over the length of the member then we should write

$$U = \sum \int_0^L \frac{F^2 dx}{2AE} \quad \ldots (3.5(a))$$

Exactly the same reasoning can be used in the case of beams subjected to bending and shearing actions only, as follows.

### 3.3.2. Strain Energy due to Bending and Shearing

If a beam is subjected to bending and shearing action only, axial forces being excluded, the relevant part of equation 3.3 is

$$\overline{P}\Delta = \int \frac{\overline{M}M dx}{EI} + \kappa \int \frac{\overline{Q}Q dx}{AN}$$

We again choose as the displacement system the deflexion $\delta\Delta$, the bending moment $\delta\overline{M}$ and the shearing force $\delta\overline{Q}$ produced by $\delta\overline{P}$, a small increment of $\overline{P}$.

Then
$$\overline{P}.\delta\Delta = \int \frac{\overline{M}.\delta\overline{M}.dx}{EI} + \kappa \int \frac{\overline{Q}.\delta\overline{Q}.dx}{AN}$$

The left-hand side of this equation again represents the work done by $\overline{P}$ in moving its point of application through the distance $\delta\Delta$. Integrating to obtain the total work done, or strain energy stored, as $\overline{P}$ increases from 0 to $\overline{P}$, and dropping the distinguishing bar

$$U = \int_0^\Delta P.d\Delta = \int\int \frac{M.dM.dx}{EI} + \kappa \int\int \frac{Q.dQ.dx}{AN}$$
$$= \int \frac{M^2 dx}{2EI} + \kappa \int \frac{Q^2 dx}{2AN} \quad \ldots (3.6)$$

### 3.3.3. Strain Energy under Combined Stress

If the beam is subjected to axial force as well as to bending moment and shearing force the strain energy is obtained by combining equations 3.5 and 3.6. Moreover, none of the force of the argument is lost if, instead of a single force $P$ acting on a single beam, there are several forces and couples applied to an interconnected framework. Thus

$$U = \sum\int P.d\Delta = \sum\int \frac{F^2 dx}{2AE} + \sum\int \frac{M^2 dx}{2EI} + \sum \kappa \int \frac{Q^2 dx}{2AN} \quad \ldots (3.7)$$

It is often convenient to have the strain energy expression in terms of stress rather than total forces and moments. For this purpose we use equation 3.2

$$\Sigma \overline{P}\Delta = \iiint [\overline{f}_x.\epsilon_x + \overline{f}_y.\epsilon_y + \overline{f}_z.\epsilon_z + \overline{q}_x \phi_x + \overline{q}_y \phi_y + \overline{q}_z \phi_z]\, dx.dy.dz \quad \ldots (3.2)$$

## 3.3. GENERAL EXPRESSIONS FOR STRAIN ENERGY    3.3.4

Proceeding as before we choose a small increment of $\overline{P}$ as the displacement system which produces an external deflexion $\delta\Delta$ and internal stresses $\delta f_x$, $\delta f_y$, $\delta f_z$, $\delta q_x$, $\delta q_y$, $\delta q_z$. The strains which accompany these stresses are, for example

$$\delta\epsilon_x = \frac{1}{E}\left\{\delta f_x - \gamma\nu\left(\delta f_y + \delta f_z\right)\right\} \quad \text{and} \quad \delta\phi_x = \frac{\delta q_x}{N}$$

The appropriate expressions in $y$- and $z$- are obtained by cyclic variation. Substituting in equation 3.2 and noting that $dx.\,dy.\,dz = dV$ we finally obtain by integration

$$U = \sum \int P.d\Delta = \int \left\{ \frac{1}{2E}(f_x^2 + f_y^2 + f_z^2) - \frac{\gamma\nu}{E}(f_x f_y + f_y f_z + f_z f_x) \right.$$
$$\left. + \frac{1}{2N}(q_x^2 + q_y^2 + q_z^2) \right\} dV \quad \ldots\,(3.8)$$

### 3.3.4. ALTERNATIVE DERIVATIONS OF STRAIN ENERGY EXPRESSIONS

A more familiar method of arriving at equation 3.7 is to begin with the expression for the strain energy of direct stress

$$U = \frac{f^2 AL}{2E}$$

which was obtained from first principles as equation 2.2a. The strain energy

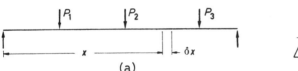

(a)   (b)

*Figure 3.4.* The strain-energy of an element, length $\delta x$ and cross-sectional area $\delta A$ is integrated to give the total strain-energy of the beam

in an element of a beam, such as that indicated in FIGURE 3.4 is therefore

$$\delta^2 U = \frac{f^2\,\delta A.\,\delta x}{2E}$$

and, recalling that

$$f = \frac{My}{I}$$

$$\delta^2 U = \frac{M^2 y^2}{I^2} \cdot \frac{\delta A.\,\delta x}{2E}$$

Integrating first across the section and then along the length, the strain energy of the whole beam is

$$U = \int_0^L \frac{M^2 y^2 dA}{2EI^2}\,dx$$

$$= \int_0^L \frac{M^2 dx}{2EI} \quad \text{for } y^2 dA = I$$

To obtain the strain energy due to the shearing stress consider FIGURE 3.5 which represents an elementary prism, of unit thickness perpendicular to

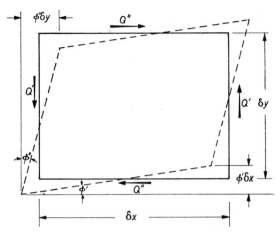

Figure 3.5. This prism, of unit thickness perpendicular to the paper, is under pure shearing stress

the diagram, under the action of pure shear. The shearing forces $Q'$ and $Q''$ correspond to the equal complementary shearing stresses $q' = q'' = q$, and
$$Q' = q' \, \delta y; \qquad Q'' = q'' \, \delta x \qquad \ldots \text{(i)}$$

As the material follows Hooke's Law the relation between shearing deflexion and shearing force is linear, and the work done by $Q'$ and $Q''$ is

$$\delta U = \tfrac{1}{2} Q' \phi' \delta x + \tfrac{1}{2} Q'' \phi'' \delta y \qquad \ldots \text{(ii)}$$

Now the sum of $\phi'$ and $\phi''$ is equal to $\phi$ the angular deformation of the prism or shearing strain and

$$\phi = q/N \qquad \ldots \text{(iii)}$$

Combining (i) and (ii) we have

$$\delta U = \tfrac{1}{2} q' \, \delta y . \phi' \, \delta x + \tfrac{1}{2} q'' \, \delta x . \phi'' \, \delta y$$
$$= \tfrac{1}{2} q \phi \, \delta x \, \delta y$$

and, introducing the relationship (iii)

$$\delta U = \frac{q^2}{2N} \, \delta x . \delta y \qquad \ldots \text{(iv)}$$

The final integration to obtain the shearing strain energy in the whole beam follows the same lines as in Section 3.2.2. Referring to FIGURE 3.3(b) and (c) and introducing the relationship

$$q = \frac{QA^* \bar{y}}{bI}$$

## 3.4. GENERAL EXPRESSIONS FOR DEFLEXION  3.4.1

we obtain, for a rectangular section

$$U = \int\int_{-d/2}^{+d/2} \frac{Q^2}{2b^2I^2N} \cdot \left\{\frac{b}{2}\left(\frac{d^2}{4} - y^2\right)\right\}^2 dy.dx$$

$$= \frac{6}{5}\int \frac{Q^2 dx}{2AN}$$

### 3.4. GENERAL EXPRESSIONS FOR THE DEFLEXION OF A STRUCTURE UNDER AXIAL AND SHEARING FORCES AND BENDING MOMENTS

#### 3.4.1. Derivation from the Virtual Work Equation 3.3

FIGURE 3.6 is intended to represent a rather general structure whose members, as a result of the action of the loads $P$, $P_2$ ...., are subjected to axial and shearing stresses and bending moments. The problem is to find the deflexion $\Delta_j$ of the joint $j$ in the direction of the dotted line that is to say in the line of action of the load $P_j$; we include the possibility that $P_j$ may be zero.

On examining equation 3.3

$$\sum \overline{P}\Delta = \sum \int \frac{\overline{M}M dx}{EI} + \sum \int \frac{\overline{F}F dx}{AE} + \sum \kappa \int \frac{\overline{Q}Q dx}{AN}$$

which has now been made more general by the addition of the summation signs, it becomes apparent that we can obtain the desired result by choosing

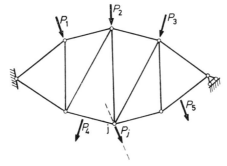

Figure 3.6. The members of this structure are under the action of axial and shearing stresses and bending moments

$P_j = 1$ as the force system. This is consistent with a displacement system comprising the required deflexion $\Delta_j$ (in the direction of $P_j$) and the distortions of the structure caused by the actual loading.

i.e. $$1 \times \Delta_j = \sum \int \frac{M'M dx}{EI} + \sum \int \frac{F'F dx}{AE} + \sum \kappa \int \frac{Q'Q dx}{AN} \quad \ldots \quad (3.9)$$

Here $M'$, $F'$, and $Q'$ are the bending moment, axial force and shearing force due to $P_j = 1$, while $M$, $F$ and $Q$ are the corresponding effects due to the actual loading which also produces the deflexion $\Delta_j$.

It remains to evaluate the integrals; an example of this is given in Section 3.4.3 below.

### 3.4.2. Derivation from the Strain Energy Equation

In Chapter 2, Section 2.6.6, we saw that Castigliano's Theorem, Part II, provided a convenient means of finding the deflexion of a specific point of a linear elastic structure. Applying this to the general strain energy expression 3.7, we have

$$\Delta_j = \frac{\partial U}{\partial P_j} = \frac{\partial}{\partial P_j} \left\{ \sum \int \frac{F^2 dx}{2AE} + \sum \int \frac{M^2 dx}{2EI} + \sum \kappa \int \frac{Q^2 dx}{2AN} \right\}$$

$$= \sum \int \frac{\partial F}{\partial P_j} \cdot \frac{F dx}{AE} + \sum \int \frac{\partial M}{\partial P_j} \cdot \frac{M dx}{EI} + \sum \kappa \int \frac{\partial Q}{\partial P_j} \cdot \frac{Q dx}{AN} \quad \ldots \text{(i)}$$

Now in a linear elastic structure, to which the principle of superposition applies,

$$\frac{\partial F}{\partial P_j} = F'; \qquad \frac{\partial M}{\partial P_j} = M'; \qquad \frac{\partial Q}{\partial P_j} = Q' \quad \ldots \text{(ii)}$$

The introduction of (ii) into (i) yields equation 3.9.

### 3.4.3. Example of General Deflexion Calculation

FIGURE 3.7(a) shows a simple beam and column structure and we seek the deflexion of the point $C$ where the load $P$ is applied. The shearing force and bending moment diagrams for the beam are shown; there is a thrust of $-2P$ in the column $BD$.

In applying equation 3.9 to this problem each of the terms is dealt with in turn. The bending term applies only to $AC$ and advantage is taken of the symmetry of the bending moment diagram to obtain the result for $AC$ by doubling that for $AB$: this is tantamount to integrating along $AB$ and $BC$ separately, taking the origin at $A$ and $C$ respectively.

i.e. $\displaystyle\sum \int \frac{M'M dx}{EI} = \left[ \int \frac{M'M dx}{EI} \right]_{AC} = 2 \left[ \int \frac{M'M dx}{EI} \right]_{AB}$

$$= \frac{2}{EI} \int_0^L (+x)(+Px) dx \quad \begin{array}{l} \text{for } M = +Px \\ \text{and } M' = M/P \\ = +x \end{array}$$

$$= \left[ \frac{2}{3} \frac{PL^3}{EI} \right]_{AC}$$

$\displaystyle\sum \int \frac{F'F dx}{AE} = \frac{F'FL}{AE} = \frac{(-2)(-2P)L}{AE} \quad \begin{array}{l} \text{for the column } BD \text{ only; there} \\ \text{is no direct stress in } AC. \end{array}$

$$= \left[ \frac{4PL}{AE} \right]_{BD}$$

## 3.4. GENERAL EXPRESSIONS FOR DEFLEXION  3.4.3

$$\sum_{\kappa} \int \frac{Q'Q\,dx}{AN} = \frac{2K}{AN}\int_0^L (\pm 1)(\pm P)\,dx$$

for the beam $AC$. This shear term is dealt with in the same manner as the bending term above.

$$= \left[\frac{2\kappa PL}{AN}\right]_{AC}$$

Hence
$$\Delta_C = \left[\frac{2PL^3}{3EI}\right]_{AC} + \left[\frac{4PL}{AE}\right]_{BD} + \left[\frac{2KPL}{AN}\right]_{AC}$$

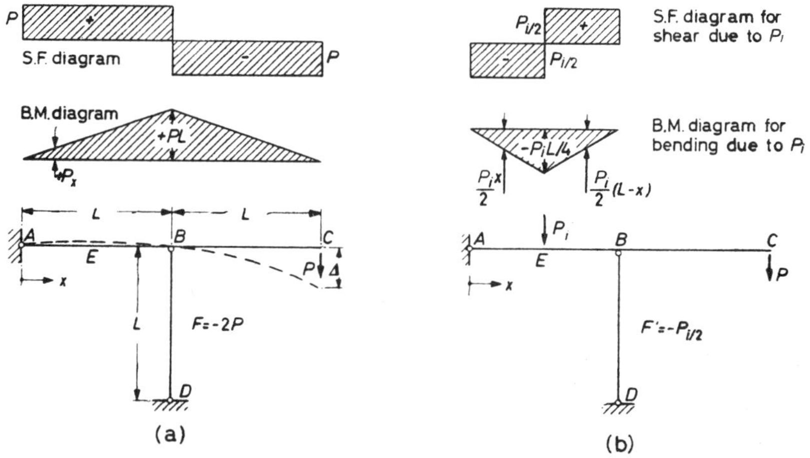

*Figure 3.7.* (a) The beam $ABC$ is hinged to the column $BD$ at $B$; the deflexion of $C$ is required; (b) In order to find the deflexion at the unloaded point $E$ an imaginary load $P_i = 1$ is applied there

If the cross-section of both beam and column is a rectangle of breadth $b$ and depth $d$ then

$$A_{AC} = A_{BD} = bd \; ; \quad I_{AC} = \frac{bd^3}{12} \; ; \quad \kappa_{AC} = 6/5$$

Taking $\qquad N = 2E/5$

$$\Delta_c = \frac{PL}{bdE}\left\{\frac{8L^2}{d^2} + 4 + 6\right\}$$

From this it appears that, while the deflexion due to shear in the beam is rather greater than that due to the compression of the column, the bending deflexion is by far the largest of the three, being proportional to the cube of the beam length. For this reason it is common to neglect any but bending deflexion in cases where beams act in combination with struts or ties.

In order to find the deflexion at the unloaded point $E$ we apply there temporarily an imaginary or 'dummy' unit load, as in FIGURE 3.7(b). The resulting B.M. and S.F. diagrams, which concern $AB$ only, are shown and it should also be noticed that the thrust in $BD$ will be $-P_i/2$.

The appropriate terms of equation 3.9 are now:

$$\sum \int \frac{M'M dx}{EI} = \frac{1}{EI}\left\{\int_0^{L/2}\left(-\frac{P_i x}{2}\right)(+Px)\,dx + \right.$$
$$\left. \int_{L/2}^L \left(-\frac{P_i L}{2} + \frac{P_i x}{2}\right)(+Px)\,dx\right\} \quad \text{for } AB$$

$$= \frac{1}{EI}\left\{\int_0^{L/2}\left(-\frac{Px^2}{2}\right)dx + \int_{L/2}^L\left(-\frac{PxL}{2} + \frac{Px^2}{2}\right)dx\right\} \quad \text{as } P_i = 1$$

$$= -\frac{PL^3}{48EI} - \frac{PL^3}{24EI} = -\frac{PL^3}{16EI}$$

The negative sign here means that the deflexion due to bending is opposite in direction to $P_i$, i.e. upwards.

$$\sum \int \frac{F'F dx}{AE} = \frac{F'FL}{AE} = \frac{(-P_i/2)(-2P)L}{AE} \quad \text{for } BD$$
$$= \frac{PL}{AE}$$

$$\sum \kappa \int \frac{Q'Q dx}{AN} = \frac{\kappa}{AN}\left\{\int_0^{L/2}\left(-\frac{P_i}{2}\right)(+P)dx + \int_{L/2}^L\left(+\frac{P_i}{2}\right)(+P)dx\right\} \quad \text{for } AB$$

$$= \frac{\kappa}{AN}\left\{-\frac{PL}{4} + \frac{PL}{4}\right\} = 0$$

Hence $\quad \Delta_E = -\dfrac{PL^3}{16EI} + \dfrac{PL}{AE}$

## 3.5. RECIPROCAL RELATIONSHIPS

Equation 3.9 can be used to demonstrate two important laws which express the reciprocal relationships existing between the forces which act on a structure and the corresponding deflexions they produce. For simplicity the proof is given for a pin-jointed structure whose members are subjected to axial forces only, but it will be seen that the same relationship must hold when the structure is more complex.

### 3.5.1. Maxwell's Reciprocal Theorem

In Figure 3.8 a framework is shown with unit load acting first at point $a$, and then at point $b$, the directions of these forces being specified by chain lines.

The resulting deflexions are conveniently described in the following way.

$\Delta_{ab}$ = Component of deflexion of point $a$, in the specified direction, caused by unit load at $b$.

$\Delta_{ba}$ = Component of deflexion of point $b$, in the specified direction, caused by unit load at $a$.

## 3.5. RECIPROCAL RELATIONSHIPS 3.5.1

This nomenclature will be found to be very convenient; it is to be remembered that the first suffix defines the point at which the deflexion is measured

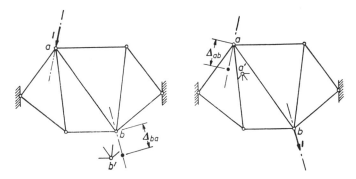

*Figure 3.8.* Maxwell's Reciprocal Theorem states that $\Delta_{ab} = \Delta_{ba}$

while the second defines the point at which unit load is applied, loads and deflexions being in the specified directions.

From the second term of equation 3.9

$$\Delta_j = \sum \int \frac{F'F dx}{AE}$$

If the members are uniform this becomes

$$\Delta_j = \sum \frac{F'FL}{AE}$$

Hence

$\Delta_{ab} = \Sigma$ (force due to unit load at $a$) (force due to unit load at $b$) $L/AE$

and

$\Delta_{ba} = \Sigma$ (force due to unit load at $b$) (force due to unit load at $a$) $L/AE$

Hence $\Delta_{ab} = \Delta_{ba}$

Although forces and linear deflexions only have been used in this proof exactly the same arguments apply to couples and angular deflexions (i.e. rotations). For example, consider the beam shown in FIGURE 3.9(a); at point $a$ the deflexion is specified to be vertically downwards while at $b$ it is a clockwise rotation. In FIGURE 3.9(b) the beam carries unit point load at $a$ which produces a slope of $-i$ (i.e. anti-clockwise) at $b$. When unit clockwise couple acts at $b$, as in FIGURE 3.9(c), the deflexion at $a$ is found to be $-i$ (i.e. upwards).

Maxwell's Reciprocal Theorem can be used to demonstrate the identity of the 'flexural centre' and the 'centre of twist' of an unsymmetrical section. For example, FIGURE 3.10 shows a channel section which is supposed to be rigidly held at one end while loads are applied at the other. It is found

101

that unless the load acts through a certain point $C'$, called the 'flexural centre', the channel will twist as well as bend. If now the channel is twisted

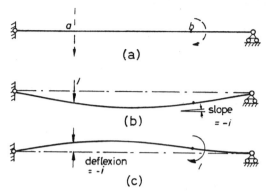

*Figure 3.9.* Directions are specified at $a$ and $b$, as shown at (a); Maxwell's Reciprocal Theorem states that the slope at $b$ due to unit force at $a$ is equal to the deflexion at $a$ due to unit couple at $b$

then the section will rotate about a point $C''$, called the 'centre of twist', which we assume for the moment to be different from $C'$.

The definition of the flexural centre implies that the rotation of $C''$ caused by unit force at $C'$ is zero and therefore, by Maxwell's Theorem, the deflexion

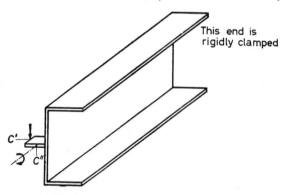

*Figure 3.10.* Unless an unsymmetrical section is loaded through its 'flexural centre' $C'$, it will twist as well as bend. If the channel is twisted it will rotate about the 'centre of twist' $C''$. Maxwell's Reciprocal Theorem can be used to show that $C'$ and $C''$ coincide

of $C'$ caused by unit couple at $C''$ is also zero. But the only point which does not deflect when the channel is twisted is $C''$ itself and so $C'$ and $C''$ coincide.

### 3.5.2. Betti's Reciprocal Theorem

Maxwell's Theorem can be generalized in the following way. At each of a series of points on a structure, such as that shown in Figure 3.11, a direction

## 3.6. DEFLEXION OF PIN-JOINTED FRAMEWORKS

is specified. Some of these points are loaded by the forces of Group 1 acting in the specified directions; others are loaded by the forces of Group 2.

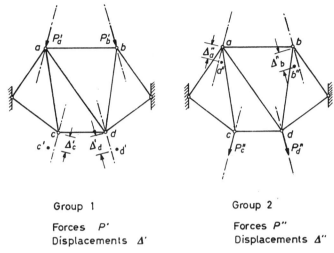

Group 1

Forces $P'$
Displacements $\Delta'$

Group 2

Forces $P''$
Displacements $\Delta''$

*Figure 3.11.* At each of a series of points on a structure a direction is specified. Some of these points are loaded, in the specified directions, by the forces of Group 1; the remainder are loaded by the forces of Group 2. Betti's Reciprocal Theorem relates these groups of forces and the corresponding deflexions

Betti's Theorem states that the product of the forces of Group 1 and the corresponding displacements caused by Group 2 is equal to the product of the forces of Group 2 and the corresponding displacements caused by Group 1.

i.e. $\quad P'_a \Delta''_a + P'_b \Delta''_b = P''_c \Delta'_c + P''_d \Delta'_d \quad \ldots$ (i)

Now $\quad \Delta''_a = P''_c \Delta_{ac} + P''_d \Delta_{ad}$

$\quad \Delta''_b = P''_c \Delta_{bc} + P''_d \Delta_{bd}$ etc.

i.e. (i) becomes

$$P'_a(P''_c \Delta_{ac} + P''_d \Delta_{ad}) + P'_b(P''_c \Delta_{bc} + P''_d \Delta_{bd})$$
$$= P''_c(P'_a \Delta_{ca} + P'_b \Delta_{cb}) + P''_d(P'_a \Delta_{da} + P'_b \Delta_{db}) \quad \ldots \text{(ii)}$$

Comparing the two sides of this equation term by term we find, for example

$$P'_a P''_c \Delta_{ac} = P''_c P'_a \Delta_{ca} \quad \text{by Maxwell's Theorem}$$

and this proves the validity of Betti's Theorem.

In Betti's Theorem, as in Maxwell's, the symbols $P$ and $\Delta$ can denote couples and rotations respectively, as well as forces and linear deflexions.

## 3.6. THE DEFLEXION OF PIN-JOINTED FRAMEWORKS

Calculations of the deflexion of structures are very often required, especially in bridge engineering. The proper control of erection procedure, for example, may require detailed and accurate deflexion figures at each

## GENERAL THEOREMS FOR LINEAR ELASTIC STRUCTURES

successive stage and under a variety of loadings. The proper design and location of bearings is obviously affected by the deflexions to be expected, and there are many other situations where this information is required.

In the following Sections two methods, one analytical and one graphical, are described and one or the other will be found the most convenient in any particular case.

### 3.6.1. THE DUMMY UNIT LOAD METHOD

The most convenient method of calculating the deflexion, in a specified direction, of a point on a truss follows at once from Mohr's Equation of Virtual Work, equation 2.6.

$$\Sigma \overline{P}\{\varDelta\} = \Sigma \overline{F}\{e\} \qquad \ldots \ (2.6)$$

FIGURE 3.12(b) for example, represent a truss whose members have undergone changes of length $e_A$, $e_B$ ... and it is required to find the component of

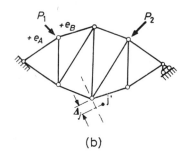

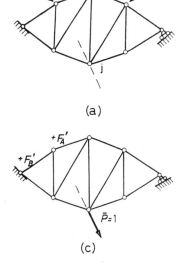

*Figure 3.12.* The members of this truss have extended by amounts $e_A$, $e_B$, ... as shown at (b) either because of the forces $P_1, P_2$ ... shown at (a) or because of a rise of temperature. The deflexion of the joint $j$, in the direction of the dotted line, is obtained by applying an imaginary unit load there as at (c)

the resulting deflexion $\varDelta_j$ of the joint $j$ to $j'$ along the dotted line. The changes of length $e$ could have been due to external loads $P_1, P_2$ ... as at (a), they could have resulted from a change of temperature, or they could have been produced by deliberately making certain members longer, or shorter, than required by the geometry of the truss in its unstrained state. This is done when it is desired to camber the truss, that is to give it an artificial upward deflexion greater than the downward deflexion produced by the worst combination of live loads†.

---

† Cambering is really an admission that the untrusting mechanism of the human mind regards a sagging girder as highly unreliable. The steps which have promoted a concession to psychology into a principle of aesthetics are fascinating but, in the present context, irrelevant.

## 3.6. DEFLEXION OF PIN-JOINTED FRAMEWORKS  3.6.1

It is apparent that, if the left-hand side of equation 2.6 is applied to the single joint $j$, the actual distortion of the truss, $\Delta_j$ and $e_A$, $e_B$ ..., defines a suitable displacement system. The force system is then conveniently selected so as to make the left-hand side of the equation numerically equal to $\Delta_j$. Hence we make $\bar{P}$ equal to unit load acting at $j$ in the specified direction, as in FIGURE 3.12(c); $\bar{F}$ is the member force produced by $\bar{P}$ and is equal to $F'_A, F'_B \ldots$

Equation 2.6 then becomes

$$1 \times \Delta_j = F'_A \cdot e_A + F'_B \cdot e_B + \ldots$$

i.e.
$$\Delta_j = \Sigma F'e \qquad \ldots (3.12)$$

In order to use this equation to find $\Delta_j$, each force $F'$ due to the 'dummy unit load' is computed and multiplied by the corresponding change of length $e$. The summation of the products is facilitated by tabulation.

If the changes of length $e$ are due to stress then

$$e = \frac{FL}{AE} = F\rho$$

where $F$ = actual force in member due to the loads acting on the structure.

Hence
$$\Delta_j = \Sigma F'F\rho \qquad \ldots (3.13)$$

This result is of course the same as that derived, in more general terms, as equation 3.9. The present treatment, based directly on the Virtual Work equation, emphasizes that the cause of the change of length of the members is unrestricted.

The use of this method is illustrated by calculating the deflexion of the joints $H$, $J$, $K$ and $L$ of the truss shown in FIGURE 3.13(a) under a load $P$ at $J$. The members are such that $\rho$ is constant.

Deflexion of $L = \Sigma F'F\rho = P\Sigma F_L F_J \rho$

where $F$ = Force in member due to load $P$ at $J = P \times F_J$ say

and $F'$ = Force in member due to unit load at $L = F_L$ say.

Similarly the deflexions of $K$, $J$ and $H$ are

$$P\Sigma F_K F_J \rho, \qquad P\Sigma F_J^2 \rho, \qquad P\Sigma F_H F_J \rho$$

The first step is evidently to calculate the forces $F_L$, $F_K$, $F_J$ and $F_H$ and then to arrange them systematically so that multiplication and summation can be conveniently performed. The symmetry of the truss suggests that some of the figures can be obtained by inspection; this can in fact be done by regarding $F_J$ and $F_H$ as the mirror images of $F_K$ and $F_L$ respectively.

FIGURES 3.13(b) and (c) show the forces in the members under unit loads at $L$ and $K$ respectively, taking tensile forces as positive. These appear in TABLE 3.1 under the columns headed $F_L$ and $F_K$; the figures under $F_J$ and $F_H$ have been written down by inspection in the manner just indicated.

Columns 5–8 show the products $F_L F_J$, $F_K F_J$ ... and these are summed at the foot of each column. The required deflexions are therefore

$$\frac{350P}{75}\rho, \quad \frac{620P}{75}\rho, \quad \frac{730P}{75}\rho \quad \text{and} \quad \frac{425P}{75}\rho$$

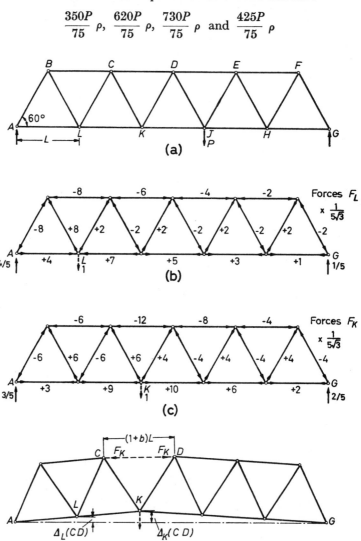

*Figure 3.13.* In order to find the deflexions of the joints of the lower chord of this truss, under the load $P$ at $J$, the bar forces are evaluated for unit load placed, in turn, at $L$ and $K$. The necessary calculations are set out in Table 3.1. (d) The effect of a change of length of a single member $CD$ is found by the direct use of Virtual Work

It is only necessary to add that if the members had been of different size the various values of $\rho$ would have been tabulated and included in the multiplication before summation.

TABLE 3.1

DEFLEXION CALCULATIONS FOR TRUSS OF FIGURE 3.13

| Member | 1 | 2 | 3 | 4 | 5 | 6 | 7 | 8 | 9 | 10 | 11 | 12 | 13 |
|---|---|---|---|---|---|---|---|---|---|---|---|---|---|
| | $F_L$ | $F_K$ | $F_J$ | $F_H$ | $F_L F_J$ | $F_K F_J$ | $F_J^2$ | $F_H F_J$ | $e$ (temp.) | $F_J e$ | $e$ (camber) | $F_L e$ | $F_K e$ |
| | | all by $1/5\sqrt{3}$ | | | all by $P/75$ | | | | $\times L.10^{-4}$ | $\times \dfrac{L.10^{-4}}{5\sqrt{3}}$ | $\times L$ | $\times L/5\sqrt{3}$ | $\times L/5\sqrt{3}$ |
| BC | −8 | −6 | −4 | −2 | 32 | 24 | 16 | 8 | 2·4 | −9·6 | a | −8a | −6a |
| CD | −6 | −12 | −8 | −4 | 48 | 96 | 64 | 32 | 2·4 | −19·2 | b | −6b | −12b |
| DE | −4 | −8 | −12 | −6 | 48 | 96 | 144 | 72 | 2·4 | −28·8 | b | −4b | −8b |
| EF | −2 | −4 | −6 | −8 | 12 | 24 | 36 | 48 | 2·4 | −14·4 | a | −2a | −4a |
| AB | −8 | −6 | −4 | −2 | 32 | 24 | 16 | 8 | 1·8 | −7·2 | | | |
| BL | 8 | 6 | 4 | 2 | 32 | 24 | 16 | 8 | 1·8 | 7·2 | | | |
| LC | 2 | −6 | −4 | −2 | −8 | 24 | 16 | 8 | 1·8 | −7·2 | | | |
| CK | −2 | 6 | 4 | 2 | −8 | 24 | 16 | 8 | 1·8 | 7·2 | | | |
| KD | −2 | 4 | 4 | −2 | −8 | −16 | 16 | 8 | 1·8 | 7·2 | | | |
| DJ | −2 | 4 | 4 | 2 | −8 | −16 | 16 | 8 | 1·8 | 7·2 | | | |
| JE | 2 | −4 | 6 | −2 | 12 | 24 | 36 | −12 | 1·8 | 10·8 | | | |
| EH | −2 | 4 | −6 | 2 | 12 | 24 | 36 | −12 | 1·8 | −10·8 | | | |
| HF | 2 | −4 | 6 | 8 | 12 | 24 | 36 | 48 | 1·8 | −10·8 | | | |
| FC | −2 | −4 | −6 | −8 | 12 | 24 | 36 | 48 | 1·8 | −10·8 | | | |
| AL | 4 | 3 | 2 | 1 | 8 | 6 | 4 | 2 | 1·2 | 2·4 | | | |
| LK | 7 | 9 | 6 | 3 | 42 | 54 | 36 | 18 | 1·2 | 7·2 | | | |
| KJ | 5 | 10 | 10 | 5 | 50 | 100 | 100 | 50 | 1·2 | 12·0 | | | |
| JH | 3 | 6 | 9 | 7 | 27 | 54 | 81 | 63 | 1·2 | 10·8 | | | |
| HB | 1 | 2 | 3 | 4 | 3 | 6 | 9 | 12 | 1·2 | 3·6 | | | |
| Σ | | | | | 350 | 620 | 730 | 425 | | −36 | | −10(a+b) | −10(a+2b) |

In order to illustrate the method of computing the effect of temperature, let it be required to find the vertical deflexion of the point $J$ if the temperature of the upper chord members be raised by 40°, the web members by 30° and the lower chord members by 20°. In the general expression for deflexion the term $e$ is now the thermal expansion, $L\,\alpha_t\,t$, and we have

$$\Delta_J = \Sigma F'e = \Sigma F_J . L\,\alpha_t\,t$$

Hence in the upper chord members, $e = L.6 \times 10^{-6}.\ 40 = 2\cdot 4\ L.10^{-4}$
in the upper web members, $e = L.6 \times 10^{-6}.\ 30 = 1\cdot 8\ L.10^{-4}$
in the lower chord members, $e = L.6 \times 10^{-6}.\ 20 = 1\cdot 2\ L.10^{-4}$

In TABLE 3.1, these figures are entered in column 9, and the product $F_J L\,\alpha_t\,t$ in column 10; the summation at the foot of the column gives the required deflexion, namely

$$-36 \times \frac{L.10^{-4}}{5\sqrt{3}}$$

The negative sign shows that the deflexion is opposite to the direction of the unit load applied to give the forces $F_J$, that is upwards.

Finally the device of cambering can be investigated. Suppose that it is arbitrarily decided that in the unloaded state the joints $K$ and $J$ should be 4 per cent of the span above the line joining $A$ and $B$, while $L$ and $H$ are 3 per cent above the line. This can be achieved by making the upper chord members longer than the others by an amount to be determined. Suppose that the length of $BC$ and $EF$ is $(1+a)L$ and that of $CD$ and $DE$ is $(1+b)L$. We have to find the values of $a$ and $b$ such that the upward deflexions of $L$ and $K$ are $\dfrac{3 \times 5L}{100}$ and $\dfrac{4 \times 5L}{100}$ respectively, that is $-0\cdot 15L$ and $-0\cdot 2L$.

In column 11 the extensions $aL$ are recorded for members $BC$ and $EF$, $bL$ for members $CD$ and $DE$, and zero for all the other members. In column 12 the product $F_L e$ is entered, to give the deflexion of point $L$, and the products $F_K e$ in column 13 give the deflexion of $K$. The summations giving the final deflexions are equated to the desired values, that is

$$\frac{-10\,(a+b)\,L}{5\sqrt{3}} = -0\cdot 15\,L$$

$$\frac{-10\,(a+2b)\,L}{5\sqrt{3}} = -0\cdot 2L$$

and, solving simultaneously for $a$ and $b$,

$$a = 0\cdot 09 \qquad b = 0\cdot 04$$

It is to be noticed that these results can be obtained by the direct use of the Virtual Work principle. FIGURE 3.13(d) shows the truss in the position it would take up if a member of length $(1+b)L$ were forced into the position $CD$. Clearly the elevation of $K$ and the other lower chord panel points above the horizontal through $A$ and $G$ could be found geometrically, but if an imaginary unit load were applied at $K$ giving rise to a force $F_K$ in $CD$, we should have, from Virtual Work

$$1 \times \{\Delta_{K(CD)}\} = F_K\{bL\}$$

i.e.
$$\Delta_{K(CD)} = F_K \cdot bL = -\frac{12}{5\sqrt{3}} bL$$

Here $\Delta_{K(CD)}$ means the deflexion of $K$ due to the elongation of $CD$; its final negative sign confirms that it is opposite in direction to the unit load at $K$. $F_K$ is a compressive force and so is negative, while the extension $bL$ is positive.

Similarly $\Delta_{L(CD)}$ can be found by putting unit load at $L$ and finding the force $F_L$ in $CD$.

$$1 \times \{\Delta_{L(CD)}\} - F_L\{bL\} = 0$$

i.e.
$$\Delta_{L(CD)} = F_L \cdot bL = -\frac{6}{5\sqrt{3}} bL$$

Proceeding next to consider the effect of a member of length $(1 + b)L$ in $DE$, we find that

$$\left.\begin{array}{l}\Delta_{L(DE)} = -4 \\ \Delta_{K(DE)} = -8\end{array}\right\} \times \frac{bL}{5\sqrt{3}}$$

Next consider a member of length $(1 + a)L$ in $BC$

$$\left.\begin{array}{l}\Delta_{L(BC)} = -8 \\ \Delta_{K(BC)} = -6\end{array}\right\} \times \frac{aL}{5\sqrt{3}}$$

Finally, consider a member of length $(1 + a)L$ in $EF$

$$\left.\begin{array}{l}\Delta_{L(EF)} = -2 \\ \Delta_{K(EF)} = -4\end{array}\right\} \times \frac{aL}{5\sqrt{3}}$$

Adding the appropriate terms together,

$$\Delta_L = -(6b + 4b + 8a + 2a)\frac{L}{5\sqrt{3}} = -\frac{10L}{5\sqrt{3}}(a + b)$$

$$\Delta_K = -(12b + 8b + 6a + 4a)\frac{L}{5\sqrt{3}} = -\frac{10L}{5\sqrt{3}}(a + 2b)$$

These are the same results as were obtained above; in this example the advantage clearly lies with the former method because the work is shorter and more systematic, but in other examples the direct application of Virtual Work will be found to be the better.

### 3.6.2. Williot-Mohr Construction for Truss Deflexion

#### 3.6.2.1. *Williot Displacement Diagrams*

FIGURE 3.14(a) shows an elementary triangular framework $ABC$; it is required to find the displacement of $B$ caused by given displacements of $A$ to $A'$, and $C$ to $C'$, together with a given extension $e_2$ of $BC$ and a contraction $e_1$ of $AB$. A simple geometrical construction will suffice: with centre $A'$ and radius $AB - e_1$ a circular arc is described cutting at $b$, the required new position of $B$, a second arc centered on $C'$ and of radius $CB + e_2$.

An approximation to the same result can be reached in the following manner. Starting at $B$ let $Ba$ be drawn parallel and equal to $AA'$, and $BC$ parallel and equal to $CC'$; $ab_a$ and $cb_c$ represent $e_1$ and $e_2$ respectively in direction and magnitude. The points $b_a$ and $b_c$ evidently lie on the arcs previously described about centres $A'$ and $C'$ respectively; if these points had been more remote from $B$ than is shown in the diagram then $b$ could have been found by representing the arcs by straight lines $b_a b$ and $b_c b$ perpendicular to $AB$ and $BC$ respectively.

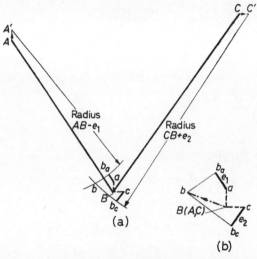

*Figure 3.14.* The movement of $B$ caused by displacements of $A$ and $C$ and by changes of length of $AB$ and $CB$ can be found by a simple geometrical construction; this can be drawn to any scale in the Williot diagram (b)

This is quite a satisfactory approximation for the present purpose because the extensions and contractions of the members of a structural framework are always small compared with the dimensions of the framework.

This construction can be performed independently, as at (b), and to any convenient scale. This is the Williot diagram for $B$; the displacement of $B$ relative to its original position is given in direction and to scale by $Bb$. As the displacement of $A$ to $A'$ has been represented by $Ba$, and of $C$ to $C'$ by $Bc$ it appears that the original position of the three points $ABC$ is represented by the single starting point $B$, while $a$ corresponds to $A'$ and $c$ to $C'$. The notation is self-explanatory, for $b_a a$ represents the displacement (by the amount $e_1$) of $B$ relative to $A$, and $b_c c$ that of $B$ relative to $C$.

A simple extension of this process enables the displacement of a number of joints forming a series of triangles to be measured, provided that two adjacent joints are fixed in position as in FIGURE 3.15(a), where a cantilever truss is shown; the figures alongside each member denote the change of length in arbitrary units. The arrows on the members follow the usual

## 3.6. DEFLEXION OF PIN-JOINTED FRAMEWORKS

convention for indicating tension or compression. If a force polygon is used with Bow's notation for determining the member forces, these arrows are inserted as the forces are found. They now serve an additional useful purpose.

Beginning with the triangle $ABC$, which should be compared with $ABC$ in FIGURE 3.14(a), we notice first that as $A$ and $C$ are fixed their position is shown by a single point $(A\,C)$ on the Williot diagram (b). Relative to $A$, joint $B$ moves 2 units downwards and to the right to $b_a$, and relative to $C$ this joint moves 5 units horizontally to the left to $b_c$. The final position of $B$ is given by $b$, which is obtained by drawing $b_a b$ and $b_c b$ perpendicular to $BA$ and $BC$ respectively. These movements of $B$ represent the extension of $AB$ and the compression of $BC$ and it will be noticed that they follow the direction of the arrows at $A$ and $C$ respectively.

Having found the new position of $B$ we can proceed to deal with $D$; $(A\,C) - d_a$ and $b - d_b$ represent the extension and contraction of $DA$ and $DB$ respectively, and $d$ is obtained by drawing appropriate perpendiculars as before. The remaining points are found in a similar manner, and the absolute displacement of each point relative to its original position is given by $(A\,C) - b$, $(A\,C) - d$, etc.

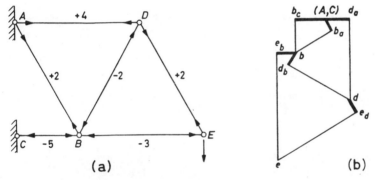

Figure 3.15. Williot diagram for a single cantilever truss

Graphical inaccuracies are inevitable, of course, but skill, care and the choice of a suitable scale will usually enable an acceptable degree of precision to be reached. Errors arise more from misalignment of the perpendiculars than from inexact plotting of the changes of length of the members, especially when the triangles are ill-conditioned. Unfortunately the errors are cumulative.

### 3.6.2.2. Mohr Rotation Diagrams

If we attempt to apply the above construction to a simply supported truss such as that shown in FIGURE 3.16(a) we at once encounter the difficulty that the two support joints are not adjacent; we begin, therefore, by assuming that some convenient member, say $AE$, is fixed in direction. This defines the direction $Ae$ on the Williot diagram and $e$ can be located. The rest of the construction is as before and is given at (b).

The final displacement of each joint relative to $A$ is shown, from which it appears that $B$ is displaced upwards, although in reality we know that it necessarily remains on the same level as $A$. Evidently the assumption that $AE$ is fixed in direction is not justified, and a moment's reflection will show that this must be the case.

The symmetry of the truss deformations indicates that $CD$ will deflect without rotating: if we had started the construction from $C$, assuming $CD$ to be fixed in direction, the Williot diagram would have given $A$ and $B$ on the same level, as in *Figure* 3.15(c). The displacement of any joint from its original position is given on this diagram by lines radiating from $a$, such as $ab$ and $ac$.

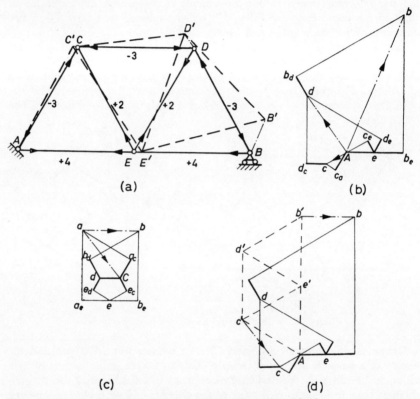

*Figure 3.16.* (a) Superimposed on the truss $AB$ are the points $B'$, $C'$ ... in the deflected position given by diagram (b); (b) Williot diagram starting at $A$ and assuming $AE$ fixed; (c) Williot diagram starting at $C$ and assuming $CD$ fixed; (d) Williot diagram (b) with Mohr diagram superimposed

In general, however, it is not possible to select a member which will not rotate and some method of correcting the Williot diagram must be devised. In FIGURE 3.16(a) the apparent position of the joints of the truss, as given by the Williot diagram, is shown in dotted lines, the displacements being exaggerated for clarity; the whole truss must now be rotated about $A$ until $B'$ is on the level of $B$. The final movement of any point will then be its displacement as given by the Williot diagram together with the displacement involved in the correcting rotation which is finally given.

## 3.6. DEFLEXION OF PIN-JOINTED FRAMEWORKS

Mohr's construction for this rotation is shown in FIGURE 3.17. The truss $ABCDE$, in diagram (a), is rotated through a small angle $\theta$ so that it takes up the position $AB'C'D'E'$. Each joint moves through a small arc, of length $r\,\theta$, which can be represented by a line perpendicular to the line joining the joint to $A$. Thus $CC' = r_4\,\theta_E$ is perpendicular to $AC$.

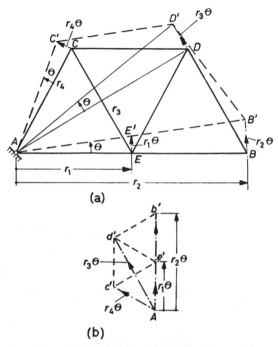

Figure 3.17. The rotation of a truss from one position to another is represented to any desired scale by the Mohr rotation diagram (b)

If the displacements relative to $A$ are represented in direction and to scale on a second diagram (b) it is found that the points $b'\ c'\ d'$ and $e'$ so obtained lie on a diagram of the original truss rotated through 90°. This is the Mohr rotation diagram.

Referring again to FIGURE 3.16(a) it is clear that the necessary corrections to the Williot diagram (b) can be obtained either by rotating the deflected truss $A$—$B'$ so that $B'$ is brought back to its original level on $A$—$B$, or by rotating the undeformed truss until $B'$ is at the level given by the Williot diagram. The latter turns out to be the more convenient and it has the effect of making the Mohr diagram the base from which deflexions due to strain are measured.

The complete process is demonstrated in FIGURE 3.16(d) where the Williot diagram (b), namely $Abcde$, is shown again. Since it is known that the level of $B$ is unchanged (and therefore that the movement of $B$ is horizontal) the point $b'$ on the Mohr diagram must lie on the horizontal through $b$; it must also lie on the perpendicular through $A$. These considerations fix $b'$ and enable the Mohr diagram $A\ b'\ c'\ d'\ e'$ to be drawn. The final movement of any point is then obtained by joining corresponding points on

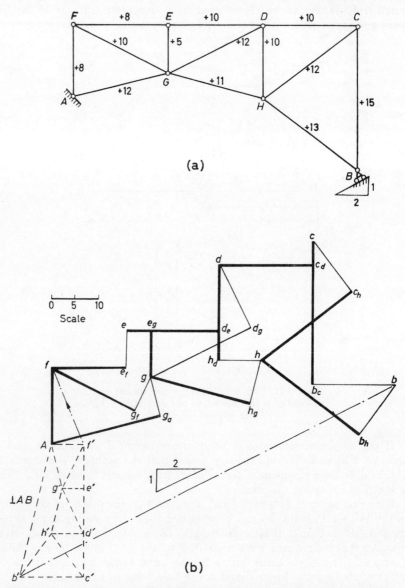

*Figure 3.18.* The Williot deflexion diagram and Mohr rotation diagram are combined at (b) to give the absolute deflexion of all the joints of the truss shown at (a)

the Mohr and Williot diagrams. Thus the movement of $C$ is given by $c'\ c$, in direction and to scale; $c'c$ will be found to be identical with $ac$ on FIGURE 3.16(c).

A second example is given in FIGURE 3.18. Diagram (a) shows an unsymmetrical arch having joint $A$ pinned to its support while $B$ is constrained to slide on a 1 : 2 slope. The arbitrary changes of the length of each member are shown. The Williot diagram, at (b), is started at $A$ with $AF$ assumed to remain vertical; construction proceeds in the usual way until $b$ is reached. Now the final movement of $B$ can only be along the 1 : 2 slope while line $Ab'$ on the Mohr diagram must be perpendicular to $AB$. The point $b'$ is thus obtained and enables the Mohr diagram $A$-$f'$-$c'$-$b'$ to be drawn. The resultant displacement is obtained, as before, by joining corresponding points on the Mohr and Williot diagrams. Thus $B$ moves from $b'$ to $b$ and $F$ from $f'$ to $f$.

The dummy unit load method described in Section 3.6.1 gave the components of the joint displacements in a specified direction. A comparison between the results obtained by the analytical and graphical methods for the truss of FIGURE 3.13(a) is made in TABLE 3.2; here the figures marked 'Dummy unit load' have been extracted from TABLE 3.1 while those marked 'Williot-Mohr' were obtained by measuring the vertical components of the deflexions given by a Williot-Mohr diagram (not shown).

TABLE 3.2
VERTICAL COMPONENTS OF DISPLACEMENT
OF JOINTS OF TRUSS SHOWN IN FIGURE 3.13(a)

| Method | Joint Displacement × $PL/75AE$ | | | | | |
|---|---|---|---|---|---|---|
| | $A$ | $L$ | $K$ | $J$ | $H$ | $G$ |
| Dummy unit load | 0 | 350 | 620 | 730 | 425 | 0 |
| Williot-Mohr | 0 | 345 | 625 | 735 | 430 | 0 |

Agreement of this order can be achieved regularly by competent draughtsmanship, and in much less time than is required for a full analytical solution. Mistakes can usually be noticed, but in the absence of any independent information on the expected order of the results, the only method of checking is to repeat the construction from a new starting point.

The following should be considered in choosing a method to use in a specific instance:

1. If the displacement of a single joint in a specified direction is required the Dummy unit load method is very suitable.
2. The Williot-Mohr is the only method that gives the absolute displacement of the joints directly. With reasonable care it gives results of quite acceptable accuracy, although errors accumulate.

Finally it should be mentioned that Chu[2] has given an analytical equivalent to the Williot-Mohr construction in a recent paper.

## 3.7. THE DEFLEXION OF BEAMS

An important problem in the theory of structures is the calculation of the deflexion of bars or beams bent by transverse loads. The results of such calculations are sometimes of intrinsic value, as when the maximum permissible deflexion of a beam is reached at a lower load than the maximum permissible stress. Deflexion calculations, however, as we saw in Chapter 1, play an essential part in the analysis of hyperstatic structures and for that reason a rather extensive account of the various methods is now given. It will be noticed that the methods described are for finding the deflexion due to bending. We saw previously that the deflexion due to shear is not often significant in structural work and is commonly ignored; it can, if necessary, be estimated by the method given in Section 3.4.3.

The student may wonder, at first, why so many different methods need be described, but he will find, as his experience grows, that each method has its own special field of utility.

### 3.7.1. SIGN CONVENTIONS

The sign convention which is here adopted is shown in FIGURE 3.19. The bending moment is taken as positive for a cantilever fixed at one end and carrying positive loads acting downwards, that is to say for a 'hogging' beam with its convex side uppermost. (If when plotting the bending moments positive moments are shown above the datum, in the usual manner, while negative moments are below the datum, it will be found that this convention agrees with that used in Chapter 4 where bending moments are plotted on the ' tension ' side of the members.)

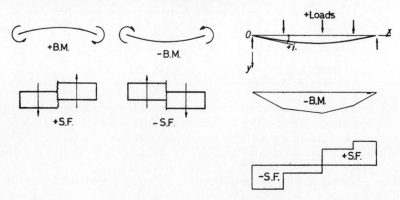

*Figure 3.19.* Convention of signs for beams

The shearing force is positive when that part of the beam lying to the right of a section tends to move upwards relative to the part lying to the left.

The coordinate axes are chosen so that $y$ is positive downwards and $x$ positive to the right; when the bending moment is positive $dy/dx$ tends to increase as $x$ increases and so $d^2y/dx^2$ is positive.

Now the radius of curvature at a point on any curve $y=f(x)$ is given by the expression

$$\frac{1}{R} = \frac{d^2y/dx^2}{\left[1 + \left(\frac{dy}{dx}\right)^2\right]^{3/2}}$$

## 3.7. THE DEFLEXION OF BEAMS

If the curvature is slight, as is usually the case in beam problems, then the above expression reduces to

$$\frac{1}{R} = \frac{d^2y}{dx^2}$$

and, from the beam equation

$$\frac{f}{y} = \frac{M}{I} = \frac{E}{R}$$

we finally have

$$\frac{d^2y}{dx^2} = +\frac{M}{EI} \qquad \ldots\ldots (3.14)$$

By considering the equilibrium of an element of a beam under the action of positive load, bending moment and shearing force, as in FIGURE 3.20, we find, in the limit, that

$$\frac{dQ}{dx} = +w \quad \text{and} \quad \frac{dM}{dx} = +Q$$

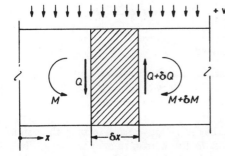

Figure 3.20. This beam element is under positive load, shearing force and bending moment

Hence, by means of equation 3.14, we arrive at the series of results

$$\left. \begin{array}{l} \dfrac{dy}{dx} = +i \\[6pt] \dfrac{d^2y}{dx^2} = +\dfrac{M}{EI} \\[6pt] \dfrac{d^3y}{dx^3} = +\dfrac{Q}{EI} \\[6pt] \dfrac{d^4y}{dx^4} = +\dfrac{w}{EI} \end{array} \right\} \qquad \ldots\ldots (3.15)$$

### 3.7.2. DEFLEXIONS BY INTEGRATION OF EQUATION 3.14

Equation 3.14 is $\quad \dfrac{d^2y}{dx^2} = \dfrac{M}{EI}$

GENERAL THEOREMS FOR LINEAR ELASTIC STRUCTURES

Integrating once $\quad\dfrac{dy}{dx} = \int \dfrac{M.dx}{EI} + C_1$

Integrating again $\quad y = \int\int \dfrac{M.dx.dx}{EI} + C_1 x + C_2$

In principle, then, the deflexion can be obtained by direct integration, but unless $\dfrac{M}{EI}$ is an integrable function of $x$ this process will have to be performed graphically or by an approximate method such as Simpson's Rule; the constants $C_1$ and $C_2$ can not be easily found except in simple cases.

The use of the method in several standard cases is given below for beams of uniform cross-section.

**3.7.2.1.** *Standard cases when the constants of integration can be obtained directly*
*Case* 1. (FIGURE 3.21) *Cantilever with end point load*

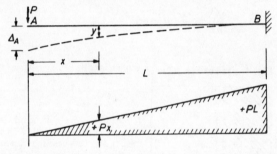

*Figure 3.21.* Deflexion Case 1. Cantilever with end point load

In this case, and in that which follows, it is convenient to take the origin at the free end. In order to conform to the convention of signs the fixed end has therefore been taken at the right-hand end. This is not essential but it is a good plan to adhere strictly to the sign convention at first.

We then have $\quad EI\dfrac{d^2y}{dx^2} = +M = +Px$

$$EI\dfrac{dy}{dx} = \dfrac{Px^2}{2} + C_1$$

At the end $B$, where $x=L$ $\quad \dfrac{dy}{dx} = 0 \quad$ i.e. $C_1 = -\dfrac{PL^2}{2}$

$$EIy = \dfrac{Px^3}{6} - \dfrac{PL^2 x}{2} + C_2$$

At the end $B$, where $x = L$ $\quad y = 0 \quad$ i.e. $C_2 = \dfrac{PL^3}{3}$

Hence, finally $\quad y = \dfrac{P}{EI}\left\{\dfrac{x^3}{6} - \dfrac{L^2 x}{2} + \dfrac{L^3}{3}\right\} \quad \ldots\ldots(3.16)$

In particular $$y_{max.} = \Delta_A = \frac{PL^3}{3EI} \quad \ldots(3.16(a))$$

Case 2. (FIGURE 3.22) *Cantilever with uniformly distributed load*

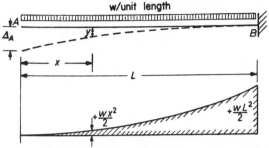

Figure 3.22. Deflexion Case 2. Cantilever with uniformly distributed load

Starting from $$EI\frac{d^2y}{dx^2} = +M = +\frac{wx^2}{2}$$

We arrive, by similar steps to the above, at

$$y = \frac{w}{EI}\left\{\frac{x^4}{24} - \frac{L^3x}{6} + \frac{L^4}{8}\right\} \quad \ldots(3.17)$$

and therefore $$y_{max.} = \Delta_A = \frac{wL^4}{8EI} \quad \ldots(3.17(a))$$

Case 3. (FIGURE 3.23) *Simply supported beam with central point load*

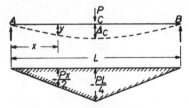

Figure 3.23. Deflexion Case 3. Simply supported beam with central point load

Since the symmetry of the beam and its loading implies that the central slope, at $C$, is zero the two constants of integration are known and the calculation can proceed as before. We finally have

$$y = \frac{P}{EI}\left\{-\frac{x^3}{12} + \frac{L^2x}{16}\right\} \quad \ldots(3.18)$$

and $$y_{max.} = \Delta_C = \frac{PL^3}{48EI} \quad \ldots(3.18(a))$$

# GENERAL THEOREMS FOR LINEAR ELASTIC STRUCTURES

*Case* 4. (FIGURE 3.24) *Simply supported beam with uniformly distributed load*

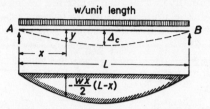

*Figure 3.24.* Deflexion Case 4. Simply supported beam with uniformly distributed load

Proceeding as before, we have

$$y = \frac{w}{EI}\left\{ -\frac{Lx^3}{12} + \frac{x^4}{24} + \frac{L^3x}{24} \right\} \quad \ldots (3.19)$$

and 
$$y_{max.} = \Delta_C = \frac{5\,w\,L^4}{384\,EI} \quad \ldots (3.19(a))$$

### 3.7.2.2. *Integration by Macaulay's Method*

In each of the above cases the expression for the bending moment was a continuous integrable function of $x$ which gave no difficulty. The constants of integration were found at once because values of the slope and deflexion were known *a priori* at a sufficient number of points. In general, however, this is not so and the previous method breaks down. It is then best to make use of the ingenious device introduced by Macaulay[3] for systematizing the procedure for dealing with the constants of integration. Several examples follow.

(a) *Beam with single eccentric point load.* FIGURE 3.25

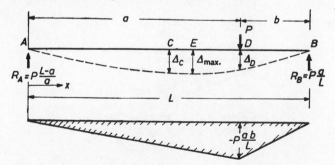

*Figure 3.25.* Deflexion of simply supported beam with single eccentric point load

The bending moment for points in $DB$ is

$$M = -R_A x + P(x-a)$$

and hence for $x > a$, 
$$EI\frac{d^2y}{dx^2} = -R_A x + P(x-a) \quad \ldots (i)$$

120

## 3.7. THE DEFLEXION OF BEAMS

In $AD$, where $x < a$, the second term in the equation disappears; this is represented conventionally in Macaulay's method by the use of distinguishing brackets $P[x-a]$, on the understanding that terms so written are ignored when $x < a$.

Thus
$$EI\frac{dy}{dx} = -R_A\frac{x^2}{2} + \frac{P}{2}[x-a]^2 + C_1 \qquad \ldots \text{(ii)}$$

and
$$EIy = -R_A\frac{x^3}{6} + \frac{P}{6}[x-a]^3 + C_1 x + C_2 \qquad \ldots \text{(iii)}$$

This expression is applicable to the whole beam so long as the term $[x-a]$ has the special meaning given to it, and the constants of integration can now be determined; since the beam must follow a continuous curve at the load point these constants have the same value over the whole length.

At $A$; $x = 0$ and $y = 0$. Hence, ignoring the term $\frac{P}{6}[x-a]^3$, $C_2 = 0$

At $B$; $x = L$ and $y = 0$. Hence

$$C_1 = R_A\frac{L^2}{6} - \frac{P}{6L}(L-a)^3$$

$$= \frac{PL}{6}(L-a) - \frac{P}{6L}(L-a)^3$$

$$= \frac{Pa}{6L}(L-a)(2L-a)$$

Hence, substituting in (iii) and rearranging

$$EIy = \frac{P}{6}\frac{x(L-a)}{L}(2aL - a^2 - x^2) + \frac{P}{6}[x-a]^3 \quad \ldots (3.20)$$

In particular, at $D$, when $x = a$

$$EI\Delta_D = \frac{Pa^2(L-a)^2}{3L} = \frac{Pa^2 b^2}{3L} \qquad \ldots (3.20(a))$$

By making $a = b = L/2$ we see that this expression coincides with equation 3.18(a) for the central deflexion of a beam carrying a central point load.

The point of maximum deflexion $E$ is obtained by setting $dy/dx$ equal to zero. Hence differentiating (iii) and writing $b = L - a$.

$$EI\frac{dy}{dx} = \frac{Pb}{6L}\{a^2 + 2ab - 3x^2\}$$

$$= 0 \text{ if } x = \sqrt{\frac{a^2 + 2ab}{3}} = \sqrt{\frac{L^2 - b^2}{3}} \qquad \ldots \text{(iv)}$$

The position of $E$ varies with the position of the load and from (iv) it is seen that $x = AE$ reaches its maximum value of $L/\sqrt{3} = \cdot 577L$ when $b = 0$, that is when the load is at one end of the beam. As the load travels across the span the point of maximum deflexion moves only over the central 15·4 per cent of the span. It appears therefore that the central deflexion will not differ much from the maximum deflexion.

(b) *Beam with several Point Loads.* FIGURE 3.26—Macaulay's method evidently improves the differential equation method considerably when the

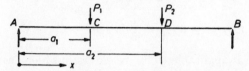

*Figure 3.26.* Deflexion of simply supported beam with several point loads

loading is not continuous. The case of a beam with several point loads is best treated by superposing the expressions for each individual load, keeping strictly to the convention as to the meaning of $[x - a]$. Thus referring to FIGURE 3.26 and making use of equation 3.20

In $AC$ $\quad (EIy)_{AC} = \dfrac{P_1}{6} \dfrac{x(L-a_1)}{L} (2a_1 L - a_1^2 - x^2)$

$\qquad \qquad \qquad + \dfrac{P_2}{6} \dfrac{x(L-a_2)}{L} (2a_2 L - a_2^2 - x^2)$

In $CD$ $\quad (EIy)_{CD} = (EIy)_{AC} + \dfrac{P_1}{6} [x - a_1]^3$

In $DB$ $\quad (EIy)_{DB} = (EIy)_{AC} + \dfrac{P_1}{6} [x - a_1]^3 + \dfrac{P_2}{6} [x - a_2]^3$

(c) *Beam with Overhang.* FIGURE 3.27—The symbols used in the figure are exactly as in FIGURE 3.25. The term $(L - a)$ is now negative so that $R_A$ is negative and therefore really acts downwards as we expect.

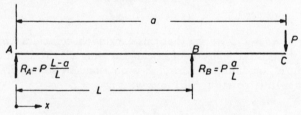

*Figure 3.27.* Deflexion of beam with overhang

In this case the discontinuity in the bending moment diagram occurs at $B$ when $x$ and $L$ are equal and hence, by Macaulay's convention, the term $[x - L]$ is now ignored when it is negative.

On this understanding

$$EI \dfrac{d^2 y}{dx^2} = M = -R_A x - R_B [x - L]$$

## 3.7. THE DEFLEXION OF BEAMS

$$EIy = -R_A \frac{x^3}{6} - R_B \frac{[x-L]^3}{6} + C_1 x + C_2$$

At A; $x = 0$, $y = 0$ and $[x - L]$ is ignored, being negative.

i.e. $C_2 = 0$

At B; $x = L$, $y = 0$, and writing $R_A = P \dfrac{L-a}{L}$

$$C_1 = \frac{PL(L-a)}{6}$$

Hence finally

$$EIy = -\frac{P(L-a)}{L}\frac{x^3}{6} - \frac{Pa[x-L]^3}{6L} + \frac{PL(L-a)x}{6}$$

(d) *Beam with Uniformly Distributed Load over part of the Span.* FIGURE 3.28

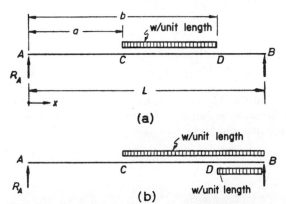

Figure 3.28. Distributed loads are most easily dealt with if they extend to the right hand end of the span

Distributed loads are most easily dealt with, by Macaulay's method, if they extend to the right-hand end of the span. Accordingly the beam shown at (a) in FIGURE 3.28 is modified by the addition of equal and opposite distributed loads over the part $DB$, as shown at (b)

The bending moment at a section in $DB$ is then

$$M = -R_A x + \frac{w[x-a]^2}{2} - \frac{w[x-b]^2}{2}.$$

The terms in brackets are ignored when $x < a$ and $x < b$ respectively. Integration of the equation

$$EI \frac{d^2 y}{dx^2} = -R_A x + \frac{w[x-a]^2}{2} - \frac{w[x-b]^2}{2}$$

can then be performed as usual and the constants of integration found from the consideration that the deflexion is zero at $A$ and $B$.

(e) *Beam acted on by a Couple.* FIGURE 3.29—This problem can be solved by

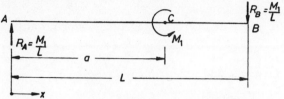

*Figure 3.29.* Beam with a couple $M_1$ applied at $C$

an extension of Macaulay's method said by Case[4] to be due to Webb. The expression for the bending moment can be written

$$M = -R_A x + [M_1] \text{ for } x > a.$$

Hence

$$EI\frac{d^2y}{dx^2} = -\frac{M_1 x}{L} + [M_1]$$

$$EI\frac{dy}{dx} = -\frac{M_1 x^2}{2L} + M_1 [x-a] + C_1$$

$$EIy = -\frac{M_1 x^3}{6L} + \frac{M_1}{2}[x-a]^2 + C_1 x + C_2$$

The integration of the $M_1$ term has been carried out as indicated in order to bring the expression within the Macaulay convention. As before $[x-a]$ is ignored when it is negative.

At A;  $x = 0$   $y = 0$   and   $[x - a]$  is ignored.

Hence  $C_2 = 0$

At B;  $x = L$   $y = 0$

Hence  $C_1 = \dfrac{M_1}{6L}\{L^2 - 3(L-a)^2\}$

### 3.7.3. Deflexions by the Dummy Unit Load Method

(a) In Section 3.4.3 of this chapter equation 3.9 was used to find the deflexion of a specific point of a structure under the influence of bending, shearing and direct forces. If the deflexion due to bending only is required equation 3.9 can be simplified to

$$\Delta_j = \int \frac{M'M dx}{EI} \qquad \ldots \ldots (3.9(a))$$

It will be recalled that $M'$ is the bending moment due to an imaginary unit load acting at $j$, the point whose deflexion is required. Integration extends, in principle, over the whole length of the beam but if either $M$ or $M'$ is zero over part of the beam the limits of integration must be suitably adjusted.

To find the deflexion of the point $C$ on the cantilever $AB$ of FIGURE 3.30(a) we apply there an imaginary unit load.

## 3.7. THE DEFLEXION OF BEAMS

Then $\qquad M = Px$

and $\qquad M' = 1 \times [x - a] \qquad$ using Macaulay's notation.

Hence $\qquad \Delta_C = \int \dfrac{M'M dx}{EI} = \int_a^L \dfrac{[x - a] \cdot Px \cdot dx}{EI}$

The lower limit is fixed by the consideration that $\{x - a\}$ is to be ignored when $x < a$.

Integrating, and putting in limits,

$$\Delta_C = \dfrac{P}{EI}\left\{\dfrac{L^2}{3} - \dfrac{aL^2}{2} + \dfrac{a^3}{6}\right\}$$

This result coincides with equation 3.16 except that $a$ is now used for the abscissa of $C$.

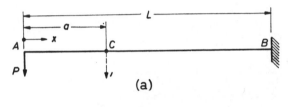

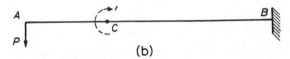

*Figure 3.30.* In the dummy unit load method the deflexion of a point on a beam is found by applying unit load there while the slope is found by applying unit couple

(b) The derivation of equation 3.9 from the Virtual Work equation made it clear that $\Delta_j$ represented the *slope* at a point on a beam provided that unit *couple* was applied there. We can therefore write

$$i_j = \int \dfrac{M'_c M dx}{EI} \qquad \ldots \ldots (3.9(b))$$

When $M'_c$ is the bending moment produced by unit couple.

Referring to FIGURE 3.30(b)

$$M = + Px$$
$$M'_c = -1 \text{ for } x > a$$

Hence $i_c = -\displaystyle\int_a^L \dfrac{Px \, dx}{EI} = -\dfrac{P}{2EI}(L^2 - a^2)$

Although this method is quite straightforward in principle, it is all too easy to make a slip in assigning the limits of integration and there is no obvious check on this; in suitable cases, especially when the deflexion of a specific point is required, the method is very useful.

## GENERAL THEOREMS FOR LINEAR ELASTIC STRUCTURES

### 3.7.4. The Moment-Area Method

In many structural problems it is sufficient to know the deflexion at one or two special points on a beam; in such cases a semi-graphical procedure, originally due to Mohr, called the moment-area method can be employed

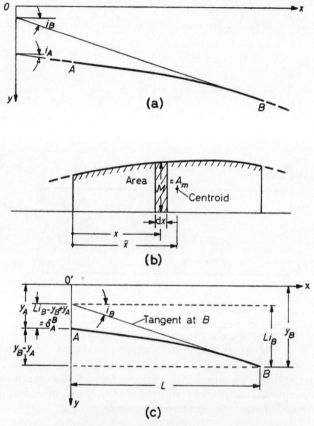

*Figure 3.31.* Derivation of the moment-area relationships. (a) $A$ and $B$ are any two points on a beam bent by transverse loads; (b) The corresponding part of the bending moment diagram; (c) The origin of coordinates shifted so that $x$ is now the horizontal distance from $A$

to good advantage. In FIGURE 3.31(a) $A$ and $B$ are any two points on a beam which is bent by a system of loads that produce the bending moment diagram shown at (b). Integrating equation 3.14,

$$\frac{d^2y}{dx^2} = \frac{M}{EI} \qquad \ldots\ (3.14)$$

we have

$$\left[\frac{dy}{dx}\right]_A^B = i_B - i_A = \int_A^B \frac{M\,dx}{EI} \qquad \ldots\ (3.21)$$

The integral term evidently represents the area of that part of the $\dfrac{M}{EI}$

## 3.7. THE DEFLEXION OF BEAMS

diagram which lies between $A$ and $B$. If $EI$ is constant we can write

$$i_B - i_A = \left[\frac{A_m}{EI}\right]_A^B \qquad \ldots \ (3.21(a))$$

where $A_m$ is the area of that part of the bending moment diagram which lies between $A$ and $B$.

If, before integrating equation 3.14, we multiply both sides by $x$ we have

$$x\frac{d^2y}{dx^2} = \frac{Mx}{EI}$$

from which

$$x\frac{dy}{dx} - y = \frac{Mx\,dx}{EI} + C$$

This equation can best be interpreted if the origin is shifted to $A$; hence, introducing the limits $x = 0$, $x = L$ at $A$ and $B$ respectively,

$$Li_B - y_B + y_A - \int_{A\leftarrow}^{B} Mx\,\frac{dx}{EI}$$

where the arrow serves as a reminder that $x$ is now measured from $A$. The terms on the left-hand side of this equation can be seen from FIGURE 3.31(c) to represent the distance from $A$ from the tangent at $B$, which can be written $\delta_A^{\tan B}$ or $\delta_A^B$.

Hence

$$\delta_A^B = \int_{A\leftarrow}^{B} Mx\,\frac{dx}{EI} \qquad \ldots \ (3.22)$$

The integral represents the sum of the moments of the elements $\frac{M\,dx}{EI}$ of the $\frac{M}{EI}$ diagram about $A$, and if $EI$ is constant we can write

$$\delta_A^B = \left[\frac{A_m\bar{x}}{EI}\right]_A^B \qquad \ldots \ (3.22(a))$$

$A_m\bar{x}$ is the moment, about $A$, of that part of the bending moment diagram which lies between $A$ and $B$.

In FIGURE 3.31(a) and (c) the slopes and curvatures have been considerably exaggerated for convenience, and it is clear that $\delta_A^B$, shown as the vertical displacement of $A$ from the tangent at $B$, differs considerably from the perpendicular distance of $A$ from the tangent. In most calculations it is assumed that these two dimensions are identical; for this to be true the tangent must not be far from the horizontal. Gross distortion is therefore excluded.

In many actual problems the sign of the solution is self-evident but it should be noticed that the equations we have derived conform strictly to our sign convention. Thus, in equations 3.21 and 3.21(a), when $x$ increases from $A$ to $B$ $i_B$ is greater than $i_A$ when the area of the bending moment $\left(\text{or } \dfrac{M}{EI}\right)$ diagram is positive.

The form of equations 3.22 and 3.22(a) provides a useful mnemonic for the position of $A$ relative to the tangent at $B$. Thus in FIGURE 3.32 we see that when the bending moment is positive $A$ lies *below* the tangent at $B$, corresponding to the symbol $\delta_A^B$ ($A$ *below* $B$); a negative bending moment means that $A$ lies *above* the tangent at $B$.

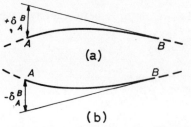

(a) Curvature convex upwards. Bending moment positive $A$ *below* the tangent at $B$. $\delta_A^B$ positive

(b) Curvature concave upwards. Bending moment negative $A$ *above* the tangent at $B$. $\delta_A^B$ negative

*Figure 3.32*

Finally we note that although $A$ has been taken at the left-hand end of the portion of the beam under consideration, none of the above would have been changed if it had been taken at the right-hand end provided $x$ had then been measured from right to left.

An essential preliminary to moment-area calculations is to make a sketch of the beam in its deflected condition. This need only give a rough idea of the state of affairs, provided that the tangents at salient points are shown.

(a) *Cantilever with end point load.* FIGURE 3.33

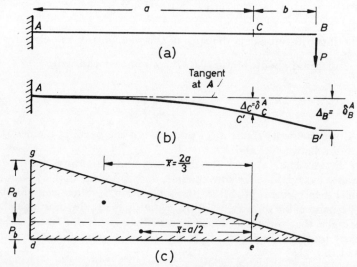

*Figure 3.33.* Diagram for moment area calculations for the deflexion of a cantilever with an end point load

Examination of the sketch at (b) shows that, because the tangent at $A$ is horizontal, the displacement of $C$ from the tangent at $A$ is the same as the absolute displacement of $C$ from its original position on the horizontal $ACB$.

## 3.7. THE DEFLEXION OF BEAMS

Hence
$$\Delta_C = \delta_C^A = \left[\frac{A_m \bar{x}}{EI}\right]_C^A = +\frac{1}{EI}\left[Pb.a.\frac{a}{2} + \frac{Pa^2}{2}\cdot\frac{2a}{3}\right]$$
$$= +\frac{P}{6EI}\left[3a^2 b + 2a^3\right]$$

Note that in order to find $A_m\bar{x}$ the trapezium *defg* has been split into a rectangle and a triangle and that $\bar{x}$ has in each case been measured from the centroid to $C$, the point whose displacement from the tangent is sought. The positive sign signifies that $C$ is below the tangent at $A$, and hence that the deflexion is downwards.

It will now be apparent that the selection of a suitable tangent, from which displacements can be measured, is the geometrical equivalent of the procedure by which the constants of integration were found in Section 3.7.2.1.

(b) *Beam with single eccentric point load.* FIGURE 3.34.

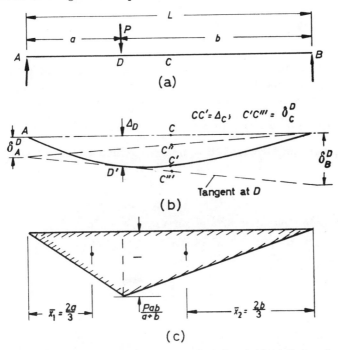

*Figure 3.34.* Diagram for moment area calculations for the deflexion of a simply supported beam with a single eccentric load

Here the problem is first to find the deflexion of $D$, and we at once encounter the difficulty that as the beam is unsymmetrically loaded there is no point whose slope is known. This same difficulty arose in the solution of the problem by the differential equation method.

It is convenient to sketch the tangent at $D'$, as in (b),

Then
$$\delta_A^D = \left[\frac{A_m \bar{x}}{EI}\right]_A^D = -\frac{1}{EI}\left[\frac{1}{2}\frac{Pa^2 b}{a+b}\frac{2a}{3}\right] = -\frac{Pa^3 b}{3EIL}$$

Similarly $$\delta_B^D = -\frac{Pab^3}{3EIL}$$

At this stage, having confirmed that $A$ and $B$ are above the tangent at $D$, we abandon the sign convention and proceed to find $D'$ from the geometry of the figure.

$$\Delta_D = \frac{b}{L}\delta_D^A + \frac{a}{L}\delta_D^B = \frac{Pa^3b^2}{3EIL^2} + \frac{Pa^2b^3}{3EIL^2} = \frac{Pa^2b^2}{3EIL}$$

We also have

$$i_D = \frac{1}{L}(\delta_B^D - \delta_A^D) = \frac{Pab}{3EIL^2}(b^2 - a^3) = \frac{Pab(b-a)}{3EIL}$$

If the deflexion of some other point, say $C$, is required the method is to find $\delta_C^D = C'C'''$ by a moment-area calculation while $CC'''$ is obtained by proportion.

Then $\Delta_C = CC' = CC''' - C'C'''$

In the two examples just worked point loads only have been considered, and it has been convenient to compute the area of the bending moment diagram and the position of the centroid from simple geometry. If the loading is continuous it is best to use equation 3.22, as follows

(c) *Continuous loading on beam with overhang.* FIGURE 3.35.

Taking moments about $A$ and $B$, in turn, we find that

$$R_A = \frac{wL(L-2a)}{2(L-a)} \quad ; \quad R_B = \frac{wL^2}{2(L-a)}$$

As $EI$ is constant, equation 3.22 becomes

$$\left|EI\,\delta_A^B\right| = \int_A^B Mx_1\,dx_1 = \int_0^{L-a}\left(R_A x_1 - \frac{wx_1^2}{2}\right)x_1\,dx_1$$

$$= \left[R_A\frac{x_1^3}{3} - \frac{wx_1^4}{8}\right]_0^{L-a} = \frac{wL(L-2a)}{6}(L-a)^2 - \frac{w(L-a)^4}{8}$$

$$\left|EI\,\delta_C^B\right| = \int_C^B Mx_2\,dx_2 = \int_0^a \frac{wx_2^2}{2}x_2\,dx_2 = \frac{wa^4}{8}$$

$$\therefore EI\,\Delta_C = \left\{\frac{a}{L-a}\delta_A^B - \delta_C^B\right\}$$

$$= \frac{waL(L-2a)(L-a)}{6} - \frac{wa(L-a)^3}{8} - \frac{wa^4}{8}$$

i.e. $\Delta_C = \frac{wa}{24EI}\{L^3 - 3aL^2 - a^2L\}$

## 3.7. THE DEFLEXION OF BEAMS

Beams with continuous loading can also be dealt with, if preferred, by splitting the bending moment diagram up into its component parts and using the geometrical properties of the parabola to evaluate equation 3.22(a).

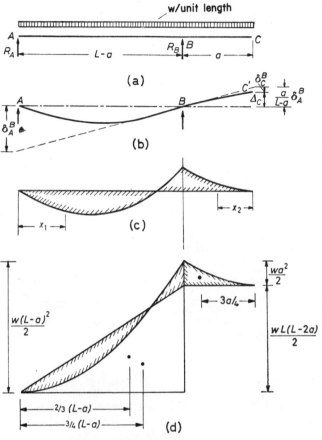

*Figure 3.35.* Diagram for moment area calculations for the deflexion of a beam with an overhang

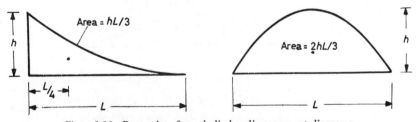

*Figure 3.36.* Properties of parabolic bending moment diagrams

FIGURE 3.36 gives the necessary information about parabolas and FIGURE 3.35(d) the bending moment diagram of the beam $ABC$ considered as two cantilevers projecting from $B$.

Then

$$\left| EI\, \delta_A^B \right| = \left[ A\bar{x} \right]_A^B$$

$$= \left[ \frac{1}{2} \cdot \frac{wL(L-2a)}{2} \cdot \frac{2(L-a)^2}{3} - \frac{1}{3} \cdot \frac{w(L-a)^2}{2} \cdot \frac{3(L-a)^2}{4} \right]$$

$$= \frac{wL(L-2a)}{6}(L-a)^2 - \frac{w(L-a)^4}{8} \quad \text{as before}$$

and $\quad \left| EI\, \delta_C^B \right| = \frac{1}{3} \cdot \frac{wa^2}{2} \cdot \frac{3a}{4} = \frac{wa^4}{8}$

In the above examples we have considered the beams to be uniform. In cases where the flexural rigidity $EI$ is not constant it is necessary to take its variation into account. This can be done quite simply by working with the $M/EI$ diagram instead of the bending moment diagram, as follows.

(d) *Non-uniform cantilever.* FIGURE 3.37.

The cantilever has a change of section at its mid-point $B$ and it is required to find the deflexion of the free end $C$. The bending

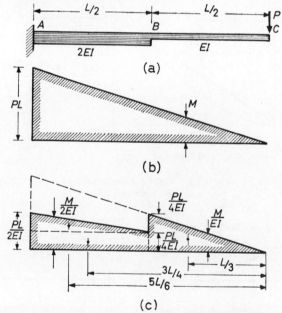

*Figure 3.37.*(a) Cantilever with non-uniform cross-section; (b) Bending moment diagram; (c) $M/EI$ diagram

moment diagram, at (b), is independent of the cross-section of the beam; when each ordinate of this diagram is divided by the corresponding value of $EI$ we get the $M/EI$ diagram shown at (c). Equation 3.22(a) now becomes

$$\Delta_c = \delta_C^A = \left[\frac{A_m}{EI} \bar{x}\right]_C^{\_\downarrow A}$$

$$= \frac{1}{EI}\left\{\frac{PL}{4} \cdot \frac{L}{2} \cdot \frac{3L}{4} + \frac{1}{2} \cdot \frac{PL}{4} \cdot \frac{L}{2} \cdot \frac{5L}{6} + \frac{1}{2} \cdot \frac{PL}{2} \cdot \frac{L}{2} \cdot \frac{L}{3}\right\}$$

$$= \frac{3PL^3}{16EI}$$

The problem of non-uniform beams is dealt with in more detail in Appendix A.

### 3.7.5. HOADLEY'S METHOD

A method of finding the deflexions of beams, arches and rigidly jointed frames which combines the Williot-Mohr and Moment-Area methods has been given by Hoadley[5]. (See also Stevens[6].) Small elements of the beam or arch are considered and the displacements caused by bending and rotation are calculated. These displacements are then plotted perpendicular to the element in a displacement diagram.

In FIGURE 3.38(a) an element $pq$ of length $l$ is shown together with the corresponding bending moment diagram at (b). A rotation $i_p$ of the end $p$

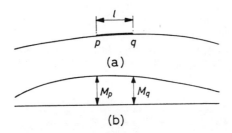

*Figure 3.38.*(a) Element $pq$ of a bent beam; (b) Corresponding bending moment diagram

this element will cause a displacement of $li_p$ of the point $q$ relative to $p$. There will also be a displacement $\delta_q^p$ due to bending within the length. The total relative displacement is therefore

$$\Delta_q^p = li_p + \delta_q^p \qquad \ldots (3.23)$$

where

$$i_p = \sum_0^{ip} di \qquad \ldots (i)$$

$di$ and $\delta_q^p$ are conveniently found by the moment area method; thus if the length $l$ is small enough for the variation of bending moment to be considered as linear

$$di = i_q - i_p = \frac{l}{EI}\left[\frac{M_p + M_q}{2}\right] \qquad \ldots (ii)$$

and
$$\delta_q^p = \frac{l^2}{EI}\left[\frac{2M_p + M_q}{6}\right] \quad \ldots \text{(iii)}$$

(a) As a simple example of Hoadley's method the cantilever shown in FIGURE 3.39(a) will be examined. It is divided into ten equal parts, the corresponding values of the bending moment being shown on diagram (b).

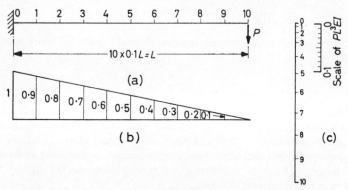

*Figure 3.39.*(a) Cantilever with end load $P$; (b) Bending moment diagram; (c) Williot diagram

TABLE 3.3 summarizes the necessary calculations. Column 3 of this table gives the increment of slope in each element, computed from (ii) above. Column 4 gives the slope at successive points on the beam obtained by cumulatively summing the figures of column 3. In column 5 the deflexions $li$ of each point, caused by the slopes so far calculated, are listed. In column

TABLE 3.3

DEFLEXION CALCULATIONS FOR BEAM OF FIGURE 3.29

| 1 | 2 | 3 | 4 | 5 | 6 | 7 | 8 |
|---|---|---|---|---|---|---|---|
| Point | $M$ $\times PL$ | $di$ (ii) $\times PL^2/EI$ | $i$ $\times PL^2/EI$ | $li$ $\times PL^3/EI$ | $\delta_q^p$ (iii) $\times PL^3/EI$ | $\Delta_q^p = 5+6$ $\times PL^3/EI$ | $\Delta$ $\times PL^3/EI$ |
| 0 | 1 | | | | | | |
| 1 | 0·9 | 0·095 | 0·095 | | 0·00483 | 0·00483 | 0·00483 |
| 2 | 0·8 | 0·085 | 0·180 | 0·0095 | 0·00433 | 0·01383 | 0·01866 |
| 3 | 0·7 | 0·075 | 0·255 | 0·0180 | 0·00383 | 0·02183 | 0·04049 |
| 4 | 0·6 | 0·065 | 0·320 | 0·0255 | 0·00333 | 0·02883 | 0·06932 |
| 5 | 0·5 | 0·055 | 0·375 | 0·0320 | 0·00283 | 0·03483 | 0·10415 |
| 6 | 0·4 | 0·045 | 0·420 | 0·0375 | 0·00233 | 0·03983 | 0·14398 |
| 7 | 0·3 | 0·035 | 0·455 | 0·0420 | 0·00183 | 0·04383 | 0·18781 |
| 8 | 0·2 | 0·025 | 0·480 | 0·0455 | 0·00133 | 0·04683 | 0·23464 |
| 9 | 0·1 | 0·015 | 0·495 | 0·0480 | 0·00083 | 0·04883 | 0·28347 |
| 10 | 0 | 0·005 | 0·500 | 0·0495 | 0·00033 | 0·04983 | 0·33330 |

6 are the terms $\delta_q^p$ for successive points calculated from (iii). Column 7, obtained by adding columns 5 and 6, gives the value of $\Delta_q^p$ for each point.

At this stage, if an arch or rigid frame were being calculated, a Williot-Mohr diagram would be drawn. This has in fact been done, at (c), but it will be seen that in this case the final deflexion of each point on the beam is

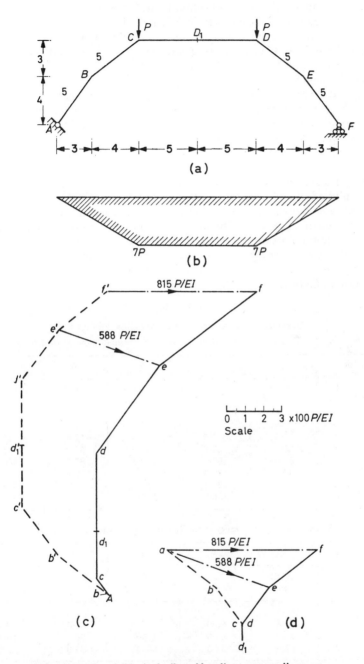

*Figure 3.40.*(a) and (b) Arch rib and bending moment diagram; (c) Williot-Mohr diagram starting at $A$; (d) Williot diagram starting at $D_1$

given by cumulative addition of column 7 with the result shown in column 8.

It will be noticed that the calculations have been carried to five places of decimals; this was to give a check against results calculated from equation 3.16 and would not normally be necessary. It is to be noticed, however, that errors accumulate in this method.

(b) A second example is shown in FIGURE 3.40(a) which represents a symmetrical uniform arch rib; the bending moment diagram is shown at (b). The variation of bending moment is everywhere linear and the length '$l$' in equations (ii) and (iii) above can conveniently be taken as 5. The appropriate calculations starting at $A$, are set out in TABLE 3.4(a). FIGURE 3.40(c) shows the Williot-Mohr diagram obtained by plotting each of the $\Delta_q^p$ terms, from column 7, at right angles to $pq$ on the diagram. Thus, starting at $A$ and assuming $AB$ fixed in direction, $bA$ is made equal to 12·5 units and is plotted at right angles to $AB$, and so on. The Mohr diagram is constructed so as to make the final movement of $F$ horizontal; its value, obtained by measurement, is 815 $P$. The movement of other points is given by joining corresponding points on the Mohr and Williot diagrams.

In this case advantage can be taken of the symmetry of the arch by starting the calculations of the Williot diagram at $D_1$, as in TABLE 3.4(b) and FIGURE 3.40(d).

TABLE 3.4

DEFLEXION CALCULATIONS FOR ARCH OF FIGURE 3.40

(a) Starting at $A$ with tangent to $AB$ at $A$ assumed fixed in direction

| 1 | 2 | 3 | 4 | 5 | 6 | 7 |
|---|---|---|---|---|---|---|
| Point | $M$ $\times P$ | $di$ (ii) $\times P/EI$ | $i$ $\times P/EI$ | $li$ $\times P/EI$ | $\delta_q^p$(iii) $\times P/EI$ | $\Delta_q^p = 5+6$ $\times P/EI$ |
| $A$ | 0 | | | | | |
| $B$ | 3 | 7·5 | 7·5 | | 12·5 | 12·5 |
| $C$ | 7 | 25·0 | 32·5 | 37·5 | 54·2 | 91·7 |
| $D_1$ | 7 | 35·0 | 67·5 | 162·5 | 87·5 | 250·0 |
| $D$ | 7 | 35·0 | 102·5 | 337·5 | 87·5 | 425·0 |
| $E$ | 3 | 25·0 | 127·5 | 512·5 | 70·8 | 583·3 |
| $F$ | 0 | 7·5 | 135·0 | 637·5 | 25·0 | 662·5 |

(b) Starting at $D_1$ with tangent to $D_1 D$ at $D_1$ assumed fixed in direction

| 1 | 2 | 3 | 4 | 5 | 6 | 7 |
|---|---|---|---|---|---|---|
| Point | $M$ $\times P$ | $di$ (ii) $\times P/EI$ | $i$ $\times P/EI$ | $li$ $\times P/EI$ | $\delta_q^p$ (iii) $\times P/EI$ | $\Delta_q^p = 5+6$ $\times P/EI$ |
| $D_1$ | 7 | | | | | |
| $D$ | 7 | 35·0 | 35·0 | | 87·5 | 87·5 |
| $E$ | 3 | 25·0 | 60·0 | 175·0 | 70·8 | 245·8 |
| $F$ | 0 | 7·5 | 67·5 | 300·0 | 25·0 | 325·0 |

In this case no Mohr diagram is necessary; the left-hand side of the Williot diagram, $a$-$b$-$c$-$d_1$, is put in by inspection, and the final movement of the joints is given by lines radiating from $a$.

# REFERENCES

[1] TIMOSHENKO, S. *Strength of Materials.* Part I
[2] KUANG-HAN CHU. 'Truss Deflexions by the Coordinate Method'. *Trans. Amer. Soc. Civ. Engrs.* Vol. 97. 1952. p. 317
[3] MACAULAY, W. H. 'Note on the Deflection of Beams'. Messenger of Mathematics. Vol. 48. 1919
[4] CASE, J. *Strength of Materials.* Arnold. 3rd edn. 1938
[5] HOADLEY, A. 'Deflexion of members of variable moment of inertia'. *Trans. Amer. Soc. Civ. Engrs.* Vol. 119. 1954. p. 499
[6] STEVENS, L. K. 'Carrying capacity of mild steel arches'. *Proc. I.C.E.* Vol. 6. 1956–7. p. 512

CHAPTER 4

# GENERAL METHODS FOR THE ANALYSIS OF LINEAR HYPERSTATIC STRUCTURES

## 4.1. INTRODUCTION

IN Chapter 2 the fundamental energy theorems of structural analysis were described and, in the course of that chapter, different methods for setting up equations of compatibility and of equilibrium were expounded. It is perhaps already apparent that two distinct systems of hyperstatic analysis can be identified by the *order* in which the two sets of equations are formulated and solved.

In the first of these, exemplified by the Principle of Minimum Potential Energy and called by Charlton[1] the Equilibrium Method, the compatibility conditions are used first to give the forces in the members in terms of unit joint displacements. The equilibrium equations are then expressed in terms of these displacements and the external loads and solved to give the actual joint displacements from which the member forces can finally be computed.

The second system, which has already been used in an elementary way in Chapter 1, is called the Compatibility Method by Charlton[1]. Engesser's and Castigliano's Theorems of Compatibility employ this system of procedure and are, in fact, simply formal statements of it in terms of complementary and strain energies respectively. The principle is now to use the equations of equilibrium first to find the forces in the members in terms of the external loads and of the redundants. The compatibility conditions are then expressed in terms of these member forces which are obtained directly on solution.

In the present Chapter general equations based on these two systems are developed in terms of stiffness and flexibility coefficients respectively. Braced frameworks are dealt with by the Equilibrium Method first directly and then by the Relaxation technique. The Compatibility Method is illustrated by the Maxwell-Mohr equations for braced structures, by the $\varDelta_{ik}$ method for rigidly jointed frames and by the use of Castigliano's Theorem of Compatibility.

## 4.2. THE STIFFNESS COEFFICIENT EQUATIONS (EQUILIBRIUM METHOD)

### 4.2.1. DERIVATION OF THE EQUATIONS

The procedure is illustrated by reference to FIGURE 4.1(a) which shows a hyperstatic braced framework in which only one joint—*A*—is free to move. This is essentially the same problem as was dealt with by the Method of

## 4.2. STIFFNESS COEFFICIENT EQUATIONS

Minimum Potential Energy in Chapter 2, Section 2.6.1.1 and the present treatment should be compared with the previous one. The first step is to

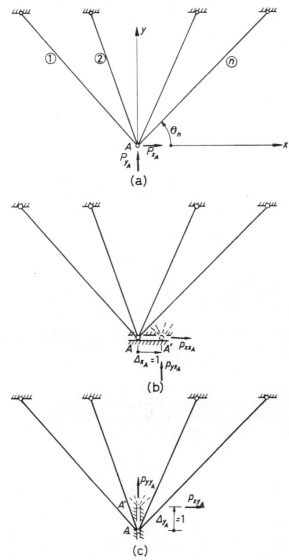

*Figure 4.1.* (a) In this hyperstatic framework only the joint $A$ is free to move; (b) When unit horizontal movement is imposed at $A$ a horizontal force $p_{xx_A}$ and a vertical force $p_{yx_A}$ act on $A$ as a result of the forces induced in the members; (c) When unit vertical movement is imposed at $A$ a horizontal force $p_{xy_A}$ and a vertical force $p_{yy_A}$ act on $A$

write down the equations of compatibility which relate the extension of each member to the horizontal and vertical components of the displacement of $A$. This is given below in equation 4.3.

## METHODS FOR THE ANALYSIS OF LINEAR STRUCTURES

The next step is to apply the conditions of equilibrium which are

$$\left.\begin{array}{l}\Sigma \text{ Horizontal component of forces acting at } A = 0 \\ \Sigma \text{ Vertical } \quad\quad \text{,,} \quad\quad \text{,, ,, ,, ,, } A = 0\end{array}\right\} \quad \ldots\text{(i)}$$

The external loads are simply dealt with by resolution and the combined effect of the member forces is expressed in terms of the *stiffness coefficients at the joint* $A$; the significance of these stiffness coefficients will be readily appreciated from FIGURE 4.1(b). Here the joint $A$ is constrained to move in a purely horizontal direction by suitable guides and it will be clear that when unit horizontal displacement has occurred forces $p_{xxA}$ and $p_{yxA}$ will be acting on $A$ in the horizontal and vertical directions respectively. Similarly at (c) a unit vertical displacement has been imposed which gives rise to forces $p_{xyA}$ and $p_{yyA}$ respectively. These forces acting *on* $A$ as a result of unit displacements are the stiffness coefficients and they can readily be found in terms of the dimensions of the members of the frame.

The equilibrium equations (i) then take the form

$$\left.\begin{array}{l}P_{xA} + p_{xxA}\,\varDelta_{xA} + p_{xyA}\,\varDelta_{yA} = 0 \\ P_{yA} + p_{yxA}\,\varDelta_{xA} + p_{yyA}\,\varDelta_{yA} = 0\end{array}\right\} \quad \ldots\text{(4.1)}$$

These are the stiffness coefficient equations for the joint $A$. The solution of these equations gives the component displacements of the joint $A$ from which the member forces can be found.

We note finally that from Maxwell's Reciprocal Theorem

$$p_{xyA} = p_{yxA}\,; \quad p_{xyB} = p_{yxB}\,, \text{etc.} \quad \ldots\text{(4.2)}$$

Although the above explanation was illustrated by reference to a simple plane braced framework it will be appreciated that the argument can readily be extended to more complicated structures having more than one joint that is free to move; an example is included in Vol. II, Chapter 4. Frames with rigid joints can also be analysed in this way; the Slope-Deflexion equations, given in Chapter 6, are in fact examples of the stiffness coefficient equations written for rigidly jointed frames in which bending stiffness only is taken into account.

### 4.2.2. Details of Application to Pin-Jointed Plane Braced Frameworks

FIGURE 4.2 shows a single member $AB$ of a pin-jointed plane framework. When displacements $\varDelta_{xA}$, $\varDelta_{yA}$, $\varDelta_{xB}$ and $\varDelta_{yB}$ are arbitrarily imposed at the ends its elongation is

$$e_{AB} = (\varDelta_{xB} - \varDelta_{xA})\cos\theta_{AB} + (\varDelta_{yB} - \varDelta_{yA})\sin\theta_{AB} \quad \ldots\text{(4.3)}$$

if its inclination is not materially changed by these displacements. The tension induced is thus

$$F_{AB} = \frac{1}{\rho_{AB}}\left[(\varDelta_{xB} - \varDelta_{xA})\cos\theta_{AB} + (\varDelta_{yB} - \varDelta_{yA})\sin\theta_{AB}\right] \quad \ldots\text{(4.4)}$$

## 4.2. STIFFNESS COEFFICIENT EQUATIONS 4.2.2

If a group of members is attached to $A$ then the stiffness coefficients corresponding to unit movements of $A$ only are easily obtained from equation 4.4. Thus

$$\begin{aligned} p_{xxA} &= \Sigma F_{AB} \cos \theta_{AB} = -\Sigma \cos^2 \theta_{AB}/\rho_{AB} \\ p_{yxA} &= \Sigma F_{AB} \sin \theta_{AB} = -\Sigma \sin \theta_{AB} \cos \theta_{AB}/\rho_{AB} \\ p_{yyA} &= \Sigma F_{AB} \sin \theta_{AB} = -\Sigma \sin^2 \theta_{AB}/\rho_{AB} \\ p_{xyA} &= \Sigma F_{AB} \cos \theta_{AB} = -\Sigma \cos \theta_{AB} \sin \theta_{AB}/\rho_{AB} = p_{yxA} \end{aligned} \quad \ldots (4.5)$$

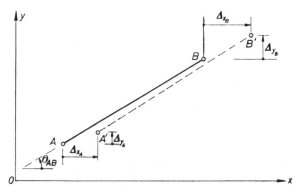

*Figure 4.2.* Derivation of the compatibility relationship

The simple example shown in FIGURE 4.3 will now be worked numerically.

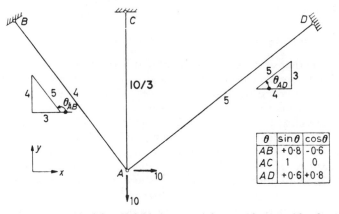

*Figure 4.3.* The joint $A$ of this framework has two degrees of freedom so that two stiffness coefficient equations must be solved to find its final position

The value of $AE$ is the same for all three members, so that the relative values of $\rho$ are

$$\rho_{AB} = \left(\frac{L}{AE}\right)_{AB} = 4; \quad \rho_{AC} = 10/3; \quad \rho_{AD} = 5$$

141

Hence the stiffness coefficients are

$$p_{xxA} = -\left(\frac{0\cdot36}{4} + 0 + \frac{0\cdot64}{5}\right) = -0\cdot218$$

$$p_{xyA} = -\left(-\frac{0\cdot48}{4} + 0 + \frac{0\cdot48}{5}\right) = +0\cdot024$$

$$p_{yyA} = -\left(\frac{0\cdot64}{4} + \frac{1}{10/3} + \frac{0\cdot36}{5}\right) = -0\cdot532$$

Substituting in the stiffness coefficient equation 4.1 we have

$$\left.\begin{array}{l} 10 - 0\cdot218\,\varDelta_{xA} + 0\cdot024\,\varDelta_{yA} = 0 \\ -10 + 0\cdot024\,\varDelta_{xA} - 0\cdot532\,\varDelta_{yA} = 0 \end{array}\right\} \text{i.e.} \quad \begin{array}{l} \varDelta_{xA} = +44\cdot0 \\ \varDelta_{yA} = -16\cdot8 \end{array}$$

Hence in equation 4.4

$$F_{AB} = \frac{1}{4}\left\{(-44\cdot0)(-0\cdot6) + 16\cdot8 \times 0\cdot8\right\} = +9\cdot96$$

$$F_{AC} = \frac{3}{10}\left\{\quad 0 \quad + 16\cdot8 \times 1 \quad\right\} = +5\cdot04$$

$$F_{AD} = \frac{1}{5}\left\{(-44\cdot0)(0\cdot8) + 16\cdot8 \times 0\cdot6\right\} = -5\cdot02$$

The correctness of these results is is confirmed by resolution.

$$9\cdot96\,(-0\cdot6) - 5\cdot02\,(0\cdot8) + 10 = +0\cdot008$$
$$9\cdot96\,(40\cdot8) + 5\cdot01\,(1) - 5\cdot02\,(0\cdot6) - 10 = -0\cdot004$$

It will now perhaps be clear that the only difference between this method and the Minimum Potential Energy method lies in the technique used to obtain the equilibrium equations; both methods then lead to a series of as many simultaneous equations as there are degrees of freedom of the joints. It is therefore clear that these methods show to best advantage when used for hyperstatic frames with many redundant members but having comparatively few joints that are free to move. Compatibility methods, by contrast, lead to as many simultaneous equations as there are redundants and are therefore most advantageous where there are few redundants even if the joints have many degrees of freedom.

Since simultaneous equations play an important part in most methods of hyperstatic analysis we now digress to discuss the relaxation method for solving them.

## 4.3. RELAXATION

### 4.3.1. INTRODUCTION

Although the examples of hyperstatic analysis so far given, in this and preceding Chapters, have for the most part been rather simple it will already be clear that the solution of as many simultaneous equations as there are redundants or degrees of freedom of the joints can be quite tedious. As the number of equations increases it becomes necessary to work with more significant figures than are finally required because the solution sought depends on the differences between comparatively large numbers. In real engineering problems there are many uncertainties in the physical quantities involved in the problem: the loads which the structure is to carry are seldom known precisely while the elastic properties of the materials used and the actual, as distinct from the nominal, dimensions of the members are subject to some uncertainty. In this situation it will obviously suffice if the solution is finally obtained to within a few per cent and we are led to enquire whether accuracies of this order can be obtained more easily than by carrying out the formal solution of a series of simultaneous equations.

So far as hyperstatic structures are concerned we are in effect seeking to find the final positions taken up by the joints after the loads have been applied. We could proceed, in theory, by making a guess at these positions and then testing the compatibility and equilibrium equations in order to see how successful we had been. If it were then possible to deduce what adjustments should be made to the first estimate in order to improve it then we should have a procedure which would lead us closer and closer to the correct position.

This is essentially the basis of Southwell's[2] method of 'Stress-calculation in frameworks by the method of systematic relaxation of constraints', which is the arithmetical equivalent of the following imaginary physical process. FIGURE 4.4(a) shows a framework under the action of loads of 3, 4 and 5 which cause the members to change length and the joints $B$, $C$, $D$ and $E$ to move to new positions. At (b) the joints are shown supported by artificial constraints in the form of screw jacks which are supposed to be capable both of moving measured distances and of recording the loads they exert. The application of the loads to the structure thus constrained produces no stress in the members, in the absence of joint displacement, and jacks $B_1$, $B_2$ and $C_2$ will record 3, 4 and 5 respectively.

If jack $C_1$ for example, is now relaxed, so that its reading falls to zero, the corresponding horizontal movement of $C$ induces some stress in the members meeting there so that the reading of the jacks at the far ends of these members will also change. The other jacks, except for $A_1$, $A_2$ and $D_2$, are now relaxed in turn until, eventually, all of them read zero. Before the unwanted jacks are discarded a note can be taken of the distance each has moved so that the final positions of the joints, and hence the load in each member, can be calculated.

Jacks $A_1$, $A_2$ and $D_2$, however, which correspond to the real constraints on the structure are not relaxed and they finally record the reactions at these points.

It is to be observed that, in general, the relaxation of one jack will alter the readings in all the others so that each jack may need several adjustments in the course of the whole operation. Progress will probably be most rapid, however, if the jack with the highest reading is the one to be relaxed.

In principle it is not difficult to derive an arithmetical equivalent to the above procedure and this is explained in Section 4.3.3 below. Such a relaxation process, however, really amounts to a step-by-step solution of the equations of compatibility and equilibrium and in this form it can be used to solve any set of linear simultaneous equations. This has proved to be an extraordinarily powerful and versatile development of the relaxation

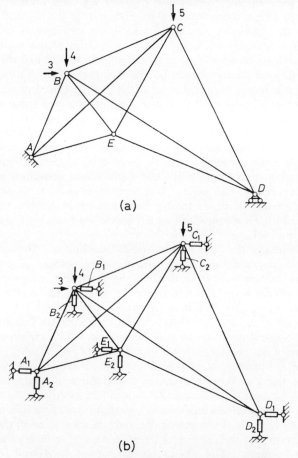

*Figure 4.4.* (a) The effect of external loads on the structure is to cause the joints to move to new positions as the members stretch or shorten under stress; (b) When the joints are restrained by jacks no stress is induced in the members until relaxation of the jacks permits movement of the joints. In this model of the relaxation process the jacks are supposed to be relaxed in sequence until they all exert zero restraint and can be discarded as superfluous

method of analysing structures which has now been applied to a wide variety of physical problems. The reader is referred to specialist works[3, 4] for a full account of the present state of the art of relaxation; in the following section the treatment of linear equations is described as a convenient preliminary to the account of the method of dealing with structures.

## 4.3.2. The Solution of Linear Simultaneous Equations

### 4.3.2.1. *Principle of the Method*

The method of relaxation can be applied to structural problems either by using it to solve the equations which have been obtained by one or other of the foregoing methods or, more directly, by operating with the actual displacement relationships. The technique can perhaps be followed more easily by reference to algebraic equations, and we begin by considering the pair of equations

$$\left.\begin{array}{l} 10x - 3y = 18 \\ 2x + 7y = 34 \end{array}\right\} \quad \ldots \text{(i)}$$

We begin by rewriting the equation in the following manner

$$\left.\begin{array}{l} R_1 = 10x - 3y - 18 \\ R_2 = 2x + 7y - 34 \end{array}\right\} \quad \ldots \text{(ii)}$$

The idea is to try successive values of $x$ and $y$ until the ' residuals ' $R_1$ and $R_2$ have been reduced either to zero or to values which are judged to be sufficiently small. In order to get some idea of how $x$ and $y$ should be varied so that the resulting changes in the residuals are in the right direction the effect of unit change in $x$ and $y$ is found by differentiating. Thus

$$\left.\begin{array}{l} dR_1 = 10dx - 3dy \\ dR_2 = 2dx + 7dy \end{array}\right\} \quad \ldots \text{(iii)}$$

Hence, if $dx = 1$ and $dy = 0$, $dR_1 = +10$ and $dR_2 = +2$

While if $dx = 0$ and $dy = 1$, $dR_1 = -3$ and $dR_2 = +7$

This information can be displayed more conveniently in an operations table, as follows:

| Operation | Change in residual | |
|---|---|---|
| | $dR_1$ | $dR_2$ |
| $dx = 1$ | $+10$ | $+2$ |
| $dy = 1$ | $-3$ | $+7$ |

It is to be noticed that changes in $x$ have most effect on $R_1$ while changes in $y$ have most effect on $R_2$; this gives a valuable guide to the order of procedure. The equations can now be set out in a relaxation table as follows.

Step 1. A first estimate is made of the values of $x$ and $y$ and, in default of a better idea, these are taken as zero; the residuals then have the initial values, obtained from (ii) of $-18$ and $-34$.

Step 2. Since $R_2$ is the larger residual and since the operations table shows that $y$ is more effective than $x$ in reducing $R_2$, a correction is made to the value of $y$ which approximately liquidates $R_2$. Thus the increment $dy = 5$ is introduced and this changes $R_1$ and $R_2$ by $-15$ and $+35$ in accordance with the operations tables.

| Step | Operation | | Residual | |
|---|---|---|---|---|
| | $dx$ | $dy$ | $R_1$ | $R_2$ |
| 1 | 0 | 0 | $-18$ | $-34$ |
| 2 | | $+5$ | $-15$ | $+35$ |
| | | | $-33$ | $+1$ |
| 3 | $+3$ | | $+30$ | $+6$ |
| | | | $-3$ | $+7$ |
| 4 | | $-1$ | $+3$ | $-7$ |
| | | | 0 | 0 |
| $\Sigma$ | $+3$ | $+4$ | 0 | 0 |

Step 3. The next line shows the new values of $R_1$ and $R_2$ ($-33$ and $+1$). Since $R_1$ is now the larger it is approximately liquidated by introducing a correction of $+3$ to the value of $x$; following this the residuals are $-3$ and $+7$ respectively.

Step 4. A final correction of $-1$ to the value of $y$ eliminates the residuals completely. The final values of $x$ and $y$ are obtained by adding up the operations columns; the values $x = +3$ and $y = +4$ are inserted in (ii) in order to check that the arithmetic is correct.

This is the basis process. It can be used, exactly as above, to solve much bigger groups of equations but in such cases it will usually be advantageous to employ special devices in order to hasten the conveyance. Before describing some of these devices, however, we draw attention to the following remarks:

(a) The largest residual should be attacked at each stage and should normally be reduced to a value as near to zero as working in whole numbers will allow.

(b) If it is noticed that an operation to reduce one residual always increases another then it may be helpful to over-relax by, say, approximately 50 per cent.

(c) It is not essential to achieve absolute arithmetical accuracy at each stage. At suitable stages in a long calculation, and in any case at its end, the values of the residuals should be checked by substitution in the original equations. If the residuals prove to be correct, all well and good; if not, their revised values are used as the starting point of a new series of relaxations.

(d) The example given above had an exact solution but this is not usual in real structural problems. The solution can best be carried out to the

## 4.3. RELAXATION

required degree of precision by working in whole numbers as far as possible, then working to one decimal place for the next phase, to two decimal places for the next phase and so on. A second example is now given to illustrate these points. The equations are

$$R_1 = -9x + 2y + 161 = 0$$
$$R_2 = 3x - 10y + 147 = 0$$

.... (iv)

The operation table is:

| Operation | $dR_1$ | $dR_2$ |
|---|---|---|
| $dx = 1$ | $-9$ | $+3$ |
| $dy = 1$ | $+2$ | $-10$ |

The relaxation table is:

| Step | Operation | | Residual | |
|---|---|---|---|---|
| | $dx$ | $dy$ | $R_1$ | $R_2$ |
| 1 | 0 | 0 | $+161$ | $+147$ |
| 2 | $+26$ | | $-234$ | $+78$ |
| | | | $-73$ | $+225$ |
| 3 | | $+22$ | $+44$ | $-220$ |
| | | | $-29$ | $+5$ |
| 4 | $-3$ | | $+27$ | $-9$ |
| | | | $-2$ | $-4$ |
| Check | $+23$ | $+22$ | $-2$ | $-4$ |
| 5 | | $-0.5$ | $-1.0$ | $+5.0$ |
| | | | $-3.0$ | $+1.0$ |
| 6 | $-0.3$ | | $+2.7$ | $-0.9$ |
| | | | $-0.3$ | $+0.1$ |
| 7 | $-0.03$ | | $+0.27$ | $-0.09$ |
| | | | $-0.03$ | $+0.01$ |
| $\Sigma$ | $+22.67$ | $+21.5$ | $-0.03$ | $+0.01$ |

Step 2. The signs of $dR_1$ and $dR_2$ in the operations table are such that when one residual is reduced the other is increased, and it will probably be profitable to over-relax. Instead of putting $x = +18$, which would have liquidated $R_1$ almost exactly, we put $x = +26$, but later work shows this to have been an overestimate.

Step 4. At this stage no progress can be made by working in whole numbers. Before introducing decimals a check is made by adding up the increments of $x$ and $y$ so far and inserting in the original equations.

METHODS FOR THE ANALYSIS OF LINEAR STRUCTURES

Step 5. $y$ has been slightly overrelaxed.

Step 7. Following step 6 it is necessary to introduce a second decimal place. The final step is to add up the $dx$ and $dy$ columns and check the residuals.

4.3.2.2. *Improving the rate of Convergence*

While the basic procedure described above is quite satisfactory in favourable circumstances it may well happen that the rate at which the residuals diminish is intolerably slow. The only tools so far at our disposal are the unit operators $dx = 1$ and $dy = 1$ and, since these are seen to be identical with the coefficients of $x$ and $y$ in the original equations their effectiveness as liquidators of the residuals is a matter of chance. These operators can be improved, however, by combining them in various ways.

*Unit Block Operators* are obtained by adding together the original unit operators. If there are only two equations then only one new operator is thus obtained but with three equations four such operators are available, three by adding the unit operators together in pairs and the fourth by adding all of them. This corresponds, of course, to the simultaneous adjustment of all the jacks shown in FIGURE 4.4.(b).

Applying this method to equations (iv) of the previous section we get a new operator $dx = dy = 1$.

| Operation | $dR_1$ | $dR_2$ |
|---|---|---|
| $dx = dy = 1$ | $-7$ | $-7$ |

The relaxation table would then begin as follows:

| Step | Operation | | Residual | |
|---|---|---|---|---|
| | $dx$ | $dy$ | $R_1$ | $R_2$ |
| 1 | 0 | 0 | $+161$ | $+147$ |
| 2 | 22 | 22 | $-154$ | $-154$ |
| | | | $+\ 7$ | $-\ 7$ |
| 3 | 0·8 | | $-7·2$ | $+2·4$ |
| | | | $-0·2$ | $+4·6$ |
| | | | etc. | |

*Group Operators* are obtained by combining the unit operators in what are judged to be helpful proportions. The three equations

$$R_1 = 10x + y + 2z - 348 = 0$$
$$R_2 = x + 2y - 2z - 6 = 0$$
$$R_3 = -2x + 3y + 8z - 354 + 0$$

## 4.3. RELAXATION

give the following operations table.

| Operation | $dR_1$ | $dR_2$ | $dR_3$ |
|---|---|---|---|
| $dx = 1$ | 10 | 1 | $-2$ |
| $dy = 1$ | 1 | 2 | 3 |
| $dz = 1$ | 2 | $-2$ | 8 |

It can be seen at once that while the operations $dx = 1$ and $dz = 1$ are very effective in liquidating the residuals $R_1$ and $R_3$ we have no very effective means for dealing with $R_2$. However, a group operator can be constructed by subtracting $dz$ from $3dy$, viz :

| $dy' =$ | $3dy - dz$ | 1 | 8 | 1 |
|---|---|---|---|---|

which will be particularly useful in liquidating $R_2$. The relaxation operation is now carried out using this operator together with $dx = 1$ and $dz = 1$.

| Step | Operation | | | Residual | | |
|---|---|---|---|---|---|---|
| | $dx$ | $dy'$ | $dz$ | $R_1$ | $R_2$ | $R_3$ |
| 1 | 0 | 0 | 0 | $-348$ | $-6$ | $-354$ |
| 2 | | | $+44$ | $+88$ | $-88$ | $+352$ |
| | | | | $-260$ | $-94$ | $-2$ |
| 3 | $+26$ | | | $+260$ | $+26$ | $-52$ |
| | | | | 0 | $-68$ | $-54$ |
| 4 | | $+8$ | | $+8$ | $+64$ | $+8$ |
| | | | | $+8$ | $-4$ | $-46$ |
| 5 | | | $+6$ | $+12$ | $-12$ | $+48$ |
| | | | | $+20$ | $-16$ | $+2$ |
| 6 | $-2$ | | | $-20$ | $-2$ | $+4$ |
| | | | | 0 | $-18$ | $+6$ |
| 7 | | $+2$ | | $+2$ | $+16$ | $+2$ |
| | | | | $+2$ | $-2$ | $+8$ |
| 8 | | | $-1$ | $-2$ | $+2$ | $-8$ |
| | | | | 0 | 0 | 0 |
| Σ | $+24$ | $+10$ | $+49$ | 0 | 0 | 0 |

The group operator $y' = 10$ hence, as a result of these operations, $y = 30$ and $z = -10$. Finally we have

$$x = 24$$
$$y = 30$$
$$z = 49 - 10 = 39$$

### 4.3.3. Direct Relaxation of Framework Problems

#### 4.3.3.1. *Principle*

We are now ready to see how the ideas developed in the preceding sections can be applied directly to frameworks. It is, perhaps, now clear that the fundamental conception of liquidating residuals by a series of relaxing operations is analagous to the process of releasing the imaginary jacks that initially constrained the structure shown in Figure 4.4. Thus we identify the *residuals* with the forces in the constraints while the controlled movement of a joint in such a direction as will reduce a residual constraining force corresponds to a relaxation operation; such an operation necessarily increases the forces in the members meeting at the joint so that there is a gradual transference of load from the constraints to the members.

#### 4.3.3.2. *Forces in terms of Displacements*

In Figure 4.5 a single member of a pin-jointed plane framework is shown. External loads $P_{xA}$ and $P_{yA}$ act a $A$ and displacements $\Delta_{xA}$, $\Delta_{yA}$, $\Delta_{xB}$ and $\Delta_{yB}$ are arbitrarily imposed at the ends $A$ and $B$ respectively. If the inclination

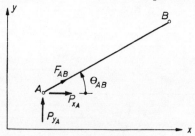

Figure 4.5. Nomenclature for the relaxation analysis of frames

of the member is not substantially changed by these displacements the tension induced in it, as we saw in Section 4.2.2, is

$$F_{AB} = \frac{1}{\rho_{AB}}\left[(\Delta_{xB} - \Delta_{xA})\cos\theta_{AB} + (\Delta_{yB} - \Delta_{yA})\sin\theta_{AB}\right] \quad \ldots \quad (4.4)$$

#### 4.3.3.3. *Equilibrium Equations and Operations Table*

If a series of members $AB$, $AC$, etc. meet at the joint $A$ the forces in them —$F_{AB}$, $F_{AC}$, etc.—can be expressed by means of equation 4.4 in terms of the displacements to be imposed at the ends; compatibility is thus ensured.

At each joint the $x$- and $y$- components of the member forces can be combined with the components of the external loads to give the equilibrium equations. For example, at joint $A$

$$P_{xA} + \Sigma F_{AB} \cos \theta_{AB} = 0 \\ P_{yA} + \Sigma F_{AB} \sin \theta_{AB} = 0 \Bigg\} \quad \ldots (i)$$

where the summation extends over all the members meeting at joint $A$. Writing these equations as in Section 4.3.2.1

$$R_{xA} = P_{xA} + \sum \frac{\cos \theta_{AB}}{\rho_{AB}} \Big[ (\Delta_{xB} - \Delta_{xA}) \cos \theta_{AB} + (\Delta_{yB} - \Delta_{yA}) \sin \theta_{AB} \Big] \\ R_{yA} = P_{yA} + \sum \frac{\sin \theta_{AB}}{\rho_{AB}} \Big[ (\Delta_{xB} - \Delta_{xA}) \cos \theta_{AB} + (\Delta_{yB} - \Delta_{yA}) \sin \theta_{AB} \Big] \Bigg\} \quad \ldots (4.6)$$

The residuals $R_{xA}$ and $R_{yA}$ represent the amounts by which the equilibrium equations are not satisfied and are thus equal to the forces in the imaginary restraining jacks. They are liquidated by imposing suitable increments of the displacements of the joints of the framework and an operations table can easily be constructed so that these displacements can be imposed systematically.

For example, if $d\Delta_{xA} = 1$, $\quad dR_{xA} = -\sum \frac{\cos^2 \theta_{AB}}{\rho_{AB}}$

and $dR_{yA} = -\sum \frac{\sin \theta_{AB} \cos \theta_{AB}}{\rho_{AB}}$

The actual process of relaxation is carried out exactly as in Section 3.2 with block and group operations introduced if it is thought that they will speed up the rate of convergence. This corresponds to the simultaneous movement of two or more joints.

### 4.3.3.4. *Example*

The simple framework of FIGURE 4.3 will now be analysed again in order to demonstrate the procedure. $AE$ is the same for all the bars. The operators are obtained from equations 4.6 as follows:

$$AEd\Delta_{xA} = 1 \; dR_{xA} = -\sum \frac{1}{L} \cos^2 \theta \;\; = -\left[ \frac{0.36}{4} + 0 + \frac{0.64}{5} \right] = -0.218$$

$$dR_{yA} = -\sum \frac{1}{L} \sin \theta \cos \theta = -\left[ -\frac{0.48}{4} + 0 + \frac{0.48}{5} \right] = +0.024$$

$$AEd\Delta_{yA} = 1 \; dR_{xA} = -\sum \frac{1}{L} \cos \theta \sin \theta = -\left[ -\frac{0.48}{4} + 0 + \frac{0.48}{5} \right] = +0.024$$

$$dR_{yA} = -\sum \frac{1}{L} \sin^2 \theta \;\; = -\left[ \frac{0.64}{4} + \frac{3}{10} + \frac{0.36}{5} \right] + -0.532$$

## METHODS FOR THE ANALYSIS OF LINEAR STRUCTURES

The operations table is therefore†

| Operation | $dR_{xA}$ | $dR_{yA}$ |
|---|---|---|
| $AEd\Delta_{xA} = 1$ | $-0\cdot218$ | $+0\cdot024$ |
| $AEd\Delta_{yA} = 1$ | $+0\cdot024$ | $-0\cdot532$ |

It will be noticed that a horizontal adjustment of $A$ is useful for liquidating a horizontal restraint, which was to be expected, while the equality of two of the operators follows from Maxwell's Reciprocal Theorem.

The relaxation is carried out as follows:

| Step | Operation | | Residual | |
|---|---|---|---|---|
| | $AEd\Delta_{xA}$ | $AEd\Delta_{yA}$ | $R_{xA}$ | $R_{yA}$ |
| 1 | 0 | 0 | $+10\cdot0$ | $-10\cdot0$ |
| 2 | | $-19$ | $-0\cdot5$ | $+10\cdot1$ |
| | | | $+9\cdot5$ | $+0\cdot1$ |
| 3 | $+44$ | | $-9\cdot6$ | $+1\cdot1$ |
| | | | $-0\cdot1$ | $+1\cdot2$ |
| 4 | | $+2\cdot3$ | $+0\cdot5$ | $-1\cdot2$ |
| | | | $+0\cdot4$ | $0$ |
| Check $\Sigma$ | $+44\cdot0$ | $-16\cdot7$ | $0$ | $-0\cdot05$ |

Finally, forces in the members are obtained by substituting in equation 4.4

$$F_{AB} = \frac{1}{4}\left[(-44\cdot0)(-0\cdot6) - (-16\cdot7)(0\cdot8)\right] = +9\cdot97$$

$$F_{AC} = \frac{3}{10}\left[(-44\cdot0)(0) - (-16\cdot7)(1)\right] = +5\cdot01$$

$$F_{AC} = \frac{1}{5}\left[(-44\cdot0)(0\cdot8) - (-16\cdot7)(0\cdot6)\right] = -5\cdot04$$

These results are close to those previously obtained.

### 4.3.4. Conclusion

It must finally be pointed out that relaxation is an art which is not seen to the best advantage in the very simple examples given above. The construction and use of unit, block and group operators has been demonstrated in principle, to be sure, but their effective use in complex problems requires a skill which can come only from practice. The speed with which the solution of a group of many simultaneous equations is reached depends very much

---

† It will be observed that these operators are equal to the stiffness coefficients of Section 4.2.2

on the intuition and experience of the analyst; the plain fact is that some people are better at it than others.

It will have been noticed that the direct relaxation analysis of structural frameworks is similar to the Minimum Potential Energy method, given in Chapter 2, Section 6, and to the method involving stiffness coefficients given in Section 2 of this chapter in that the solution is obtained in terms of displacements. The calculation of the operators and the final evaluation of the bar forces involve work which is avoided, for example, in the procedures given later in this chapter so that relaxation only saves time in very complicated cases, especially in those where the proportion of members to movable joints is high. On the other hand the actual arithmetic is simple and the inevitable errors can be eliminated as they occur without starting the calculations afresh. A still more important point is that it is not necessary to work to more significant figures than are finally required.

Although examples have not been given structures in which bending and direct stresses must be simultaneously considered can be dealt with by comparatively simple extensions of the procedure. The most important case, however, is in dealing with rigid frames in which bending stress predominates. This technique, known as Moment Distribution, was discovered independently by Hardy Cross; it is described in some detail in Chapter 6.

## 4.4. THE FLEXIBILITY COEFFICIENT EQUATIONS (COMPATIBILITY METHOD)

### 4.4.1. INTRODUCTION

In Chapter 1, Section 1.3, a general method for evaluating the force in the single redundant member of a simple hyperstatic structure was described. The idea was to compute the deflexions of the primary structure—the statically determinate structure remaining when the redundant has been temporarily removed—first under the prescribed external loading and secondly under the force, $R$ or $F$ say, in the redundant. These two deflexions were then combined to give the equation of compatibility which expressed in symbolic form the known information about the deflexion of the complete structure. This equation was then solved to give the required force in the redundant.

This procedure can now be recognized as being the same as the compatibility method mentioned in the Introduction to the present chapter. We now develop and systematize this method so that it can be used conveniently in more complicated structures. With the results of previous chapters at our disposal we are able to move quite rapidly towards the necessary equations. The original development did not proceed quite so smoothly, however. In an interesting paper Niles[5] shows that Maxwell's original announcement[6] of the method of analysing over-stiff braced frameworks was not in general form, was not presented in such a way as to attract attention and was not entirely free from error. Certainly it had no apparent influence even on those, like Rankine for instance, who might have been expected both to see the paper in the Philosophical Magazine and to appreciate its significance.

It was not until Mohr[7], working apparently without knowledge of Maxwell's writings, presented an account which was both complete and manifestly relevant to practice that engineers began to study and use the new theory. It seems appropriate, then, to link the names of these two men and to refer now to the Maxwell-Mohr Equations.

These equations, however, are a special case, applicable to braced frames, of the more general equations which were apparently first stated by Müller-Breslau[8] and which apply to all hyperstatic structures. In the following Section, therefore, we begin with these general equations and then go on to deduce from them the particular equations needed to deal with braced and rigidly-jointed frameworks respectively. The general equations, called the Flexibility Coefficient Equations, will also be found to lead directly to the graphical and experimental techniques described in the next Chapter.

### 4.4.2. Redundant Reactive Restraints

The conception of flexibility coefficients can best be illustrated by reference to structures with external redundant restraints, such as those shown at (a) and (e) in Figure 4.6; structures with redundant internal forces and moments can then be brought within the scope of the method by a simple extension of the underlying principle.

The two frames mentioned each have four redundant restraints which can be chosen as at (b) and (f). Different choices could, of course, have been made without affecting the following discussion.

The recognition and identification of these redundants, $X_a \ldots X_d$, automatically identifies both their point of application and their direction so that the suffixes $a \ldots d$ can be used to describe deflexions as well as forces and moments. A second suffix is used to denote the force producing the deflexion in the following way.

$\Delta_{ab}$ = Component of deflexion, in the $X_a$ − direction, of the point on the primary structure where the redundant $X_a$ acts, caused by the force $X_b = 1$.

$\Delta_{bo}$ = Component of deflexion, in the $X_b$ − direction, of the point on the primary structure where the redundant $X_b$ acts, caused by the actual loads $P_1$, $P_2$ acting on the structure.

Since we are now restricting the discussion to linear elastic structures we can superimpose forces and deflexions so that $X_c \Delta_{cc}$ = Component of deflexion, in the $X_c$ − direction, of the point on the primary structure where the redundant $X_c$ acts, caused by the force $X_c$.

The deflexions correspond completely to the redundants so that if $X_a$ happens to be a clockwise couple acting at a certain point then $\Delta_a$ means a clockwise rotation of the same point. Examples of the use of the notation are shown in Figure 4.6 at (c), (d), (g) and (h). It will be clear that the notation is essentially the same as was used in Chapter 3, Section 3.5.1, so that Maxwell's Reciprocal Theorem

$$\Delta_{ab} = \Delta_{ba}$$

can be used.

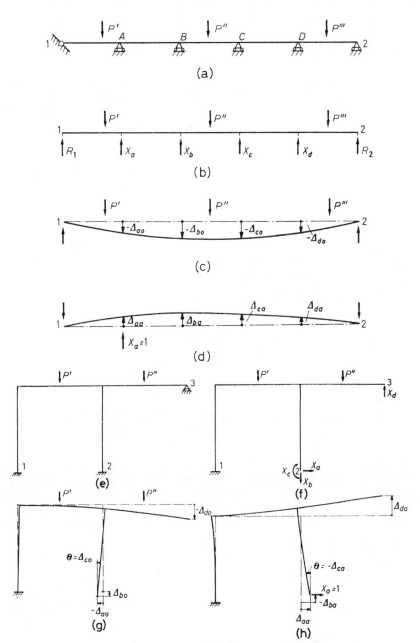

*Figure 4.6.* (a) and (b) The 5-span continuous beam, under vertical loads, has four redundant reactions which we take to be the four interior reactions labelled $X_a \ldots X_d$; (c) The primary structure is the simply-supported beam 1–2. The loads $P' - P'''$ produce deflexions $-\Delta_{ao}, -\Delta_{bo}, \ldots$ corresponding to $X_a, X_b, \ldots$. The negative sign indicates that the deflexions are opposite in direction to the forces $X_a, X_b \ldots$; (d) When the force $X_a = 1$ is applied to the primary structure 1–2 deflexions $\Delta_{aa}, \Delta_{ba} \ldots$ are produced, corresponding to $X_a, X_b, \ldots$; (e) and (f) This rigidly-jointed frame is built-in at 1 and 2 and supported on a roller joint at 3. There are four redundants which are considered to be the reactive restraints shown, namely $X_a \ldots X_d$; (g) The loads $P', P''$ acting on the primary structure produce deflexions $\Delta_{ao}, \Delta_{bo} \ldots$ It will be seen that $\Delta_{co}$ is a rotation corresponding to the couple $X_c$; (h) The force $X_a = 1$ acting on the primary structure produces deflexions $\Delta_{aa}, \Delta_{ba} \ldots$ corresponding to $X_a, X_b$

If it is known that there is no reaction displacement the compatibility conditions are that when the applied loads $P'$, $P''$ .. are acting on the primary structure together with the redundants $X_a$, $X_b$, .. the various component deflexions $\Delta_a$, $\Delta_b$ .. are zero. We thus have as many equations as there are redundants, as follows:

$$\begin{aligned}
\Delta_a &= 0 = \Delta_{ao} + X_a\Delta_{aa} + X_b\Delta_{ab} + X_c\Delta_{ac} + X_d\Delta_{ad} \\
\Delta_b &= 0 = \Delta_{bo} + X_a\Delta_{ba} + X_b\Delta_{bb} + X_c\Delta_{bc} + X_d\Delta_{bd} \\
\Delta_c &= 0 = \Delta_{co} + X_a\Delta_{ca} + X_b\Delta_{bc} + X_c\Delta_{cc} + X_d\Delta_{cd} \\
\Delta_d &= 0 = \Delta_{do} + X_a\Delta_{da} + X_b\Delta_{db} + X_c\Delta_{dc} + X_d\Delta_{dd}
\end{aligned} \quad \ldots (4.7)$$

These equations were first given by Müller-Breslau[8] and they form the basis for several different methods of analysing hyperstatic structures. The coefficients $\Delta_{aa}$, $\Delta_{ab}$, ... of the redundants are, as we have seen, the deflexions of the primary structure under one or other unit load; they can therefore be termed 'flexibility coefficients' or 'influence coefficients'[9].

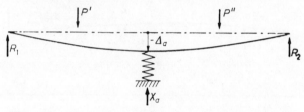

*Figure 4.7.* The spring, which exerts the single redundant reactive force on this propped beam, deflects an amount $k$ under unit load

If known reaction displacements are permitted then the left-hand side of the equations can be suitably adjusted. If the supports are elastic and deflect an amount proportional to the load on them the deflexion on the left-hand side becomes a function of the appropriate redundant. For example, the central support of the beam shown in FIGURE 4.7 is a spring, of stiffness $k$ (or flexibility $1/k$), which deflects an amount $\dfrac{1}{k}$ under unit load. Equations 4.7 then reduce to the single equation

$$-\Delta_a = -\frac{1}{k}X_a = -\Delta_{ao} + X_a\Delta_{aa}$$

when the appropriate signs are inserted.

Solving this equation, we have,

$$X_a = \frac{\Delta_{ao}}{1/k + \Delta_{aa}}$$

so that the value of $X_a$ diminishes as the flexibility of the spring increases; this is what we should expect.

This procedure has been systematized by Wilson[10] for use with continuous beams on elastic supports.

## 4.4.3. REDUNDANT INTERNAL FORCES AND MOMENTS

Although we chose to regard certain of the reactive restraints as redundant in the first two examples of the preceding Section it would have been quite legitimate, and it might have been advantageous, to use appropriate internal forces and moments in this way. It often happens, too, that such a choice is inevitable as in the case of overstiff braced frameworks where one or more of the bars is redundant.

In such a case the meaning of the symbols used in the preceding Section has to be extended; the redundant forces and couples now act on this primary structure in equal and opposite pairs and the corresponding deflexions are the relative displacements of parts of the structure.

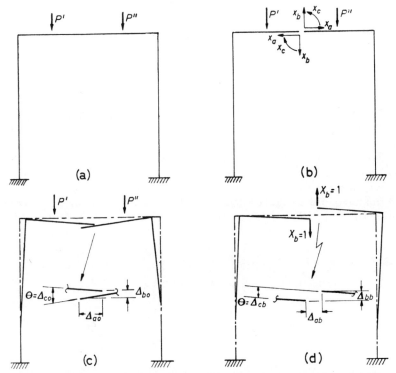

*Figure 4.8.* (a) This rigidly-jointed portal frame has three redundants; (b) The direct and shearing forces and the bending moment acting at mid-span of the beam have been chosen as the redundants; (c) Relative displacements $\Delta_{ao}$, $\Delta_{bo}$, $\Delta_{co}$ are produced by the loads $P'$, $P''$ acting on the primary structure, which in this case consists of two cranked cantilevers; (d) Relative displacements $\Delta_{ab}$, $\Delta_{bb}$, $\Delta_{cb}$ are produced in the primary structure by the pair of equal and opposite shearing forces $X_b = 1$

For example, the fixed-ended portal frame shown in FIGURE 4.8(a) has three redundants as we can see by making a cut at some point in order to reduce the structure to two determinate cantilevers. The redundant direct and shearing forces and bending moment which must have acted at the cut section are shown at (b) while at (c) are shown the corresponding relative displacements $\Delta_{ao}$, $\Delta_{bo}$ and $\Delta_{co}$, produced by the loads acting on the primary structure (the two cantilevers). Compatibility requires that the redundants $X_a$, $X_b$ and $X_c$ be suitably adjusted to reduce these displacements to zero and this is done by finding the relative displacements produced by pairs of unit forces $X_a = 1$, $X_b = 1$, $X_c = 1$, as shown, for example, at (d).

With this understanding of the extended meaning now given to the various symbols, equations 4.7 follow as before. It is necessary, of course, to be quite consistent in the signs used: as soon as the redundant forces have been selected, as at (b), the corresponding displacements are defined. Relative movement in the same sense is given a positive sign.

The methods used to compute the flexibility coefficients distinguish the different systems of analysis to be described in subsequent sections, the fundamental objective in each case being to systematize the calculations and to deal with the sign question as automatically as possible.

## 4.5. THE MAXWELL-MOHR EQUATIONS FOR PIN-JOINTED STRUCTURES

### 4.5.1. Derivation of the Equations

When equations 4.7 are applied to a braced pin-jointed hyperstatic structure they yield the equations, to which reference has already been made in Section 4.4.1, which were given first by Maxwell and then by Mohr. Consider, for example, the two-hinged braced arch shown in FIGURE 4.9(a). Application of equation 1.7 shows that there are two redundants which can conveniently be taken as the horizontal component of reaction at $A$ ($X_a$) and the force in the member $CD$. ($X_b$). The primary structure, obtained by cutting $CD$ and replacing the pin at $A$ by a roller is shown in FIGURE 4.9(b). The quasi-external forces $X_a$ and $X_b$ will be equal to the redundant forces they replace when the compatibility equations 4.7 are satisfied. These equations express the following conditions

(i) The horizontal displacement of $A$ is zero,

i.e. $\quad \Delta_a = 0 = \Delta_{ao} + X_a \Delta_{aa} + X_b \Delta_{ab}$ .... (i)

(ii) The 'gap' in the cut redundant member $CD$ is zero,

i.e. $\quad \Delta_b = 0 = \Delta_{bo} + X_a \Delta_{ba} + X_b \Delta_{bb}$ .... (ii)

The flexibility coefficients $\Delta_{aa}$, $\Delta_{ab}$ can be found by making use of equation 3.13,

$$\Delta_j = \Sigma F' F \rho$$

although the method of use of the equation must be slightly developed in order to give the relative displacement now required. The symbol $F$ denotes the force in a member of the framework under the loading which produces the desired deflexion $\Delta_j$; and in the present case it evidently corresponds to the second suffix of the flexibility coefficient. $F'$, it will be remembered, is the member force produced by unit load corresponding to the desired deflexion $\Delta_j$. The displacement of $C'$ towards $D'$, for instance, is found by applying unit load at $C'$ acting in the direction $C'D'$, while the displacement of $D'$ towards $C'$ is found by applying unit load at $D'$ in the direction $D'C'$.

## 4.5. MAXWELL-MOHR EQUATIONS FOR PIN-JOINTED STRUCTURES 4.5.1

The sum of these displacements, which gives the relative movement of $C'$ and $D'$ towards each other, can conveniently be found by applying these

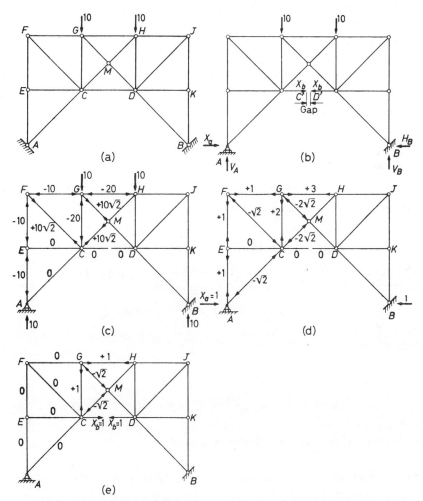

*Figure 4.9.* (a) This two-hinged braced arch has two redundants which can be taken as the force in the member $CD$ and the horizontal component of the reaction at $A$; (b) The redundant member $CD$ has been cut and the pin at $A$ replaced by a roller. Quasi-external forces $X_a$ and $X_b$ replace the redundant forces formerly acting. $X_a$ and $X_b$ will just equal the redundant forces if there is zero displacement at $A$ and if the gap $C'D'$ is zero; (c) The forces $F_o$ are those acting in the primary structure when the external loads only are acting; (d) The forces $F_a$ are produced in the primary structure when $X_a = 1$ and $X_b = 0$; (e) The forces $F_b$ are produced in the primary structure when $X_b = 1$ and $X_a = 0$

two unit loads simultaneously. $F'$ evidently corresponds to the first suffix of the flexibility coefficient.

We now introduce the following symbols to designate the member forces under the various force systems.

## METHODS FOR THE ANALYSIS OF LINEAR STRUCTURES

$F$ = Force in a member of the complete hyperstatic structure under the actual applied loads.

$F_o$ = Force in a member of the primary structure under the actual applied loads.

$F_a$ = Force in a member of the primary structure acted on by unit load in redundant *a*, i.e. $X_a = 1$.

$F_b$ = Force in a member of the primary structure acted on by unit load in redundant *b*, i.e. $X_b = 1$.

Evidently
$$F = F_o + X_a F_a + X_b F_b \qquad \ldots\text{(iii)}$$

Applying equation 3.13, with its extended meaning, to the terms of equations (i) and (ii) we have,

$$\Delta_{ao} = \Sigma F_a F_o \rho \qquad \Delta_{aa} = \Sigma F_a^2 \rho \qquad \Delta_{ab} = \Sigma F_a F_b \rho \qquad \text{and so on.}$$

The two equations (i) and (ii) thus become

$$\left. \begin{array}{l} \Delta_a = 0 = \Sigma F_o F_a \rho + X_a \Sigma F_a^2 \rho + X_b \Sigma F_a F_b \rho \\ \Delta_b = 0 = \Sigma F_o F_b \rho + X_a \Sigma F_b F_a \rho + X_b \Sigma F_b^2 \rho \end{array} \right\} \qquad \ldots\text{(iv)}$$

The same arguments can clearly be used for several redundants which will be given by the series of simultaneous equations

$$0 = \Sigma F_n (F_o + X_a F_a + X_b F_b + \text{---} X_n F_n) \rho \qquad \ldots\text{(4.8)}$$

where $n$ is given in turn the values 1, 2, etc. up to $n$, the total number of redundants.

These are the Maxwell-Mohr Equations; it will be noticed that they could have been derived by the direct application of equation 3.13. Thus:

$$\Delta_n = 0 = \Sigma F_n F \rho = \Sigma F_n (F_o + X_a F_a + \text{---} X_n F_n) \rho$$

It is a more instructive approach, however, to regard them as a special case of the general flexibility coefficient equations.

### 4.5.2. Completion of the Analysis of the Framework of Figure 4.9

The forces $F_o, F_a$ and $F_b$ in the members are obtained by analysing the primary structure under the three relevant force systems. In simple cases this can be done by resolution at the joints and this method has been used to give the member forces shown in FIGURE 4.9(c), (d) and (e); in more complicated cases it may be better to use a force diagram.

The actual calculations are best carried out, as far as possible, in tabular form as shown in TABLE 4.1. The cross-sectional area of the members has been assumed to be constant and it will be noticed that a constant factor $L/AE$ runs right through the calculation and is finally cancelled. From the summations shown at the feet of the last five columns we have:

$$\Sigma F_o F_a \rho = -(200 + 120\sqrt{2}) \; L/AE = -369 \cdot 70 \; L/AE$$
$$\Sigma F_o F_b \rho = -(60 + 40\sqrt{2}) \; L/AE = -116 \cdot 57 \; L/AE$$

## 4.5. MAXWELL-MOHR EQUATIONS FOR PIN-JOINTED STRUCTURES 4.5.2

$$\Sigma F_a F_b \rho = +(\ 7\ +\ 8\sqrt{2})\ L/AE = +\ 18\cdot314\ L/AE$$

$$\Sigma F_a^2 \rho = +(\ 23\ +\ 24\sqrt{2})\ L/AE = +\ 56\cdot941\ L/AE$$

$$\Sigma F_b^2 \rho = +(\ 4\ +\ 4\sqrt{2})\ L/AE = +\ 9\cdot6568\ L/AE$$

Substituting in equations (iv) we have

$$\left. \begin{array}{l} 0 = -369\cdot70 + 56\cdot941 X_a + 18\cdot314 X_b \\ 0 = -116\cdot57 + 18\cdot314 X_a + 9\cdot6568 X_b \end{array} \right\} \text{ i.e. } \begin{array}{l} X_a = 6\cdot6924 \\ X_b = -0\cdot6208 \end{array}$$

The significance of the signs of these values is that while the horizontal component of thrust at the hinges is directed inwards, as expected, the force in member $CD$ is compressive rather than tensile as assumed.

TABLE 4.1
ANALYSIS OF FRAMEWORK OF FIGURE 4.9

| Member | Length $\times L$ | $\rho$ $\times L/AE$ | $F_o$ | $F_a$ | $F_b$ | $F_o F_a \rho$ $\times L/AE$ | $F F_b \rho$ $\times L/AE$ | $F_a F_b \rho$ $\times L/AE$ | $F_a^2 \rho$ $\times L/AE$ | $F_b^2 \rho$ $\times L/AE$ |
|---|---|---|---|---|---|---|---|---|---|---|
| AE | 1 | 1 | $-10$ | $+1$ | 0 | $-10$ | 0 | 0 | $+1$ | 0 |
| BK | 1 | 1 | $-10$ | $+1$ | 0 | $-10$ | 0 | 0 | $+1$ | 0 |
| AC | $\sqrt{2}$ | $\sqrt{2}$ | 0 | $-\sqrt{2}$ | 0 | 0 | 0 | 0 | $+2\sqrt{2}$ | 0 |
| BD | $\sqrt{2}$ | $\sqrt{2}$ | 0 | $-\sqrt{2}$ | 0 | 0 | 0 | 0 | $+2\sqrt{2}$ | 0 |
| EC | 1 | 1 | 0 | 0 | 0 | 0 | 0 | 0 | 0 | 0 |
| DK | 1 | 1 | 0 | 0 | 0 | 0 | 0 | 0 | 0 | 0 |
| EF | 1 | 1 | $-10$ | $+1$ | 0 | $-10$ | 0 | 0 | $+1$ | 0 |
| KJ | 1 | 1 | $-10$ | $+1$ | 0 | $-10$ | 0 | 0 | $+1$ | 0 |
| FC | $\sqrt{2}$ | $\sqrt{2}$ | $+10\sqrt{2}$ | $-\sqrt{2}$ | 0 | $-20\sqrt{2}$ | 0 | 0 | $+2\sqrt{2}$ | 0 |
| JD | $\sqrt{2}$ | $\sqrt{2}$ | $+10\sqrt{2}$ | $-\sqrt{2}$ | 0 | $-20\sqrt{2}$ | 0 | 0 | $+2\sqrt{2}$ | 0 |
| FG | 1 | 1 | $-10$ | $+1$ | 0 | $-10$ | 0 | 0 | $+1$ | 0 |
| HJ | 1 | 1 | $-10$ | $+1$ | 0 | $-10$ | 0 | 0 | $+1$ | 0 |
| GC | 1 | 1 | $-20$ | $+2$ | $+1$ | $-40$ | $-20$ | $+2$ | $+4$ | $+1$ |
| HD | 1 | 1 | $-20$ | $+2$ | $+1$ | $-40$ | $-20$ | $+2$ | $+4$ | $+1$ |
| GM | $1/\sqrt{2}$ | $1/\sqrt{2}$ | $+10\sqrt{2}$ | $-2\sqrt{2}$ | $-\sqrt{2}$ | $-20\sqrt{2}$ | $-10\sqrt{2}$ | $+2\sqrt{2}$ | $+4\sqrt{2}$ | $+\sqrt{2}$ |
| HM | $1/\sqrt{2}$ | $1/\sqrt{2}$ | $+10\sqrt{2}$ | $-2\sqrt{2}$ | $-\sqrt{2}$ | $-20\sqrt{2}$ | $-10\sqrt{2}$ | $+2\sqrt{2}$ | $+4\sqrt{2}$ | $+\sqrt{2}$ |
| CM | $1/\sqrt{2}$ | $1/\sqrt{2}$ | $+10\sqrt{2}$ | $-2\sqrt{2}$ | $-\sqrt{2}$ | $-20\sqrt{2}$ | $-10\sqrt{2}$ | $+2\sqrt{2}$ | $+4\sqrt{2}$ | $+\sqrt{2}$ |
| DM | $1/\sqrt{2}$ | $1/\sqrt{2}$ | $+10\sqrt{2}$ | $-2\sqrt{2}$ | $-\sqrt{2}$ | $-20\sqrt{2}$ | $-10\sqrt{2}$ | $+2\sqrt{2}$ | $+4\sqrt{2}$ | $+\sqrt{2}$ |
| GH | 1 | 1 | $-20$ | $+3$ | $+1$ | $-60$ | $-20$ | $+3$ | 9 | $+1$ |
| CD | 1 | 1 | 0 | 0 | $+1$ | 0 | 0 | 0 | 0 | $+1$ |
| $\Sigma$ | | | | | | $-200$ $-120\sqrt{2}$ | $-60$ $-40\sqrt{2}$ | $+7$ $+8\sqrt{2}$ | $+23$ $+24\sqrt{2}$ | $+4$ $+4\sqrt{2}$ |

The final forces in the members of the frame are then as computed in TABLE 4.2.

Two comments on these calculations should be made. The first is that it is essential that the redundant members themselves be included in the calculations in order that their own changes of length are not left out of account. Since these members have been cut they are, of course, only stressed when unit load is applied to them.

The second comment is that these particular calculations have been carried out to five significant figures in order to give a reliable comparison if alternative solutions, using other redundants, are attempted.

## METHODS FOR THE ANALYSIS OF LINEAR STRUCTURES

TABLE 4.2

CALCULATION OF FINAL FORCES IN THE MEMBERS OF THE STRUCTURE OF FIGURE 4.9

| Member | $F_o$ | $F_a$ | $X_a F_a$ | $F_b$ | $X_b F_b$ | $F$ |
|---|---|---|---|---|---|---|
| AE | − 10 | + 1 | + 6·692 | 0 | 0 | − 3·308 |
| AC | 0 | − $\sqrt{2}$ | − 9·464 | 0 | 0 | − 9·464 |
| EC | 0 | 0 | 0 | 0 | 0 | 0 |
| EF | − 10 | + 1 | + 6·692 | 0 | 0 | − 3·308 |
| FC | + 14·142 | − $\sqrt{2}$ | − 9·464 | 0 | 0 | + 4·678 |
| FG | − 10 | + 1 | + 6·692 | 0 | 0 | − 3·308 |
| GC | − 20 | + 2 | + 13·385 | + 1 | − 0·621 | − 7·236 |
| GM | + 14·142 | − $2\sqrt{2}$ | − 18·929 | − $\sqrt{2}$ | + 0·878 | − 3·909 |
| CM | + 14·142 | − $2\sqrt{2}$ | − 18·929 | − $\sqrt{2}$ | + 0·878 | − 3·909 |
| GH | − 20 | + 3 | + 20·077 | + 1 | − 0·621 | − 0·544 |
| CD | 0 | 0 | 0 | + 1 | − 0·621 | − 0·621 |

### 4.5.3. COMMENTS ON THE USE OF THE MAXWELL-MOHR EQUATIONS

In Chapter 1, Section 1.2.2, we observed that there is often some freedom in the choice of the redundant and that one choice may well be more favourable than another. Unfortunately it is not possible to give explicit rules for the selection of the redundants but two useful suggestions emerge from the calculations just concluded.

The first is that symmetry should be preserved if at all possible. It will be noticed that TABLE 4.1 has been set out so that corresponding members, on either side of the truss, follow one another. In this way almost half the figures can be written in immediately; it is indeed hardly necessary to write them in at all if care is taken to double the results emanating from members which have a mirror image on the other side of the truss.

The second point is that redundants should be chosen so that as many as possible of the $F_a$, $F_b$ terms are zero. In this way the number of terms in the product columns is kept to a minimum.

In choosing one internal member and one external component of reaction as redundants we made what is perhaps the most obvious selection but we can profitably enquire whether a better selection could have been made. Certainly, by choosing CD rather than one of the diagonal members of the over-stiff central panel, we have preserved symmetry. But it will be noticed that none of the $F_a$ terms is zero except those corresponding to EC and DK, members which are soon seen to be superfluous anyway. By contrast the work involved in evaluating the $F_b$ terms was quite trivial and this makes one wonder whether it would not have been better to take GH and CD as two redundant members; the primary structure would then have been an arch with three hinges, at A, M and B respectively.

The final step in the calculation of the redundants is the solution of as many simultaneous equations as there are redundants. The accuracy that is eventually attained is apt to be lower than expected because the solution of the equations involves the differences between comparatively large numbers. This is an inherent difficulty in the solution of simultaneous equations that may be quite troublesome if the equations are ill-conditioned. Southwell[11] has pointed out that ill-conditioned equations are unlikely to occur if a

## 4.5. MAXWELL-MOHR EQUATIONS FOR PIN-JOINTED STRUCTURES 4.5.4

well-designed structure is analysed by the Principle of Minimum Potential Energy where the unknowns are the joint displacements. On the other hand it is not easy to select the redundant forces so as to ensure that the Maxwell-Mohr Equations are well-conditioned.

This difficulty can always be circumvented by working to a larger number of significant figures than is finally required although, as Goldstein[12] remarked when discussing the analysis of shell roofs, this may call for the use of a desk calculating machine [13, 14, 15].

### 4.5.4. PRE-STRESSING

It has already been remarked that a characteristic of hyperstatic structures is that stresses are induced in them if, at assembly, the redundant members are of incorrect length and have to be forced into position. Advantage can sometimes be taken of this to induce a favourable initial stress pattern which in combination with the stresses produced by the applied loads, is less severe

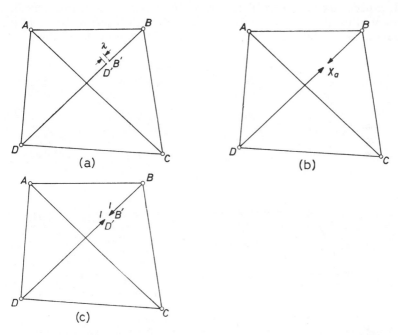

*Figure 4.10.* (a) The member $BD$ is too short by an amount $\lambda$ to fit into position (The gap $B'D'$ has been shown towards the middle of the member for clarity.); (b) The force $X_a$ in $BD$ must be sufficient to close the gap by stretching the member $BD$ and by distorting the rest of the frame; (c) Forces $F_a$ are produced by unit tensile forces applied at $B'$ and $D'$

than would otherwise have been the case. The advantages accruing to this procedure are perhaps less dramatic in steel than in concrete structures but the idea would seem to be worth pursuing further than has so far been the case.

It is always possible to design a structure to carry a given set of loads so that, by means of appropriate prestressing, *all* the members are fully stressed.

Consider first the quadrilateral frame shown in FIGURE 4.10(a), the diagonal $BD$ being an amount $\lambda$ too short to fit into its position in the unloaded frame. The tension induced in this member when it is forced into position is $X_a$; it is evident from (b) that $X_a$ must be just sufficient to stretch member $BD$ and distort the frame so that $B'$ and $D'$ are brought together.

i.e. $$\lambda = \Sigma F' F \rho$$
$$= \Sigma F_a (X_a F_a) \rho = X_a \Sigma F_a^2 \rho$$

i.e. $$X_a = \lambda / \Sigma F_a^2 \rho$$

The summation extends over all the members of the frame including the redundant $BD$.

Unit loads applied at $B'$ and $D'$ produce the forces $F_a$ as in the previous example. The final force in any member when the frame is assembled is, of course, $X_a F_a$.

It is necessary to be careful over signs, but it will be clear from the above example that if the member is too short, so that it has to be *extended* on assembly, it will be in *tension*. In our convention tension and extension are positive.

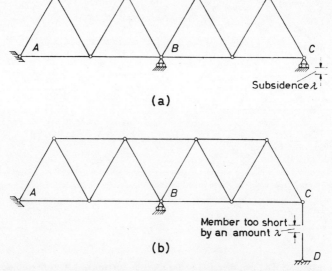

*Figure 4.11.* (a) One support of this truss is supposed to have subsided by an amount $\lambda$; (b) This situation is the same as would have arisen if the support $C$ had been replaced by a very stiff pin-ended member $CD$ which was too short by an amount $\lambda$

Reaction displacements will similarly produce forces in the framework if there are redundant reactive restraints. The truss shown in FIGURE 4.11(a) has one redundant reaction, say $C$, and we consider the effect of a downward displacement, $\lambda$, at $C$. It is easy to see that there is really no difference between this problem and the last if we replace the roller at $C$ by an imaginary member $CD$, very stiff in comparison with the other members of the truss,

## 4.5. MAXWELL-MOHR EQUATIONS FOR PIN-JOINTED STRUCTURES

which is $\lambda$ too short to fit into place. Evidently, as before, a tensile force will be induced in the imaginary member, corresponding to a downward force on the reaction, which is what would be expected. As before, then

$$X_a = \lambda / \Sigma F_a^2 \rho$$

The summation will presumably have to include the imaginary member $CD$, but as the stiffness is very great $\rho$ is correspondingly small and its contribution to the sum can be ignored.

We can now work out the member forces produced in the frame of FIGURE 4.9(a) if the redundant member $CD$ is 1 per cent too short on assembly while $A$ is displaced an amount $3L/100$ away from $B$. The corresponding values of $X_a$ and $X_b$ are given by

$$\left. \begin{aligned} -\frac{3L}{100} &= X_a \Sigma F_a^2 \rho + X_b \Sigma F_a F_b \rho = X_a 56 \cdot 941 \frac{L}{AE} + X_b 18 \cdot 314 \frac{L}{AE} \\ \frac{L}{100} &= X_a \Sigma F_a F_b \rho + X_b \Sigma F_b^2 \rho = X_a 18 \cdot 314 \frac{L}{AE} + X_b 9 \cdot 6568 \frac{L}{AE} \end{aligned} \right\}$$

Solving these equations gives

$$X_a = -0 \cdot 2205 \frac{AE}{100} \qquad X_b = +0 \cdot 5217 \frac{AE}{100}$$

Inspection shows that the signs of $X_a$ and $X_b$ are as was to be expected.

### 4.5.5. TEMPERATURE STRESSES

Evidently a change of temperature can be expected to introduce stresses into the members of a hyperstatic framework. Dealing first with external redundant restraints we see that a uniform change of temperature will tend to lengthen or shorten all the members in the same proportion without changing the shape of the structure. Stress is induced in the members when this change in the size of the structure is resisted by reactions, the situation being essentially the same as that produced by a reaction displacement. For example, if the frame of FIGURE 4.9(a) is cooled the members will shrink relative to the hinges at $A$ and $B$, and this is clearly equivalent to the outward displacement of $A$ postulated in the last example. Thus if the temperature falls by $t°$ and the coefficient of expansion is $\alpha_t$, $X_a$ and $X_b$ are given by:

$$\left. \begin{aligned} -\alpha_t (3L) &= X_a \Sigma F_a^2 \rho + X_b \Sigma F_a F_b \rho \\ 0 &= X_a \Sigma F_a F_b \rho + X_b \Sigma F_b^2 \rho \end{aligned} \right\}$$

The second of these equations expresses the fact that a *uniform* change of temperature has no effect if the structure is only redundant internally and is of the same material throughout. A change of the temperature of a redundant member, relative to that of the rest of the frame, has the same effect as if the member were of incorrect length.

A slightly more complicated problem arises, for example, if the members of the top chord $FJ$ are heated relative to the rest of the frame. It is now necessary first to work out the relative movements of $A$ and $B$, and $C$ and $D$, using the method of Section 3.6.1 of Chapter 3. These are then the values of $\Delta_a$ and $\Delta_b$ to be used in equations (iv) above (Section 4.5.1) to give the corresponding values of $X_a$ and $X_b$.

### 4.5.6. Hyperstatic Space Frames

#### 4.5.6.1. *Tension Coefficients*

The Maxwell-Mohr Equations can be used to analyse space frames as well as plane frames but the calculation of the bar forces under the different load-systems is apt to be more complex. The computations can be systematized by a useful procedure, first expounded in English by Southwell[9],

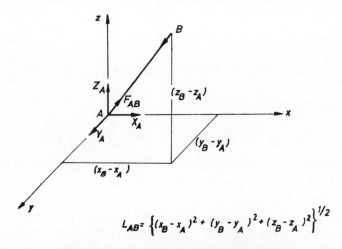

*Figure 4.12.* Nomenclature for tension coefficients

which amounts to the application of the equations of equilibrium to the joints of the frame in terms of direction cosines. In FIGURE 4.12 a single bar $AB$, having a tension $F_{AB} = t_{AB} L_{AB}$ where $t_{AB}$ is the 'tension coefficient', is shown. External forces $X_A$ $Y_A$ and $Z_A$ act on the joint $A$ in the directions of the coordinate axes.

Resolving in the $x$- direction at $A$,

$$X_A + F_{AB} \frac{x_B - x_A}{L_{AB}} = X_A + t_{AB}(x_B - x_A) = 0$$

Similarly, resolving in the $y$- and $z$- directions,

$$Y_A + t_{AB}(y - y_A) = Z_A + t_{AB}(z_B - z_A) = 0$$

If several bars—$AB$, $AC$, $AD$, etc.—meet at $A$, we have

$$\left. \begin{array}{l} t_{AB}(x_B - x_A) + t_{AC}(x_C - x_A) + t_{AD}(x_D - x_A) + \ldots + X_A = 0 \\ t_{AB}(y_B - y_A) + t_{AC}(y_C - y_A) + t_{AD}(y_D - y_A) + \ldots + Y_A = 0 \\ t_{AB}(z_B - z_A) = t_{AC}(z_C - z_A) + t_{AD}(z_D - z_A) + \ldots + Z_A = 0 \end{array} \right\} \ldots (4.9)$$

## 4.5. MAXWELL-MOHR EQUATIONS FOR PIN-JOINTED STRUCTURES 4.5.6.2

Similar equations are written for each joint† in turn and solved for $t_{AB}$, $t_{AC}$, etc. Finally we have

$$\left. \begin{aligned} F_{AB} &= t_{AB} L_{AB} = t_{AB} \{(x_B - x_A)^2 + (y_B - y_A)^2 + (z_B - z_A)^2\}^{\frac{1}{2}} \\ F_{AC} &= t_{AC} L_{AC} = t_{AC} \{(x_C - x_A)^2 + (y_C - y_A)^2 + (z_C - z_A)^2\}^{\frac{1}{2}} \end{aligned} \right\} \quad \ldots (4.10)$$

### 4.5.6.2. *Example of a Hyperstatic Space Frame*

The framework shown in FIGURE 4.13(a), whose members all have the same value of $\rho = L/AE$, has a single redundant member $AC$. This can be seen to be the case if it is observed that joint $D$ is adequately supported by the three bars $DA$, $DE$ and $DF$, while joint $C$ is supported by the bars $CD$, $CB$ and $CE$.

The $F_o$-system is shown at (b) while the forces in the bars are calculated by means of tension coefficients as follows:

$F_o$-system:

At $C$:  
x-dir$^n$ · $t_{CB}(0 - 3) + t_{CD}(3 - 3) + t_{CE}(\ \ 0 - 3) = 0$  
y-dir$^n$ · $t_{CB}(0 - 0) + t_{CD}(4 - 0) + t_{CE}(\ \ 0 - 0) = 0$  $\quad \ldots$ (i)
z-dir$^n$ · $t_{CB}(0 - 0) + t_{CD}(0 - 0) + t_{CE}(-4 + 0) = 0$

Solving these equations we have

$$t_{CB} = t_{CD} = t_{CE} = 0 \quad \ldots \text{(ii)}$$

This result could, in such a simple case, have been obtained by inspection. Indeed it is to be observed that in any space frame, if three members not in one plane meet at an unloaded joint the members are unstressed.

At $D$:  x-dir$^n$ · $t_{DA}(0 - 3) + t_{DC}(3 - 3) + t_{DE}(\ \ 0 - 3)$  
$\qquad\qquad\qquad\qquad\qquad\qquad\qquad + t_{DF}(\ \ 0 - 3) = 0$  
y-dir$^n$ · $t_{DA}(4 - 4) + t_{DC}(0 - 4) + t_{DE}(\ \ 0 - 4)$  
$\qquad\qquad\qquad\qquad\qquad\qquad\qquad + t_{DF}(\ \ 4 - 4) = 0$  $\quad \ldots$ (iii)
z-dir$^n$ · $t_{DA}(0 - 0) + t_{DC}(0 - 0) + t_{DE}(-4 - 0)$  
$\qquad\qquad\qquad\qquad\qquad\qquad\qquad + t_{DF}(-4 - 0) = P$

Solving these equations, we have

$t_{DC} = 0 \quad$ (from (ii) above)

$t_{DE} = -\ t_{DC} = 0$

$t_{DF} = -\ P/4 \qquad \therefore\ F_{DF} = -\ L_{DF} P/4 = -\ 5P/4$

$t_{DA} = +\ P/4 \qquad \therefore\ F_{DA} = +\ L_{DA} P/4 = +\ 3P/4$

---

† In order to ensure that the equations are written down correctly in the first instance, it is convenient to write the tension coefficient of a member, say $CB$, as $t_{BC}$ when the joint $B$ is being considered and $t_{CB}$ when $C$ is being considered.

Thus at $B$,  $\quad t_{BC}\ (x_C - x_B) + \ldots$
at $C$,  $\quad t_{CB}\ (x_B - x_C) + \ldots$

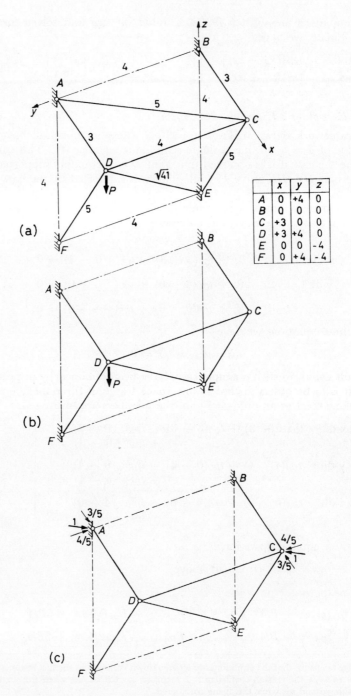

*Figure 4.13.* (a) This space frame has one redundant member, $AC$; (b) $F_o$ forces; (c) $F_a$ forces

## 4.5. MAXWELL-MOHR EQUATIONS FOR PIN-JOINTED STRUCTURES 4.5.6.2

Again we could have foreseen that there would be zero force in $DE$ for if, at a joint in a space frame, the external load and all the members except one lie in a plane that member is unstressed. In this case we know, from (ii) above, that $CD$ is unstressed and can be removed; this leaves $AD$, $DF$ and the force $P$ acting in one plane while $DE$ is now the only member not in that plane.

We now analyse the $F_a$-system, shown at (c), beginning by resolving the unit loads acting along $A - C$ in the directions of the coordinate axes.

$F_a$-system:

The rule just given is seen to apply to joint $C$ and we can state at once that $CE$ is unstressed. The forces in $CB$ and $CD$ are $-3/5$ and $-4/5$ respectively.

Hence
$$t_{DC} = F_{DC}/L_{DC} = -\frac{4/5}{4} = -\frac{1}{5}$$

At $D$:

x-dir$^n$ · $t_{DA}(0 - 3) + t_{DC}(3 - 3) + t_{DE}(\ 0 - 3) + t_{DF}(\ 0 - 3) = 0$

y-dir$^n$ · $t_{DA}(4 - 4) + t_{DC}(0 - 4) + t_{DE}(\ 0 - 4) + t_{DF}(\ 4 - 4) = 0$

z-dir$^n$ · $t_{DA}(0 - 0) + t_{DC}(0 - 0) + t_{DE}(-4 - 0) + t_{DF}(-4 - 0) = 0$

Hence  $t_{DA} = 0$  $\qquad F_{DA} = 0$

$t_{DE} = +1/5$ $\qquad F_{DE} = 1/5 \cdot \sqrt{41} = \sqrt{41}/5$

$t_{DF} = -1/5$ $\qquad F_{DF} = -1/5 \cdot 5 = -1$

The Maxwell-Mohr Equation is

$$0 = \Sigma F_o F_a \rho + X_a \Sigma F_a^2 \rho \qquad \ldots \text{(4.8)}$$

The results of the previous calculations are summarized in TABLE 4.3.

TABLE 4.3

| Member | $F_o$ | $F_a$ | $F_o F_a$ | $F_a^2$ |
|---|---|---|---|---|
| AC | 0 | $+1$ | 0 | $+1$ |
| AD | $+3P/4$ | 0 | 0 | 0 |
| BC | 0 | $-3/5$ | 0 | $+9/25$ |
| CD | 0 | $-4/5$ | 0 | $+16/25$ |
| CE | 0 | 0 | 0 | 0 |
| DE | 0 | $+\sqrt{41}/5$ | 0 | $+41/25$ |
| DF | $-5P/4$ | $-1$ | $+5P/4$ | $+1$ |
| $\Sigma$ | | | $+1 \cdot 25P$ | $+116/25$ $= +4 \cdot 64$ |

Hence $\qquad 0 = 1 \cdot 25P + 4 \cdot 64 X_A$

i.e. $\qquad X_A = -0 \cdot 27P$

## 4.6. THE $\Delta_{ik}$ METHOD FOR FRAMES IN WHICH BENDING STRESS PREDOMINATES

### 4.6.1. Derivation of the Method

The discussion of Section 4.4.2 of this chapter can, of course, be applied to so-called 'rigid' frames which rely on the rigidity of their joints for stability. The deflexion of such a frame depends mainly on the flexural stiffness of the beams and columns, axial and shearing effects being usually small in comparison.

The flexibility coefficients $\Delta_{ab}$, $\Delta_{aa}$, etc. are now found by computing the deflexion, due to bending, of the primary structure under the relevant force systems, equation 3.9(a).

$$\Delta_j = \int \frac{M'M dx}{EI} \qquad \ldots \ldots (3.9(a))$$

being particularly suitable for this purpose.

We now develop a geometrical procedure for evaluating this integral which is often to be preferred to the direct integration used in Chapter 3, Section 3.7.3. Here $M'$ is the bending moment due to unit load acting at $j$ in the direction of the desired deflexion and it is therefore represented by straight lines. $M$ is the actual bending moment producing the deflexion; if the loading is a series of point loads then $M$ will be represented by a series of straight lines, while if the loading is continuous $M$ will be represented by parabolic curves. These are the two most common types of loading and for these cases the integrations can be performed by splitting the frame up into suitable lengths and applying the following procedure.

Suppose that a single beam $AB$, Figure 4.14(a), is acted on by certain loads which produce bending moments $M_k$ and that it is required to find the resulting deflexion $\Delta_{ik}$ at $E$. We know that

$$\Delta_{ik} = \frac{1}{EI}\int_A^B M_i M_k \, dx \qquad \ldots \ldots (4.11)$$

where $M_i$ is the bending moment, shown at (c), corresponding to unit load acting at $E$. We concentrate first on a length of the beam $CD$ over which one of the bending moments, say $M_i$, varies linearly. The bending moment diagrams for the length $CD$ are shown at (d) and (e), $G$ being the centroid of this part of the $M_k$ diagram.

Now at (e)

$$M_i = c\frac{L-x}{L} + d\frac{x}{L}$$

Hence $$\int_C^D M_i M_k dx = \int_C^D M_k c \frac{L-x}{L} dx + \int_C^D M_k d \frac{x}{L} dx$$

Now $$\int_C^D M_k(L-x)dx = A_k(L-\bar{x})$$

## 4.6. $\Delta_{ik}$ METHOD FOR FRAMES

and
$$\int_C^D M_k x\,dx = A_k \bar{x}$$

Hence
$$\int_C^D M_i M_k\,dx = A_k \left( c\frac{L - \bar{x}}{L} + d\frac{\bar{x}}{L} \right)$$
$$= \eta_i A_k \qquad \ldots\ (4.12)$$

where $\eta_i$ is the value of $M_i$ at the centroid of the $M_k$ diagram.

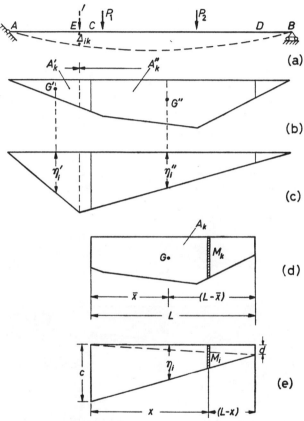

*Figure 4.14.* (a) It is required to find the deflexion of the point $E$ on this beam produced by the loads $P_1$ and $P_2$; (b) Bending moments $M_k$ due to the loads $P_1$ and $P_2$; (c) Bending moments $M_i$ due to unit load at $E$; (d) $M_k$ for a short length $CD$ of the beam; (e) $M_i$ for $CD$

Applying this result to $AB$, then, we have

$$EI.\Delta_{ik} = \eta'_i A'_k + \eta''_i A''_k$$

Evidently this method is quite convenient when both diagrams are built up of straight lines although a sufficiently close approximation can often be found when $M_k$ is parabolic. In FIGURE 4.15(a) two trapezoidal diagrams are shown from which all the straight-line cases can be deduced.

Here we have,

$$\Delta_{ik} = \eta_i A_k = \frac{aL}{2}\left(\frac{d}{3} + \frac{2c}{3}\right) + \frac{bL}{2}\left(\frac{2d}{3} + \frac{c}{3}\right)$$

$$= \frac{L}{6}\left\{ a\,(2c + d) + b\,(c + 2d) \right\} \quad \ldots(4.13)$$

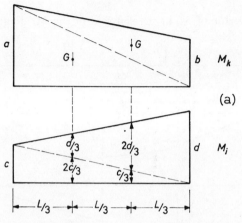

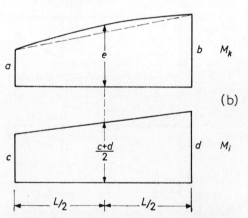

*Figure 4.15.* (a) Diagram for the calculation of $\Delta_{ik}$ when both $M_i$ and $M_k$ are trapezoidal; (b) Diagram for the calculation of $\Delta_{ik}$ when $M_i$ is trapezoidal while $M_k$ is bounded by a parabolic curve

In FIGURE 4.15(b) $M_k$ is bounded by part of a parabola. Dividing the diagram as shown and taking the centroid of the segmental part to be half way along, we have

$$\eta_i A_k = \frac{L}{6}\left\{ a\,(2c + d) + b\,(c + 2d) \right\} + \frac{2L}{3}\left\{ \left(e - \frac{a+b}{2}\right)\left(\frac{c+d}{2}\right) \right\}$$

$$= \frac{L}{6} \left\{ c\,(a + 2e) + d\,(b + 2e) \right\} \qquad \ldots (4.14)$$

This method has the advantage that the sign of $\Delta_{ik}$ is obtained automatically. In FIGURE 4.14 both $M_i$ and $M_k$ were negative according to our usual convention, so that their product was positive. In rigid frame analysis the same convention is used in a slightly different form—moments are

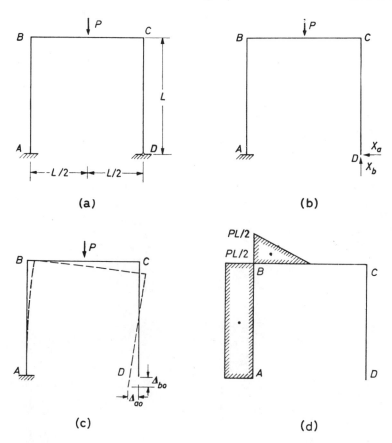

*Figure 4.16.* (a) and (b) This portal frame has two redundants which are taken to be the horizontal and vertical components of the reaction at $D$; (c) System $o$; the primary structure carrying the actual load $P$. The deflexions of $D$ are $\Delta_{ao}$ and $\Delta_{bo}$ in the directions of action of $X_a$ (horizontal) and $X_b$ (vertical); (d) System $o$ bending moments $M_o$

plotted on the tension sides of the members. Hence if $M_i$ and $M_k$ are on the same side of the member then $\Delta_{ik}$ is positive while if they are on the opposite sides it is negative.

Finally it should be noted that we have tacitly assumed in the above that $EI$ is constant. If it is not then the calculations should be based on $M/EI$ diagrams.

### 4.6.2. *Application of the method*

We now apply this technique to the portal frame illustrated in Figure 4.16(a) which has two redundant restraints; the members have been taken

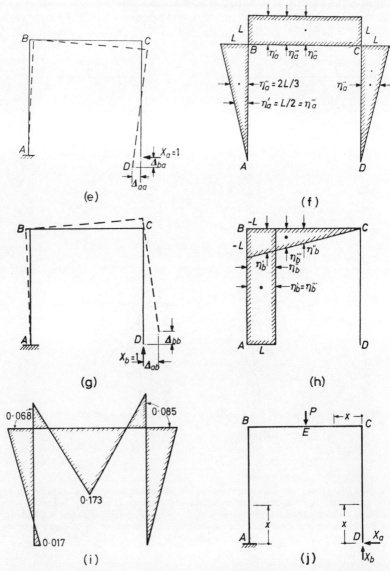

*Figure 4.16.* (e) System *a*; the primary structure loaded by $X_a = 1$; (f) System *a* bending moments $M_a$; (g) System *b*; the primary structure loaded by $X_a = 1$; (h) System *b* bending moments $M_b$; (i) Final bending moments ($\times PL$); (j) Symbols for strain energy solution of this problem

to be uniform for simplicity. We choose as redundants the horizontal and vertical components of the reaction at the hinged abutment *D*. In the

## 4.6. $\Delta_{ik}$ METHOD FOR FRAMES    4.6.2

absence of reaction displacements the compatibility equations 4.7 are:

$$0 = \Delta_{ao} + X_a\Delta_{aa} + X_b\Delta_{ab}$$
$$0 = \Delta_{bo} + X_a\Delta_{ab} + X_b\Delta_{bb}$$
.... (i)

The primary structure under the appropriate load systems is shown at (c), (e) and (g), the corresponding bending moment diagrams being given alongside at (d), (f) and (h). The various flexibility coefficients are now computed by the $\Delta_{ik}$ method, the details being given in TABLE 4.4. The primes against the $\eta$ terms simply serve to distinguish one from the other and help to identify them on FIGURE 4.16. The positive sign of $\eta$ and $A$ is used when the bending moment, plotted on the tension side of the members, is on the *outside* of the frame.

### TABLE 4.4
$\Delta_{ik}$ CALCULATIONS FOR THE FRAME OF FIGURE 4.16

| | | Member | $\eta$ | $A$ | $\eta A$ | $\Sigma \times L^3$ |
|---|---|---|---|---|---|---|
| $\Delta_{ao}$ | $\eta'_a A_o$ | AB<br>BC<br>CD | $+L/2$<br>$+L$<br>— | $+PL^2/2$<br>$+PL^2/8$<br>$0$ | $+PL^3/4$<br>$+PL^3/8$<br>$0$ | $\dfrac{3P}{8}$ |
| $\Delta_{bo}$ | $\eta'_b A_o$ | AB<br>BC<br>CD | $-L$<br>$-5L/6$<br>— | $+PL^2/2$<br>$+PL^2/8$<br>$0$ | $-PL^3/2$<br>$-5PL^3/48$<br>$0$ | $-\dfrac{29P}{48}$ |
| $\Delta_{aa}$ | $\eta''_a A_a$ | AB<br>BC<br>CD | $+2L/3$<br>$+L$<br>$+2L/3$ | $+L^2/2$<br>$+L^2$<br>$+L^2/2$ | $+L^3/3$<br>$+L^3$<br>$+L^3/3$ | $+\dfrac{5}{3}$ |
| $\Delta_{ab}$ | $\eta'''_a A_b$ | AB<br>BC<br>CD | $+L/2$<br>$+L$<br>— | $-L^2$<br>$-L^2/2$<br>$0$ | $-L^3/2$<br>$-L^3/2$<br>$0$ | $-1$ |
| $\Delta_{ba}$ | $\eta''_b A_a$ | AB<br>BC<br>CD | $-L$<br>$-L/2$<br>$0$ | $+L^2/2$<br>$+L^2$<br>$+L^2/2$ | $-L^3/2$<br>$-L^3/2$<br>$0$ | $-1$ |
| $\Delta_{bb}$ | $\eta'''_b A_b$ | AB<br>BC<br>CD | $-L$<br>$-2L/3$<br>$0$ | $-L^2$<br>$-L^2/2$<br>$0$ | $+L^3$<br>$+L^3/3$<br>$0$ | $+\dfrac{4}{3}$ |

It was to be expected that $\Delta_{ab}$ and $\Delta_{ba}$ would turn out to be equal.

Substituting these values in equations (i)

$$0 = \frac{3PL^3}{8} + X_a\frac{5L^3}{3} - X_bL^3 \quad\quad X_a = +0\cdot0853P$$

$$0 = \frac{-29PL^3}{48} - X_aL^3 + X_b\frac{4L^3}{3} \quad\quad X_b = +0\cdot517P$$

METHODS FOR THE ANALYSIS OF LINEAR STRUCTURES

The positive signs of $X_a$ and $X_b$ signify that the directions arbitrarily assigned to these redundant forces in FIGURE 4.16(b) are correct. It is now possible to compute the actual bending moments in the frame. For example

$$M_{BA} = M_o + X_a M_a + X_b M_b$$
$$= PL\{0 \cdot 5 + 0 \cdot 0853 - 0 \cdot 517\} = 0 \cdot 0683 PL$$

The complete bending moment diagram, with the moments calculated in this way, is shown in FIGURE 4.16(i).

This method has a good deal to commend it from a practical point of view as the arithmetic is simple and easily checked and signs are self-evident.

## 4.7. APPLICATIONS OF CASTIGLIANO'S THEOREM OF COMPATIBILITY

### 4.7.1. DERIVATION OF THE MAXWELL-MOHR EQUATIONS

In Chapter 2, Section 2.6.8, we saw that when the deflexions used in the compatibility equations for a hyperstatic structure were calculated by means of Castigliano's Theorem, Part II, a general relationship

$$\frac{\partial U}{\partial X_i} = \lambda_i \qquad \ldots (2.14)$$

was obtained to which we gave the title Castigliano's Theorem of Compatibility. Here $X_i$ stands for one of a series of redundant forces and moments while $\lambda_i$ is the corresponding lack of fit. When this is zero we have

$$\frac{\partial U}{\partial X_i} = 0 \qquad \ldots (2.14(a))$$

a series of equations which is commonly used for solving hyperstatic problems of many kinds. It is interesting to begin by showing that the equations derived in earlier Sections of this Chapter follow at once from appropriate application of equations 2.14 and 2.14(a).

If we consider a braced framework, such as that shown in FIGURE 4.9(a) and (b), having redundants $X_a$ and $X_b$ then, if there is no lack of fit,

$$\frac{\partial U}{\partial X_a} = \frac{\partial U}{\partial X_b} = 0 \qquad \ldots \text{(i)}$$

The strain energy due to direct stress only is given by equation 3.5

$$U = \sum \frac{F^2 L}{2AE} \qquad \ldots (3.5)$$

Hence
$$\frac{\partial U}{\partial X_a} = \frac{\partial}{\partial X_a} \sum \frac{F^2 L}{2AE} = \sum F \frac{\partial F}{\partial X_a} \rho = 0$$
$$\qquad \ldots \text{(ii)}$$
$$\frac{\partial U}{\partial X_b} = \frac{\partial}{\partial X_b} \sum \frac{F^2 L}{2AE} = \sum F \frac{\partial F}{\partial X_b} \rho = 0$$

## 4.7. APPLICATIONS OF CASTIGLIANO'S THEOREM

Using the notation of Section 4.5.1, equation (iii),

$$F = F_o + X_a F_a + X_b F_b$$

i.e. $\quad \dfrac{\partial F}{\partial X_a} = F_a \qquad \dfrac{\partial F}{\partial X_b} = F_b \qquad \ldots$ (iii)

Inserting these in (ii) gives

$$\left. \begin{array}{l} 0 = \Sigma F_o F_a \rho + X_a \Sigma F_a^2 \rho + X_b \Sigma F_a F_b \rho \\ 0 = \Sigma F_o F_a \rho + X_a \Sigma F_a F_b \rho + X_b \Sigma F_b^2 \rho \end{array} \right\} \quad \ldots \text{(iv)}$$

This argument can clearly be generalized to apply to a frame with $n$ redundants, giving equation 4.8

$$0 = \Sigma F_n (F_o + X_a F_a + X_b F_b + \ldots + X_n F_n) \rho \quad \ldots \text{(4.8)}$$

It follows that the application of Castigliano's Theorem of Compatibility to a braced structure leads to calculations which are identical with those arising from the use of the Maxwell-Mohr Equations.

### 4.7.2. Derivation of the $\Delta_{ik}$ Method

In the same way it will be found that the $\Delta_{ik}$ method emerges from the application of the Theorem of Compatibility to frames in which bending stress predominates. Here the strain energy of the members is given by the first term of equation 3.6,

$$U = \sum \int \frac{M^2 dx}{2EI}$$

Hence

$$\left. \begin{array}{l} \dfrac{\partial U}{\partial X_a} = \sum \int M \dfrac{\partial M}{\partial X_a} \dfrac{dx}{EI} = 0 \\ \dfrac{\partial U}{\partial X_b} = \sum \int M \dfrac{\partial M}{\partial X_b} \dfrac{dx}{EI} = 0 \end{array} \right\} \quad \ldots \text{(4.15)}$$

The bending moment in a member is

$$M = M_o + X_a M_a + X_b M_b$$

i.e. 
$$\left. \begin{array}{l} 0 = \sum \int \dfrac{M_o M_a dx}{EI} + X_a \sum \int \dfrac{M_a^2 dx}{EI} + X_b \sum \int \dfrac{M_a M_b dx}{EI} \\ 0 = \sum \int \dfrac{M_o M_b dx}{EI} + X_a \sum \int \dfrac{M_a M_b dx}{EI} + X_b \sum \int \dfrac{M_b^2 dx}{EI} \end{array} \right\} \quad \ldots \text{(4.16)}$$

Although these equations were not given explicitly in Section 5 they are in fact the same as those which are given when equation 4.11 is applied to the flexibility coefficient equations 4.7. The essential feature of the $\Delta_{ik}$ method, of course, is simply that the integrals in equations 4.16 are evaluated by a semi-graphical method.

It is usually simpler, however, when it is desired to use Castigliano's Theorem of Compatibility in rigid frame problems to proceed directly, in

METHODS FOR THE ANALYSIS OF LINEAR STRUCTURES

the following manner, rather than to start from equation 4.16. We now analyse the frame of FIGURE 4.16 in this way, using the symbols shown at $(j)$.

Member $DC$:  $\quad M = X_a x \qquad\qquad\qquad\qquad \dfrac{\partial M}{\partial X_a} = x \quad \dfrac{\partial M}{\partial X_b} = 0$

Member $CB$ ($C$ to $E$):

$$M = X_a L - X_b x \qquad\qquad \dfrac{\partial M}{\partial X_a} = L \quad \dfrac{\partial M}{\partial X_b} = -x$$

Member $CB$ ($E$ to $B$):

$$M = X_a L - X_b x + P\left(x - \dfrac{L}{2}\right) \qquad \dfrac{\partial M}{\partial X_a} = L \quad \dfrac{\partial M}{\partial X_b} = -x$$

Member $AB$:  $\quad M = X_a x - X_b L + \dfrac{PL}{2} \qquad \dfrac{\partial M}{\partial X_a} = x \quad \dfrac{\partial M}{\partial X_b} = -L$

Substituting in equations 4.15, we have

$$0 = \dfrac{\partial U}{\partial X_a} = \int_0^L X_a x^2 dx + \int_0^{L/2} (X_a L - X_b x) L\, dx$$

$$+ \int_{L/2}^L \left\{ X_a L - X_b x + P\left(x - \dfrac{L}{2}\right) \right\} L\, dx + \int_0^L \left( X_a x - X_b L + \dfrac{PL}{2} \right) x\, dx$$

$$= \dfrac{5}{3} X_a L^3 - X_b L^3 + \dfrac{3 PL^3}{8}$$

$$0 = \dfrac{\partial U}{\partial X_b} = \int_0^{L/2} (X_a L - X_b x)(-x)\, dx$$

$$+ \int_{L/2}^L \left\{ X_a L - X_b x + P\left(x - \dfrac{L}{2}\right) \right\}(-x)\, dx$$

$$+ \int_0^L \left( X_a x - X_b L + \dfrac{PL}{2} \right)(-L)\cdot dx = -X_a L^3 + \dfrac{4}{3} X_b L^3 - \dfrac{29}{48} PL^3$$

These equations are identical with those following TABLE 4.4.

### 4.7.3. EXAMPLES OF THE DIRECT USE OF CASTIGLIANO'S THEOREM OF COMPATIBILITY

In actual practice hyperstatic braced frames are most often dealt with by the direct application of the Maxwell-Mohr Equations, while rigid frames are analysed by one or other of the methods of Chapter 6. Strain energy methods show to best advantage when the frame to be analysed is subject to a combination of stresses, direct, shearing and bending. Two examples now illustrate the procedure.

(a) The roof principal illustrated in FIGURE 4.17(a) has a rigid joint at $B$; the tie $AC$ is connected to the rafters $AB$ and $BC$ by frictionless pins at

## 4.7. APPLICATIONS OF CASTIGLIANO'S THEOREM 4.7.3

$A$ and $C$. The tension $X$ in the tie can conveniently be regarded as the redundant quantity.

Since from symmetry the joint $B$ does not rotate each half of the frame is statically equivalent to the arrangement shown at (b). The expression for the strain energy of the whole structure is

$$U = \sum \int \frac{F^2 dx}{2AE} + \sum \int \frac{M^2 dx}{2EI} + \sum_\kappa \int \frac{Q^2 dx}{2AN} \qquad \ldots (3.7)$$

the first term referring to direct axial forces the second to bending moments and the third to shearing forces; so far as the tie $AC$ is concerned the first term only need be considered. The rafters $AB$ and $BC$, however, are definitely subjected to bending and shear as well as to direct force and a complete solution requires the inclusion of all three terms of equation 3.7.

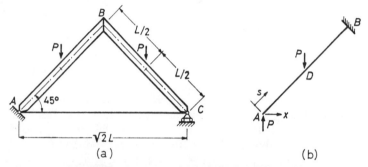

*Figure 4.17.* (a) This roof principal has a rigid joint at $B$ so that the tie $AC$, which is not subject to bending, is redundant; (b) Each half of the frame is statically equivalent to the arrangement shown here

It is often sufficiently accurate, at least for a preliminary analysis, to ignore direct and shearing effects in members whose primary function is to resist bending. The justification for this was given in Chapter 3, Section 3.4.3, when it was found that the bending deflexion far outweighed that due to direct or shearing stress. Making this simplification, and applying Castigliano's Theorem of Compatibility, we have:

$$0 = \frac{\partial U}{\partial X} = \frac{\partial}{\partial X} \left\{ \left[ \frac{F^2 L}{2AE} \right]_{AC} + \left[ \int \frac{M^2 ds}{2EI} \right]_{AB \text{ and } BC} \right\} \qquad \ldots \text{(i)}$$

In the rafters $AB$ and $BC$ we have

$$\left. \begin{array}{ll} M = (P - x)\,s/\sqrt{2} & \text{for } 0 < s < \dfrac{L}{2} \\[4pt] M = (P - x)\,s/\sqrt{2} - P/\sqrt{2}\,(s - L/2) & \text{for } \dfrac{L}{2} < s < L \end{array} \right\} \quad \therefore \frac{\partial M}{\partial x} = -\frac{s}{\sqrt{2}}$$

In the tie $AC$, which is $\sqrt{2}L$ in length, we have

$$F = X \qquad \frac{\partial F}{\partial X} = 1$$

## METHODS FOR THE ANALYSIS OF LINEAR STRUCTURES

Hence (i) becomes
$$0 = F \frac{\partial F}{\partial X} \frac{\sqrt{2}L}{AE} + 2 \int_0^L M \frac{\partial M}{\partial X} \frac{ds}{EI}$$

$$= \frac{\sqrt{2}\,XL}{AE} - 2\left\{ \int_0^{L/2} (P-X) \frac{s}{\sqrt{2}} \left(\frac{s}{\sqrt{2}}\right) \frac{ds}{EI} \right.$$

$$\left. + \int_{L/2}^L \left\{(P-X)\frac{s}{\sqrt{2}} - \frac{P}{\sqrt{2}}\left(s - \frac{L}{2}\right)\right\} \left(\frac{s}{\sqrt{2}}\right) \frac{ds}{EI} \right\}$$

$$= \frac{\sqrt{2}\,XL}{AE} - \frac{2L^3}{EI}\left\{\frac{11P}{96} - \frac{X}{6}\right\}$$

i.e.
$$X = \frac{\dfrac{11PL^2}{48I}}{\dfrac{\sqrt{2}}{A} + \dfrac{L^2}{3I}}$$

Here $A$ refers to the cross-sectional area of the tie while $I$ is the second moment of area of the rafters; $E$ has been assumed to be the same for tie and rafters.

This completes the main work of analysis. It should be remembered, however, that although the *deflexion* of the rafters under direct and shearing forces has been neglected in finding the tension in the tie the *stress* at any point is due to the combined effect of all the forces. Thus at $D$

$$\text{the bending moment is } (P-X)\,L/2\sqrt{2}$$
$$\text{the shearing force is } (P-X)/\sqrt{2}$$
$$\text{and the thrust is } \quad (P+X)/\sqrt{2}$$

(b) The beam shown in FIGURE 4.18(a) is strengthened by the strut $CD$ and the ties $AD$ and $BD$, which are adjusted to be taut before the uniformly distributed load of $w$ per foot is applied. The system is equivalent to that shown at (b). As in the previous example the direct compression of the beam and any deflexion caused by shear are neglected; in addition the strut $CD$ is supposed to be so stiff, compared with the rest of the structure, that its compression is negligible. The corresponding strain energy terms are consequently omitted from the analysis†.

Taking $X$, the thrust in the strut $CD$, as the redundant, Castigliano's Theorem of Compatibility gives

$$0 = \frac{\partial U}{\partial X} = \frac{\partial}{\partial X}\left\{\left[\frac{F^2 L}{2AE}\right]_{AD \text{ and } BD} + \left[\int \frac{M^2 dx}{2EI}\right]_{ACB}\right\}$$

$$= \left[F \frac{\partial F}{\partial X} \frac{L}{AE}\right]_{AD \text{ and } BD} + \left[\int M \frac{\partial M}{\partial X} \frac{dx}{EI}\right]_{ACB} \quad \ldots\ldots (i)$$

---

† This is only done here for simplicity. The reader can satisfy himself, by a numerical test, that a reasonable approximation is thus obtained.

## 4.7. APPLICATIONS OF CASTIGLIANO'S THEOREM

Dealing first with the ties $AD$ and $BD$, which are $\dfrac{L}{2\cos\theta}$ in length;

$$F = \frac{X}{2\sin\theta} \quad \therefore \quad \frac{\partial F}{\partial X} = \frac{1}{2\sin\theta}$$

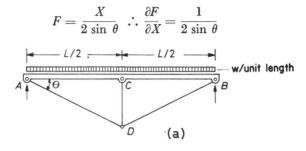

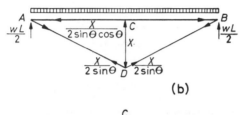

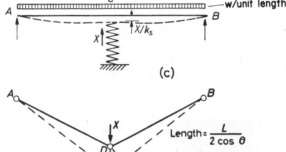

*Figure 4.18.* (a) The beam $ABC$ is strengthened by the bracing $ADB$ which is adjusted to be taut before the uniformly distributed load is applied; (b) The direct forces in the members are as shown; (c) The analogous problem in which the beam is carried on a spring of stiffness $k_s$. If the force in the spring is $X$ the deflexion at $C$ must be $X/k_s$; (d) The displacement $DD'$ of $D$ is $\dfrac{XL}{4AE\sin^2\theta\cos\theta}$ and hence the stiffness of the bracing, that is the force required to give unit displacement of $D$, is $\dfrac{4AE\sin^2\theta\cos\theta}{L}$

In the beam $ACB$, the two halves $AC$ and $CB$ are identical, and we have

$$M = \left(\frac{wL-X}{2}\right)x - \frac{wx^2}{2} \qquad \frac{\partial M}{\partial X} = -\frac{x}{2}$$

where $x$ is measured inwards from $A$ and $B$.

181

Hence (i) becomes

$$0 = \frac{X}{2\sin\theta} \frac{1}{2\sin\theta} \frac{L}{2\cos\theta.AE} \times 2 + \frac{2}{EI}\int_0^{L/2} \left\{\left(\frac{wL-X}{2}\right)x - \frac{wx^2}{2}\right\}\left(\frac{-x}{2}\right)dx$$

$$= \frac{XL}{4AE\sin^2\theta\cos\theta} + \frac{L^3}{384EI}(8X - 5wL)$$

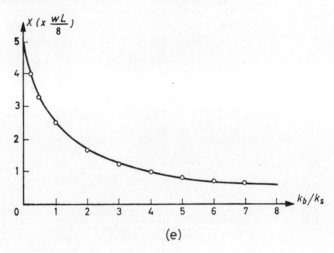

(e)

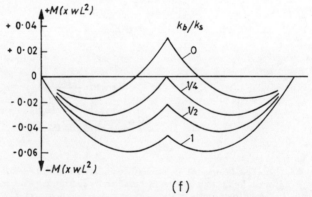

(f)

*Figure 4.18 (contd.).* (e) The load in the strut as a function of the ratio of the stiffness of the beam to that of the bracing; (f) The bending moments in the beam for different values of the ratio $k_b/k_s$

Hence

$$X = \frac{5wL^4/384EI}{L^3/48EI + L/4AE\sin^2\theta\cos\theta} = \frac{5wL/8}{1 + \dfrac{48EI}{L^3}\bigg/\dfrac{4AE\sin^2\theta\cos\theta}{L}} \quad \ldots \text{(ii)}$$

One difficulty about this method is that, being an essentially algebraic procedure, it conceals the underlying physical principles. We digress for a moment, therefore, to consider the analogous problem shown in FIGURE 4.18(c) which we now deal with by the method of Chapter 1, Section 1.3. The spring, which requires a force $k_s$ to compress it unit amount, is initially unstressed; when the uniformly distributed load of $w$ per unit length is applied a force $X$ is induced in the spring whose compression is evidently then $X/k_s$. The beam will finally take up the shape indicated by the dotted line.

Now the central deflexion of the simply-supported beam $AB$ under $w$ is

$$5wL/384EI$$

while its central deflexion under a central load $X$ is

$$XL^3/48EI$$

The equation of compatibility, which expresses the known information about the final nett deflexion of $C$ is

$$\frac{5wL^4}{384EI} - \frac{XL^3}{48EI} = \frac{X}{k_s}$$

i.e.
$$X = \frac{5wL/8}{1 + \dfrac{48EI}{L^3}/k_s} = \frac{5wL/8}{1 + k_b/k_s} \qquad \ldots \text{(iii)}$$

Here $k_b = \dfrac{48EI}{L^3}$ is the central force required to give unit central deflexion to the beam $AB$ and it can therefore be regarded as the stiffness of the beam.

Comparison with (ii) shows an immediate similarity; indeed reference to FIGURE 4.18(d) shows that the ties require a force $\dfrac{4AE \sin^2 \theta \cos \theta}{L}$ to depress them unit amount, so that this expression corresponds to the stiffness of the spring. This example has been worked in full in order to emphasize the underlying physical significance of Castigliano's Theorem of Compatibility†. The form of equation (iii) shows how the load in the strut, and hence the bending moments in the beam, depend on the relative stiffnesses of beam and bracing system. It is, indeed, characteristic of all hyperstatic systems that the partition of load between the component parts depends on this relative stiffness. FIGURE 4.18(e) shows how the load in the strut varies with the ratio $k_b/k_s$; if the bracing is very stiff compared with the beam, so that the ratio is small, its effect approximates to that of an immovable support whose reaction is $5wL/8$. If the beam stiffness is increased the force in the strut is correspondingly reduced. The bending moment diagrams for several values of $k_b/k_s$ are shown in FIGURE 4.18(f) from which it can be seen that the most favourable value of $X$ occurs when the maximum positive and negative values of the bending moment are numerically equal. It seems that this will occur when $k_b/k_s$ is slightly greater than zero, so that the most beneficial effect is felt when the bracing is very stiff compared with the beam.

---

† It may perhaps be thought that the second, more direct, method is simpler and less liable to error than the first. The point is, of course, that the Castigliano method shows to best advantage in more complex problems than this one.

If the load always acts downwards, as on a floor, it may be advantageous to prestress the structure by making CD a small amount too long so that it is forced into initial compression when the structure is first assembled. We can investigate how this is to be done as follows: we consider the case when $k_b/k_s = 1/2$. We notice first that the maximum negative moment will occur when $x$ is about $3L/16$; equating the numerical values of the bending moment at this point and at the centre, we have

$$\left\{\frac{wL}{2} \cdot \frac{3L}{16} - \frac{w}{2}\left(\frac{3L}{16}\right)^2\right\} - \frac{3}{8} \cdot \frac{XL}{4} = \frac{XL}{4} - \frac{wL^2}{8}$$

i.e. $\qquad\qquad\qquad X = 0\cdot585wL \qquad\qquad \ldots.\text{(iv)}$

Now Castigliano's Theorem of Compatibility, when there is an initial lack of fit, is

$$\lambda = \frac{\partial U}{\partial X} \qquad\qquad \ldots.\text{(2.14)}$$

so that (iii) becomes $X = \dfrac{5wL/8 + \lambda 48EI/L^3}{1 + \frac{1}{2}} \qquad\qquad \ldots.\text{(v)}$

Equating (iv) and (v) $\qquad \lambda = 0\cdot255\,\dfrac{wL^4}{48EI}$

It is obviously rather difficult to ensure that the structure is fabricated and assembled with the necessary accuracy to give the desired value of $X$. An alternative method is to introduce some device for altering the length of CD, or of the ties AD and BD, so that the value of $X$, as measured by a tensometer, can be suitably adjusted.

## 4.8. DEFLEXION CALCULATIONS FOR HYPERSTATIC STRUCTURES

Following the determination of the redundants and the subsequent evaluation of the forces and bending moments in a hyperstatic structure it may be necessary to compute its deflexion at one or more points. If this is to be done by means of equation 3.9, or by the use of Castigliano's Theorem, Part II, it will be necessary to find the stresses in the structure when unit load acts at the point whose deflexion is required. It might perhaps be thought that this would involve new calculations to give the values of the redundants under these unit loads, but, as already indicated in Chapter 2, Section 2.6.2.5, Example (a), and Section 2.6.5.2, this is not necessary. In calculating the values of the redundants the conditions of compatibility, in one form or another, have been used to relate the deflexion of the primary structure to the distortion (zero or otherwise) of the redundant members or restraints. Compatibility is, in fact, assured when the deflected shape of the primary structure is the same as that of the complete structure. We can take advantage of this to find deflexions by making our calculations for the primary rather than for the complete structure.

## 4.8. DEFLEXION CALCULATIONS FOR HYPERSTATIC STRUCTURES

A simple example will perhaps make the process clear, and will lead on to a demonstration that the principle is sound. FIGURE 4.19(a) shows a simple overstiff frame; we choose $BC$ as the redundant member. In FIGURE 4.19(b) the primary structure is shown in its strained position with the

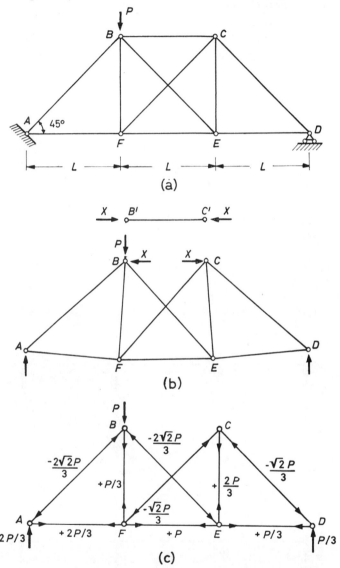

*Figure 4.19.* (a) This framework has one redundant member, taken here to be $BC$; (b) The redundant member alongside the distorted primary structure. Equilibrium and compatibility are both satisfied since the forces $X$ are equal and opposite while $B'C' = BC$. The member $BC$ can be restored to its proper position without further distortion of the primary structure which is thus the same shape as the complete structure; (c) Forces $F_o$

185

redundant member alongside. It is obvious that equilibrium is satisfied since the same forces $X$ act on the primary structure as on the redundant member, while compatibility requires that $BC$ shall be equal to $B'C'$. If both these conditions obtain then the redundant member can be restored to its proper position without further distortion of the structure. The primary structure, then, when stressed by the external load $P$ and by quasi-external loads $X$ representing the effect of the redundant member, occupies the same position as the complete loaded structure.

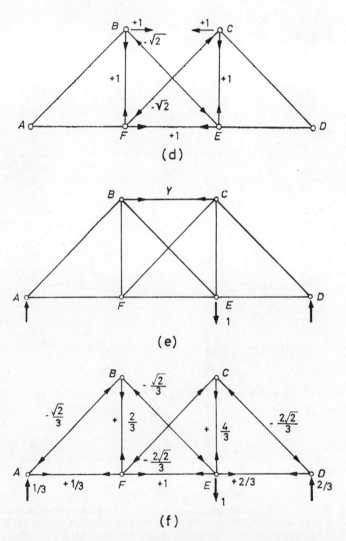

*Figure 4.19 (contd.).* (d) Forces $F_a$; (e) In order to find the deflexion of $E$ we apply unit load there; but this causes a force $Y$ in $BC$ which we do not know; (f) Unit load acting on the primary structure at $E$ produces forces $F_2$

## 4.8. DEFLEXION CALCULATIONS FOR HYPERSTATIC STRUCTURES

We begin the analysis by recalling the Maxwell-Mohr equation 4.8 which, for a single redundant $X$, is

$$0 = \Sigma F_o F_a \rho + X \Sigma F_a^2 \rho \qquad \ldots \text{(i)}$$

i.e.
$$X_a = -\frac{\Sigma F_o F_a \rho}{\Sigma F_a^2 \rho} \qquad \ldots \text{(ii)}$$

FIGURES 4.19(c) and (d) show the forces $F_o$ and $F_a$ respectively, while the necessary multiplications and additions are performed in TABLE 4.5. The cross-sectional area of the diagonal members $BE$ and $FC$ is $\dfrac{3A}{2}$ while that of all the other members is $A$.

TABLE 4.5

SOLUTION OF PROBLEM OF FIGURE 4.19

| Member | $\rho$ $\times L/AE$ | $F_o$ | $F_a$ | $F_o F_a \rho$ $\times L/AE$ | $F_a^2 \rho$ $\times L/AE$ | $F_2$ | $F_o F_2 \rho$ $\times L/AE$ | $F_a F_2 \rho$ $\times L/AE$ |
|---|---|---|---|---|---|---|---|---|
| AB | $\sqrt{2}$ | $-2\sqrt{2}P/3$ | 0 | 0 | 0 | $-\sqrt{2}/3$ | $+4\sqrt{2}P/9$ | 0 |
| BC | 1 | 0 | $+1$ | 0 | $+1$ | 0 | 0 | 0 |
| CD | $\sqrt{2}$ | $-\sqrt{2}P/3$ | 0 | 0 | 0 | $-2\sqrt{2}/3$ | $+4\sqrt{2}P/9$ | 0 |
| DE | 1 | $+P/3$ | 0 | 0 | 0 | $+2/3$ | $+2P/9$ | 0 |
| EF | 1 | $+P$ | $+1$ | $+P$ | $+1$ | $+1$ | $+P$ | $+1$ |
| AF | 1 | $+2P/3$ | 0 | 0 | 0 | $+1/3$ | $+2P/9$ | 0 |
| BF | 1 | $+P/3$ | $+1$ | $+P/3$ | $+1$ | $+2/3$ | $+2P/9$ | $+2/3$ |
| CE | 1 | $+2P/3$ | $+1$ | $+2P/3$ | $+1$ | $+4/3$ | $+8P/9$ | $+4/3$ |
| BE | $2\sqrt{2}/3$ | $-2\sqrt{2}P/3$ | $-\sqrt{2}$ | $+8\sqrt{2}P/9$ | $+4\sqrt{2}/3$ | $-\sqrt{2}/3$ | $+8\sqrt{2}P/27$ | $+4\sqrt{2}/9$ |
| FC | $2\sqrt{2}/3$ | $-\sqrt{2}P/3$ | $-\sqrt{2}$ | $+4\sqrt{2}P/9$ | $+4\sqrt{2}/3$ | $-2\sqrt{2}/3$ | $+8\sqrt{2}P/27$ | $+8\sqrt{2}/9$ |
| $\Sigma \times \dfrac{L}{AE}$ | | | | $P\left(2+\dfrac{4\sqrt{2}}{3}\right)$ | $\left(4+\dfrac{8\sqrt{2}}{3}\right)$ | | $P\left(\dfrac{23}{9}+\dfrac{40\sqrt{2}}{27}\right)$ | $\left(3+\dfrac{4\sqrt{2}}{3}\right)$ |

From the summations in TABLE 4.5 we have

$$X = -P\frac{2 + 4\sqrt{2}/3}{4 + 8\sqrt{2}/3} = -\frac{P}{2} \quad \text{(i.e. compression)}$$

Suppose now that the vertical deflexion of $E$ is required. This is given by equation 3.13.

$$\Delta = \Sigma F' F \rho \qquad \ldots \text{(iii)}$$

where $F'$ is the force produced by unit vertical load at $E$, as at (e). As we do not know the force $Y$ in $BC$ under this loading we proceed to expand (iii) as follows:

$$\Delta_E = \Sigma F' F \rho$$
$$= \Sigma(F_2 + Y F_a)(F_o + X F_a)\rho$$

where $F_2 =$ Force in member of primary structure under unit vertical load at $E$

i.e.
$$\Delta_E = \Sigma F_2(F_o + X F_a)\rho + Y[\Sigma F_o F_a \rho + X \Sigma F_a^2 \rho] \qquad \ldots \text{(iv)}$$

METHODS FOR THE ANALYSIS OF LINEAR STRUCTURES

It will be noticed that the terms in square brackets, being identical with the right-hand side of (i), are zero. Hence

$$\Delta_E = \Sigma F_o F_2 \rho + X \Sigma F_a F_2 \rho \qquad \ldots \text{(v)}$$

The forces $F_2$ are shown in FIGURE 4.19(f) and are tabulated in TABLE 4.5 from which we eventually obtain

$$\Delta_E = \frac{PL}{AE}\left(\frac{23}{9} + \frac{40\sqrt{2}}{27}\right) - \frac{PL}{2AE}\left(3 + \frac{4\sqrt{2}}{3}\right)$$

$$= 2.22 \frac{PL}{AE}$$

The first part of expression (iv) is exactly the same as we should have obtained if we had set out to compute the deflexion of the primary structure loaded, as at (b), with the actual external loads $P$ and quasi-external forces $X$ replacing the redundants. A similar demonstration can be given when bending and shearing effects are involved as well as direct stress and we therefore conclude that this procedure is valid in all cases.

## REFERENCES

[1] CHARLTON, T. M. 'Statically Indeterminate Frames: the two basic approaches to analysis'. *Engineering.* Dec. 28, 1956. p. 822
[2] SOUTHWELL, R. V. *Relaxation Methods in Engineering Science.* Oxford. 1940
[3] SHAW, F. S. *An Introduction to Relaxation Methods.* Dover Publications Inc. 1953
[4] ALLEN, D. N. de G. *Relaxation Methods.* McGraw-Hill. 1954
[5] NILES, A. S. 'Clerk Maxwell and the Theory of Indeterminate Structures'. *Engineering.* September 1, 1950
[6] MAXWELL, J. C. 'On the Calculation of the Equilibrium and Stiffness of Frames'. The Scientific Papers of James Clerk Maxwell. London. 1890. Vol. L. pp. 598–604
[7] MOHR, O. *Abhandlung aus dem Gebiete der Technischen Mechanik.* 1906 and later editions
[8] MÜLLER-BRESLAU, H. *Die neueren Methoden der Festigkeitslehre.* 1886 and later editions
[9] SOUTHWELL, R. V. *Theory of Elasticity.* Oxford. 1936
[10] WILSON, G. 'On a method of determining the reactions at the points of support of continuous beams'. *Proc. Roy. Soc.* Vol. 62. 1897–8
[11] SOUTHWELL, R. V. 'Current trends in Structural Research'. *Research.* Engineering Structures Supplement. Butterworth. 1949. p. 6
[12] GOLDSTEIN, A. *Symposium on Concrete Shell Roof Construction.* Cement and Concrete Association. 1954. p. 97
[13] MAGNEL, G. 'Prestressed Steel Structures'. *Structural Engineer.* 1950. p. 285
[14] SEFTON JENKINS. 'Prestressed Steel Lattice Girders'. *Structural Engineer.* 1954. p. 43
[15] SAMUELY, F. J. 'Structural Pre-stressing'. *Structural Engineer.* 1955. p. 41

# Chapter 5

# MOVING LOADS ON STRUCTURES

## 5.1. INTRODUCTION

In previous Chapters we have been concerned to find the effects of specified load systems on hyperstatic structures. The determination of the appropriate load system to be used in any given case is an important preliminary step in the design stage although the designer of building frames, being bound by the terms of Building Regulations, may have little choice in the matter. Many structures are liable to be subjected to different load systems which can act either independently or in combination: for example the important variable loads on a power station frame are likely to be horizontal wind and crane surge forces and vertical crane and coal bunker loads in addition to the permanent weight of the building and of the machinery supported by it. The effects of these will usually have to be analysed separately so that the critical combinations can be recognized by trial.

In bridge design, however, while it is possible to specify the maximum train or road traffic loads to be resisted the critical positions of these loads can often not be readily found by inspection. It is then necessary to employ special methods in order to find out how the stresses in the members vary as the loads traverse the structure. This is especially true when the stresses are liable to be of different sign for different positions of the same load.

The most powerful and versatile of these methods depend on the conception of 'influence lines', first introduced by Winkler in 1868, which are essentially graphs showing how the various stress functions—shear, bending moment, etc.—vary under the influence of a moving unit load. This idea was developed and generalized by various workers among whom Müller-Breslau was pre-eminent. In the following Section some of the important ideas about influence lines are reviewed in preparation for the Sections dealing with the construction of these diagrams for hyperstatic structures.

## 5.2. INFLUENCE LINES

### 5.2.1. Influence Lines for Beams

An influence line is a graph showing how the value of a particular stress-function varies with the position of a single unit moving load. Thus in Figure 5.1(a) a simply-supported beam $AB$ is shown with unit load at a movable point $m$ distant $x$ from $B$. The vertical reaction $R_A$ obtained by taking moments about $B$, is

$$R_A = 1\frac{x}{L} \qquad \ldots\ldots \text{(i)}$$

while $R_B$ is similarly

$$R_B = \frac{L-x}{L} \qquad \ldots\ldots \text{(ii)}$$

These two expressions are shown graphically at (b) and (c) which are the influence lines for $R_A$ and $R_B$ respectively.

The bending moment at $C$ is best found by considering unit load in two positions in turn; when it lies in $CB$, to the right of $C$,

$$M_C = R_A \cdot a = \frac{ax}{L} \qquad \ldots\ldots \text{(iii)}$$

When it lies in $AC$, to the left of $C$,

$$M_C = R_B \cdot b = \frac{b(L-x)}{L} \qquad \ldots\ldots \text{(iv)}$$

These two expressions are shown graphically at (d) and it will be noticed that the two straight lines represented by (iii) and (iv) cross at $C$ where they have the common value $\frac{ab}{L}$. Since expression (iii) is only valid in the region $CB$, while (iv) applies only to $AC$, the effective part of the diagram is that shown shaded, which is thus the required influence line for bending moment at $C$. The superficial resemblance to a bending moment diagram must be ignored; such a diagram shows the bending moments at every point on the beam for a fixed load while an influence line shows the value of the bending moment (say) at a single fixed point for the various positions of a *moving* unit load.

Finally FIGURE 5.1(e) shows the influence line for shearing force at $C$, obtained in much the same way as that just described.

It is not necessary to employ a formal convention of signs for influence lines but it will be found convenient to mark the diagram, after construction, with positive or negative signs according to the shearing force and bending moment convention.

When the influence lines have been constructed it is a simple matter to find the total effect of a train of wheel loads. Thus at (e) the ordinates corresponding to a certain position of a series of wheel loads are $\alpha$, $\beta$, $\gamma$, $\delta$ and $\epsilon$, so that the shearing force at C corresponding to this position of the loads is

$$4\alpha + 11\beta - 5(\gamma + \delta + \epsilon)$$

This operation is obviously assisted by drawing the diagrams to a natural vertical scale.

The designer is interested in the maximum numerical values of this expression and, although it is possible to devise rules to aid the identification of such critical positions, these can usually be found very quickly by inspection and trial. It is obvious, for example, that the 11 load will have a dominating effect, so that the maximum bending moment is likely to occur when this load is at $C$. Similarly the largest negative shear will occur when this load is just to the right of $C$.

5.2.2. INDIRECT LOADING THROUGH FLOOR SYSTEM

In many bridges the loads actually reach the main girders through a system of subsidiary longitudinal and cross-beams which support the rails or roadway. In order to see the effect of such an arrangement we consider

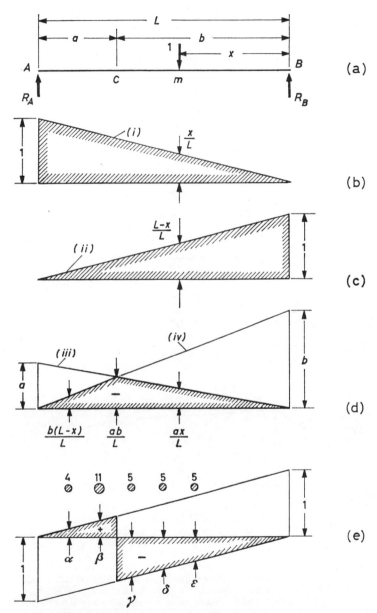

*Figure 5.1.* (a) Simply-supported beam carrying a moving unit point load; (b) Influence line for the reaction at $A$; (c) Influence line for the reaction at $B$; (d) Influence line for the bending moment at $C$; (e) Influence line for shear at $C$

191

again the beam of FIGURE 5.1 which is shown again at (a) in FIGURE 5.2 with a simple indirect floor system added. The influence line for bending moment at $C$, FIGURE 5.1(d), is reproduced again at (b) together with the modification which we now find to be necessary. It is clear that when the

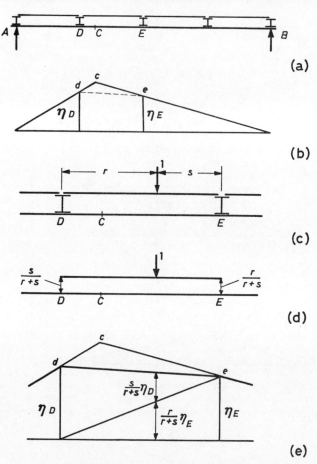

*Figure 5.2.* (a) The beam $ADCEB$ is loaded indirectly through a floor-system consisting of a series of simply-supported beams resting on cross-girders; (b) Influence line for bending-moment at $C$, showing the modification necessitated by the indirect method of loading; (c) Detail of one unit of the floor-system; (d) The load is actually transferred to the main beam through the cross-girders; (e) Detail of the influence line showing that $d$ and $e$, which correspond to the load points $D$ and $E$, must now be connected by a straight line

moving unit load is actually over one of the cross girders, say $D$ or $E$, the floor system has no effect and the original influence line applies. When the load is between the cross-girders, as at (c), the load is divided between them in proportion to its distance from them, as at (d), and the actual effect on the beam is given by the expression

$$\frac{s}{r+s}\eta_D + \frac{r}{r+s}\eta_E$$

This is the required influence line ordinate and we see from FIGURE 5.2(e) that it means that between $d$ and $e$ the influence line is a straight line. This is a general result which applies to trusses as well as to beams and which can be stated as follows:

When a structure is indirectly loaded, through a floor system consisting of a series of simply-supported beams, the influence line ordinates for the points of attachment of the floor system are computed as though the structure were directly loaded; the influence diagram is completed by connecting the points so obtained by straight lines.

### 5.2.3. INFLUENCE LINES FOR TRUSSES

The principles outlined in previous Sections can easily be extended to give influence lines for trusses by using the method of sections to isolate the individual members under examination. FIGURE 5.3(a) shows a rather general truss with an independent floor system carried by columns; the arrangement is analogous to that of FIGURE 5.2 and the influence diagrams will consist of a series of straight lines connecting the points corresponding to the loaded joints $B, C, D$, etc. Consideration of the equilibrium of the truss as a whole shows that the reactions at $A$ and $G$ will be the same as if these points were connected by a simply-supported beam so that the influence lines for $R_A$ and $R_G$ are as shown at (b) and (c). The procedure for dealing with individual members will be sufficiently illustrated by deriving influence lines for one web member, say $MC$, and one chord member, say $ML$; the same section, shown by the wavy line, serves to isolate both these members.

*Web Member MC*—FIGURES 5.3(d) and (e) show the truss separated into two portions by the section, the equilibrium of each section being preserved by the forces in the cut members, now regarded as quasi-external forces, which we seek to find; for the time being these are all assumed to be tensile forces. As in the previous Section it is best to eliminate the moving load from the equations by considering, in turn, that portion of the sectioned truss which does not carry the unit load.

When the load is to the right of the section, in the region $C'G'$, we consider the equilibrium of the left-hand portion of the truss shown at (d). The force in $MC$, which we seek, is conveniently found by taking moments about $P_1$, the point of intersection of the other two cut members, which is called the ' centre of moments ' for $MC$.

Thus $\qquad R_A.a_1 + F_{MC}.p_1 = 0$

$$F_{MC} = -R_A \frac{a_1}{p_1} \qquad \dots\text{(i)}$$

The negative sign signifies that $F_{BC}$ is, in fact, compressive when the load is to the right of the section.

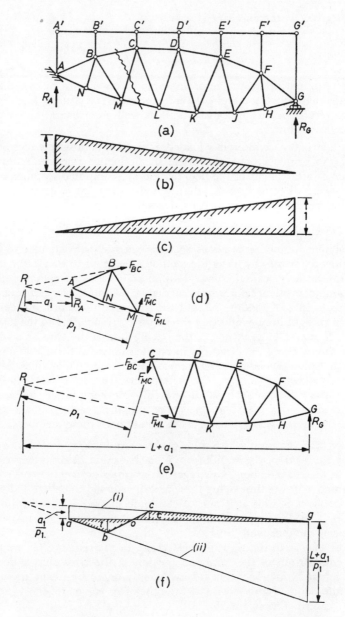

*Figure 5.3.* (a) The truss $A - G$ is loaded by means of columns connecting the top chord joints to the independent floor-system shown diagrammatically at $A' - G'$; (b) Influence line for the reaction at $A$; (c) Influence line for the reaction at $G$; (d) The truss is divided into two portions by a suitable section. When the load is to the right of the section the force in $MC$ is obtained by considering the equilibrium of the portion $BAM$; (e) When the load is to the left of the section $F_{MC}$ is obtained from the equilibrium equations for $CGL$; (f) Influence line for $F_{MC}$

## 5.2. INFLUENCE LINES  5.2.3

When the load is to the left of the section we consider the right-hand portion of the truss, shown at (e), and taking moments about $P_1$ as before we have

$$R_G(L + a_1) = F_{MC} \cdot p_1$$

$$F_{MC} = + R_G \frac{L + a_1}{p_1} \quad \ldots \text{(ii)}$$

The two expressions (i) and (ii) have been plotted, at (f), with due regard to sign and it can easily be shown that the two lines intersect under $P_1$. It now remains to complete the diagram by recalling the range over which (i) and (ii) are valid and by joining $b$ and $c$, the points corresponding to the joints on the loaded chord on either side of the section, by a straight line.

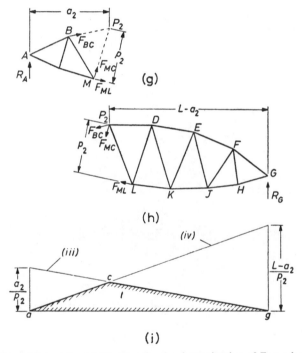

*Figure 5.3 (contd.)*; (g) Diagram for the determination of $F_{ML}$ when the load is to the right of the section; (h) Diagram for the determination of $F_{ML}$ when the load is to the left of the section; (i) Influence line for $F_{ML}$

This has been done in FIGURE 5.3(f). It will be noticed that the force in member $MC$ changes sign as the load passes from one side to the other of the point $O$, which is sometimes called the 'load-divide'. It is possible to devise constructions for giving this point directly[1].

*Chord Member ML*—The centre of moments for this member is evidently $C$ which has been marked $P_2$ on FIGURES 5.3(g) and (h). The dimensions $a_2$

195

and $p_2$ are the appropriate lever arms and by taking moments about $P_2$ we have

Load to right: $\quad F_{ML} = + R_A \dfrac{a_2}{p_2}$ .... (iii)

Load to left: $\quad F_{ML} = + R_A \dfrac{L - a_2}{p_2}$. .... (iv)

These expressions are shown on the diagram (i) from which it appears that the required influence line is $acg$; no modification is required here since $c$, the point of intersection of the lines (iii) and (iv), corresponds to a join on the loaded chord.

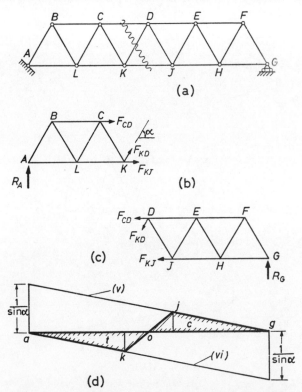

*Figure 5.4.* (a) This Warren truss has parallel chords so that the centre of moments for the inclined web members is at infinity; (b) Diagram for finding $F_{KD}$ when the load is to the right of the section; (c) Diagram for finding $F_{KD}$ when the load is to the left of the section; (d) Influence line for $F_{KD}$

The above procedure has to be modified slightly when the chords of the truss are parallel, as in FIGURE 5.4(a), for the centre of moments for the inclined web members is now at infinity: by resolving vertically for the left- and right-hand portions of the truss in turn we obtain

## 5.2. INFLUENCE LINES

Load to right (FIGURE 4.4(b)):  $F_{KD} = -\dfrac{R_A}{\sin \alpha}$  .... (v)

Load to left (FIGURE 4.4(c)):  $F_{KD} = +\dfrac{R_G}{\sin \alpha}$  .... (vi)

These expressions are shown at (d); the points $k$ and $j$, corresponding to joints $K$ and $J$ on the indirectly loaded chord $AG$, have been joined by a straight line to give the influence line for $F_{KD}$.

### 5.2.4. MÜLLER-BRESLAU'S PRINCIPLE

In subsequent Sections, where influence lines for hyperstatic structures are discussed, it will be demonstrated that as the expression for the stress function under consideration can always be obtained as a ratio of deflexions, the corresponding influence line is given by plotting the deflected shape of the loaded chord of the structure, under the appropriate loading, to the correct scale. Although this fact, first established by Müller-Breslau, is not particularly important in dealing with determinate structures it is of interest to notice that it can be used in such cases. This has actually already been demonstrated in Chapter 2, Section 2.6.2.1, although not in the present context of moving load analysis.

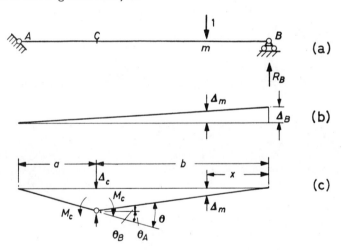

*Figure 5.5.* (a) The influence lines for this beam can be obtained by imposing suitable imaginary displacements; (b) That for $R_B$ is found by giving a displacement corresponding to $R_B$, that is a bodily rotation about $A$; (c) That for $M_C$ involves the insertion of an imaginary hinge at $C$ so that rotation corresponding to $M_C$ can occur

Suppose that it is required to find the influence line for the right-hand reaction of the beam $AB$ in FIGURE 5.5. In order to employ the principle of virtual work we imagine a small rotation about $A$ involving displacements $\Delta_m$ and $\Delta_B$ at the load point $m$ and the end $B$ respectively. Then, from virtual work, we have

$$R_B . \Delta_B - 1 . \Delta_m = 0$$

i.e.
$$R_B = \frac{\Delta_m}{\Delta_B} \qquad \ldots \text{(i)}$$

This expression implies that the deflexion diagram (b) is actually an influence line; it can be converted to a more useful form by replotting so that the end ordinate is 1 unit. It will then be seen to be identical with FIGURE 5.1(c).

The influence line for bending moment at $C$ is obtained by inserting an imaginary hinge at that point and permitting a small relative rotation $\theta$ of the two portions of the beam meeting these as shown at (c). This rotation can only be accommodated by vertical movement of the beam, the unit load, in particular, being displaced by an amount $\Delta_m$.

Applying the principle of virtual work,

$$-M_C \cdot \theta + 1 \cdot \Delta_m = 0 \qquad \ldots \text{(ii)}$$

Now if $\theta$ is small
$$\theta = \frac{\Delta_C}{a} + \frac{\Delta_C}{b}$$

While
$$\Delta_m = \Delta_C x/b$$

Substituting in (ii)
$$M_c = \frac{\Delta_m}{\theta} = \frac{\Delta_C x/b}{\Delta_C/a + \Delta_C/b} = \frac{ax}{a+b} \qquad \ldots \text{(iii)}$$

This expression is coincident with the expression (iii) of Section 5.2.1; it is also clear that it is represented by the displacement diagram FIGURE 5.5(c).

The influence line for shear at $C$ involves cutting the beam there and imposing a relative displacement of the cut ends in such a way that no relative rotation occurs. It will be found that this produces the diagram FIGURE 5.1(e).

FIGURE 3.13(d) shows how the truss $A—G$ is distorted by the change of length of a single member $CD$ and it will now be clear that the deflexion diagram for the lower chord is in fact an influence line for the force in member $CD$. When unit load is at $K$, for example, we have

$$1 \times \{\Delta_K(CD)\} = F_K\{bL\}$$

i.e.
$$F_K = \frac{\Delta_K(CD)}{bL}$$

and a similar equation applies when the unit load acts at the other lower chord joints; the fact that these joints move upwards, that is in the opposite direction to the unit load, implies that the sign of $F_K$ will be negative and $CD$ in compression, as expected.

The influence line for the force in any truss member, then, is given by the deflected shape of the loaded chord when the member is given a small increase in length. Cornish[2] gives several examples of the use of Williot-Mohr diagrams for this purpose. Alternatively, if the general shape of the

## 5.3. INFLUENCE LINES FOR HYPERSTATIC STRUCTURES

diagram can be visualized its scale can be found by applying unit load at one lower chord joint, and finding the resulting force in the member by means of the method of sections.

### 5.3. INFLUENCE LINES FOR HYPERSTATIC STRUCTURES

#### 5.3.1. Structures with a Single Redundant Reaction

##### 5.3.1.1. *Beams*

The two-span continuous beam $ADCB$ shown in Figure 5.6(a) has a single redundant reaction which can be taken to be $R_C$. Applying the flexibility coefficient equation 4.7 to this example, in order to express the fact that the deflexion of $C$ is zero, we have

$$0 = \Delta_{co} + R_C \Delta_{cc}$$

In this case the actual loading on the structure which produces the deflexion $\Delta_{co}$ is the unit load at the movable point $m$ so that the first term can be written $\Delta_{cm}$;

Hence
$$R_C = -\frac{\Delta_{cm}}{\Delta_{cc}} \qquad \ldots \text{(i)}$$

This expression can be made much more useful if Maxwell's Reciprocal Theorem is used to transpose the numerator.

We then have
$$R_C = -\frac{\Delta_{mc}}{\Delta_{cc}} \qquad \ldots \text{(5.1)}\dagger$$

This is another example of Müller-Breslau's Principle. The advantage of this expression lies in the fact that the deflexions appearing in both numerator and denominator are due to the same load, namely unit vertical load acting at $C$. Any of the methods given in Chapter 3, Section 3.7, can be used to find the shape of the deflected beam under unit load at $C$, which is what $\Delta_{mc}$ signifies; this is indicated at (b) while (c) shows the influence line for $R_C$ obtained by dividing each deflexion ordinate by $\Delta_{cc}$.

As soon as the reaction chosen as the redundant has been found the others follow at once from statics. Thus in Figure 5.6(a), taking moments about $B$, we have

$$R_A = \frac{x}{L} - R_c \frac{b}{L} \qquad \ldots \text{(ii)}$$

and
$$R_B = \frac{L-x}{L} - R_C \frac{a}{L} \qquad \ldots \text{(iii)}$$

These two diagrams have been plotted at (d) and (e) respectively. It appears that $R_A$ and $R_B$ are now given by the vertical intercept between the straight line and the curve; for actual use it may be better to redraw these diagrams on a horizontal base as at (f) and (g).

---

† It may seem odd that a ratio of deflexions should be equal to a *force* but $\Delta_{cc}$, of course, is really a deflexion per force.

## MOVING LOADS ON STRUCTURES

When the influence lines for the reactions have been obtained those for shear and bending moment can be derived in much the same way as in

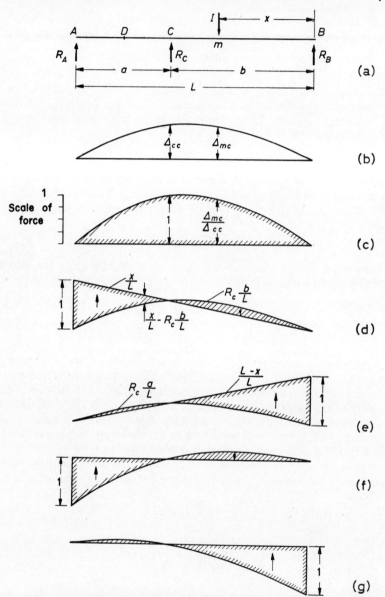

*Figure 5.6.* (a) Any one of the three reactions can be taken as the redundant: $R_C$ has been chosen here; (b) Deflexion diagram for the primary structure—the simply-supported beam $AB$—under unit load at $C$; (c) Influence line for $R_C$, obtained by dividing the ordinates of diagram (b) by $\Delta_{CC}$, plotted to a vertical scale of 1 unit; (d) Influence line for $R_A$; (e) Influence line for $R_B$; (f) Influence line for $R_A$ plotted on a horizontal base; (g) Influence line for $R_B$ plotted on a horizontal base

## 5.3. INFLUENCE LINES FOR HYPERSTATIC STRUCTURES  5.3.1.2

Section 5.2.1. Suppose that influence lines for shear and bending moment at the mid-point $D$ of $AC$ are required. Referring to FIGURE 5.6(h), which shows the relevant portion of the beam isolated by means of a section through $D$, we see that so long as the load is to the right of the section,

$$\text{Shear at } D = R_A \qquad \ldots\ldots \text{(iv)}$$

$$\text{B.M. at } D = R_A\, a/2 \qquad \ldots\ldots \text{(v)}$$

while when the load comes on to the portion $AD$, as at (i), we have

$$\text{Shear at } D = 1 - R_A \qquad \ldots\ldots \text{(vi)}$$

$$\text{B.M. at } D = 1 \times x' - R_A\, a/2 \qquad \ldots\ldots \text{(vii)}$$

These expressions are shown graphically at (j) and (k).

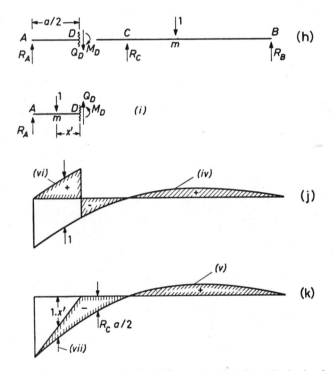

Figure 5.6 (cont.); (h) & (i) The beam sectioned at $D$ to obtain the shear and bending moment influence lines there; (j) Influence line for shear at $D$; (k) Influence line for bending moment at $D$

### 5.3.1.2. *Trusses with a Single Redundant Reaction*

(a) The arguments of the previous Section apply equally well to a truss continuous over two spans. Thus in FIGURE 5.7(a) the deck-type Pratt girder is loaded indirectly, by means of stringers and cross-girders, at the upper chord joints $B—K$ and is supported at the lower chord joints $A$, $R$ and $L$. Equations 5.1 and (i) to (iii) of the previous Section are perfectly relevant, and it is only in finding the shape of the deflected truss, $\Delta_{mr}$, that

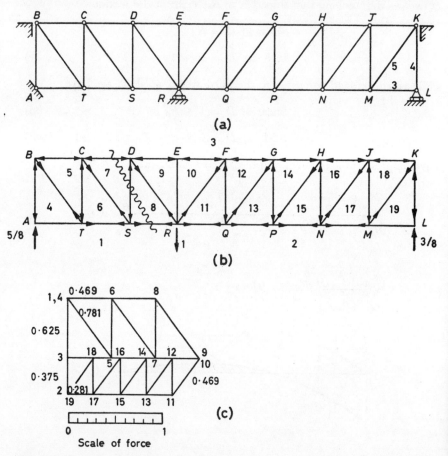

*Figure 5.7.* (a) The Pratt truss is supported at the lower chord joints $A$, $R$ and $L$. The loads are transmitted to the upper chord joints $B — K$ through a deck system; (b) Taking the truss, simply-supported at $A$ and $L$, as the primary structure, the influence line for $R_R$ is obtained by applying unit load at $R$ and finding the deflected shape of the loaded chord; (c) Force diagram for the primary structure under unit load at $R$. The nodes of the diagram have been *numbered* to identify the members according to Bow's notation

any difference between the two problems arises. Although it is possible to use the dummy unit load method of Chapter 3, Section 3.6.1, to find the deflexion, in turn, of each of the lower chord joints this is a rather laborious procedure and it will usually be found better to construct a Williot-Mohr deflexion diagram. This has been done, at (d), following the determination of the forces in and hence the changes of length of the members. The relative values of these changes of length are shown in TABLE 5.1; the lengths of the members have been taken as 3, 4 and 5 units and their cross-sectional areas as 2, 3 and 4 units. $E$ has been taken to be unity.

202

## 5.3. INFLUENCE LINES FOR HYPERSTATIC STRUCTURES  5.3.1.2

The vertical components of the deflexions of the top chord joints, $\Delta_{mr}$, are shown to the right of the Williot-Mohr diagram. The required ordinates of the influence line for $R_R$, $\Delta_{mr}/\Delta_{rr}$, have been evaluated and plotted at (e); it will be noticed that the end ordinates, instead of being zero as might

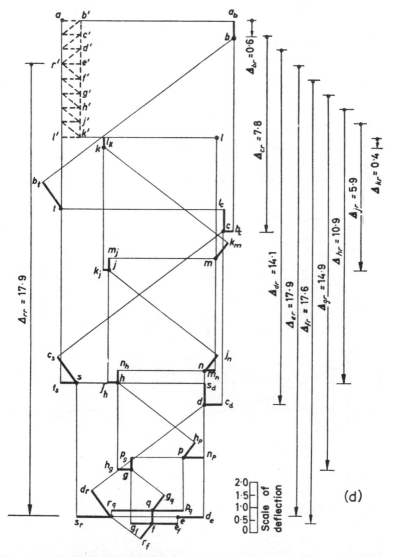

*Figure 5.7 (cont.)*; (d) Williot-Mohr deflexion diagram for the primary structure beginning at joint $Q$ and assuming $QF$ to remain vertical. The changes of length of the members, extracted from TABLE 5.1, are shown as thick lines. Note that an unstressed member is dealt with by treating its change of length as very small. The Mohr rotation diagram displays the effect of rotating the truss about $A$ until $L$ and $A$ are at the same level. The deflexions, $\Delta_{mn}$, of the loaded joints are the vertical components of the deflexions given by the distance from the point on the Mohr rotation diagram to the corresponding point on the Williot deflexion diagram

MOVING LOADS ON STRUCTURES

be expected, have small values corresponding to the compression of the end posts $AB$ and $KL$. Influence lines for the other reactions can be developed

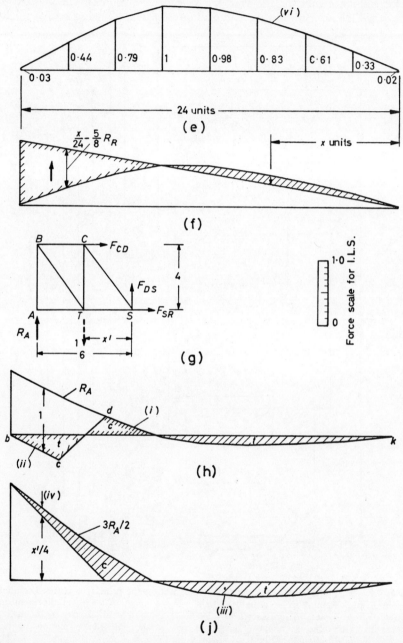

*Figure 5.7 (cont.)*; (e) Influence line for $R_R$. Each ordinate $= \Delta_{mr}/\Delta_{rr}$; (f) Influence line for $R_A$; (g) Part of the truss isolated, by the section shown at (b), to enable the forces $F_{CD}$, $F_{DS}$ and $F_{SR}$ to be evaluated; (h) Influence line for $F_{DS}$; (j) Influence line for $F_{CD}$

## 5.3. INFLUENCE LINES FOR HYPERSTATIC STRUCTURES 5.3.1.2

exactly as in FIGURES 5.6(d) and (e) and that for $R_A$ is shown at (f) in FIGURE 5.7. In order to find the influence lines for individual members appropriate sections, as shown at (g) for $CD$ and $DS$, are made in the truss. Then for $F_{DS}$, considering the equilibrium of the isolated portion of the truss,

Load to right of section: $F_{DS} = -R_A$ .... (i)

Load to left of section: $F_{DS} = 1 - R_A$ .... (ii)

These two expressions are shown at (h); the influence line for $R_A$ has been replotted, from (f), on a horizontal base and, with due regard to sign, is

TABLE 5.1
COMPUTATIONS FOR TRUSS, FIGURE 5.7(b)

Top chords  $L=3$  $A=4$  
End posts  $L=4$  $A=4$  Bottom chords  $L=3$  $A=2$

| Member | | $\dfrac{L}{A}$ | $F$ | $\dfrac{FL}{A}$ | Member | | $\dfrac{L}{A}$ | $F$ | $\dfrac{FL}{A}$ |
|---|---|---|---|---|---|---|---|---|---|
| 3– 4 | AB | 1   | −0·62 | −0·62 | 1– 4  | AT | 3/2 | 0     | 0     |
| 3– 5 | BC | 3/4 | −0·47 | −0·35 | 1– 6  | TS | ,,  | +0·47 | +0·70 |
| 3– 7 | CD | ,,  | −0·94 | −0·70 | 1– 8  | SR | ,,  | +0·94 | +1·41 |
| 3– 9 | DE | ,,  | −1·41 | −1·05 | 2–11  | RQ | ,,  | +1·12 | +1·68 |
| 3–10 | EF | ,,  | −1·41 | −1·05 | 2–13  | QP | ,,  | +0·84 | +1·26 |
| 3–12 | FG | ,,  | −1·12 | −0·84 | 2–15  | PN | ,,  | +0·56 | +0·84 |
| 3–14 | GH | ,,  | −0·84 | −0·63 | 2–17  | NM | ,,  | +0·28 | +0·42 |
| 3–16 | HJ | ,,  | −0·56 | −0·42 | 2–19  | ML | ,,  | 0     | 0     |
| 3–18 | JK | ,,  | −0·28 | −0·21 |       |    |     |       |       |
| 3–19 | KL | 1   | −0·37 | −0·37 |       |    |     |       |       |

Verticals  $L=4$  $A=3$     Diagonals  $L=5$  $A=3$

| Member | | $\dfrac{L}{A}$ | $F$ | $\dfrac{FL}{A}$ | Member | | $\dfrac{L}{A}$ | $F$ | $\dfrac{FL}{A}$ |
|---|---|---|---|---|---|---|---|---|---|
| 5– 6  | CT | 4/3 | −0·62 | −0·83 | 4– 5  | BT | 5/3 | +0·78 | +1·30 |
| 7– 8  | DS | ,,  | ,,    | ,,    | 6– 7  | CS | ,,  | ,,    | ,,    |
| 9–10  | ER | ,,  | 0     | 0     | 8– 9  | DR | ,,  | ,,    | ,,    |
| 11–12 | FQ | ,,  | −0·37 | −0·5  | 10–11 | FR | ,,  | +0·47 | +0·78 |
| 13–14 | GP | ,,  | ,,    | ,,    | 12–13 | GQ | ,,  | ,,    | ,,    |
| 15–16 | HN | ,,  | ,,    | ,,    | 14–15 | HP | ,,  | ,,    | ,,    |
| 17–18 | JM | ,,  | ,,    | ,,    | 16–17 | JN | ,,  | ,,    | ,,    |
|       |    |     |       |       | 18–19 | KM | ,,  | ,,    | ,,    |

equal to the above expression (i). The expression (ii) is obtained by plotting a line unit distance below the $R_A$ line. The transition from one line to another is accomplished by joining the points $c$ and $d$, corresponding to $C$ and $D$, by a straight line; the panel $CD$ is that panel of the loaded chord which is cut by the section.

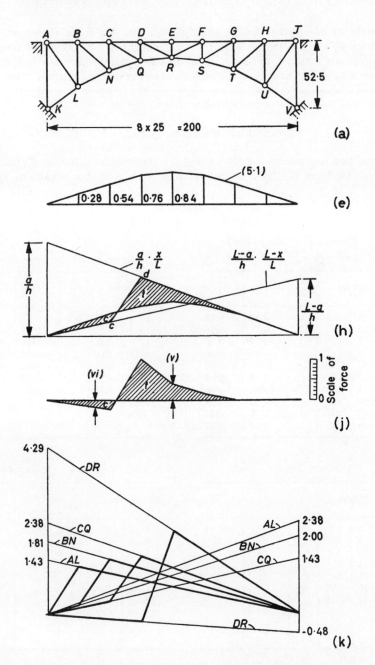

*Figure 5.8.* (a) All the members of this spandrel-braced two-hinged arch have the same value of $\rho$. If the horizontal component of the reactions is taken as the redundant the primary structure is obtained by replacing the hinge at $K$ by horizontal rollers; (e) Influence line for $H = -\Delta_{mk}/\Delta_{kk}$; (h) Construction of the influence line for $F_{CQ}$; (j) Influence line for $F_{CQ}$ replotted from (h) to the proper scale; (k) Construction of the influence lines for the family of diagonals

## 5.3. INFLUENCE LINES FOR HYPERSTATIC STRUCTURES 5.3.1.2

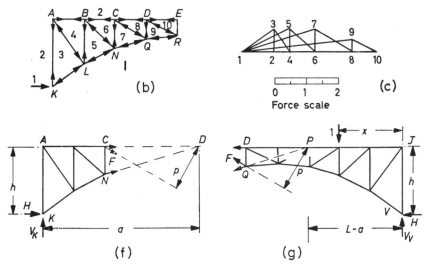

Figure 5.8 (cont.); (b) & (c) When unit horizontal load is applied at K the force diagram need be drawn for one half of the truss only; the other half is known from symmetry to be identical; (f) & (g) Diagrams for the calculation of the influence line for a diagonal member, CQ

As for the force $F_{CD}$ in the member $CD$, we see from FIGURE 5.7(g) that the centre of moments is $S$. Then

Load to right of section: $F_{CD} = -3R_A/2$ .... (iii)

Load to left of section: $F_{CD} = x'/4 - 3R_A/2$ .... (iv)

These expressions, which are shown in FIGURE 5.7(j), define the influence line for $F_{CD}$.

(b) FIGURE 5.8(a) shows a two-hinged arch of the spandrel-braced type. The horizontal component $H$ of the reactions at the hinges $K$ and $V$ can conveniently be taken as the single redundant, and we therefore have

$$0 = \Delta_{km} + H\Delta_{kk}$$

i.e. $$H = -\frac{\Delta_{km}}{\Delta_{kk}} = -\frac{\Delta_{mk}}{\Delta_{kk}}$$

These deflexions refer, of course, to the primary structure; this is obtained by replacing the pin at $K$ by rollers which permit movement along, but not perpendicular to, the line $KV$. $\Delta_{kk}$ is evidently the *horizontal* movement of $K$ under unit horizontal load while $\Delta_{mk}$ is the *vertical* movement of one of the top chord joints, corresponding to the actual loading that comes on to the arch.

The symmetry of the arch can be exploited, as at (b) and (c), by drawing a force diagram for one half only while if the deflexion diagram is started at $E$, assuming $EP$ to remain vertical as shown at (d), only one half need be drawn and no rotation diagram will be required. The relevant values of $\Delta_{mk}$ are shown on FIGURE 5.8(d) so that $H$ can be worked out for the successive positions of the moving unit load.

When the load is at $E$, for example,

$$H = \frac{\Delta_{ek}}{\Delta_{kk}} = \frac{84}{100} = 0 \cdot 84$$

The values of $H$ obtained in this way are plotted at (e) to give the required influence line for $H$.

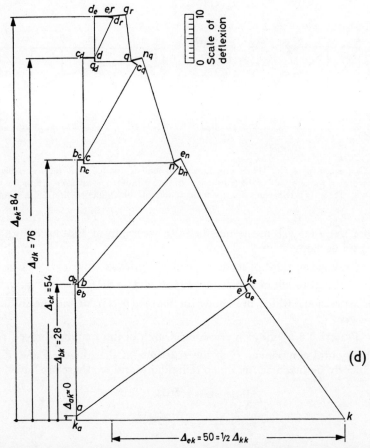

*Figure 5.8 (cont.)*; (d) The deflexion diagram is started at $E$ assuming $ER$ to be vertical. Since the assumption is correct the final position of $K$ is obtained directly without the need for a Mohr rotation diagram. The position of $V$ can be deduced and $\Delta_{kk}$ obtained from the deflexion diagram for one half of the arch. As $\rho$ is constant the changes in length of the members are directly proportional to the forces in them as scaled from diagram (c)

The construction of the influence lines for the individual members will be sufficiently illustrated by dealing with the diagonals. FIGURES 5.8(f) and (g) show the arch separated at a vertical section through one of the panels to give the force $F$ in one diagonal, $CQ$. When the load is to the right of the section we consider diagram (f) and, taking moments about $P$,

$$F.p = V_K.a - H.h$$

## 5.3. INFLUENCE LINES FOR HYPERSTATIC STRUCTURES 5.3.2

Now $V_K$ is the left-hand reaction of a simply-supported beam $KV$, carrying unit load a distance $x$ from $V$.

i.e. $$V_K = 1 \cdot x/L$$

Hence $$F = \frac{h}{p}\left\{\left[\frac{a}{h} \cdot \frac{x}{L}\right] - H\right\} \quad \ldots (v)$$

Similarly, considering the equilibrium of the part of the truss shown at (g) for the case when the load is to the left of the section,

$$F = \frac{h}{p}\left\{\left[\frac{L-a}{h} \cdot \frac{L-x}{L}\right] - H\right\} \quad \ldots (vi)$$

In both the expressions (vi) and (vii) the terms in square brackets represent straight lines, as at (h), intersecting under the centre of moments $P$. One of these lines is operative so long as the load is in $AC$ and the other when it is in $DJ$, the transition from one to another being accomplished by joining $cd$ by a straight line. The diagram is completed by superimposing the influence line for $H$ and the scale is determined from the multiplying factor $h/p$. The influence line is made more suitable for actual use by redrawing it, as at (j), on a horizontal base. It is advantageous, however, to treat all similar members as a family. The figures listed in TABLE 5.2 have been used to prepare the preliminary diagram for the diagonals which is shown at (k). This diagram would be completed by superimposing the influence line for $H$, after which the influence line for each diagonal would be constructed by scaling the ordinates and multiplying by the factor $h/p$.

TABLE 5.2
COMPUTATIONS FOR DIAGONAL MEMBERS. FIGURE 5.8(k)

| Member | $a$ | $L-a$ | $a/h$ | $\dfrac{L-a}{h}$ | $p$ | $h/p$ |
|---|---|---|---|---|---|---|
| AL | 75' | 125' | 1·43 | 2·38 | 61' | 0·86 |
| BM | 95' | 105' | 1·81 | 2·00 | 47' | 1·12 |
| CQ | 125' | 75' | 2·38 | 1·43 | 39' | 1·35 |
| DR | 225' | −25' | 4·29 | −0·48 | 66' | 0·79 |

### 5.3.2. TRUSSES WITH A SINGLE REDUNDANT MEMBER

FIGURE 5.9(a) shows a Pratt truss with a counterbraced central panel. The present method can be employed to find the influence line for the force $X$ in the redundant diagonal, say $LE$, as follows: The diagonal is supposed to be severed, as at (b), and to have unit load applied at its ends. The resulting forces in and changes of length of the members are tabulated in TABLE 5.3, and the Williot-Mohr diagram (c) is constructed from these figures.

The required force, $X$, in the redundant is given by equation 4.7:

$$0 = \Delta_{xm} + X\Delta_{xx} \quad \ldots (i)$$

Where the suffixes $x$ now refer to displacements or forces along $LE$ while $m$ refers, as usual, to a unit vertical load acting at a variable point on the loaded chord $AH$. Hence

$$X = -\frac{\Delta_{xm}}{\Delta_{xx}} = -\frac{\Delta_{mx}}{\Delta_{xx}} \qquad \ldots\ (\text{ii})$$

The various values of $\Delta_{mx}$ are shown alongside the Williot-Mohr diagram; it is to be remembered, however, that $\Delta_{xx}$ measures the movement at the

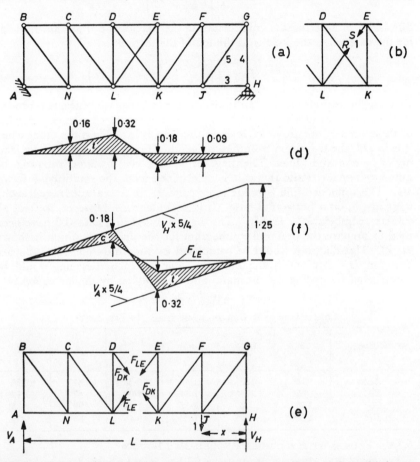

*Figure 5.9.* (a) This truss is statically determinate externally but the central panel is hyperstatic because of the counterbracing; (b) The member $LE$, chosen as the redundant, is severed and unit loads are applied at the cut ends $R$ and $S$ of the member. The resulting forces in and changes of length of the members are recorded in TABLE 5.3; (d) Influence line for $F_{LE}$; (e) A section through the central panel enables an expression for the force in $DK$ to be found; (f) Influence line for $F_{DK}$

'gap' in the severed member. It therefore comprises the component, along $L - E$, of the relative movement of $E$ and $L$ ($e''1 = 5{\cdot}56$ units on the Williot-Mohr diagram) together with the stretch of $LE$ under unit load (2 units, from TABLE 5.3). The required ordinates of the influence line for $LE$ are thus obtained by dividing each of the $\Delta_{mx}$ terms by $7{\cdot}56$. This has been done and the results plotted at (d).

## 5.3. INFLUENCE LINES FOR HYPERSTATIC STRUCTURES 5.3.2

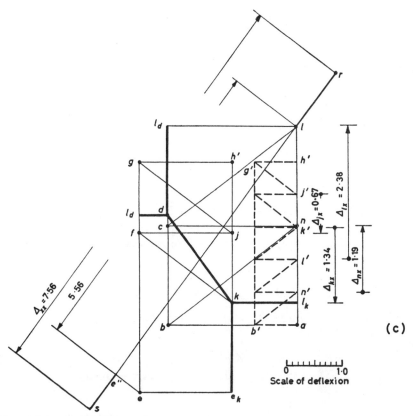

(c)

*Figure 5.9 (cont.)*; (c) Williot-Mohr diagram for the truss under unit load in *LE* starting from *D* with *DK* fixed in direction. The relative movement of *E* and *L* is *el* and this has the component $e''l$ along the line *L–E*. In order to obtain the relative movement of the cut ends, *R* and *S*, of *LE* the stretch of *LR* and *SE*, represented by *lr* and $se''$ (totalling 2 units) must be added. The final movement of *R* relative to *S* is *sr* which has a scaled length of 7·56 units. Note that the cut can be made at any point in *LE* without affecting this result

TABLE 5.3

COMPUTATIONS FOR TRUSS FIGURE 5.9(a)

| Member | L | A | F | $\dfrac{FL}{A}$ |
|---|---|---|---|---|
| LE | 5 | 2·5 | +1 | +2·0 |
| DE | 3 | 3·6 | −0·6 | −0·5 |
| DL | 4 | 2 | −0·8 | −1·6 |
| EK | 4 | 2 | −0·8 | −1·6 |
| LK | 3 | 1·5 | −0·6 | −1·2 |
| DK | 5 | 2·5 | +1 | +2·0 |

The influence lines for the remaining members can now be obtained. $F_{DK}$, for example, is given by cutting the truss through panel *LK*, as at (e), and considering the equilibrium, in the usual way, of the portions thus isolated. This yields the following expressions:

Load to right: equilibrium of $A$—$L$; $F_{DK} = \dfrac{5}{4} \cdot \dfrac{x}{L} - F_{LE}$ .... (iii)

Load to left:  equilibrium of $K$—$H$; $F_{DK} = \dfrac{5}{4} \cdot \dfrac{L-x}{L} - F_{LE}$ .... (iv)

These expressions are shown at (f) from which it appears, as was to be expected, that the influence lines for $F_{LE}$ and $F_{DK}$ are mirror images. Such checks on the work are always to be sought.

The influence lines for the remaining members of the counter-braced panel follow in a similar manner; those for the members of the other panels are obtained by the methods of Section 5.2.3 since the truss is statically determinate externally.

### 5.3.3. Structures with Two Redundant Reactions

Figure 5.10(a) shows a beam on four supports two of which, say $B$ and $C$, are redundant. Applying equation 4.7 to $B$ and $C$ in turn, we have

$$0 = \Delta_{bm} + R_B \Delta_{bb} + R_C \Delta_{bc}$$
$$0 = \Delta_{cm} + R_B \Delta_{cb} + R_C \Delta_{cc}$$  .... (i)

Solving these for $R_B$ and $R_C$, and replacing $\Delta_{bm}$ and $\Delta_{cm}$ by $\Delta_{mb}$ and $\Delta_{mc}$ respectively, we have

$$\left. \begin{array}{l} R_B = \Delta_{mb} \dfrac{\Delta_{cc}}{\Delta_{bc}^2 - \Delta_{bb}\Delta_{cc}} - \Delta_{mc} \dfrac{\Delta_{bc}}{\Delta_{bc}^2 - \Delta_{bb}\Delta_{cc}} \\[2mm] R_C = \Delta_{mc} \dfrac{\Delta_{bb}}{\Delta_{bc}^2 - \Delta_{bb}\Delta_{cc}} - \Delta_{mb} \dfrac{\Delta_{bc}}{\Delta_{bc}^2 - \Delta_{bb}\Delta_{cc}} \end{array} \right\} \quad \ldots\ldots (5.2)$$

In these equations the terms $\Delta_{mb}$ and $\Delta_{mc}$ represent the deflected shape of the primary structure (the simply-supported beam $AD$) under unit load at $B$ and $C$ respectively while $\Delta_{bc}$, $\Delta_{bb}$ and $\Delta_{cc}$ are the deflexions of specific points under these loads. The equations can therefore be written

$$R_B = k_1 \Delta_{mb} - k_2 \Delta_{mc}$$
$$R_C = k_3 \Delta_{mc} - k_2 \Delta_{mb}$$  .... (5.2(a))

in order to bring out the fact that the influence line for each reaction is the difference between the deflexion diagrams each multiplied by a certain factor. $\Delta_{mb}$ and $\Delta_{mc}$ have been calculated by Macaulay's method and are shown in Figure 5.10 at (b) and (c) respectively; the multiplying factors are

$$k_1 = \dfrac{85}{87^2 - 147 \times 85} = -0\cdot0173$$

$$k_2 = \dfrac{87}{87^2 - 147 \times 85} = -0\cdot0177$$

$$k_3 = \dfrac{147}{87^2 - 147 \times 85} = -0\cdot0299$$

## 5.3. INFLUENCE LINES FOR HYPERSTATIC STRUCTURES 5.3.3

The expressions 5.2 (or 5.2a) have been plotted at (d) and (e); the shaded areas represent $R_B$ and $R_C$ respectively and it will be noticed that they have the expected values of 1 and 0 at $B$ and $C$. This follows not only

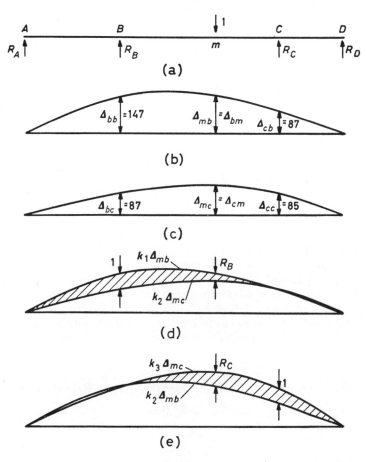

*Figure 5.10.* (a) Three-span beam continuous over four supports; (b) Deflexion diagram for the primary structure—the simply-supported beam $AB$ of span $L$—under unit load at $B$. $\left(\times \dfrac{L^3 \cdot 10^{-4}}{EI}\right)$; (c) Deflexion diagram for $AB$ under unit load at $C$. $\left(\times \dfrac{L^3 \cdot 10^{-4}}{EI}\right)$; (d) The influence line for $R_B$ is the difference between $k_1 \Delta_{mb}$ and $k_2 \Delta_{mc}$; (e) The influence line for $R_C$

from the basic properties of influence lines but directly from equations 5.2. Considering $R_B$, for example; when the moving point $m$ coincides with $B$, we have

$$R_B = - \Delta_{bb} \frac{\Delta_{cc}}{\Delta_{bc}^2 - \Delta_{bb} \Delta_{cc}} + \Delta_{bc} \frac{\Delta_{bc}}{\Delta_{bc}^2 - \Delta_{bb} \Delta_{cc}}$$

$$= 1$$

While when $m$ coincides with $C$,

$$R_B = - \Delta_{cb} \frac{\Delta_{cc}}{\Delta_{bc}^2 - \Delta_{bb} \Delta_{cc}} + \Delta_{cc} \frac{\Delta_{bc}}{\Delta_{bc}^2 - \Delta_{bb} \Delta_{cc}}$$

$$= 0$$

This concludes the introduction to the special features arising when there are two redundant reactions. The influence lines for the remaining reactions and for bending moment and shearing force follow in the same manner as in Section 5.3.1. The treatment for trusses continuous over three spans differs from the above only in the method of computing the deflected shape, when the dummy unit load or the Williot-Mohr methods are used rather than Macaulay's.

It will be seen that the influence line, being obtained from the difference between two deflexion diagrams, is likely to be less accurate than either. Fortunately the region of least accuracy occurs in the least important part of the diagram—in the neighbourhood of $C$, for example, in the case of $R_B$ —and in the important region the accuracy is likely to be quite acceptable. Nevertheless this consideration moved Müller-Breslau to search for a method of choosing the redundants so that they are independent of one another. His method is described in the following Section.

In principle the methods described above can be extended to beams and trusses continuous over many spans. It will be found in practice, however, that the graphical method for trusses is inadequate for more than three spans and the necessary deflexions would have to be calculated, laborious as this may well be. So far as beams are concerned special methods have been developed for continuous beams of many spans which are mentioned in Chapter 6.

### 5.3.4. Special Selection of Redundants: Müller-Breslau's Technique

Examination of equations 5.2 shows that if $R_B$ and $R_c$ are such that $\Delta_{bc}$ is zero the equations reduce to

$$R_B = - \frac{\Delta_{mb}}{\Delta_{bb}} \qquad \ldots \text{(5.2(b))}$$

$$R_C = - \frac{\Delta_{mc}}{\Delta_{cc}}$$

If these equations are to be used, rather than 5.2, the redundants must be so selected that neither produces a deflexion corresponding to the other; this can always be accomplished if the two act at the same joint. For example the truss shown in Figure 5.11(a) has two redundant reactions which we take to be components of the reaction at $A$. The primary structure is thus the truss supported by a pin at $E$ and by rollers at $H$.

We arbitrarily choose the vertical component of the reaction at $A$, $X_x$, to be one of these and (b) shows the primary structure with unit load acting in place of $X_x$. The Williot-Mohr diagram, at (c), has been constructed on the assumption that the members are identical; the Mohr rotation diagram

## 5.3. INFLUENCE LINES FOR HYPERSTATIC STRUCTURES 5.3.4

gives effect to the fact that the movement of $H$, which is on rollers, is horizontal while $E$ is not free to move. The movement of $A$ is from $a'$ to $a$ along the chain line so that if the direction of the second redundant $X_y$ is perpendicular to this line $\Delta_{yx}$ is zero.

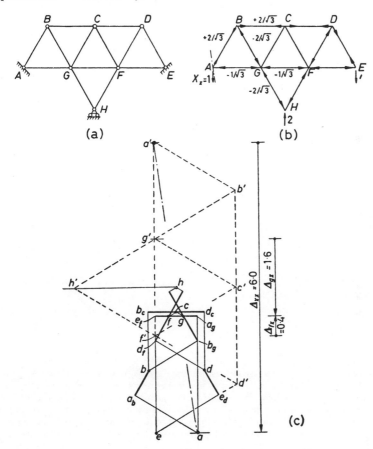

*Figure 5.11.* (a) The components of the reaction at $A$ are selected as the redundants for this structure, which has two redundant reactive restraints; (b) The primary structure, obtained by removing the pin at $A$, is here shown under unit vertical load at $A$ corresponding to one redundant $X_x$; (c) The corresponding Williot-Mohr diagram is started from $F$, assuming $FG$ to be horizontal. Since the movement of $H$ is horizontal, while $E$ is stationary, the Mohr diagram is so constructed that $h'h$ is horizontal. The resultant movement of $A$ is given by the chain line $a'a$

FIGURE 5.11(d) shows the primary structure with unit load acting in this direction at $A$ and the corresponding force diagram is shown at (e). The Williot diagram, at (f), requires no rotation diagram since, because of symmetry, the movement of $H$ turns out to be horizontal which accords with the support condition. It is to be noted that the movement of $A$ is also horizontal so that $\Delta_{xy}$ is zero; this result satisfactorily confirms the accuracy of the graphical work.

215

MOVING LOADS ON STRUCTURES

The appropriate deflexions can now be scaled from diagrams (c) and (f) to give the influence lines for $X_x$ and $X_y$; when the moving unit load is at $G$, for example,

$$X_x = -\frac{\Delta_{gx}}{\Delta_{xx}} = -\frac{1\cdot 6}{6\cdot 0} = -0\cdot 23$$

and

$$X_y = -\frac{\Delta_{gy}}{\Delta_{yy}} = -\frac{1\cdot 7}{10} = -0\cdot 17$$

The influence lines for the remaining reactions and for the forces in the individual members now follow in the same manner as before but it is inevitable that these involve both $X_x$ and $X_y$.

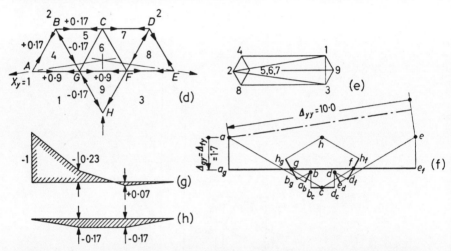

*Figure 5.11 (cont.)*; (d) The second redundant, $X_y$, is now applied to the primary structure in the direction perpendicular to the chain line $a'a$ on (c); (e) The corresponding force diagram; (f) The Williot deflexion diagram starting from $F$ with $FG$ assumed to be horizontal. No Mohr rotation diagram is required since the movement of $H$, given by $he$, is found to be horizontal as required. The movement of $A$, given by $ae$, is also horizontal and the component parallel to $X_y$ is measured; (g) Influence line for $X_x$; (h) Influence line for $X_y$

It is possible to devise a technique analogous to the above for use with trusses continuous over three spans, but a more interesting and important example is to be found in the case of the braced arch shown in FIGURE 5.12(a). Here there are three redundant restraints which could be taken to be the three shown. Application of the flexibility coefficient equations would give three equations which would have to be solved simultaneously for $X_1$, $X_2$ and $X_3$.

Müller-Breslau's device for obtaining independent equations involves finding three redundants such that each of them, when applied to the primary structure, produces zero displacement in the directions of the other two. This can hardly be achieved if all three redundants are linear forces but if one is a couple then the required solution can be found. At (b) a supplementary rigid arm is shown attached to the primary structure and the redundants, two forces and a couple, are applied at the end $O$ of this arm. If the required conditions are fulfilled then $O$ will simply rotate, without translation, when the couple $M$ is applied to the arch. Recalling that the forces induced in the members are independent of the point of application of $M$, we find these forces and construct the corresponding Williot-Mohr diagram to find the displacements of $A$ and $B$. The required

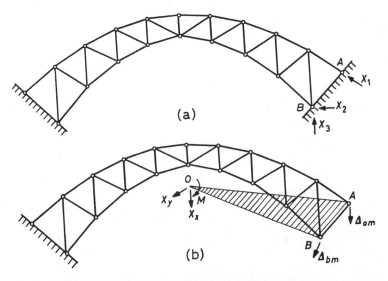

*Figure 5.12.* (a) This braced arch has three redundants which could be the restraints $X_1$ $X_2$ and $X_3$ shown; (b) In order to find independent redundants a rigid arm is imagined. The point $O$ is the instantaneous centre of rotation of this arm when a couple $M$ is applied to the primary structure. The redundant $X_x$ is then arbitrarily chosen while $X_y$ acts at right angles to the direction of movement of $O$ when $X_x$ is applied alone

position of $O$ is then the instantaneous centre of rotation of the rigid arm, obtained by constructing lines perpendicular to $\Delta_{am}$ and $\Delta_{bm}$ the directions of movement of $A$ and $B$ respectively. $X_x$ can then be arbitrarily chosen and $X_y$ is finally directed, as in the case of FIGURE 5.11 so that $\Delta_{xy} = \Delta_{yx} = 0$.

The importance of the above is that it is essentially the forerunner of the elastic centre method of analysing arched ribs and hence of the column analogy for rigidly-jointed frames generally.

## 5.4. STRUCTURAL MODELS

The fundamental principle, established in the previous sections, that influence lines are always obtainable from modifications and combinations of deflexion diagrams suggests the possibility of measuring the appropriate deflexions on a model of the structure as an alternative to calculating them. Such a diagram as FIGURE 5.6(b), for instance, could surely be obtained by imposing a suitable deflexion, $\Delta_{cc}$, on a model of the structure and then measuring the corresponding deflexions, $\Delta_{mc}$, at other points. This will only be a valid procedure if the model is elastically similar to its prototype and in Section 5.4.2.1 below we discuss how this similarity can be ensured.

## MOVING LOADS ON STRUCTURES

Several different procedures are now available for the detailed design and manipulation of models in order to obtain the desired information but the basic requirement of similarity must never be overlooked.

### 5.4.1. MODELS OF STATICALLY DETERMINATE STRUCTURES

The above remarks on elastic similarity apply essentially to hyperstatic structures. Similarity between model and prototype must be ensured in all cases but influence lines for determinate structures can be found from models which satisfy the requirements of kinematic similarity only.

Reference to FIGURE 5.5(c) suggests that a simple cardboard model, with a hinge in the form of a paper clip inserted at the point corresponding to $C$, can be used to represent the influence line for the bending moment at that point; the diagram for shear can be obtained by a relative displacement, with the portions of the beam kept parallel with each other, as in FIGURE 5.1(e).

Influence lines for the forces in truss members can be obtained in a similar manner. The principle has already been established in the concluding remarks of Section 5.2.4 and FIGURE 5.13(a) shows a simple model, made of cardboard strips and drawing pins, arranged to give the influence line for the force in member $CD$; the influence line for the force in $KD$ is shown at (b) In each case the member under examination has been replaced by a slightly longer member, thus distorting the truss into the required shape.

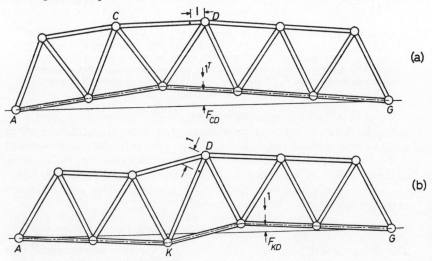

*Figure 5.13.* Model of a Warren girder made of cardboard strips and drawing pins. If a member, such as $CD$ or $KD$, is made 1in. too long the truss is distorted and the chord which carries the deck system, in this case $AG$, takes up the shape of the influence line for the member

These particular influence lines can be obtained so readily by other means that even such easily made models are hardly worth while except to demonstrate a principle. In more complex cases the analyst may find that a simple model will help him to visualize the general shape of the required influence line; it is then only necessary to calculate a single ordinate.

5.4.2. Models of Hyperstatic Structures

5.4.2.1. *Laws of Elastic Similarity*

The deflexion of a point $j$ of a structure was derived in Chapter 3, Section 3.4, in the following form:

$$\Delta_j = \sum \int \frac{M'M dx}{EI} + \sum \int \frac{F'F dx}{AE} + \sum \kappa \int \frac{Q'Q dx}{AN} \quad \ldots (3.9)$$

where the first term describes the effect of bending, the second that of direct stress and the third that of shearing stress.

(A fourth term, describing torsional effects, can also be included.)

Integration extends along the length of each member and the summations include all the members of the structure. The deflexion of a point $j$ can therefore be written

$$\Delta_{jF} = \beta_1 \frac{FL^3}{EI} + \beta_2 \frac{FL}{AE} + \beta_3 \frac{FL}{AN} \quad \ldots (i)$$

Where $F$ denotes a single load typifying the whole pattern of loads acting on the frame; in the same way $L$ is the length of a typical member and $EI$, $AE$ and $AN$ its flexural, extensional and shearing rigidities respectively, $\beta_1$, $\beta_2$ and $\beta_3$ are numerical coefficients that depend on the geometrical layout of the structure while $\beta_3$ also includes $\kappa$, the shape factor for shearing deflexion.

In order that the model shall correctly reproduce the behaviour of its prototype it is necessary that the deflexions at corresponding points shall bear a constant ratio to one another,

i.e. $$\frac{(\Delta_{jF})_p}{(\Delta_{jF})_m} = \text{constant} = \frac{\left(\beta_1 \frac{FL^3}{EI} + \beta_2 \frac{FL}{AE} + \beta_3 \frac{FL}{AN}\right)_p}{\left(\beta_1 \frac{FL^3}{EI} + \beta_2 \frac{FL}{AE} + \beta_3 \frac{FL}{AN}\right)_m} \quad \ldots (5.3)$$

5.4.2.1.1. *Similarity of extensional deformation only.*—In a pin-jointed structure the bending and shearing terms are zero and 5.3 reduces to

$$\frac{\Delta_p}{\Delta_m} = \frac{\left(\beta_2 \frac{FL}{AE}\right)_p}{\left(\beta_2 \frac{FL}{AE}\right)_m} = \text{constant} \quad \ldots (5.3(a))$$

For this relationship to be satisfied at all points of the structure the coefficients $\beta_2$ must have the same value in model and prototype and so their layouts must be geometrically similar. On the other hand it is permissible for the length scale $\frac{L_p}{L_m}$ to be different from the force scale $\frac{F_p}{F_m}$, the area scale $\frac{A_p}{A_m}$ and the modular ratio $\frac{E_p}{E_m}$.

We can thus write

$$s_\Delta = \frac{s_F s_L}{s_A s_E} \qquad \dots (5.4)$$

Where $s_\Delta$ and $s_F$, for example, denote the deflexion and force scales respectively. It is particularly to be noted that since $s_\Delta$ is constant the deflexions of two points $a$ and $b$ on the model will now be in the same ratio as those of the corresponding points $A$ and $B$ on the prototype
so that

$$\frac{\Delta_a}{\Delta_b} = \frac{\Delta_A}{\Delta_B} \qquad \dots (v)$$

This is the basis of the indirect model technique described below.

5.4.2.1.2. *Similarity of bending deformation only.*—Calculations of the action of rigidly jointed frames and arch ribs are generally based on the assumption that the deflexions due to bending far outweigh those due to axial and shearing stress. When a model is to be constructed on the same basis we must have

$$\frac{\Delta_p}{\Delta_m} = \frac{\left(\beta_1 \frac{FL^3}{EI}\right)_p}{\left(\beta_1 \frac{FL^3}{EI}\right)_m} = \text{constant} \qquad \dots (5.3(b))$$

This again requires similarity of geometrical layout in order that $\beta_1$ shall be the same for model and prototype but it is still possible to select the force, length and second moment of area scales and the modular ratio independently.

Thus $$s_\Delta = \frac{s_F s_L{}^3}{s_E s_I} \qquad \dots (5.5)$$

In numerical work it is usual to assume that the members coincide with their centre lines and are of negligible width and thickness. As the bending deflexion is proportional to the cube of the length of the members the fact that the effective lengths of the members differ from their assumed lengths, because of the finite size of the joints, can introduce appreciable errors. By means of a suitable choice of scales this difficulty can be avoided in a model analysis.

In the model, too, some deflexion will inevitably occur because of axial and shearing stresses and, although it will probably not be quite accurate, the model may well be more nearly correct in this respect† than the calculation which assumed that these stresses were negligible. In a recent paper, describing experiments conducted under the direction of the Author, Tse[3] makes some interesting comments on this point.

---

† The reader who is familiar with hydraulic model theory will recognise here a situation analogous to that where both viscous effects (Reynolds Number) and gravitational effects (Froude Number) occur ; in that case it is not usually possible to represent both effects to scale and the model is built so that the major effect is represented while the 'scale effects' due to the minor one are either simply accepted or estimated by some other means.

5.4.2.1.3. *Similarity of bending and extensional deformations.*—Equation 5.3 now becomes

$$\frac{\Delta_p}{\Delta_m} = \frac{\left(\beta_1 \frac{FL^3}{EI} + \beta_2 \frac{FL}{AE}\right)_p}{\left(\beta_1 \frac{FL^3}{EI} + \beta_2 \frac{FL}{AE}\right)_m} \quad \ldots (5.3(c))$$

and it is now necessary, in general, to make the cross-sections of the model members geometrically† similar to those of the prototype and to the same linear scale as the prototype.

If this is done we shall have

$$\frac{\Delta_p}{\Delta_m} = \frac{aL_p{}^2}{aL_m{}^2} = s_L^2 \; ; \; \frac{I_p}{I_m} = \frac{bL_p{}^4}{bL_m{}^4} = s_L^4$$

where $a$ and $b$ are numerical constants which depend on the shape and size of the cross-sections of the members. Then

$$\frac{\Delta_p}{\Delta_m} = \frac{F_p}{F_m} \cdot \frac{E_m}{E_p} \frac{\left(\beta_1 \frac{L_p^3}{I_p} + \beta_2 \frac{L_p}{A_p}\right)}{\left(\beta_1 \frac{L_m^3}{I_m} + \beta_2 \frac{L_m}{A_m}\right)}$$

$$s_\Delta = \frac{s_F}{s_E} \frac{\left(\frac{\beta_1}{bL_p} + \frac{\beta_2}{aL_p}\right)}{\left(\frac{\beta_1}{bL_m} + \frac{\beta_2}{aL_m}\right)} = \frac{s_F}{s_E \, s_L} \quad \ldots (5.6)$$

5.4.2.1.4. *Complete similarity.*—Finally we note that if shearing deformations are to be correctly reproduced equation 5.3 must be satisfied. This means that in addition to the requirements of the previous Section the ratio $E/N$ must be the same in both model and prototype: Poisson's ratio must therefore be the same in both.

The scales are again connected by equation 5.6.

5.4.2.2. *The indirect method for models of structures with a single redundant.*— The principle of the indirect method is to impose a displacement, corresponding to the redundant, on an elastically similar model of the prototype and to measure the resulting deflexion of the joints of the loaded chord. This is strictly analogous to the analytical or graphical procedure used in earlier sections of this chapter and yields an influence line for the redundant. The method is indirect in the sense that an influence line for a redundant force, external or internal as the case may be, is obtained without actually measuring any force at all.

---

† Pippard and Baker[4] show that by a suitable choice of scales it is sometimes possible to evade this restriction.

5.4.2.2.1. *Models of structures in which axial stress predominates.*—Suppose that a model is to be constructed of the truss of FIGURE 5.7(a), which has a single redundant reaction. The similarity relationship 5.3(a) of Section 5.4.2.1.1 indicates that the linear scale $L_p/L_m$ and the extensional rigidity scale $(AE)_p/(AE)_m$ can be chosen independently when axial stress predominates, so that the cross-section of the members need not be exactly reproduced. Indeed it is most convenient to use screwed rods for the model members and to simulate their rigidity by introducing springs of appropriate stiffness within the length of each member. Helical springs have been used for this purpose by Pippard[5] while Charlton[6] advocates leaf springs; both writers give descriptions of the mechanical design of joints and other details which have proved successful in practice.

It remains to devise a suitable means of imposing a known displacement corresponding to the redundant and of measuring the resulting deflexions of the joints of the loaded chord; there are obviously many ways in which this can be done.

When the influence line for the redundant has been obtained from the model those for the members of the truss can be deduced from statics or, if suitable provisions has been made in the design, by imposing further displacements on the model. This would involve the possibility of shortening or lengthening the members in turn, by known amounts, in a manner analogous to that of Section 5.4.1.

5.4.2.2.2. *Models of structures in which bending stress predominates.*—A model of a continuous beam, that shown in FIGURE 5.6(a) for example, can be made of any convenient size; the flexural rigidity $EI$ must everywhere bear a constant ratio, not necessarily the same as the linear scale, to that of the prototype. These considerations make it possible to construct the model of sheet material of uniform thickness and to adjust the depth of the members to give the correct second moment of area.

Since

$$\frac{E_m I_m}{E_p I_p} = \frac{E_m b_m d_m^3}{12 E_p I_p} = \text{constant}$$

we have

$$d_m \propto \sqrt[3]{I_p} \qquad \ldots (5.7)$$

In accordance with the above models have often been cut out of cardboard, sheet metal or, more frequently, of perspex or xylonite. The last-named plastic materials are very convenient but the fact that they 'creep' under stress must not be overlooked. Fortunately their creep properties are such as not to invalidate the indirect model technique for, if a constant displacement is imposed corresponding to a redundant, the resulting deflexions at other points also remain constant. Although the force required to hold the model in its displaced position will change with time the results are not affected in consequence; an important advantage of the indirect system is its independence of the actual magnitudes of the forces involved.

## 5.4. STRUCTURAL MODELS

When he originally described this technique, Beggs[7] insisted that the deflexions must be very small. He therefore devised a system of accurately made gauges for imposing suitable small displacements and a micrometer microscope was necessary for measuring the resulting deflexions. Although the non-linearity due to gross distortion was eliminated in this way the errors caused by temperature changes could be quite serious. Pippard and others have since shown that quite large displacements can be tolerated and it is now common to mount the models on a drawing board, preferably held vertical to eliminate friction, and to measure the displacements against squared paper or with a rule.

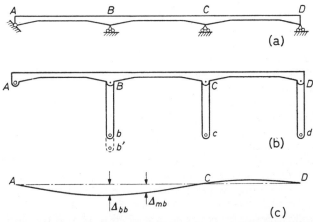

*Figure 5.14.* (a) Three-span continuous beam with variable second moment of area; (b) A model of the beam, cut out of sheet metal or perspex, with the depth everywhere proportional to the cube root of the second moment of area of the prototype. The links $Bb$, $Cc$ and $Dd$ are stiff and are hinged to the model at $B$, $C$ and $D$. Pins $A$, $b$, $c$ and $d$ are used to attach the model to the baseboard; (c) When the pin $b$ is moved to $b'$ the centre line of the model takes up the shape indicated by the full line and this is the influence line for the vertical reaction at $B$

The influence line for the redundant is then given by the shape of the deflected beam, the scale being determined by the displacement imposed at $B$. The influence lines for the bending moment and shear at points within the span can then be calculated or, if desired, obtained from the model by the technique given in Section 5.4.2.3 below.

5.4.2.2.3. *Models of structures in which complete similarity is important.*—The conditions of similarity now demand that the model be constructed of a material with the same value of Poisson's ratio as the prototype and that the flexural, extensional and shearing rigidities of model and prototype shall be in the same ratio. These requirements can hardly be satisfied except by reproducing the whole structure in miniature. Although it may be essential to do this, especially to investigate complex and important parts of aircraft frames, for example, the procedure is usually to measure the stresses directly rather than to attempt to apply the indirect method which is better suited to problems, like those described above, in which simplifying assumptions could be made.

**5.4.2.3.** *The Application of the Indirect Method to Structures having Several Redundants.*—In Section 5.3.3 we showed that influence lines for two or more redundant reactions could be obtained by imposing corresponding displacements, in turn, on the statically determinate primary structure and then combining the results. The problem was thus reduced to a series of deflexion calculations for a determinate structure. Although these deflexions can easily be measured on a model the process of combining them must then be carried out as a separate operation. This difficulty is avoided by working with a hyperstatic primary structure as described below.

**5.4.2.3.1.** *Hyperstatic primary structures.*—In Section 5.3.4 we showed that equations 5.2 were greatly simplified if the interaction term $\Delta_{bc}$ could be eliminated and we went on to explain how to select the redundants so that this came about. A model, however, has the advantage that an unwanted deflexion can be suppressed by mechanical means. FIGURE 5.14(a) shows a three-span continuous beam and we take $R_B$ and $R_C$ as the two redundant reactions. If a model is constructed with suitable linear and flexural rigidity scales and supported so that $A$, $C$ and $D$ are free to rotate, while $C$ and $D$ can accommodate small lateral movements, as sketched at (b), a displacement imposed at $B$ will produce the deflexion indicated at (c). It will now be seen that this is indeed the required influence line for $R_B$ for if the flexibility coefficient equation is applied to joint $B$ we have

$$0 = \Delta_{bm} + R_B \Delta_{bb}$$

i.e.
$$R_B = -\frac{\Delta_{bm}}{\Delta_{bb}} = -\frac{\Delta_{mb}}{\Delta_{bb}}$$

In this case the deflexions $\Delta_{mb} = \Delta_{bb}$ refer to the hyperstatic primary structure $ACD$. These deflexions could, of course, be calculated but the work would be rather tedious. On the other hand they can be measured on a model just as easily as if the primary structure were statically determinate. The number of redundant reactions that can be dealt with in this way is unlimited provided the model is so arranged that each support point can be released in turn and given a suitable displacement.

FIGURE 5.15(a) shows a structure with three redundants which can best be taken, for the present purpose, to be the horizontal and vertical forces and the restraining moment acting at $D$. The model must now be displaced in the direction of each redundant in turn, any displacement corresponding to the other redundants being prevented. A simple method of achieving this is described below; for the present we observe that the resulting distorted shapes of the structure will be as indicated at (b), (c) and (d). Referring first to diagram (b) it will be noticed that the influence ordinate $\Delta_{md}$ has been indicated at three points: two of these, on the beam $ABC$, give the effect on $V_D$ of unit *vertical* load while the third gives the effect of unit *horizontal* load acting on the column. Diagram (d), giving the influence

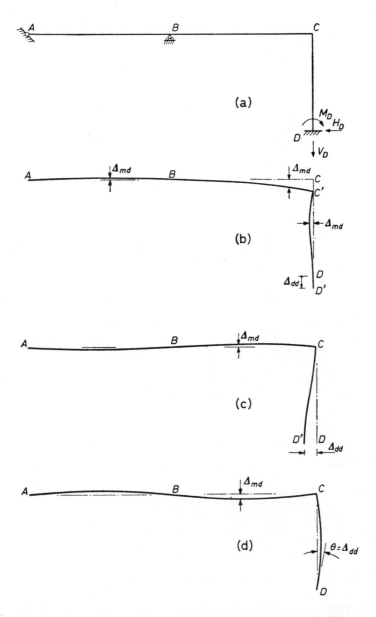

*Figure 5.15.* (a) This structure is supported by a hinge at $A$ and rollers at $B$ while $D$ is fixed; the joint $C$ is rigid; (b) In order to find the influence line for $V_D$ the model is displaced so that $D$ moves vertically to $D'$ with no accompanying rotation or lateral movement. The centre line then takes up the shape of the required influence line; (c) Lateral movement of $D$, without accompanying rotation or vertical movement, gives the influence line for $H_D$; (d) Rotational movement of $D$ gives the influence line for $M_D$

line for $M_D$, has been obtained by imposing a rotation at $D$; the flexibility coefficient equation still applies, however, and we have

$$M_D = -\frac{\Delta_{md}}{\Delta_{dd}} \text{ as before}$$

As $\Delta_{dd}$ is now an angle while $\Delta_{md}$ is a linear displacement the linear scale factor must be used to give the value of $M_D$ in the prototype: for if the prototype is elastically similar to the model its deflected shape will be

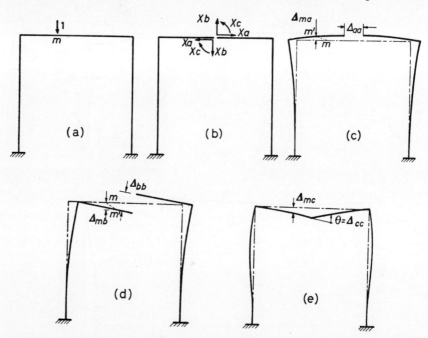

*Figure 5.16.* (a) This portal frame has three redundants; (b) The redundants are chosen as the axial and shearing forces and the bending moment at the centre of the beam; (c) Influence line for $X_a$ obtained by imposing a displacement $\Delta_{aa}$ on the model while keeping $\Delta_{bb}$ and $\Delta_{cc}$ zero; (d) Influence line for $X_b$ obtained by imposing a displaceemnt $\Delta_{bb}$ on the model while keeping $\Delta_{aa}$ and $\Delta_{cc}$ zero; (e) Influence line for $X_c$ obtained by imposing a rotational displacement $\theta = \Delta_{cc}$ on the model while keeping $\Delta_{aa}$ and $\Delta_{bb}$ zero

geometrically similar, and the angular deflexion $(\Delta_{dd})_p$ will be equal to that on the model while the linear deflexion $(\Delta_{md})_p$ will be $s_L$ times that on the model.

i.e. $$(M_D)_p = -\frac{(\Delta_{md})_m \times s_L}{\Delta_{dd}} \quad \ldots \text{(5.8)}$$

The above discussion, being based on the fundamental flexibility coefficient equation 4.7, is identical with that given in Chapter 4, Section 4.4.2 except that a hyperstatic, rather than a statically determinate, primary structure is used in the model method in order to avoid simultaneous equations. In just the same way as in Chapter 4, Section 4.4.3. we can use internal forces and moments as the redundants provided that the imposed displacements correspond to one redundant at a time. In FIGURE 16(b), for example, the axial and shearing force and the bending moment at the centre of the beam have been chosen as the three redundants. Each of the diagrams (c), (d) and (e) shows the deflected shape taken up by the portal when one of the displacements $\Delta_{aa}$, $\Delta_{bb}$ or $\Delta_{cc}$ is imposed and this is the required influence line for the corresponding redundant $X_a$, $X_b$ or $X_c$. FIGURE 5.16 (d), in particular, should be compared with FIGURE 4.8(d) when the essential difference between the model and the analytical method described in Chapter 4 will be apparent.

5.4.2.3.2. *Derivation from Betti's Reciprocal Theorem.*—While the above explanation of model theory fits well into the logical pattern of this book by linking the analytical and experimental methods it may leave some questions unanswered in the reader's mind. The part played by the other redundants, for instance, while a displacement corresponding to the one under consideration is imposed, may not immediately suggest itself. An alternative approach, using Betti's Theorem†, may therefore be of interest.

A prototype portal frame is shown at (a), in FIGURE 5.17, under the action of unit load at a point $m$; the three redundants $X_a$, $X_b$ and $X_c$ have

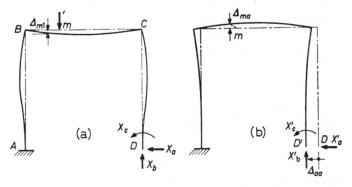

Figure 5.17. (a) The forces of Group 1 acting on this portal frame are the three redundants $X_a$ $X_b$ and $X_c$ together with a unit load; (b) The forces of Group 2 are such that a deflexion $\Delta_{aa}$ is imposed at $D$ while the unit load previously acting is removed

been adjusted until there are neither translational nor rotational displacements at $D$. These are the forces of Group 1. At (b) the structure is shown under the action of the forces of Group 2 which have been adjusted so that the only movement of $D$ is $\Delta_{aa}$ in the direction of $X_a$. This group comprises $X'_a$, $X'_b$ and $X'_c$ only, corresponding to $X_a$, $X_b$ and $X_c$ but not necessarily equal to them. Now Betti's Theorem states that the product of the forces of Group 1 and the corresponding displacements of Group 2 is equal to the product of the forces of Group 2 and the corresponding

---

† See Chapter 3, Section 3.5.2.

displacements of Group 1. We therefore have

$$1 \times \Delta_{ma} + X_a \times \Delta_{aa} + X_b \times 0 + X_c \times 0 = 0 \times \Delta_{m1} + X_a' \times 0 + X_b' \times 0 + X_c' \times 0$$

i.e. $\qquad X_a = -\dfrac{\Delta_{ma}}{\Delta_{aa}}$ as before

Although this explanation of the indirect model method may appear to be quite different from the previous one it depends, in fact, on the same principle that the unwanted forces are not themselves zero but are multiplied by zero deflexion. Their exact value is therefore immaterial and we are at liberty to make an elastically similar model on which the required deflexions, which will be in the correct ratio, can be measured.

#### 5.4.2.4. *Practical details*

Details of different techniques for model investigations have been given by many writers; the following notes give the principle of one method which has proved to be both effective and convenient in practice.

Many models have been made from sheet perspex or xylonite by cutting round the profile of the frame with a fretsaw and finishing with a fine file but it is more convenient to build up the model, especially if it is large, by using strips of material for the model members. It is essential that the joints shall be rigid and a suitable joint is described by Tse[8] and illustrated at (a) in FIGURE 5†.

It is, of course, quite possible to use wire or strip metal with soldered joints but perspex is both lighter and cleaner to use and its transparency facilitates the reading of the displacements.

The base of each column is enlarged to take screws for fastening the model to the baseboard or, if desired, arrangements can be made to impose displacements at these points. At sections where internal redundants are to be measured (or at the column feet) the member is cut as at (b); groups of holes drilled at the neutral axis, permit the insertion of displacement jigs of the kind illustrated at (c), (d) and (e) in FIGURE 5.18. It will be noticed that these restrict the relative movement of the cut ends of the member to the direction of the redundant under examination but permit bodily movement of the member as a whole: inspection of FIGURE 5.16(c), (d) and (e) shows that such a bodily movement does take place. It is obviously essential that friction be kept to a minimum and this is best done by mounting the baseboard vertically; if the horizontal position is preferred the model must be supported on balls. The vertical position has the additional advantage, if squared paper is used for reading the deflexions, that parallax errors are avoided.

Finally it should be mentioned that the best results are obtained by imposing displacements first to one side of the neutral position and then to the other; the mean of the two sets of readings is adopted. Although models are capable of yielding reliable results if carefully made and used it is always possible that a mistake has been made. It is therefore a wise precaution to impose a series of checks to see that the requirements of static equilibrium are everywhere satisfied.

---

† It is believed that these joints, while not conforming to the exact requirements of similarity, do not introduce appreciable errors.

#### 5.4.2.5. *Direct model methods*

The indirect model method described above shows to best advantage when a rather complex load system can act on a structure having a small number of redundants. If the structure is highly redundant and has a comparatively simple loading system—a tall, rigidly jointed transmission tower, for example—then the process of imposing deformations, one at a time, at a large number of points becomes very tedious. In such cases it may be better to impose forces directly on the model and to measure the resulting strains, slopes or deflexions at critical points on the members.

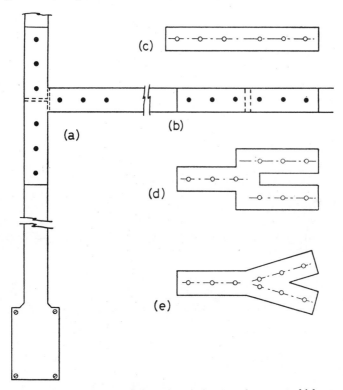

*Figure 5.18.* (a) The model is built up of strips of perspex which are joined by means of *T*-jigs; (b) At sections where internal redundants are to be measured the member is cut and joined by a jig; the one shown in place holds the member in the undisported position; (c) Jig for producing axial distortion; (d) Shear jig; (e) Moment jig

This can be done in a great many ways and it is now possible to mention only one or two points of interest. Further details can be obtained from the references listed at the end of the Chapter and from the Proceedings of the Society for Experimental Stress Analysis.

##### 5.4.2.5.1. *Scale factors.*—If the laws of elastic similarity, discussed in Section 5.4.2.1. are obeyed the various scale factors now required can easily be devised. If the main effect to be reproduced is bending the scales are

connected by equation 5.5.

$$s_A = \frac{s_F s_L^3}{s_E s_I} \qquad \ldots (5.5)$$

i.e.
$$s_F = \frac{s_A s_E s_I}{s_L^3}$$

If the model is completely similar to the prototype in all respects then

$$s_I = s_L^4 \quad \text{and} \quad s_A = s_L$$

so that
$$s_F = s_E s_L^2 \qquad \ldots (5.5(a))$$

The moment scale will be given by

$$s_M = s_F s_L = s_E s_L^3 \qquad \ldots (5.5(b))$$

While the scale of stress is given by

$$s_f = \frac{s_F}{s_L^2} = s_E \qquad \ldots (5.5(c))$$

On the other hand if the model is made of rectangular strips so proportioned that the correct second moment of area is obtained by adjusting the depths of the strips, according to equation 5.7, the stresses, which depend on the modulus of section, will not be correctly represented. If the bending moments are obtained by the use of the moment indicator (see below), or are calculated from the slopes and deflexions, the prototype stresses can always be calculated by using the moment scale

$$s_M = s_F s_L = \frac{s_A s_E s_I}{s_L^2} \qquad \ldots (5.5(d))$$

If other effects than bending are to be reproduced the appropriate scale factors are derived in a similar manner from equations 5.4 and 5.6.

5.4.2.5.2. *Measurement of slopes and deflexions.*—The slope deflexion equations 6.5 can be used to calculate the moments at the ends of the members of rigidly jointed frames under load; details of the method have been given by Pippard and Baker[4]. Briefly the system involves attaching suitable targets to the joints of the model, which is loaded according to the force scale, and measuring the rotations and displacements by means of measuring microscope or otherwise. Good results are obtainable by this means but a fair amount of calculation lies between the experimental observations and the final result.

5.4.2.5.3. *The moment indicator.*—This instrument, designed by Ruge and Schmidt[9], provides a convenient direct means of measuring bending moment in members of uniform cross-section. FIGURE 5.19 shows the principle diagrammatically, $wAx$ and $yBz$ being rigid arms attached to the members at $A$ and $B$. When the member is bent the gaps $wz$ and $xy$ change by amounts which can be measured and which are given by the intercepts $a$ and $b$ on the line diagram (b). The bending moment diagram is shown at (c) and it is easy to show that

$$M_A = 3L^2b/2EI$$
$$M_B = 3L^2a/2EI$$
.... (5.9)

provided there is no load between $A$ and $B$.

**5.4.2.5.4. Creep.**—When the model is made of metal the loads can be applied by means of dead weights or through proving rings or spring balances. If

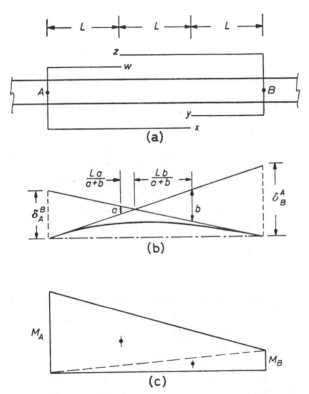

*Figure 5.19.* (a) The Ruge moment indicator consists, in principle, of two pairs of arms which are clamped to the model member at $A$ and $B$. After the application of the load the distances between $w$ and $z$, and $x$ and $y$, are measured; (b) The changes in these distances are equal to $a$ and $b$ in the diagram; (c) The bending moment diagram that is sought

this is done with a perspex or xylonite model errors will be introduced because the material creeps under load. This means that the modulus of elasticity, and hence the scale factor, is not constant.

This difficulty can be avoided by loading the model through a spring balance made of the same material. If it can be assumed that the creep properties of model and balance are identical then the modulus of elasticity will be the same for both at any instant and the behaviour of the model can be interpreted by reference to the deflexion of the balance, which is measured, and to its stiffness, which is calculated.

Details of the arrangement and of the method are given by Charlton[10] while Wright[11] describes an instrument called the isodynamometer, which enables the same principle to be used when it is desired to impose known internal forces and moments on the model.

## REFERENCES

[1] JOHNSON, J. B., BRYAN, C. W. and TURNEAURE, F. E. *Modern Framed Structures.* Wiley. 10th edn. 1946. Vol. I. p. 228

[2] CORNISH, R. J. 'Influence Lines by Williot-Mohr Diagrams'. *Engineering.* Dec. 1954. pp. 827–8

[3] TSE, K. F. 'Model Analysis of Belfast Power Station Building Frame'. *Civil Engineering and Public Works Review.* Vol. 52. 1957. pp. 660 and 792

[4] PIPPARD, A. J. S. and BAKER, J. F. *Analysis of Engineering Structures.* 3rd edn. Arnold. 1957

[5] PIPPARD, A. J. S. *The Experimental Study of Structures.* Arnold. 1947

[6] CHARLTON, T. M. *Model Analysis of Structures.* Spon. 1954

[7] BEGGS, G. E. 'The use of models in the solution of indeterminate Structures'. *J. Franklin Institute.* Vol. 203 (1927). pp. 375–386

[8] TSE, K. F. 'A practical method of analysing structures using large models'. *Civil Engineering and Public Works Review.* Vol. 51. 1956. p. 1341 and Vol. 52. 1957, p. 67

[9] RUGE, A. C. and SCHMIDT, E. C. 'Mechanical Structural Analysis by the Moment-indicator'. *Proc. Amer. Soc. Civ. Engrs.* Vol. 64. 1938. p. 1613–25

[10] CHARLTON, T. M. 'A Direct Method for the Model Analysis of Structures'. *Civil Engineering and Public Works Review.* Vol. 48. 1953. p. 51

[11] WRIGHT, J. 'The Isodynamometer'. *Civil Engineering and Public Works Review.* Vol. 48. 1953

## ADDITIONAL REFERENCES

BULL, M. G. 'Model Analysis of an Arch Span of 344 Feet'. *Proc. Inst. Engrs.* Australia. 1937. p. 58

MAKOWSKI, Z. S. and PIPPARD, A. J. S. 'Experimental Analysis of space structures'. *Proc. Inst. Civil Engineers.* 1. III. pp. 421–441

PIPPARD, A. J. S. 'Stresses by Analysis and Experiment'. *Proc. Inst. Mech. Engrs.* 1947

REDSHAW, S. C. and PALMER, P. J. 'The Construction and Testing of a Xylonite Model of a Delta Aircraft'. *Aeronautical Quarterly.* Vol. III. 1951. p. 83

SCRUTON, C. 'An investigation of the oscillations of suspension bridges in wind'. *Prelim. Pub. 4th Congress. International Association for Bridge and Structural Engineering.* 1952. pp. 329–51

LANDDECK, N. E. 'A direct method for Model Analysis'. *Proc. Amer. Soc. Civ. Engrs, J. Struct. Div.* Vol. 82. January, 1956

ROCHA, M. 'Model Tests in Portugal'. *Civil Engineering and Public Works Review.* Vol. 53. 1958. p. 49

# Chapter 6
# FRAMES WITH RIGID JOINTS
## 6.1. INTRODUCTION

The theories of rigid frame analysis developed in the first five Sections of this Chapter rely on a number of simplifications in addition to the usual assumption of linear elastic behaviour. The assumption that the joints are perfectly rigid, so that all the members meeting at a joint rotate through the same angle, is satisfactory for reinforced concrete and welded steel structures but may be appreciably in error when certain types of rivetted steel joints are used. This subject was studied in some detail by the Steel Structures Research Committee[1]. In structures of normal proportions it is quite satisfactory to neglect the size of the joints and to represent the structure by the gravity axes of the members. But if the depth of the members exceeds about 10 per cent. of their span appreciable errors are introduced because of the size of the joints. The stress distribution within the joints is complicated and difficult to include in the frame analysis but it is worth noting that a properly designed model, even if the effects of shear are only approximately represented, will take some account of the effects of the size of the joints and may well be an improvement, in this respect, on a calculation.

The deformation of members caused by shearing and axial stresses is usually ignored in the methods described below, the effect of bending being regarded as predominant. This simplification can again lead to appreciable error if the depth of the members is more than about 10 per cent. of their length. In such cases the bending moments in the frame can first be found in one of the usual ways; the axial deformations of the members are then computed and used to evaluate a second set of bending moments which are finally added to the first. The equations given in Chapter 9 include the effects of axial deformation automatically but it is hardly practicable to use these equations unless a computing machine is available.

Finally it should be noted that if the axial forces are large enough to affect the flexural stiffness the methods of Chapter 8 should be used.

The development of methods of rigid frame analysis based on the above assumptions was the main preoccupation of research engineers in the decade following the publication of Hardy Cross's famous paper[2] but in more recent years, while refinements of the moment distribution method have continued to appear at regular intervals, attention has gradually shifted to the plastic theory. It has therefore been thought appropriate to include an introduction to this topic in Section 6.6.

## 6.2. HYPERSTATIC BEAMS
### 6.2.1. The Propped Cantilever

Although its usefulness as a structural unit is limited the propped cantilever is the simplest form of hyperstatic beam and for this reason it served to illustrate the introductory remarks in Chapter 1.

An example of its use is shown diagrammatically in FIGURE 6.1(a) which shows a cantilever retaining wall in the basement of a building. If the ground floor beams and slabs frame into the top of the wall they act as props and affect the moments in the vertical stem of the retaining wall, in sign as well as in magnitude.

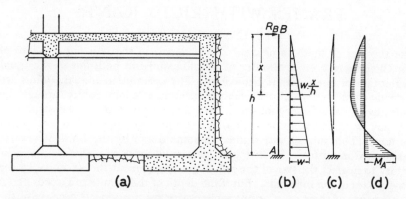

*Figure 6.1.* (a) A basement retaining wall is shown propped by the ground floor slab; (b) The structural scheme to be analysed; (c) Deflexion diagram; (d) Bending moment diagram

FIGURE 6.1(b) shows the problem schematically: a length of one foot of the wall is considered and the loading is assumed to be hydrostatic, varying linearly from zero at the top to $w$ per unit length at the base of the wall.

The stem of the wall will deflect into the shape indicated in FIGURE 6.1(c) and the bending moment diagram will be of the form shown at (d) although numerical values cannot be assigned until the single redundant $R_B$ is known. The equation of compatibility must express the fact that there is zero deflexion at $B$ and so, using equation 3.9(a), we have

$$\Delta_B = \int_0^h \frac{M'M\,dx}{EI} = 0$$

i.e.
$$0 = \int_0^h x\left(R_B x - \frac{wx^3}{6h}\right)dx = \frac{R_B h^3}{3} - \frac{wh^4}{30}$$

Hence
$$R_B = \frac{wh}{10} \quad \text{and} \quad M_A = \frac{wh^2}{15}$$

The uncertainties that surround the assumptions on which this analysis is based can be seen to turn on the nature and position of the reactive restraints: the exact point of action of the force $R_B$, for example, and the rotational restraint offered by the slab cannot be exactly known. The base of the wall, again, has been taken to be fixed against rotation but this can hardly be strictly true. At the cost of a more elaborate analysis the designer can estimate the probable effect of these uncertainties and in important and costly works he may decide that further calculations are justified; very probably he will then try to assign limits, to the position of $R_B$, for example, which will give the range that his design must cover.

Calculations of this sort can be made so easily that there is nothing to be gained by attempting any sort of general formula for propped cantilevers.

### 6.2.2. THE BEAM WITH FIXED ENDS

FIGURE 6.2(a) shows a beam $AB$ whose ends $A$ and $B$ are supposed to be completely restrained against both rotation and vertical translation. Few actual beams are so effectively gripped at their ends that the slope there is zero although light beams welded to heavy columns approximate to this

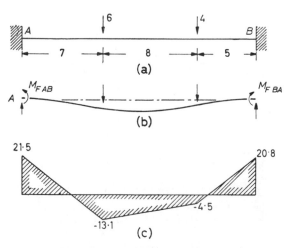

*Figure 6.2.* (a) The ends of this beam are restrained so that neither rotation nor translation occurs there; (b) The beam takes up the shape indicated and fixing moments are developed at the ends; (c) The corresponding bending moment diagram

condition. The importance of this problem lies in its being the starting point for the moment distribution and other methods of analysing rigid frames; the idea here is to calculate the moments that would develop in the beam if its ends were completely fixed and then to see what changes are brought about by the rotations which actually occur.

If vertical loads only are considered the beam has two redundant reactive restraints usually taken as the 'fixing moments' $M_{FAB}$ and $M_{FBA}$ that develop at the ends as shown at (b). A horizontal reaction will also be developed as the beam deflects but this is usually neglected as being of secondary importance[3].

In order to obtain general expressions for the fixing moments it is convenient to consider a simply-supported beam $AB$, (FIGURE 6.3(a)) under a single vertical load which produces the 'free' bending moment diagram (b) together with the end couples whose 'fixing moment' diagram is (c); for the time being the latter is assumed to be positive. If the couples are of such a value that the deflexion and slope at both ends are zero we shall have:

$$i_B = i_A = 0$$

From equation 3.21(a)

$$\left[\frac{A_m}{EI}\right]_A^B = 0$$

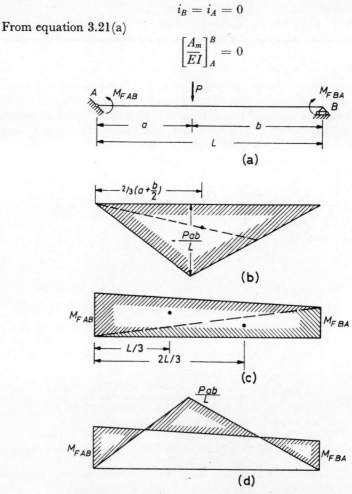

*Figure 6.3.* (a) The primary structure for the beam of FIGURE 6.2 is a simply-supported beam, here shown carrying a single point load; (b) The 'free' bending moment diagram obtained by considering the point load only; (c) The 'fixing moment' diagram is of this form though the end ordinates have yet to be found; (d) When the fixing moments are known the complete bending moment diagram is obtained by superimposing diagrams (b) and (c) with due regard to sign

Hence if the beam is uniform

$(A_m)$ fixing moments $= - (A_m)$ free moments

$$\frac{L}{2}(M_{FAB} + M_{FBA}) = +\frac{Pab}{L} \cdot \frac{L}{2}$$

## 6.2. HYPERSTATIC BEAMS

i.e.
$$M_{FAB} + M_{FBA} = \frac{Pab}{a+b} \qquad \ldots (i)$$

The displacement of $A$ from the tangent at $B$ is also zero and so, from equation 3.22(a),

$$\left[\frac{A_m \bar{x}}{EI}\right]_A^B = 0$$

i.e.
$$M_{FAB} \frac{L}{2} \cdot \frac{L}{3} + M_{FBA} \cdot \frac{L}{2} \cdot \frac{2L}{3} = + \frac{Pab}{L} \cdot \frac{L}{2} \cdot \frac{2}{3}\left(a + \frac{b}{2}\right)$$

$$M_{FAB} + 2M_{FBA} = \frac{Pab}{(a+b)^2}(2a+b) \qquad \ldots (ii)$$

solving (i) and (ii) simultaneously gives

$$\left.\begin{array}{l} M_{FAB} = + \dfrac{Pab^2}{(a+b)^2} = + \dfrac{Pab^2}{L^2} \\[2ex] M_{FBA} = + \dfrac{Pa^2b}{L^2} \end{array}\right\} \qquad \ldots (6.1)$$

The positive sign confirms that with the usual downward applied loads, giving a negative 'free' bending moment diagram, the fixing moments are positive. The two diagrams (b) and (c) can now be combined to give the

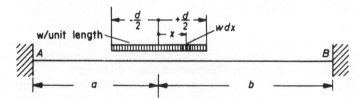

*Figure 6.4.* Beam with fixed ends carrying a load uniformly distributed over part of its length

final diagram (d) and, if desired, this can be replotted on a horizontal base. If there are several point loads acting on the beam then their effects can be superimposed by writing

$$M_{FAB} = + \sum \frac{Pab^2}{L^2}$$
$$M_{FBA} = + \sum \frac{Pa^2b}{L^2} \qquad \ldots (6.1(a))$$

These expressions have been used to work out the fixing moments for the beam if FIGURE 6.2(a) whose bending moment diagram, plotted on a horizontal base, is shown at (c). If the beam carries distributed loads the summations of equation 6.1(a) can be obtained by integration. For the beam

237

FRAMES WITH RIGID JOINTS

of FIGURE 6.4 we have, for example,

$$M_{FAB} = \int_{-\frac{d}{2}}^{+\frac{d}{2}} \frac{wdx(a+x)(b-x)}{L^2}$$

$$= \frac{wd.ab^2}{L^2} + \frac{wd^3}{12L^2}(a - 2b) \quad \ldots \quad (6.2)$$

In this equation the first term is the value that the fixing moment would have if the load were concentrated at its centre, while the second term takes account of the spread of the load. In order to find the effect of a load uniformly distributed over the whole length of the beam we can write

$$a = b = L/2 = d/2$$

i.e. $$M_{FAB} = \frac{wL^2}{12} \quad \ldots \quad (6.2(a))$$

and from symmetry, $$M_{FBA} = \frac{wL^2}{12}$$

Equations 6.1 and 6.2 only apply to beams whose cross-section is constant; the modifications which become necessary in other cases are discussed in Appendix A.

### 6.2.3. CONTINUOUS BEAMS

The beam of more than one span, carried by several supports, has already been used by way of illustration in previous Chapters; the following Sections include a more systematic account of the methods of analysis beginning with Clapeyron's well-known Theorem of Three Moments.

Continuous beams are such a valuable structural device, especially in bridge engineering, that they have been intensively studied over a long period. The obvious starting point is to choose as the redundants an appropriate number of reactions as was done, for example, in Chapter 4, Section 4.4.2; this is particularly useful if influence lines are required. Clapeyron's great contribution was to recognize that the bending moments over the supports could be used as the redundants and that the compatibility equations were then obtained in terms of the end slopes of adjacent spans. This proved to be an extraordinarily fertile idea: not only did it lead to his own equations for the continuous beam problem which, with later developments and modifications, are still in regular use a century later but it foreshadowed the solution of more complex frames with interconnected columns and beams.

When reactions are chosen as the redundants we obtain an equal number of simultaneous equations each of which contains all the unknowns. Although the same number of equations must eventually be solved when the support moments are used as the redundants each equation contains only three of the unknowns and a simple relationship between the unknowns enables the equations to be written down easily and systematically.

## 6.2.3.1. *Clapeyron's Equation of Three Moments*

In its original form[4] Clapeyron's Equation gave the relationship between the applied loading and the moments over three adjacent supports at the same level; it is more convenient, however, to include the effect of support displacement from the outset. In FIGURE 6.5(a) are shown two adjacent

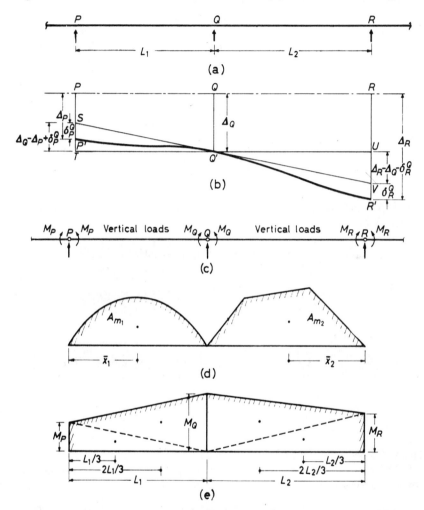

*Figure 6.5.* (a) Two adjacent spans from a beam continuous over many spans; (b) Subsidence of the supports to $P'$, $Q'$ and $R'$ together with the action of the applied loads bends the beam into the shape of the thick line; (c) The primary structure consists of a series of simply-supported beams; (d) Bending moment diagram for the applied vertical loads acting on the primary structure; (e) Bending moment diagram for the primary structure under the action of couples $M_P$, $M_Q$ and $M_R$

spans, $PQ$ and $QR$, of a continuous beam with an indefinite number of spans carrying loads defined by the bending moments they would produce in each span if it were simply supported; these are assumed to be positive

## FRAMES WITH RIGID JOINTS

and are shown at (d). Subsidence, amounting to $\varDelta_P$, $\varDelta_Q$ and $\varDelta_R$, has taken place at the supports and the beam has the deflected shape indicated by the thick line in diagram (b). The primary structure, of course, comprises the series of simply-supported beams obtained by inserting imaginary hinges over each support, as at (c), and the bending moment diagrams produced in spans $PQ$ and $QR$ by the support moments, whose magnitude is not yet known but whose sign is assumed to be positive, is shown at (e).

A compatibility equation can be written for each support to express the fact that the end slopes of adjacent spans are equal in magnitude but opposite in sign; the geometrical relationships between these slopes and the various deflexions follow from diagram (b), $SQ'V$ is the tangent to the beam at $Q'$ and in the similar triangles $STQ'$ and $VUQ'$ we have

$$\frac{ST}{TQ'} = \frac{VU}{UQ'} \quad \text{i.e.} \quad \frac{\varDelta_Q - \varDelta_P + \delta_P^Q}{L_1} = \frac{\varDelta_R = \varDelta_Q - \delta_R^Q}{L_2}$$

i.e. 
$$\frac{1}{L_1}\delta_P^Q + \frac{1}{L_2}\delta_R^Q = \frac{\varDelta_P - \varDelta_Q}{L_1} + \frac{\varDelta_R - \varDelta_Q}{L_2} \quad \ldots \text{(i)}$$

The displacements $\delta_P^Q$ and $\delta_R^Q$ are given by applying equation 3.22(a) to the two spans in turn: thus

$$\delta_P^Q = \frac{1}{E_1 I_1}\left\{A_{m_1}\bar{x}_1 + \frac{1}{2}M_P L_1 . L_1/3 + \frac{1}{2}M_Q L_1 . 2L_1/3\right\}$$

$$\delta_R^Q = \frac{1}{E_2 I_2}\left\{A_{m_2}\bar{x}_2 + \frac{1}{2}M_R L_2 . L_2/3 + \frac{1}{2}M_Q L_2 . 2L_2/3\right\} \quad \ldots \text{(ii)}$$

Combining equations (i) and (ii) we get

$$M_P \frac{L_1}{E_1 I_1} + 2M_Q\left\{\frac{L_1}{E_1 I_1} + \frac{L_2}{E_2 I_2}\right\} + M_R \frac{L_2}{E_2 I_2} + 6\left\{\frac{A_{m_1}\bar{x}_1}{E_1 I_1 L_1} + \frac{A_{m_2}\bar{x}_2}{E_2 I_2 L_2}\right\}$$

$$= 6\left\{\frac{\varDelta_P - \varDelta_Q}{L_1} + \frac{\varDelta_R - \varDelta_Q}{L_2}\right\} \quad \ldots \text{(6.3)}$$

This is Clapeyron's Equation of Three Moments[†].

In the special case when the beam is uniform and there are no reaction displacements we have

$$M_P L_1 + 2M_Q(L_1 + L_2) + M_R L_2 = -6\sum \frac{A_m \bar{x}}{L} \quad \ldots \text{(6.3(a))}$$

When this equation is applied to an actual problem it must be remembered that $\bar{x}$ is measured outwards in each span from the loads to the ends.

It is convenient to have algebraic expressions for $A_m \bar{x}$ in one or two standard cases:

---

[†] An extension of the above, known as the 'Theorem of Four Moments' can be used in the analysis of rigid frames.

*Point load:* FIGURE 6.6(a)
$$-\frac{6A_m\bar{x}}{L} = +\frac{6 \cdot 1/2\, Pa(L-a)\left\{2/3\left(a + \frac{L-a}{2}\right)\right\}}{L}$$
$$= \frac{Pa(L^2 - a^2)}{L} \qquad \ldots\ (6.4(a))$$

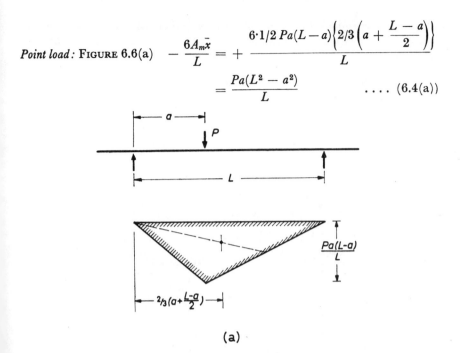

(a)

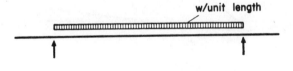

(b)

*Figure 6.6.* Point and uniformly distributed loads on one span of a continuous beam

*Uniformly distributed load:*
FIGURE 6.6(b)
$$-\frac{6A_m\bar{x}}{L} = +\frac{6.2/3.\ wL^2/8.\ L.\ L/2}{L}$$
$$= \frac{wL^3}{4} \qquad \ldots\ (6.4(b))$$

241

## FRAMES WITH RIGID JOINTS

In solving continuous beam problems the Three Moment Equation is applied in succession to pairs of adjacent spans, as in the following examples.

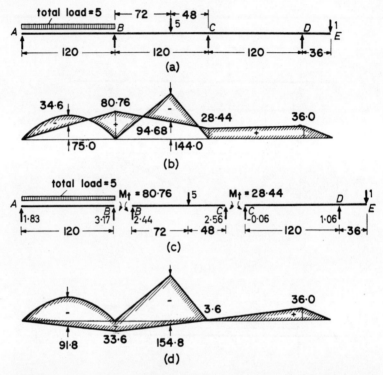

*Figure 6.7.* (a) Continuous beam of uniform cross-section; (b) Bending moment diagram obtained by superimposing the ' support ' moments on the ' free ' moments; (c) The beam subdivided into separate spans to allow the reactions to be computed; (d) Bending moment diagram for the same beam when carried on elastic supports

(a) Applying Clapeyron's Equation to the two spans $ABC$ of the beam of FIGURE 6.7(a)

$$M_A.120 + 2M_B(120 + 120) + M_C.120$$
$$= + \left(\frac{5}{120}\right)\left(\frac{120^3}{4}\right) + \frac{5 \times 48(120^2 - 48^2)}{120}$$

Since $A$ is a free end $M_A$ is zero and this reduces to
$$480M_B + 120M_C = + 42,200 \qquad \ldots \text{(i)}$$

In the two spans $BCD$:
$$M_B.120 + 2M_C(120 + 120) + M_D.120$$
$$= + \frac{5 \times 72(120^2 - 72^2)}{120} = + 27,650$$

$M_D$ is equal to $+36$, the moment of the unit load on the overhang, so
$$120M_B + 480M_C = + 23,330 \qquad \ldots \text{(ii)}$$

242

Solving these equations simultaneously gives

$$M_B = + 80.76$$
$$M_C = + 28.44$$

These moments can now be incorporated in the 'support-moment' diagram which is shown superimposed on the 'free' moment diagram in FIGURE 6.7(b).

The reactions and shearing forces are best found by dividing the beam up into its individual spans, as at (c), taking care to replace the support moments by couples acting in the correct direction; the shearing force diagram (d) can then be drawn.

The following points arising from the above solution are especially to be noted:

(i) Successive pairs of spans are selected by moving along the beam one span at a time, thus: $ABC$, $BCD$, etc.

(ii) The right-hand side of equations (i) and (ii) is positive because the free bending moments in spans $AB$ and $BC$ are negative.

(iii) The dimension 'a' in the right-hand side of equation 6.4(a) is 48 when $BC$ is the right-hand span of the pair $AB$—$BC$, but 72 when $BC$ is the left-hand span of the pair $BC$—$CD$: it is always measured outwards from the loads to the ends.

(iv) The small value of $R_D$ suggests that with different loads on the beam it might have been negative; this draws attention to the tacit assumption that the supports can provide upward or downward reactions, as necessary, and yet allow rotation of the beam at that point.

(b) A special case arises when one or both ends of the beam are fixed. The moment at each fixed end is a redundant for which a new equation must be found and it is now best to regard the support to which the

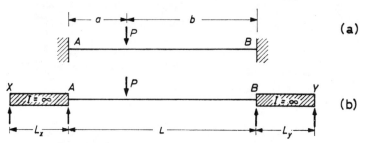

Figure 6.8. The beam with fixed ends can be regarded as a three-span continuous beam with outer spans of infinite second moment of area

beam is attached as a span of infinite second moment of area. The fixed ended beam of FIGURE 6.8 can be treated in this way. Applying the theorem of these moments to the two spans $XAB$

$$M_x \frac{L_x}{\infty} + 2M_A \left(\frac{L_x}{\infty} + \frac{L}{I}\right) + M_B \frac{L}{I} = \frac{Pb(L^2 - b^2)}{LI}$$

i.e. 
$$2M_A + M_B = \frac{Pb(L^2 - b^2)}{L^2} \quad \cdots \text{(i)}$$

Similarly for the two spans $ABY$

$$M_A + 2M_B = \frac{Pa(L^2 - a^2)}{L^2}$$ .... (ii)

Solving simultaneously for $M_A$ and $M_B$

$$M_A = \frac{Pab^2}{L^2} \;;\; M_B = \frac{Pa^2b}{L^2}$$

This will be recognized as the standard result for the fixed ended beam, previously obtained as equation 6.1.

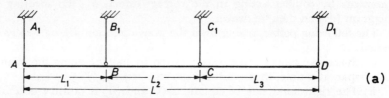

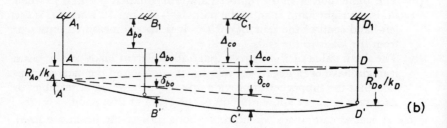

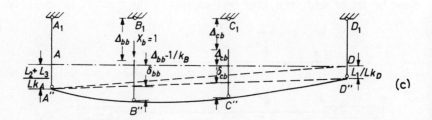

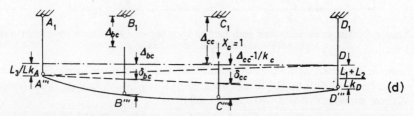

*Figure 6.9.* (a) The continuous beam $ABCD$, shown without its applied loads, is supported by elastic rods $A_1A, B_1B$ etc.; (b) The primary structure is the beam $AD$ supported by the rods $A_1A$ and $D_1D$, the deflexions shown are those due to the applied load; (c) The primary structure under the loading $X_B=1$; (d) The primary structure under the loading $X_C=1$

## 6.2. HYPERSTATIC BEAMS

### 6.2.3.2. *Continuous Beam on Elastic Supports*

If the supports of a continuous beam subside by known amounts Clapeyron's Equation can be used in a straightforward manner to find the resulting bending moments. But if the supports are elastic and deflect in proportion to the reactions the Three Moment Equation is not convenient because the reactions do not appear explicitly.

In such cases it is better to discard the argument advanced in Section 6.2.3 and to use the reactions as the redundants: the flexibility coefficient equations can then be arranged to take account of the flexibility of the supports as well as of the beam. The procedure has already been indicated in Chapter 4, Section 4.4.2, and is now expanded with reference to FIGURE 6.9(a) which shows a continuous beam carried by a series of supports whose stiffness is $k_A$, $k_B$, etc. The primary structure is the simply-supported beam obtained by removing the pins at $B_1$ and $C_1$ as shown at (b). Equations 4.7 are then

$$\left. \begin{aligned} 0 &= \varDelta_{bo} + X_b\, \varDelta_{bb} + X_c\, \varDelta_{bc} \\ 0 &= \varDelta_{co} + X_b\, \varDelta_{cb} + X_c\, \varDelta_{cc} \end{aligned} \right\} \quad \ldots \text{(i)}$$

where the terms $\varDelta_{bo}$, $\varDelta_{bb}$, etc., refer to the beam and supports combined. We can best proceed by separating the deflexions of the beam and of the supports and write:

$$\left. \begin{aligned} \varDelta_{bo} &= \frac{L_2 + L_3}{L} \cdot \frac{R_{Ao}}{k_A} + \frac{L_1}{L} \cdot \frac{R_{Do}}{k_D} + \delta_{bo} \\ \varDelta_{co} &= \frac{L_3}{L} \cdot \frac{R_{Ao}}{k_A} + \frac{L_1 + L_2}{L} \cdot \frac{R_{Do}}{k_o} + \delta_{co} \\ \varDelta_{bb} &= \left(\frac{L_2 + L_3}{L}\right)^2 \frac{1}{k_A} + \left(\frac{L_1}{L}\right)^2 \frac{1}{k_D} + \frac{1}{k_B} + \delta_{bb} \\ \varDelta_{cb} &= \frac{L_3}{L} \cdot \frac{L_2 + L_3}{L} \cdot \frac{1}{k_A} + \frac{L_1 + L_2}{L} \cdot \frac{L_1}{L} \frac{1}{k_D} + \delta_{cb} \\ \varDelta_{cc} &= \left(\frac{L_3}{L}\right)^2 \frac{1}{k_A} + \left(\frac{L_1 + L_2}{L}\right)^2 \frac{1}{k_A} + \frac{1}{k_C} + \delta_{cc} \\ \varDelta_{bc} &= \frac{L_2 + L_3}{L} \cdot \frac{L_3}{L} \cdot \frac{1}{k_A} + \frac{L_1}{L} \cdot \frac{L_1 + L_2}{L} \cdot \frac{1}{k_D} + \delta_{bc} \end{aligned} \right\} \quad \ldots \text{(ii)}$$

Where $\delta_{bo}$, $\delta_{co}$, etc., are the deflexion of the beam relative to its ends and $R_{Ao}$, $R_{Do}$ are the end reactions of the primary structure under the applied loading. These results will be applied to the beam of FIGURE 6.7(a), which is now supposed to be supported at $A$, $B$, $C$ and $D$ on springs which deflect 1 unit per unit load, tension or compression (i.e. $k_A = k_B = k_C = k_D = 1$). The beam is taken to have a flexural rigidity of 600,000. The values of $\delta_{bo}$, $\delta_{co}$, etc., obtained by any convenient method are:

## FRAMES WITH RIGID JOINTS

$$\delta_{bo} = 5·994 \times 10^6/EI = 9·99$$
$$\delta_{co} = 5·688 \times 10^6/EI = 9·48$$
$$\delta_{bb} = 0·768 \times 10^6/EI = 1·28$$
$$\delta_{bc} = \delta_{cb} = 0·672 \times 10^6/EI = 1·12$$
$$\delta_{cc} = 0·768 \times 10^6/EI = 1·28$$

.... (iii)

Also $L_1 = L_2 = L_3 = 120$; $L = 360$, $R_{Ao} = 6·4$, $R_{Do} = 4·6$.

Hence in (ii)  $\Delta_{bo} = \dfrac{240 \times 6·4}{360} + \dfrac{120 \times 4·6}{360} + 9·99 = 15·79$

Similarly  $\Delta_{co} = 14·68$

$\Delta_{bb} = 2·83 = \Delta_{cc}$

$\Delta_{cb} = 1·56 = \Delta_{bc}$

.... (iv)

Inserting equations (iv) in (i) we have:

$$0 = 15·79 + 2·83\,X_b + 1·56\,X_c$$
$$0 = 14·68 + 1·56\,X_b + 2·83\,X_c$$

i.e.  $X_B = -3·91 \quad X_C = -3·04.$

Knowledge of these two reactions enables the remaining reactions to be found from statics: they are $R_A = 2·78$; $R_D = 1·27$; the bending moment diagram can then be found and is shown at (d) on FIGURE 6.7.

Equations (i) can be used to find the reactions of the beam on rigid supports by excluding from (ii) the terms that describe the support deflexions and writing $\Delta_{bo} = \delta_{bo}$, etc.

Thus inserting equations (iii) in (i) we have

$$0 = 9·99 + 1·28\,X_b + 1·12\,X_c$$
$$0 = 9·48 + 1·12\,X_b + 1·28\,X_c$$

i.e.  $X_b = -5·65 \quad X_c = -2·46.$

These results agree closely with those obtained previously for this problem apart from small discrepancies, in the second place of decimals, due to rounding off.

Comparison of FIGURES 6.7(b) and (d) reveals that the elasticity of the supports has been responsible for a considerable change especially in the maximum bending moment and to the sign of the bending moment over support B. This draws attention to the very important feature of continuous beams, and indeed of all other structures with redundant reactions, that stresses are induced by foundation movements. There are several ways in which this difficulty can be circumvented. The first is to use the methods of soil mechanics to estimate, in terms of the soil properties and of the superimposed loading, the foundation settlement that is likely to occur; the resulting stresses can then be computed and allowed for in the design. This procedure is likely to result in a less economical beam than might have been hoped for and it may be thought worth while to make provision for adjusting the level of the supports so as to counteract the settlement. It is not easy to make these adjustments with sufficient accuracy by direct measurement of level and it may be necessary instead to weigh the reactions by means of proving rings or calibrated jacks.

If the foundation conditions are too unfavourable the intention to use a continuous beam may have to be abandoned in favour of a statically determinate cantilever arrangement. This retains the economical form of the continuous beam bending moment diagram and, by introducing hinges at the points of contraflexure, ensures that the moments are unaffected by foundation movements. This is a favourite solution in steel bridge practice but has been less often adopted by reinforced concrete designers.

### 6.2.3.3. *Influence Lines for Continuous Beams*

In Chapter 5, Section 5.3.3, a method was given for finding the influence lines for the redundant reactions of continuous beams. Although this method is quite general it is not very suitable for use with beams of many spans especially if these are of different cross-section. The following method, using Clapeyron's Equation, has been described by Cornish and Jones[5]. FIGURE 6.10(a) shows a continuous beam with a point load on each span. Then from equations 6.3 and 6.4(a) we have:

For spans $AB$ and $BC$

$$2M_B \left(\frac{L_1}{I_1} + \frac{L_2}{I_2}\right) + M_C \frac{L_2}{I_2} = \frac{P_1 n_1 L_1^2 (1 - n_1^2)}{I_1} + \frac{P_2 n_2 L_2^2 (1 - n_2)(2 - n_2)}{I_2}$$

For spans $BC$ and $CD$

$$M_B \frac{L_2}{I_2} + 2M_C \left(\frac{L_2}{I_2} + \frac{L_3}{I_3}\right) = \frac{P_2 n_2 L_2^2 (1 - n_2^2)}{I_2} + \frac{P_3 n_3 L_3^2 (1 - n_3)(2 - n_3)}{I_3}$$

....(i)

Eliminating $M_C$ from these equations, we have:

$$\begin{aligned} M_B &\frac{4(L_1 I_2 + L_2 I_1)(L_2 I_3 + L_3 I_2) - L_2^2 I_1 I_3}{I_1 I_2 I_3} \\ &= \frac{2 P_1 n_1 L_1^2 (1 - n_1^2)(L_2 I_3 + L_3 I_2)^2}{I_1 I_3} \quad \text{....(a)} \\ &+ \frac{P_2 n_2 L_2^2 \{2(1 - n_2)(2 - n_2)(L_2 I_3 + L_3 I_2) - (1 - n_2^2) L_2 I_3\}}{I_2 I_3} \quad \text{....(b)} \\ &- \frac{P_3 n_3 L_2 L_3^2 (1 - n_3)(2 - n_3)}{I_3} \quad \text{....(c)} \end{aligned}$$

....(ii)

In order to find the influence ordinates for $M_B$ for unit load in span 1, say, it is only necessary to put $P_1 = 1$ and $P_2 = P_3 = 0$ in the above expression and then to introduce successive values of $n_1$; $M_C$ can be found by transposition.

FIGURE 6.10(b) shows the influence line for $M_B$ obtained in this way for the particular case of $I_1 = I_2 = I_3$; $L_1 = L_3 = 35$; $L_2 = 40$.

## FRAMES WITH RIGID JOINTS

Influence lines for other load functions can then be derived quite easily. For example, when the unit load is in spans $BC$ and $CD$,

$$R_A = - M_B/L_1 \qquad \ldots \text{(iii)}$$

When the load is in span $AB$,

$$R_A = (1 - n_1) - M_B/L_1 \qquad \ldots \text{(iv)}$$

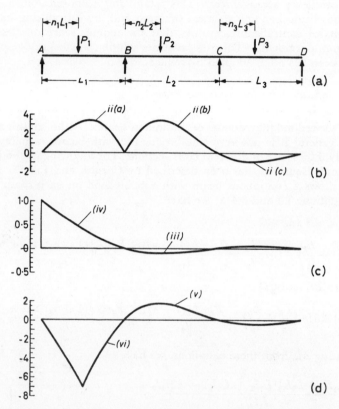

*Figure 6.10.* (a) In order to calculate influence ordinates for this continuous beam it is considered under the load system shown; (b) Influence line for $M_B$; (c) Influence line for $R_A$; (d) Influence line for bending moment at the mid-point of $AB$

In a similar way the influence line for bending moment at the mid-point of span $AB$, when the load is to the right of the point, is

$$- R_A . L_1/2 \qquad \ldots \text{(v)}$$

When the load is to the left of the point it is

$$1(L_1/2 - n_1 L_1) - R_A L_1/2 \qquad \ldots \text{(vi)}$$

These two influence lines are plotted at (c) and (d) in FIGURE 6.10.
Although this method can be extended, in principle, to continuous beams of any number of spans the solution, in algebraic terms, of more than two

## 6.3. THE SLOPE-DEFLEXION EQUATIONS

or three equations of Three Moments is tedious. In such cases it is usually better to seek a more direct solution in terms of the actual dimensions of the particular beam under consideration and to use for this purpose one or other of the methods described below: the beam-line method of Section 6.5.3 is particularly convenient here.

### 6.3. THE SLOPE-DEFLEXION EQUATIONS

In 1880 Manderla published a method of computing the secondary bending moments induced in truss members whose joints were rigid rather than pinned. His work was exact in the sense that the effect of the axial forces in the members on these bending moments was taken into account but the resulting equations were too complex to be used for design purposes. As simplified by Winkler, and again by Mohr, they have been extensively used in secondary stress calculations but their application in rigid frame analysis seems to be due to Wilson and Maney. In a well-known paper[6] these authors showed how to analyse the effect of lateral loads on buildings of the skyscraper type and in so doing gave designers, for the first time, a convenient means of dealing with welded steel and reinforced concrete structures.

A fundamental difficulty is that the sign conventions used in drawing bending moment diagrams, which distinguish the tension from the compression side of the beam, are not suitable for the analysis of frames when the rotational equilibrium of the joints must be considered. For this purpose couples which tend to rotate the joint in a certain direction must always be given the same sign regardless of the type of bending they impose on adjacent beams. Accordingly we define, in FIGURE 6.11, a convention of signs, suitable for rigid frame analysis, which must be followed quite strictly until all the unknown moments have been determined. If it is then necessary to draw a bending moment diagram these moments are finally given correct signs according to the bending moment convention of Chapter 3, Section 3.7.1.

### 6.3.1. DERIVATION OF THE EQUATIONS

The purpose of the following Section is to obtain the relationships which connect the couples acting on the ends of a single beam and the relative movements, rotational and translational, of the two ends. Accordingly we consider the beam $AB$ of FIGURE 6.11(a) whose ends are initially horizontal and at the same level. If the beam is unloaded the couples acting on the ends are zero but if it carries loads, as at (b), the couples are equal to the fixing moments and are given by equation 6.1(a). If the ends are now permitted to move, so that the beam takes up the position (c) the *changes* in the end couples, $m_{AB}$ and $m_{AC}$, are uniquely connected with the relative rotations of the ends but the angles $\theta_{AB}$ and $\theta_{BA}$, measured from the horizontal, depend also on the displacement $\delta$ of one end relative to the other.

In order to find $m_{AB}$ and $m_{AC}$ we employ the moment-area equation 3.22(a); the bending moment diagram is shown at (d) and from this we have

$$\left. \begin{array}{l} L\,\theta_{BA} = \delta + aA = \delta + \delta_A^B \\ L\,\theta_{AB} = \delta + bB' = \delta + (-\delta_B^A) \end{array} \right\} \quad \ldots\ldots \text{(i)}$$

## FRAMES WITH RIGID JOINTS

The minus sign of $\delta_B^A$ arises from the convention of signs for the moment-area method: see FIGURE 3.32(b).

Now $\quad \delta_A^B = \left[ \dfrac{A_m \bar{x}}{EI} \right]_A^B = \dfrac{1}{EI}(1/3\ m_{BA}\ L^2 - 1/6\ m_{AB}\ L^2)$

and $\quad \delta_B^A = \left[ \dfrac{A_m \bar{x}}{EI} \right]_B^A = \dfrac{1}{EI}(1/6\ m_{BA}\ L^2 - 1/3\ m_{AB}\ L^2)$ $\quad \ldots$ (ii)

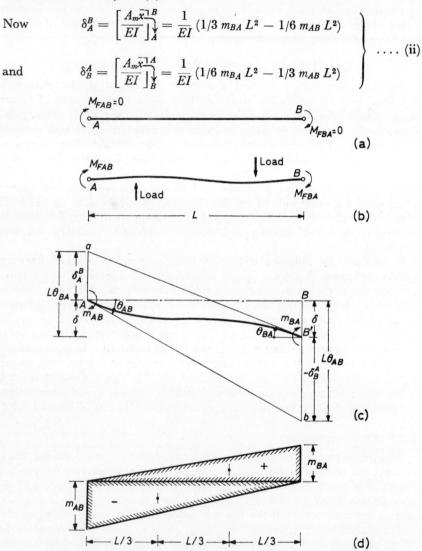

*Figure 6.11.* (a) An unloaded beam has zero fixing moments acting on its ends; (b) When loads are imposed the ends can be kept horizontal if moments equal to the fixing moments act on the ends; (c) The beam is shown in a displaced position after $A$ and $B$ have moved in a clockwise direction relative to each other; clockwise slopes, $\theta_{AB}$ and $\theta_{BA}$ have been caused at the ends and the changes in end moments, from those acting at (a) and (b) are $m_{AB}$ and $m_{BA}$, also clockwise. These are the positive directions of displacement, slope and end moment according to the slope deflexion convention; (d) Bending moment diagrams for the beam under clockwise end moments $m_{AB}$ and $m_{BA}$ which produce negative and positive diagrams according to the bending moment convention

## 6.3. THE SLOPE-DEFLEXION EQUATIONS

Hence from (i) and (ii)

$$\left.\begin{array}{l}\dfrac{6EI}{L}(\theta_{BA} - \delta/L) = 2m_{BA} - m_{AB} \\[1em] \dfrac{6EI}{L}(\theta_{AB} - \delta/L) = -m_{BA} + 2m_{AB}\end{array}\right\} \quad \ldots\text{(iii)}$$

Solving (iii) for $m_{AB}$ and $m_{BA}$, and writing $I/L = K$, we have

$$\left.\begin{array}{l}m_{AB} = 2EK(2\theta_{AB} + \theta_{BA} - 3\delta/L) \\ m_{BA} = 2EK(2\theta_{BA} + \theta_{AB} - 3\delta/L)\end{array}\right\} \quad \ldots\text{(iv)}$$

But $m_{AB}$ and $m_{BA}$ were defined as the changes in the end moments, from their initial values, resulting from the displacement of the beam from its initial position when the ends were horizontal and at the same level. The final values of the end moments, then, are given by adding these changes in moment to their initial values.

Hence
$$\left.\begin{array}{l}M_{AB} = m_{AB} + M_{FAB} = 2EK\left(2\theta_{AB} + \theta_{BA} - 3\dfrac{\delta}{L}\right) + M_{FAB} \\[1em] \text{and} \quad M_{BA} = m_{BA} + M_{FBA} = 2EK\left(2\theta_{BA} + \theta_{AB} - 3\dfrac{\delta}{L}\right) + M_{FBA}\end{array}\right\}\ldots(6.5)$$

These are the slope-deflexion equations; they are correctly signed if they refer to moments and rotations in the directions shown on FIGURE 6.11 which we now assert to be the positive directions. Our new sign convention can therefore be stated in the following terms:

Rotations of the ends of a beam from the horizontal, moments acting *on* its ends and displacements of one end relative to the other are all considered positive if they are *clockwise*.

In using the equations the first step is to estimate the fixing moments, $M_{FAB}$ and $M_{FBA}$, for each member and to allot to them a positive or negative sign according as they are clockwise or anti-clockwise. A moment equilibrium equation is then written for each joint, using the slope-deflexion equations, while the compatibility equations simply express the fact that, since they are rigidly attached to one another, all the members at a joint rotate through the same angle

e.g. $\quad \theta_{AB} = \theta_{AC} = \theta_{AD} = \ldots = \theta_A$

In the absence of joint translation or 'sway', from which we know that $\delta = 0$, enough equations are thus obtained to give $\theta$ for each joint; substitution in equations 6.5 then gives the end moments for each member.

### 6.3.2. APPLICATION OF THE SLOPE-DEFLEXION EQUATIONS TO FRAMES IN WHICH SWAY CANNOT OCCUR

Two numerical examples now follow to illustrate the above points. In each case the frame is of such a nature that there is no joint translation apart from that due to changes of length of the members.

(a) The continuous beam of FIGURE 6.7(a).

(i) Fixing moments:

$$M_{FAB} = \frac{wL^2}{12} = \left(\frac{5}{120}\right) \times 120^2 = 50 \qquad \text{(anti-clockwise, i.e. }-\text{)}$$

$$M_{FBA} = 50 \qquad \text{(clockwise, i.e. }+\text{)}$$

$$M_{FBC} = \frac{Pab^2}{L^2} = \frac{5 \times 72 \times 48^2}{120^2} = 57 \cdot 6 \qquad \text{(anti-clockwise, i.e. }-\text{)}$$

$$M_{FCB} = \frac{Pa^2b}{L^2} = \frac{5 \times 72^2 \times 48}{120^2} = 86 \cdot 4 \qquad \text{(clockwise, i.e. }+\text{)}$$

There is no load in $CD$ and hence

$$M_{FCD} = M_{FDC} = 0$$

At the end $D$, however, the overhanging load exerts a clockwise couple on the end $D$ of member $DC$ and so we know that

$$M_{DC} = +36$$

(ii) Compatibility equations:

At $B$: $\theta_{BA} = \theta_{BC} = \theta_B$

At $C$: $\theta_{CB} = \theta_{CD} = \theta_C$

(iii) Member equations:

It is easily shown that in solving for the bending moments we can let $EK$ (= constant) equal unity. Hence the slope-deflexion equations 6.5 become

$$M_{AB} = 2(2\theta_A + \theta_B) - 50$$
$$M_{BA} = 2(2\theta_B + \theta_A) + 50$$
$$M_{BC} = 2(2\theta_B + \theta_C) - 57 \cdot 6$$
$$M_{CB} = 2(2\theta_C + \theta_B) + 86 \cdot 4$$
$$M_{CD} = 2(2\theta_C + \theta_D)$$
$$M_{DC} = 2(2\theta_D + \theta_C)$$

(iv) Joint equations:

At $A$: $\Sigma M_A = 0$ i.e. $0 = M_{AB} = 4\theta_A + 2\theta_B - 50$

At $B$: $\Sigma M_B = 0$ i.e. $0 = M_{BA} + M_{BC} = 2\theta_A + 8\theta_B + 2\theta_C - 7 \cdot 60$

At $C$: $\Sigma M_C = 0$ i.e. $0 = M_{CB} + M_{CD} = 2\theta_B + 8\theta_C + 2\theta_D + 86 \cdot 4$

At $D$: $\Sigma M_D = 0$ i.e. $0 = M_{DC} + M_{DE} = 2\theta_C + 4\theta_D - 36$

We thus have four equations for the four unknown joint rotations; solving these simultaneously gives:

$$\theta_A = +11 \cdot 53; \qquad \theta_B = +1 \cdot 93; \qquad \theta_C = -15 \cdot 58; \qquad \theta_D = +16 \cdot 73$$

Substitution in the member equations gives:

$$M_{AB} = 2(23{\cdot}06 + 1{\cdot}93) - 50 \quad\ = -0{\cdot}01$$
$$M_{BA} = 2(3{\cdot}86 + 11{\cdot}53) + 50 \quad\ = +80{\cdot}8$$
$$M_{BC} = 2(3{\cdot}86 - 15{\cdot}58) - 57{\cdot}6 \quad = -80{\cdot}8$$
$$M_{CB} = 2(-31{\cdot}16 + 1{\cdot}93) + 86{\cdot}4 = +28{\cdot}4$$
$$M_{CD} = 2(-31{\cdot}16 + 16{\cdot}73) \quad\ \ \ = -28{\cdot}4$$
$$M_{DC} = 2(33{\cdot}46 - 15{\cdot}58) \quad\ \ \ \ \ = +36{\cdot}0$$

These are the required moments which act on the ends of the beams, signed according to our new convention. The signs can be translated into the bending moment convention by reference to FIGURE 6.12 where the action of the couples of different sign on the ends of a single beam is illustrated.

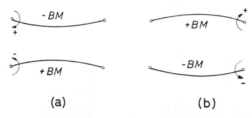

Figure 6.12. (a) Moments acting at the *left* hand end of a beam produce bending moments of *opposite* sign; (b) Moments acting at the right hand end of the beam produce bending moments of the *same* sign

We deduce from this that the couples $M_{BA}$, $M_{BC}$, $M_{CB}$ and $M_{CD}$ are all positive according to the bending moment convention so that the above results coincide with those shown in FIGURE 6.7(b), obtained by means of the three-moment equations.

(b) The portal frame of FIGURE 6.13(a)

In the previous example joint translation was excluded by the nature of the problem; in the present we can be sure that lateral movement of $B$ and $C$, although possible, will not occur because of symmetry. The structure will evidently take up the deflected shape indicated at (b) from which we see that

$$\left.\begin{array}{l}\theta_B = -\theta_C \\ \theta_E = -\theta_F\end{array}\right\} \qquad \ldots\ldots \text{(i)}$$

There are no loads within the lengths of the beams and so the fixing moments are everywhere zero.

The member equations thus become:

$$\left.\begin{array}{l}M_{BC} = 2(2\theta_B + \theta_C) = 2\theta_B \\ M_{BE} = 2(2\theta_B + \theta_E) = 4\theta_B + 2\theta_E\end{array}\right\} \qquad \ldots\ldots \text{(ii)}$$

$$M_{EB} = 2(2\theta_E + \theta_B) = 4\theta_E + 2\theta_B \qquad \ldots\ldots \text{(iii)}$$

# FRAMES WITH RIGID JOINTS

From the equilibrium of joint $B$, we have

$$\Sigma M_B = M_{BA} + M_{BC} + M_{BE} = 0$$

i.e. $\quad 0 = +35 + 2\theta_B + 4\theta_B + 2\theta_E$

i.e. $\quad 0 = 17 \cdot 5 + 3\theta_B + \theta_E \qquad \ldots (iv)$

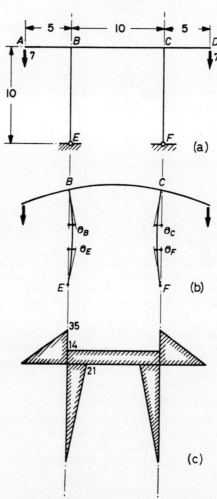

*Figure 6.13.* (a) The portal frame, which might represent a crane structure, is constructed of uniform members having the same cross-section; (b) The symmetry of the structure and its loading ensures symmetrical deflexion, so that $\theta_B = -\theta_C$ and $\theta_E = -\theta_F$; (c) The bending moments have been plotted here on the tension side of the members

From the equilibrium of joint $E$, we have

$$\Sigma M_E = M_{EB} = 0 = 4\theta_E + 2\theta_B \qquad \ldots (v)$$

## 6.3. THE SLOPE-DEFLEXION EQUATIONS

Solving (iv) and (v) simultaneously

$$\left.\begin{array}{l}\theta_B = -7\cdot 0 \\ \theta_E = +3\cdot 5\end{array}\right\}$$

Hence, substituting in equations (ii)

$$M_{BC} = -14\cdot 0$$
$$M_{BE} = -21\cdot 0$$

The bending moment diagram can then be plotted as at (c).

### 6.3.3. Application of the Slope-Deflexion Equations to Frames in which Sway can Occur

#### 6.3.3.1. *Principle of the Method*

The possibility of sway could be eliminated from the previous example only because of the symmetry both of the frame and of its loading; it seems very likely, then, that lateral movement of $BC$ will occur if the columns are made unequal or if one of the loads is changed in value. We now examine this possibility: in Figure 6.14(a) the same frame is shown with only one load acting; the member $CD$ has been removed as unnecessary.

There are now five unknowns, $\theta_B$, $\theta_C$, $\theta_E$, $\theta_F$ and $\delta$, the horizontal displacement of $B$ and $C$. Four equilibrium equations can be written down for the four joints $B$, $C$, $E$ and $F$ and a fifth for the equilibrium of the frame as a whole. This, known as a 'sway equation', is obtained by considering Figure 6.14(c) from which we see that

$$H_E + H_F = 0 \quad \ldots\text{(i)}$$

Taking moments about $B$ and $C$, in turn, to express the conditions of equilibrium of each column we have

$$\left.\begin{array}{l}10.H_E = M_{BE} \\ 10.H_F = M_{CF}\end{array}\right\} \quad \ldots\text{(ii)}$$

so that (i) becomes

$$M_{BE} + M_{CF} = 0 \quad \ldots\text{(iii)}$$

which is the required sway equation.

We can now write down the member equations **omitting the constant** term $2EK$

$$\left.\begin{array}{l}M_{EB} = (2\theta_E + \theta_B - 3\delta/10) \\ M_{BE} = (2\theta_B + \theta_E - 0\cdot 3\delta) \\ M_{BC} = (2\theta_B + \theta_C) \\ M_{CB} = (2\theta_C + \theta_B) \\ M_{CF} = (2\theta_C + \theta_F - 0\cdot 3\delta) \\ M_{FC} = (2\theta_F + \theta_C - 0\cdot 3\delta)\end{array}\right\} \quad \ldots\text{(iv)}$$

From the joint equilibrium equations we obtain

$$\left.\begin{aligned}\Sigma M_E = 0 &= \theta_B + 2\theta_E - 0\cdot3\delta \\ \Sigma M_B = 0 &= 4\theta_B + \theta_C + \theta_E - 0\cdot3\delta + 35 \\ \Sigma M_C = 0 &= \theta_B + 4\theta_C + \theta_F - 0\cdot6\delta \\ \Sigma M_F = 0 &= \theta_C + 2\theta_F - 0\cdot3\delta\end{aligned}\right\} \quad \ldots(v)$$

From the sway equation (iii) we have

$$0 = 2\theta_B + 2\theta_C + \theta_E + \theta_F - 0\cdot6\delta$$

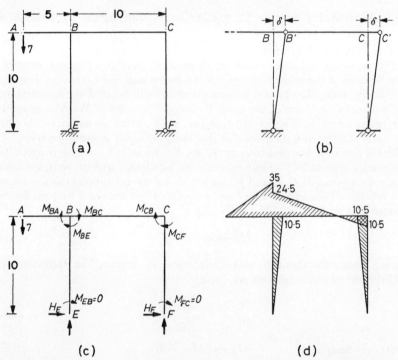

*Figure 6.14.* (a) An unsymmetrical arrangement of the frame and its loading usually results in sway if this can be accommodated by bending of the members; (b) The nature of the sway can be visualised by replacing the rigid joints by pin joints. The joints $B$ and $C$ can thus be seen to move through the same horizontal distance if the axial distortion of $BC$ is excluded from consideration; (c) In order to obtain the sway equation the equilibrium of the whole frame is studied. The end couples $M_{BA}$, $M_{BE}$ etc. are here shown acting in the positive directions; (d) The final bending moment diagram with moments plotted on the tension side of the members

These equations are now solved simultaneously and give

$$\left.\begin{aligned}\theta_B &= -5\cdot5(7/3) & \theta_F &= -4(7/3) \\ \theta_C &= +0\cdot5(7/3) & \delta &= -25(7/3) \\ \theta_E &= -1(7/3) & &\end{aligned}\right\}$$

## 6.3. THE SLOPE-DEFLEXION EQUATIONS 6.3.3.2

When these results are substituted in (iv) we obtain the following values of the end moments:

$M_{EB} = 0$  $\qquad\qquad M_{CB} = -10·5$

$M_{BE} = -10·5$  $\qquad M_{CF} = +10·5$

$M_{BC} = -24·5$  $\qquad M_{FC} = 0$

giving the bending moment diagram shown at (d).

### 6.3.3.2. *Derivation of Sway Equations*

In the previous problem the sway equation was obtained by considering the equilibrium of the frame as a whole; in more complicated cases the equilibrium of suitable parts of the frame must be examined and this is best done as follows:

(i) The appropriate number of sway equations must first be found by identifying the number of independent joint translations, $\delta$, that exist.

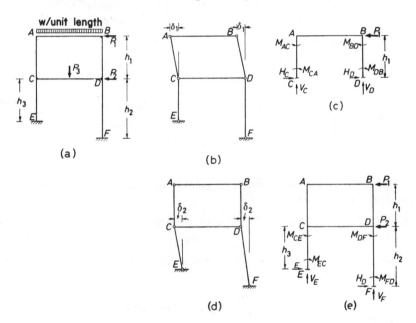

Figure 6.15. (a) The two-storey frame has two modes of sway; (b) The two modes of sway are revealed by replacing the rigid joints by pinned joints. Sway of the upper storey is called Sway I; (c) The corresponding sway equation is obtained by isolating the upper storey; (d) Sway of the lower storey is called Sway II; (e) Cuts are made in the lower-storey columns to give the second sway equation

This can be done very readily by imagining the rigid joints to be replaced by pinned joints; the number of independent joint translations is then equal to the number of kinematic degrees of freedom possessed by the mechanism which the frame has now become.

## FRAMES WITH RIGID JOINTS

(ii) The geometrical relationship between the $\delta$'s of the individual members and the independent joint translations must then be found. Here again the analogous pin-jointed mechanism will be found to be very useful.

(iii) The sway equations are now obtained by considering the equilibrium of the parts of the frame which are distorted when displacements, corresponding to each of the degrees of freedom of the pin-jointed analogue, are imposed in turn.

This procedure is now illustrated by deriving the sway equations for a number of frames.

(a) Double-storey portal frame. FIGURE 6.15

(i) When the frame is replaced by the corresponding mechanism, as at (b) and (d), the two possible modes of sway are at once revealed and the two displacement terms, $\delta_1$ and $\delta_2$, identified.

(ii) The two sway equations are obtained by considering the equilibrium of the part of the frame associated with each type of sway. Thus for Sway I the panel $ABCD$ is isolated by sections and we have

Resolving horizontally: $H_C + H_D = P_1$

In $AC$: Moments about A: $H_C = \dfrac{M_{AC} + M_{CA}}{h_1}$

In $BD$: Moments about B: $H_D = \dfrac{M_{BD} + M_{DB}}{h_1}$

i.e.
$$\frac{M_{AC} + M_{CA}}{h_1} + \frac{M_{BD} + M_{DB}}{h_2} = P_1 \qquad \ldots\text{(i)}$$

For Sway II the whole frame is isolated by sections cutting the lower panel and we have

$$\frac{M_{CE} + M_{EC}}{h_3} + \frac{M_{DF} + M_{FD}}{h_2} = P_1 + P_2 \qquad \ldots\text{(ii)}$$

Sway equations for frames with vertical columns are always of this type and can be written generally as

$$\sum \frac{M \text{ columns}}{\text{col. height}} = f(\text{lateral loads}) \qquad \ldots(6.6)$$

It will be noticed that the vertical loads do not appear in this equation, but if the columns are inclined, as in FIGURE 6.16(a) it is found that all the loads are involved and also that the displacement terms become interconnected.

(b) FIGURES 6.16(b) and (d) show the movements associated with each degree of freedom and the diagrams alongside the joints enable the '$\delta$' terms for each member to be deduced. Sway I, for example, shown at (b), involves the movement of $A$ to $A'$ and $B$ to $B'$, the horizontal components of these movements, $aA'$ and $Bb$, being equal. Now

## 6.3. THE SLOPE-DEFLEXION EQUATIONS

$\triangle AaA'$ is similar to $\triangle CGA$ so that $AA'$, the '$\delta$' term for the member $CA$ is $\dfrac{13\delta_1}{12}$; in a similar way we find that $BB'$ is $\dfrac{12\cdot 5\delta_1}{12}$, while the '$\delta$' term for the member $AB$ is $Aa + B'b$, that is $-\left(\dfrac{5\delta_1}{12} + \dfrac{3\cdot 5\delta_1}{12}\right)$

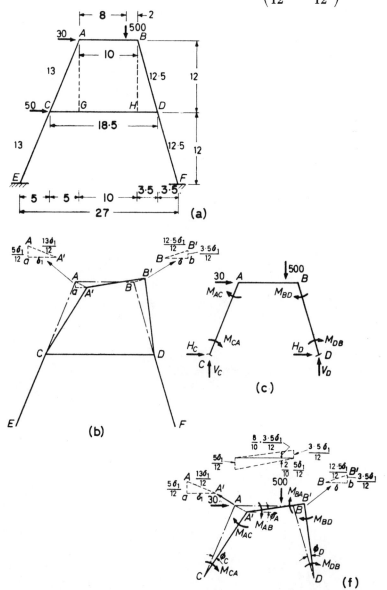

*Figure 6.16.* (a) Two-storey frame with inclined columns under vertical and horizontal loads; (b) Analogous frame with pinned joints distorted in the manner of Sway I; (c) The upper storey isolated so that the first sway equation can be derived; (f) Derivation of first sway equation by means of virtual work

Sway II, at (d), has the additional feature that the movements $CC'$ and $DD'$ of the lower joints also involve movements $AA'$ and $BB'$ so that an additional amount

$-\dfrac{8\cdot 5\delta_2}{12}$ has to be added to the '$\delta$' term for the number $AB$.

Some of the member equations are now given:

$$M_{AB} = 2EK_{AB}\left\{2\theta_{AB} + \theta_{BA} + \dfrac{3 \times 8\cdot 5\,(\delta_1 + \delta_2)}{12 \times 10}\right\} + M_{FAB}$$

$$M_{CA} = 2EK_{CA}\left(2\theta_C + \theta_A - \dfrac{3 \times 13\delta_1}{12 \times 13}\right)$$

$$M_{EC} = 2EK_{EC}\left(\theta_C - \dfrac{3 \times 13\delta_2}{12 \times 13}\right)$$

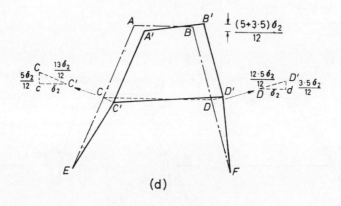

(d)

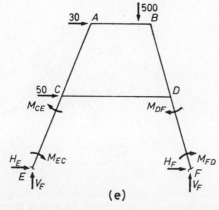

(e)

*Figure 6.16. (cont.)*; (d) Analogous pin-jointed frame distorted in the manner of Sway II; (e) Derivation of the second sway equation

## 6.3. THE SLOPE-DEFLEXION EQUATIONS

The sway equations are obtained from diagrams (c) and (e). In diagram (c), resolving horizontally,
$$H_C + H_D + 30 = 0$$
Taking moments about $D$ for the whole panel
$$V_C \cdot 18{\cdot}5 + 30.12 - 500 \cdot 5{\cdot}5 + M_{CA} + M_{DB} = 0$$
Taking moments about $C$ for the whole panel
$$V_D \cdot 18{\cdot}5 - 500 \cdot 13 - 30.12 - (M_{CA} + M_{DB}) = 0$$
Taking moments about $A$ for the column $AC$
$$M_{AC} + M_{CA} + V_C \cdot 5 - H_C \cdot 12 = 0$$
Taking moments about $B$ for the column $BD$
$$M_{BD} + M_{DB} - V_D \cdot 3{\cdot}5 - H_D \cdot 12 = 0$$
Hence
$$(M_{AC} + M_{BD}) + \frac{10}{18{\cdot}5}(M_{CA} + M_{DB}) = 292 \quad \ldots(I)$$

This is the first sway equation; the second is obtained in an exactly similar manner by considering diagram (e) and is
$$(M_{CE} + M_{DF}) + \frac{18{\cdot}5}{27}(M_{EC} + M_{FD}) = -211 \quad \ldots(II)$$

Clyde[7] and Neal[8] have pointed out that the virtual work principle can be used to combine steps (ii) and (iii) above and so to obtain the sway equation from the geometrical relationships between the displacements of the various members. Thus in FIGURE 6.16(f) the upper storey is shown in its displaced position with the end couples and external couples acting. The necessary relationship between these couples and loads is then obtained by using the virtual work principle to express the condition of equilibrium.

Thus
$$(M_{AC} + M_{CA})\phi_C - (M_{AB} + M_{BA})\phi_A + (M_{BD} + M_{DB})\phi_D + 30\delta_1$$
$$- 500\left(\frac{8}{10} \times \frac{3{\cdot}5}{12} - \frac{2}{10} \times \frac{5}{12}\right)\delta_1 = 0$$

Now
$$\phi_C = \frac{13\delta_1}{12 \times 13} \; ; \; \phi_A = \frac{5+3{\cdot}5}{12} \times \frac{\delta_1}{10} \; ; \; \phi_D = \frac{12{\cdot}5\delta_1}{12 \times 12{\cdot}5}$$

And $\quad M_{AB} + M_{AC} = 0 \; ; \; M_{BA} + M_{BD} = 0$

Hence
$$(M_{AC} + M_{BD}) + \frac{20}{37}(M_{CA} + M_{DB}) = \frac{10{,}800}{37}$$

and this is the same equation as was obtained above.

The final example is the pitched roof portal shown in FIGURE 6.17(a). There are two independent joint translations corresponding to two modes of sway as indicated at (b) and (c) from which we obtain the required geometrical relationships connecting $\phi = \dfrac{\delta}{L}$ for each member.

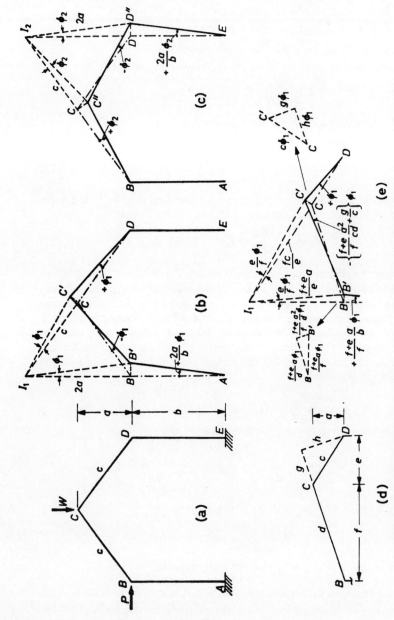

*Figure 6.17.* (a) The ridged portal frame has two modes of sway; (b) The relationship between the angles of sway for the members involved in Sway I; (c) The relationship between the angles of sway for the members involved in Sway II; (d) Dimensions for the ridge of an unsymmetrical portal; (e) Sway I for an unsymmetrical portal

## 6.3. THE SLOPE-DEFLEXION EQUATIONS

Thus for Sway I, at (b), the instantaneous centre of rotation is at $I_1$ so that

$$\phi_{DC} = \phi_{CD} = +\phi_1 \quad ; \quad \phi_{AB} = \phi_{BA} = +\frac{2a}{b}\phi_1$$

As the rafters $BC$ and $CD$ are equal in length

$$\phi_{BC} = \phi_{CB} = -\phi_{CD} = -\phi_1$$

[If the rafters are unequal, as at (d), rather more involved geometry, indicated at (e), is required to obtain the corresponding results]

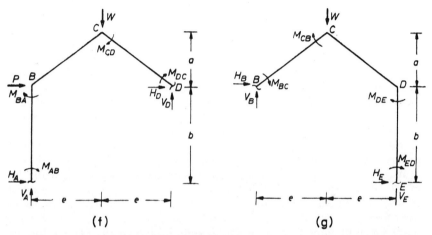

Figure 6.17 (cont.); (f) Derivation of Sway Equation I; (g) Derivation of Sway Equation II

Similar results for Sway II are obtained from (c)

i.e.
$$\phi_{BC} = \phi_{CB} = +\phi_2 = -\phi_{CD} = -\phi_{DC}$$

$$\phi_{DE} = \phi_{ED} = +\frac{2a}{b}\phi_2$$

Some of the member equations are now given:

$$M_{AB} = 2EK_{AB}\left\{\theta_B - 3\frac{2a}{b}\phi_1\right\}$$

$$M_{BC} = 2EK_{BC}\{2\theta_B + \theta_C - 3(-\phi_1 + \phi_2)\}$$

$$M_{CD} = 2EK_{CD}\{2\theta_C + \theta_D - 3(\phi_1 - \phi_2)\}$$

$$M_{DE} = 2EK_{DE}\left\{2\theta_D - 3\frac{2a}{b}\phi_2\right\}$$

The sway equation corresponding to Sway I follows from FIGURE 6.17(f) and is

$$P - W\frac{e}{2a} = -\frac{1}{b}(M_{AB} + M_{BA}) - \frac{1}{2a}(M_{BA} + M_{DC}) - \frac{1}{a}M_{CD}$$

The second sway equation, corresponding to Sway II, is obtained from (g) and is

$$W \frac{e}{2a} = -\frac{1}{2a}(M_{BC} + M_{DE}) - \frac{1}{b}(M_{DE} + M_{ED}) - \frac{1}{a} M_{CB}$$

Further examples of the derivation of sway equations will be found in a paper by Matheson[9].

The modifications which must be made to the slope-deflexion method before it can be used for non-uniform beams are described in Appendix A.

### 6.3.4. Comments on the Slope-Deflexion procedure

The reader will perhaps have noticed that in setting up the slope-deflexion and sway equations no account has been taken of the number of redundants; the procedure thus appears to have a different logical basis from other methods, especially those based on the flexibility coefficient equations, which have been described previously. In the portal frame of Figure 6.14, for example, the horizontal component of the reaction is the only external redundant; this could presumably have been found from a single flexibility coefficient equation whereas the present method involved the solution of no less than five simultaneous equations.

It can be seen, however, that the slope-deflexion procedure is really one of the 'equilibrium methods' mentioned in the introduction to Chapter 4. It will be recalled that in these methods the compatibility conditions are used first, to be followed by the equilibrium equations expressed in terms of the forces required to give the actual joint displacements. In the present method the fact that the members are rigidly attached to one another at the joints leads at once to the compatibility equations, while equations 6.5 give the moments required to produce unit joint rotation and translation.

The slope-deflexion method therefore, like other equilibrium methods, shows to best advantage when the number of joints that can move is small in comparison with the number of members. In all these methods forces, or moments, disappear from the calculations at an early stage so that the exact number of redundants is irrelevant; but the arithmetic required to translate the original loads to angles and back again to bending moments is not directly useful and, to that extent, is a waste of effort.

The slope-deflexion method is very convenient for problems, such as symmetrical closed frames, where the number of unknowns is small; but as soon as there is more than a small number of simultaneous equations to be solved some other method is usually to be preferred. The basic equations, however, are of fundamental importance; we now use them to derive the basic formulae of the moment distribution method.

### 6.4. THE MOMENT DISTRIBUTION METHOD

In 1930 Hardy Cross published his famous paper[2] and started a revolution which has transformed engineering thinking and design office practice throughout the world. Although Timoshenko[10] traces the origins of the method back through the work of Calisev and Waddell to Mohr himself,

## 6.4. THE MOMENT DISTRIBUTION METHOD

(and was sharply taken to task by Bateman[11] for so doing) there can be no doubt that the idea of successive approximations, applied directly to the structure rather than to a series of simultaneous equations, came to be widely studied and accepted as a direct result of Cross's writings. It is true that the original paper was incomplete and singularly succinct; that most of the difficulties, especially that of dealing with sway, were in fact surmounted by others; and that we can now recognize in Southwell's Relaxation method a much more powerful and adaptable technique. But Hardy Cross's contribution is secure; by bringing the esoteric mystery of rigid frame analysis within the competence of every engineer he enabled the designer to catch up with the constructor of reinforced concrete and welded steelwork, who was fast outstripping him, and in so doing he prepared the way for the tremendous development of successive approximation methods which now pervade every branch of engineering science.

The immediate success of his work is easily explained: engineers are accustomed to trying to visualize the action of the structures they are designing and in this country at least, if not on the continent of Europe, prefer such an intuitive approach to one involving elaborate mathematics. The moment distribution method lends itself admirably to this procedure for it is easy to picture the structure with its joints locked by imaginary external clamps which can be relaxed in sequence. This process is described below after the necessary basic equations have been derived.

### 6.4.1. DERIVATION OF THE BASIC EQUATIONS OF MOMENT DISTRIBUTION

The following derivations apply exclusively to uniform beams; the modifications which can be made when non-uniform beams are involved are described in Appendix A.

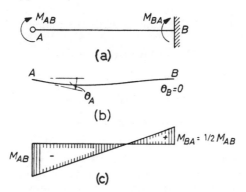

*Figure 6.18.* (a) A couple $M_{AB}$ acts at the pinned end $A$ of the beam $AB$; a couple $M_{BA}$ is induced at the fixed end $B$; (b) The deflected shape of the beam; (c) The corresponding bending moment diagram

(i) Couple acting on Beam $AB$: far end clamped, FIGURE 6.18(a). It is clear from the deflexion diagram (b) that a restraining couple $M_{BA}$ is induced at the clamped end $B$.

## FRAMES WITH RIGID JOINTS

Applying the slope-deflexion equation 6.5 to the ends $A$ and $B$ in turn, we have:

$$M_{AB} = 2EK_{AB}(2\theta_A) = 4EK_{AB}\,\theta_A \qquad \ldots\ldots(6.7)$$

$$M_{BA} = 2EK_{AB}(\theta_A)$$

i.e. $\qquad M_{BA} = \tfrac{1}{2} M_{AB} \qquad \ldots\ldots(6.8)$

The factor $\tfrac{1}{2}$ is spoken of as the 'carry-over factor', while the stiffness of the beam, that is the end couple required to give unit slope, is $4EK_{AB}$.

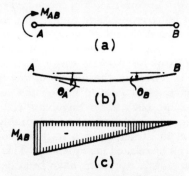

*Figure 6.19.* (a) A couple $M_{AB}$ acts at the end $A$ of the pin ended beam $AB$; (b) The deflected shape of the beam; (c) The corresponding bending moment diagram

(ii) *Couples acting on Beam AB: far end pinned.* FIGURE 6.19(a). The slope at the end $B$ is now $\theta_B$ while the moment there is zero and we have:—

$$M_{AB} = 2EK_{AB}(2\theta_A + \theta_B)$$
$$0 = M_{BA} = 2EK(2\theta_B + \theta_A)$$

i.e. $\qquad \theta_B = -\tfrac{1}{2}\theta_A \qquad \ldots\ldots(6.9)$

and $\qquad M_{AB} = 3EK_{AB}\,\theta_A \qquad \ldots\ldots(6.10)$

Comparing equations 6.7 and 6.10 we see that the stiffness of the beam has now been reduced to $\tfrac{3}{4}$ of its former value, while equations 6.8 and 6.9 show that the carry-over factor of $\tfrac{1}{2}$ applies to angles as well as to moments although with an obvious change of sign.

(iii) *Partition of moment between members meeting at a joint.* FIGURES 6.20 and 6.21.

If a couple $M$ is applied to the joint $B$ as in FIGURE 6.20(a) it will be divided into $M_{BA}$, acting on beam $BA$, and $M_{BC}$, acting on $BC$. The relationship between these two couples follows from consideration of FIGURE 6.20(b) from which we see that

$$\theta_{BA} = \theta_{BC}$$

## 6.4. THE MOMENT DISTRIBUTION METHOD

Hence from equation 6.7

$$\frac{M_{BA}}{4EK_{AB}} = \frac{M_{BC}}{4EK_{BC}}$$

From which we obtain

$$\frac{M_{BA}}{M} = \frac{K_{AB}}{K_{AB} + K_{BC}} \quad \ldots (6.11)$$

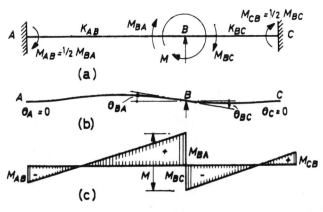

*Figure 6.20.* (a) When a couple $M$ acts at the interior support of this two-span beam it is divided between the spans $BA$ and $BC$ in proportion to the $K$-values of the two beams. Couples $M_{AB}$ and $M_{CB}$ are induced at the fixed ends $A$ and $C$; (b) The deflected shape of the beam; (c) The corresponding bending moment diagram

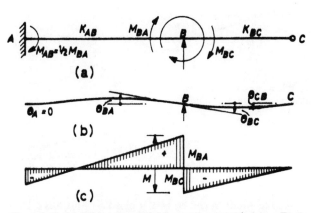

*Figure 6.21.* (a) When one of the beams is pin-ended its effective $K$-value becomes equal to $\frac{3}{4} I/L$; (b) The deflected shape of the beam; (c) The corresponding bending moment diagram

If one of the beams, say $BC$, is pinned at the far end, we shall have, as before,

$$\theta_{BA} = \theta_{BC}$$

Hence from equations 6.7 and 6.10

$$\frac{M_{BA}}{4EK_{AB}} = \frac{M_{BC}}{3EK_{BC}}$$

From which we obtain

$$\frac{M_{BA}}{M} = \frac{K_{AB}}{K_{AB} + \tfrac{3}{4} K_{BC}}$$

$$\frac{M_{BC}}{M} = \frac{\tfrac{3}{4} K_{BC}}{K_{AB} + \tfrac{3}{4} K_{BC}}$$

.... (6.12)

Equations 6.11 and 6.12, which are the basic equations of moment distribution, can readily be extended to explain the partition of moments at a joint where several members meet; at such a joint the moment is distributed in proportion to the *effective K-values* of the members meeting there. The effective $K$-value of a member which is fixed at the far end is $I/L$, and it is $\tfrac{3}{4} I/L$ for a member which is pinned at the far end.

The factors which govern this partition of moments, such as $\dfrac{K_{AB}}{K_{AB} + \tfrac{3}{4} K_{BC}}$ are known as 'distribution factors'.

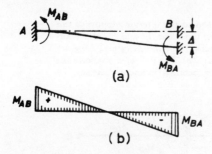

(a)

(b)

*Figure 6.22.* (a) When one end of a beam with direction fixed ends is displaced relative to the other equal couples are induced at the ends; (b) The corresponding bending moment diagram

*Figure 6.23.* (a) When one end is pinned a relative displacement induces a couple at the other; (b) The corresponding bending moment diagram

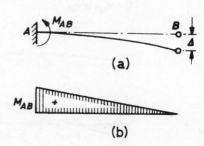

(a)

(b)

(iv) *Relative displacement of the ends of a beam.* FIGURES 6.22 and 6.23.

It is convenient to have expressions for the moments produced at the ends of beam if relative translation, without rotation, is imposed as in FIGURE 6.22(a). From the slope deflexion equation we see at once that

$$M_{AB} = M_{BA} = 2EK_{AB}\left(-\frac{3\Delta}{L}\right) = -\frac{6EK\Delta}{L} \quad \ldots (6.13)$$

## 6.4. THE MOMENT DISTRIBUTION METHOD

If one end of the beam is pinned, as in FIGURE 6.23(a) we shall have

$$0 = M_{BA} = 2EK_{AB}\left(2\theta_B - \frac{3\Delta}{L}\right)$$

i.e.
$$\theta_B = 3\Delta/2L$$

$$M_{BA} = 2EK_{AB}\left(\theta_B - \frac{3\Delta}{L}\right) = -\frac{3EK\Delta}{L} \quad \ldots (6.14)$$

These last two equations are the basis of the method of calculating the effects of sway.

### 6.4.2. MOMENT DISTRIBUTION ANALYSIS OF FRAMES IN WHICH SWAY CANNOT OCCUR

The procedure is as follows:

(i) Imaginary clamps are applied, at each joint, which are capable of applying sufficient restraint to prevent rotation.

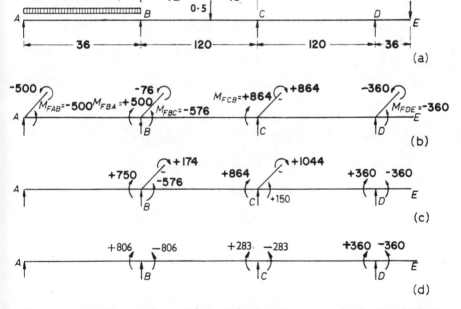

*Figure 6.24.* (a) Uniform continuous beam to be analysed by moment distribution; (b) Each joint is supposed to be initially clamped; when the loads are applied fixed end moments the developed at the ends of each loaded member. The clamps have therefore to supply are algebraic sum of the moments acting at each joint. (The external loads are omitted from this diagram); (c) The clamps have been released and the joints balanced at the pinned ends $A$ and $D$; moments have been carried over to $BA$ and $CD$, and the diagram records the state of affairs following line (f) in TABLE 6.1; (d) After a sufficient number of balances all the joints are in equilibrium and the moments at the ends of the members are as shown

## FRAMES WITH RIGID JOINTS

(ii) The loads are applied: the clamps prevent rotation of the joints by exerting restraining moments equal to the algebraic sum of the fixed end moments.

(iii) The clamps are released one by one: this means that at each joint a moment, equal and opposite to the restraining moment, is applied and distributed between the members meeting at the joint in accordance with equations 6.11 or 6.12.

(iv) The moment at the far end of each member thus released is adjusted by ' carrying over ' in accordance with equation 6.8.

(v) The process is repeated, joint by joint, until the out-of-balance moments provided by the clamps are judged to be negligibly small. The final moment acting at the end of each member is obtained by addition.

Example (a)

This process is now illustrated by reference to the continuous beam of FIGURE 6.24, which has previously been analysed in other ways. The necessary arithmetical operations can be recorded either on a sketch of the frame or in tabular form: the former is perhaps to be preferred as being the more vivid but slips can be more easily avoided with the latter. A necessary preliminary is to work out the distribution factors from equations 6.12 and, as a mistake at this stage will probably not be detected, it is worth while to set out the calculations, simple as they are, in a systematic manner. Noticing that joints $A$ and $D$ are effectively pinned ends, we have:

| Joint | Member | $I$ | $L$ | $I/L$ | Effective $K$-value | $\Sigma K$ | Distribution Factor |
|---|---|---|---|---|---|---|---|
| $B$ | $BA$ | $I$ | 120 | $I/120$ | $\tfrac{3}{4} \times I/120$ | $\tfrac{7}{4} \cdot \tfrac{I}{120}$ | 3/7 |
|  | $BC$ | $I$ | 120 | $I/120$ | $I/120$ |  | 4/7 |
| $C$ | $CB$ | $I$ | 120 | $I/120$ | $I/120$ | $\tfrac{7}{4} \cdot \tfrac{I}{120}$ | 4/7 |
|  | $CD$ | $I$ | 120 | $I/120$ | $\tfrac{3}{4} \times I/120$ |  | 3/7 |

The imaginary clamps are now applied and the fixed end moments computed. These are

$$M_{FAB} = - M_{FBA} = - wL^2/12 = - 50$$

$$M_{FBC} = - \frac{5 \times 72 \times 48^2}{120^2} = - 57 \cdot 6$$

$$M_{FCB} = + \frac{5 \times 72^2 \times 48}{120^2} = 86 \cdot 4$$

$$M_{FDE} = - 1 \times 36 = - 36$$

The last term requires some explanation: strictly speaking it is not a fixed end moment, as the term is ordinarily used, but the clamp will, in the first instance, have to support the unit load on its cantilever.

## 6.4. THE MOMENT DISTRIBUTION METHOD  6.4.2

The successive steps of the distribution are set out in Table 6.1.

Line (b) gives the fixed end moments multiplied by 10 for convenience, and the first step is to release the end joints which are pinned. Beginning at $A$, in line (c), the joint is balanced by releasing the clamp, this is equivalent to allowing a moment of $+500$ to come on to the joint. In line (d) half this moment, $+250$ (to the nearest whole number) is carried over to the other end of the member, in accordance with equation 6.8, so that $M_{AB}$ is now $+500 + 250 = +750$. Joint $D$ is now balanced, in line (e), by relaxing the clamp and allowing a moment of $+360$ to come on to the joint; in line (f) a moment of $+\frac{1}{2} \cdot 360 = +180$ is carried over to the end $C$ of member $CD$.

TABLE 6.1
MOMENT DISTRIBUTION ANALYSIS. FIGURE 6.24

| A | B | | C | | D | | Joint |
|---|---|---|---|---|---|---|---|
| AB | BA | BC | CB | CD | DC | DE | End of Member |
|  | 3/7 | 4/7 | 4/7 | 3/7 | 1 | 0† | (a) Distribution factor |
| − 500<br>+ 500 | + 500 | − 576 | + 864 |  |  | − 360 | (b) Fixed End Moments<br>(c) Balance A |
|  | + 250 |  |  |  | + 360 |  | (d) Carry-over<br>(e) Balance D |
|  |  |  | − 596 | + 180<br>− 448 |  |  | (f) Carry-over<br>(g) Balance C |
|  | + 53 | − 298<br>+ 71 |  |  |  |  | (h) Carry-over<br>(i) Balance B |
|  |  |  | + 35<br>− 21 | − 14 |  |  | (j) Carry-over<br>(k) Balance C |
|  | + 4 | − 10<br>+ 6 |  |  |  |  | (l) Carry-over<br>(m) Balance B |
|  |  |  | + 2<br>− 1 | − 1 |  |  | (n) Carry-over<br>(o) Balance C |
| 0 | + 806 | − 806 | + 283 | − 283 | + 360 | − 360 | (p) Total |
| 0 | + 80·6 | − 80·6 | + 28·3 | − 28·3 | + 36·0 | − 36·0 | (q) Total × $^{1}/_{10}$ |

† The cantilever $DE$ offers no resistance to rotation, so that its stiffness is zero.

As the end joints $A$ and $D$ are now in equilibrium the clamps are unnecessary and can be discarded. At joint $C$, however, the clamp is providing a restraining couple of $864 + 180 = +1044$; when this clamp is relaxed a moment of $-1044$ comes on to the joint and is divided between the members meeting there, $CB$ and $CD$, in accordance with equation 6.12, for $CD$ is now a pin-ended member. Thus $CB$ gets $\frac{4}{7}(-1044) = -596$ and $CD$ gets

FRAMES WITH RIGID JOINTS

$\frac{3}{7}(-1044) = -448$; these are entered in line (g), the horizontal line signifying that the joint is balanced at this stage. The moment carried over to $BC$ is $-\frac{1}{2}.596 = -298$, in line (h), but no moment is carried over to $D$ which has been permanently converted into a pinned joint (cf. FIGURE 6.21(c)).

The out-of-balance moment provided by the clamp at $B$ is now $-124$ and to balance this moments of $+53$ and $+71$, which are in the ratio of $3 : 4$, are introduced at $BA$ and $BC$ respectively.

The processes of balancing and carrying-over are repeated at joints $B$ and $C$ until the numbers involved are sufficiently small. The total moment at each end of each member is then obtained by summing the columns and dividing by 10. It will be noticed that the results coincide with those found previously by Clapeyron's method and by slope deflexion but the amount of labour required is noticeably smaller than in either of the previous methods.

Example (b):

The members of the two-storey portal frame shown in FIGURE 6.25(a)

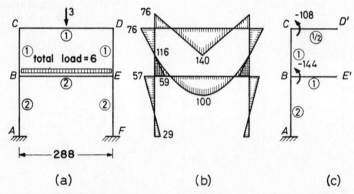

*Figure 6.25.* (a) Symmetrical frame symmetrically loaded; (b) Corresponding bending moment diagram; (c) Substitute frame which gives the same result

have the $K$-values given in circles. The distribution factors are therefore:

| Joint | Member | K | Σ | Distribution factor | |
|---|---|---|---|---|---|
| $C$ | $CB$ | 1 | 2 | $\frac{1}{2}$ | and similarly |
|  | $CD$ | 1 |  | $\frac{1}{2}$ | for joint $D$ |
| $B$ | $BA$ | 2 | 5 | 2/5 | and similarly |
|  | $BC$ | 1 |  | 1/5 |  |
|  | $BE$ | 2 |  | 2/5 | for joint $E$ |

272

## 6.4. THE MOMENT DISTRIBUTION METHOD 6.4.2

The fixed end moments are

$$M_{FCD} = -M_{FDC} = -\frac{PL}{8} = -\frac{3 \times 288}{8} = -108$$

$$M_{FBE} = -M_{FEB} = -\frac{wL^2}{12} = -\frac{(6/288) \times 288^2}{12} = -144$$

TABLE 6.2(a)

MOMENT DISTRIBUTION ANALYSIS : FIGURE 6.25(a)

| ↓CB | CD → | ← DC | | ↓DE | | |
|---|---|---|---|---|---|---|
| 1/2 | 1/2 | 1/2 | | 1/2 | | |
| | −108 | +108 | | | | Fixed end moments |
| +14 | | | | −14 | | Carry over |
| +47 | +47 | −47 | | −47 | 2 | Balance C & D |
| | −24 | +24 | | | | Carry over |
| +12 | +12 | −12 | | −12 | 4 | Balance C & D |
| | −6 | +6 | | | | Carry over |
| +3 | +3 | −3 | | −3 | 6 | Balance C & D |
| +76 | −76 | +76 | | −76 | | TOTAL |

| ↓BA | ↑BC | BE → | ← EB | ↓EF | ↑ED | | |
|---|---|---|---|---|---|---|---|
| 2/5 | 1/5 | 2/5 | 2/5 | 2/5 | 1/5 | | |
| | | −144 | +144 | | | | Fixed end moments |
| +58 | +28 | +58 | −58 | −58 | −28 | 1 | Balance B & E. |
| | +24 | −29 | +29 | | −24 | | Carry over |
| +2 | +1 | +2 | −2 | −2 | −1 | 3 | Balance B & E |
| | +6 | −1 | +1 | | −6 | | Carry over |
| −2 | −1 | −2 | +2 | +2 | +1 | 5 | Balance B & E |
| | +1 | +1 | −1 | | −1 | | Carry over |
| −1 | | −1 | +1 | +1 | | 7 | Balance B & E |
| +57 | +59 | −116 | +116 | −57 | −59 | | TOTAL |

| AB |
|---|
| +29 |
| +1 |
| −1 |
| +29 |

| FE |
|---|
| −29 |
| −1 |
| +1 |
| −29 |

} Carry over

TOTAL

# FRAMES WITH RIGID JOINTS

The distribution is carried out in TABLE 6.2(a) which has been arranged so that moments are 'carried over' to an adjacent column (either above, below, or to one side). In order to preserve symmetry, however, the joints have been balanced in pairs. Thus the first step was to balance joints $B$ and $E$; the next to carry over; the next to balance joints $C$ and $D$ and so on. The total end moments are given at the foot of each column and have been used to plot the bending moment diagram shown at (b) in FIGURE 6.25.

TABLE 6.2(b)
MOMENT DISTRIBUTION ANALYSIS: FIGURE 6.25(c)

| ↓CB | CD′ | |
|---|---|---|
| 2/3 | 1/3 | |
| +72 | −108<br>+36 | Fixed end moment<br>1. Balance $C$ |
| +13<br>−9 | −4 | Carry over<br>3. Balance $C$ |
| +76 | −76 | TOTAL |

| ↓BA | ↑BC | BE′ | |
|---|---|---|---|
| 1/2 | 1/4 | 1/4 | |
| +54 | +36<br>+27 | −144<br>+27 | Fixed end moment<br>Carry over<br>2. Balance $B$ |
| +2 | −4<br>+1 | +1 | Carry over<br>4. Balance $B$ |
| +56 | +60 | −116 | TOTAL |

| AB | |
|---|---|
| +27<br>+ 1 | } Carry over |
| +28 | TOTAL |

Examination of the distribution figures, however, reveals that when a certain moment has been allotted to one end of a beam ($+58$ to $BE$ in step 1, for example) the succeeding carry-over at once reduces that moment by one half ($-29$ has been carried over to $BE$ from $EB$). It therefore seems that we could carry out the distribution for one half of the frame only if the effective $K$-value of the beam is reduced to one half of its actual value.

FIGURE 6.25(c) shows the substitute frame that gives effect to this idea; the corresponding distribution is given in TABLE 6.2(b) and it is seen that, apart from rounding off differences, the solution coincides with the previous one.

## 6.4. THE MOMENT DISTRIBUTION METHOD    6.4.2

This is a very useful device for symmetrical frames; not only is the work reduced because only half the frame is analysed but convergence is assisted because the carry-over factor for beams is now zero.

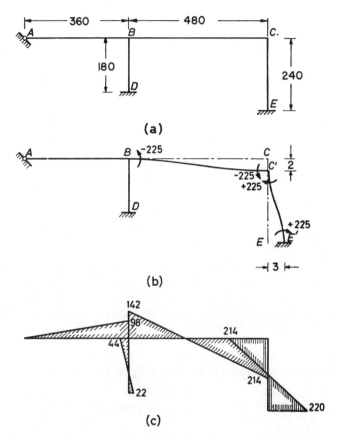

*Figure 6.26.* (a) The frame shown here is to be analysed for the effect of a downward and sideways movement of $E$; (b) The fixed end moments from which distribution starts are those produced if the frame is pushed into the distorted position with imaginary clamps preventing rotation of the joints; (c) The final bending moment diagram

**Example (c):**

The $EI$ value of the girder $ABC$ in Figure 6.26(a) is $43 \cdot 2 \times 10^9$, while that of the columns $BD$ and $CE$ is $7 \cdot 2 \times 10^9$. It is required to find the moments induced if a foundation subsidence causes $E$ to move a distance 2 downwards and 3 to the right without rotation.

The fixed end moments are due to the subsidence and are given by equation 6.13, the frame being pushed into the displaced position with all joints locked against rotation, as at (b).

275

# FRAMES WITH RIGID JOINTS

Thus

$$M_{FBC} = M_{FCB} = -\frac{6EK_{BC}\Delta}{L} = -6\left(\frac{43\cdot 2 \times 10^9}{480}\right)\frac{2}{480} = -225.10^4$$

$$M_{FCE} = M_{FEC} = +\frac{6EK_{EC}\Delta}{L} = +6\left(\frac{7\cdot 2 \times 10^9}{240}\right)\frac{3}{240} = +225.10^4$$

The distribution factors are worked out as follows:

| Joint | Member | $EI$ $\times 10^{-9}$ | $L$ | $EI/L$ $\times 10^{-6}$ | Effective $K$-value $\times 10^{-6}$ | $\Sigma K$ $\times 10^{-6}$ | Distribution Factor |
|---|---|---|---|---|---|---|---|
| B | BA | 43·2 | 360 | 120 | $\tfrac{3}{4} \times 120 = 90$ |  | 9/22 |
|  | BD | 7·2 | 180 | 40 | 40 | 220 | 4/22 |
|  | BC | 43·2 | 480 | 90 | 90 |  | 9/22 |
| C | CB | 43·2 | 480 | 90 | 90 |  | 3/4 |
|  | CE | 7·2 | 240 | 30 | 30 | 120 | 1/4 |

The distribution, given in TABLE 6.3, follows very similar lines to the previous one: on entering the fixed end moments it is noticed that joint $C$ happens to be balanced to start with so distribution begins at joint $B$. The moment of $+41$ carried over to joint $C$ causes a lack of balance there which is eventually liquidated by balancing $C$, $B$ and finally $C$, in turn. The end moments obtained by summing each column are converted to the bending moment convention and plotted at (c) in FIGURE 6.26.

The main interest of the preceding example lies in the method adopted to bring what is essentially a sway problem within the scope of moment distribution. The extension of the properties of the imaginary joint clamps

TABLE 6.3
MOMENT DISTRIBUTION ANALYSIS OF FRAME OF FIGURE 6.26(a)

| A | B | | | C | | E | D | |
|---|---|---|---|---|---|---|---|---|
| AB | BA | BD | BC | CB | CE | EC | DB | |
|  | 9/22 | 4/22 | 9/22 | 3/4 | 1/4 |  |  |  |
|  |  |  | −225 | −225 | +225 | +225 |  | F.E. Moments |
| +92 | +41 | +92 |  |  |  |  |  | Balance B |
|  |  |  |  | +41 |  |  | +20 | Carry over |
|  |  |  |  | −31 | −10 |  |  | Balance C |
|  |  |  | −15 |  |  | −5 |  | Carry over |
| +6 | +3 | +6 |  |  |  |  |  | Balance B |
|  |  |  |  | +3 |  |  | +2 | Carry over |
|  |  |  |  | −2 | −1 |  |  | Balance |
|  | +98 | +44 | −142 | −214 | +214 | +220 | +22 | Total |

so that they can accommodate a specified joint translation, without accompanying translation, gives the key to the solution of sway problems of a more general character. Although many variants of the basic procedure have been published in recent years the principles now to be described are fundamental and, in one form or another, can be recognized in all methods of sway analysis.

### 6.4.3. Moment Distribution Analysis of Frames in which Sway can occur

The moment distribution process described above differs from other methods, slope-deflexion for example, only in the method used to ensure that the equations of equilibrium and compatibility are satisfied at every joint; the equations themselves are universal. Exactly the same can be said of joint translation and its analysis. The geometrical relationships which lead to the compatibility equations and the sway equations of equilibrium are exactly the same, and can be obtained in the same way, as was described in Section 6.3.3.2 of this Chapter in connexion with the slope-deflexion method. But the solution of these equations is now reached by a series of approximations which are equivalent to the successive relaxations of appropriately designed imaginary clamps. In the following Section a method of correcting for sway by proportion is described which follows the fundamental conception rather closely; developments of this procedure which are often more convenient in actual practice are described later.

#### 6.4.3.1. *Correction for Sway Effects by Proportion*

In all methods of analysis of sway effects the first step is the recognition of the number of modes of sway that are possible, and this is most readily done by finding the number of degrees of freedom to the corresponding pin-jointed mechanism, as in Section 6.3.3.2.(i). The subsequent procedure is most conveniently introduced by reference to frames with but one mode of sway, with which we begin.

6.4.3.1.1. *Application to frames with a single mode of sway.*—(i) The sway equation is derived by describing the equilibrium of the frame in terms of the column moments and the external loads, as in Section 6.3.3.2.

(ii) The fixed end moments are distributed while sway is prevented by imaginary forces acting at the appropriate joint.

(iii) The resulting moments are substituted in the sway equation and if this is satisfied the imaginary force is zero and no further work is required; if it is not satisfied the imaginary force must be neutralized by superimposing a sway distortion, as follows.

(iv) The geometrical relationships between the sways of the individual members are used with equations 6.13 or 6.14 to give the moments produced when an arbitrary sway displacement is imposed on the frame. This arbitrary displacement is supposed to occur with the joints held against rotation by imaginary clamps exerting couples which are then neutralized by distribution. The result of this sway distribution is to give the moments produced at the ends of the members by an arbitrary sway of the frame with its joints free to rotate during the process.

# FRAMES WITH RIGID JOINTS

(v) It is now necessary to combine the original distribution of step (ii) with the sway distribution in such a way that the sway equation is satisfied. This is achieved by multiplying the sway moments by a suitable factor and adding them to those resulting from the original distribution. This step is tantamount to evaluating the imaginary force operating in step (ii) and applying an equal and opposite force.

This process is now illustrated by analysing the single storey portal frame with sloping legs, shown again in FIGURE 6.27, which was the upper storey of the frame of example (b) of Section 6.3.3.2, FIGURE 6.16; sway equation I (p. 261) applies to this storey.

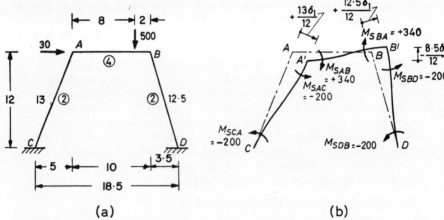

*Figure 6.27.* (a) The figures in circles give the relative $K$-values of the members of this frame, which has one mode of sway; (b) When the frame is forced into the distorted position shown, with the joints prevented from rotating, moments $S_{CA}$, $S_{AC}$ etc. are induced in the members

(i) The sway equation is

$$(M_{AC} + M_{BD}) + \frac{10}{18 \cdot 5}(M_{CA} + M_{DB}) = 292 \quad \ldots \text{(I)}$$

(ii) The fixed end moments are:

$$M_{FAB} = -\frac{500 \cdot 8 \cdot 2^2}{10^2} = -160$$

and $$M_{FBA} = +\frac{500 \cdot 8^2 \cdot 2}{10^2} = +640$$

The figures in circles are the $K$-values for the members so that the distribution factors are:

at $A$: $M_{AB} : M_{AC} = 2/3 : 1/3$

at $B$: $M_{BA} : M_{BD} = 2/3 : 1/3$

TABLE 6.4(a) gives the distribution of fixed end moments with the joints prevented from swaying by an imaginary force.

## 6.4. THE MOMENT DISTRIBUTION METHOD    6.4.3.1.1

(iii) Substitution of the results of this distribution in the sway equation gives

$$+ 140 - 259 + \frac{10}{18 \cdot 5}(70 - 129) = -151 \quad [+292]$$

As this equation is not satisfied sway correction is required.

TABLE 6.4(a)

DISTRIBUTION OF FIXED END MOMENTS WITH SWAY PRESENTED IN FIGURE 6.27

| CA | AC | AB | BA | BD | DB |
|---|---|---|---|---|---|
|  | 1/3 | 2/3 | 2/3 | 1/3 |  |
|  |  | − 160 | + 640<br>− 427 | − 213 |  |
|  | + 124 | − 213<br>+ 249 |  |  | − 107 |
| + 62 |  |  | + 124<br>− 83 | − 41 |  |
|  | + 14 | − 41<br>+ 27 |  |  | − 20 |
| + 7 |  |  | + 14<br>− 9 | − 5 |  |
|  | + 2 | − 5<br>+ 3 |  |  | − 2 |
| + 1 |  |  | + 1<br>− 1 |  |  |
| + 70 | + 140 | − 140 | + 259 | − 259 | − 129 |

(iv) When the frame is pushed into the position indicated at (b), the sway moments $M_{SAB}$, $M_{SBA}$, etc. are induced and can be found from equation 6.13 in terms of the relative displacements $\delta$ of the ends of the members; these were found previously (see FIGURE 6.16(b)) and we have

$$M_{SCA} = M_{SAC} = - \frac{6EK_{AC}\left(\frac{13\delta_1}{12}\right)}{L_{AC}} = -E\delta_1 \quad \alpha - 200$$

$$M_{SAB} = M_{SBA} = + \frac{6EK_{AB}\left(\frac{8 \cdot 5\delta_1}{12}\right)}{L_{AB}} = +1 \cdot 7E\delta_1 \quad \alpha + 340$$

$$M_{SBD} = M_{SDB} = - \frac{6EK_{BD}\left(\frac{12 \cdot 5\delta_1}{12}\right)}{L_{BD}} = -E\delta_1 \quad \alpha - 200$$

# FRAMES WITH RIGID JOINTS

These moments are distributed in TABLE 6.4(b) and the results, when substituted in the sway equation, give

$$-234 - 236 + \frac{10}{18 \cdot 5}(-217 - 218) = -705$$

TABLE 6.4(b)
DISTRIBUTION OF SWAY MOMENTS: FIGURE 6.27(b)

| CA | AC | AB | BA | BD | DB |
|---|---|---|---|---|---|
|  | 1/3 | 2/3 | 2/3 | 1/3 |  |
| − 200 | − 200 | + 340 | + 340<br>− 93 | − 200<br>− 47 | − 200 |
|  | − 31 | − 47<br>− 62 |  |  | − 23 |
| − 16 |  |  | − 31<br>+ 21 | + 10 |  |
|  | − 3 | + 10<br>− 7 |  |  | + 5 |
| − 1 |  |  | − 3<br>+ 2 | + 1 |  |
| − 217 | − 234 | + 234 | + 236 | − 236 | − 218 |

TABLE 6.4(c)
COMPUTATION OF FINAL MOMENTS

| CA | AC | AB | BA | BD | DB | |
|---|---|---|---|---|---|---|
| + 70<br>+ 136 | + 140<br>+ 147 | − 140<br>− 147 | + 259<br>− 148 | − 259<br>+ 148 | − 129<br>+ 137 | Original distribution<br>Sway distribution × − 0·629 |
| + 206 | + 287 | − 287 | + 111 | − 111 | + 8 | Final Solution |

(v) The factor by which these sway moments must be multiplied is given by

$$-151 - 705x = 292$$

i.e. $\qquad x = -0 \cdot 629$

In TABLE 6.4(c) the results of the original distribution are added to the sway moments multiplied by this factor to give the final solution. Substitution in the sway equation as a final check gives:

$$+ 287 - 111 + \frac{10}{18 \cdot 5}(+206 + 8) = +291 \quad [+292]$$

## 6.4. THE MOMENT DISTRIBUTION METHOD 6.4.3.1.1

This process may, at first sight, appear to be unnecessarily indirect but it should be realized that it is not possible to find the moments produced by a sway force in a single operation. It would, of course, be quite easy to use the sway equation to find the lateral force required to push the frame into the position shown at (b) but the frame is only in equilibrium in this position when the imaginary joint clamps are operating. Successive relaxations of

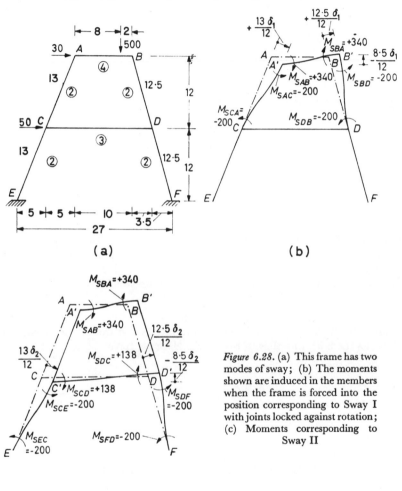

Figure 6.28. (a) This frame has two modes of sway; (b) The moments shown are induced in the members when the frame is forced into the position corresponding to Sway I with joints locked against rotation; (c) Moments corresponding to Sway II

these clamps not only bring the joints into equilibrium but also change the value of the lateral force. It is only when the clamps have been eliminated and the joints are independently in equilibrium that the sway equation can be used to find the relevant value of the lateral force. This is, by implication, given by the figure — 705 obtained when the sway correction moments are substituted in the sway equation. It is better not to evaluate the force explicitly, although this can readily be done, in order to avoid difficulty over signs which are taken care of automatically by the procedure just given.

# FRAMES WITH RIGID JOINTS

Finally we note that if the frame is subjected to lateral loads only, with no vertical loads, steps (ii) and (iii) are omitted.

**6.4.3.1.2.** *Application to frames with several modes of sway.*—It is now necessary to extend the above procedure by imposing sway distortions corresponding to each mode of sway in turn and to combine the results of these, each in its correct proportion, with the original distribution in which sway was suppressed. The factors by which each sway distribution must be multiplied are obtained from simultaneous equations constructed from the results of substitutions in the sway equation.

This process is illustrated by analysing the two storey frame of FIGURE 6.28 which has two modes of sway. The sway equations were derived previously (Section 6.3.3.2) and are:

$$M_{AC} + M_{BD} + \frac{10}{18\cdot 5}(M_{CA} + M_{DB}) = 292 \qquad \ldots \text{(I)}$$

$$M_{CE} + M_{DF} + \frac{18\cdot 5}{27}(M_{EC} + M_{FD}) = -211 \qquad \ldots \text{(II)}$$

The upper storey of this frame is that used to illustrate the previous Section, so that the fixed end moments, $M_{FAB} = -160$ and $M_{FBA} = +640$ are as before. The distribution factors are unchanged at $A$ and $B$ but at $C$ and $D$ they are as shown in TABLE 6.5(a), which gives the distribution of the fixed end moments with imaginary forces preventing sway.

TABLE 6.5(b) gives the distribution of the sway correction moments corresponding to Sway I, FIGURE 6.28(b), computed as in the previous example, while TABLE 6.5(c) applies to Sway II. Here we have:

$$M_{SEC} = M_{SCE} = -\frac{6.E.2\left(\frac{13\delta_2}{12}\right)}{13} = -E\delta_2 \qquad \alpha - 200$$

$$M_{SCD} = M_{SDC} = +\frac{6.E.3\left(\frac{8\cdot 5\delta_2}{12}\right)}{18\cdot 5} = +0\cdot 689 E\delta_2 \quad \alpha + 138$$

$$M_{SDF} = M_{SFD} = -\frac{6.E.2\left(\frac{12\cdot 5\delta_2}{12}\right)}{12\cdot 5} = -E\delta_2 \qquad \alpha - 200$$

$$M_{SAB} = M_{SBA} = +\frac{6.E.4\left(\frac{8\cdot 5\delta_2}{12}\right)}{10} = +1\cdot 7 E\delta_2 \quad \alpha + 340$$

## 6.4. THE MOMENT DISTRIBUTION METHOD   6.4.3.1.2

### TABLE 6.5(a)

DISTRIBUTION OF FIXED END MOMENTS: FIGURE 6.28(a)

|   |   | ↓AC | AB → | BA ← |   | ↓BD |   |
|---|---|---|---|---|---|---|---|
|   |   | 1/3 | 2/3 | 2/3 |   | 1/3 | Order of balancing joints |
|   |   |   | − 160 <br> − 213 | + 640 <br> − 427 |   | − 213 | 1 |
| 2 |   | + 124 | + 249 | + 124 |   |   |   |
|   |   | − 12 | − 41 | − 83 |   | − 41 | 3 |
| 6 |   | + 18 | + 35 | + 17 |   | + 18 |   |
|   |   | − 2 | − 12 | − 23 |   | − 12 | 7 |
| 10 |   | + 5 | + 9 | + 4 |   | + 3 |   |
|   |   |   | − 2 | − 5 |   | − 2 | 11 |
| 14 |   |   | + 2 |   |   |   |   |
|   |   | + 133 | − 133 | + 247 |   | − 247 | Totals |

|   | ↓CE | ↑CA | CD → | DC ← | ↓DF | ↑DB |   |
|---|---|---|---|---|---|---|---|
|   | 2/7 | 2/7 | 3/7 | 3/7 | 2/7 | 2/7 |   |
|   |   | + 62 |   |   |   | − 107 <br> − 20 |   |
|   |   |   | + 27 | + 55 | + 36 | + 36 | 4 |
| 5 | − 25 | − 25 | − 39 | − 19 |   | − 6 |   |
|   |   | + 9 | + 5 | + 11 | + 7 | + 7 | 8 |
| 9 | − 4 | − 4 | − 6 | − 3 |   | − 1 |   |
|   |   | + 2 | + 1 | + 2 | + 1 | + 1 | 12 |
| 13 |   | − 1 | − 2 |   |   |   |   |
|   | − 29 | + 43 | − 14 | + 46 | + 44 | − 90 | Totals |

|   | EC |   |   |   | FD |   |   |
|---|---|---|---|---|---|---|---|
|   | − 12 <br> − 2 |   |   |   | + 18 <br> + 4 |   |   |
|   | − 14 |   |   |   | + 22 |   | Totals |

This table has been arranged so that moments are always carried over to the column of figures immediately above, below, to the right or to the left; the figures in the margins give the order in which the joints are balanced.

## FRAMES WITH RIGID JOINTS

When the results of these distributions are substituted in the sway equations the right-hand sides are as follows:

|  | Original Distribution | Sway I | Sway II | Correct R.H.S. |
|---|---|---|---|---|
| In equation I | − 139 | − 613 | − 172 | + 292 |
| In equation II | + 20 | + 140 | − 607 | − 211 |

TABLE 6.5(b)

DISTRIBUTION OF SWAY I MOMENTS: FIGURE 6.28(b)

|  | ↓AC | AB → | BA ← |  | ↓BD |  |
|---|---|---|---|---|---|---|
|  | 1/3 | 2/3 | 2/3 |  | 1/3 |  |
| 3 | − 200<br>+ 22<br>− 54 | + 340<br>− 108 | + 340<br>− 54 |  | − 200<br>+ 28 |  |
|  | + 4 | − 38 | − 76 |  | − 38 | 4 |
| 7 | + 11 | + 23 | + 11 |  | − 2 |  |
|  |  | − 3 | − 6 |  | − 3 | 8 |
| 11 | + 1 | + 2 |  |  |  |  |
|  | − 216 | + 216 | + 215 |  | − 215 | Totals |

|  | ↓CE | ↑CA | CD → | DC ← | ↓DF | ↑DB |  |
|---|---|---|---|---|---|---|---|
|  | 2/7 | 2/7 | 3/7 | 3/7 | 2/7 | 2/7 |  |
|  |  | − 200 | + 43 | + 86 | + 57 | − 200<br>+ 57 | 1 |
| 2 | + 45 | + 45 | + 67 | + 33 |  | − 19 |  |
|  |  | − 27 | − 3 | − 6 | − 4 | − 4 | 5 |
| 6 | + 8 | + 9 | + 13 | + 6 |  | − 1 |  |
|  |  | + 5 | − 1 | − 3 | − 1 | − 1 | 9 |
| 10 | − 1 | − 1 | − 2 |  |  |  |  |
|  | + 52 | − 169 | + 117 | + 116 | + 52 | − 168 | Totals |

|  | EC |  |  |  | FD |  |  |
|---|---|---|---|---|---|---|---|
|  | + 22<br>+ 4 |  |  |  | + 28<br>− 2 |  |  |
|  | + 26 |  |  |  | + 26 |  | Totals |

### 6.4. THE MOMENT DISTRIBUTION METHOD   6.4.3.1.2

If the results of Sway I are multiplied by $x$ and those of Sway II by $y$ we have:

$$\left. \begin{array}{l} -139 - 613x - 172y = +292 \\ +20 + 140x - 607y = -211 \end{array} \right\} \text{ i.e. } \begin{array}{l} x = -0.761 \\ y = +0.205 \end{array}$$

TABLE 6.5(c)
DISTRIBUTION OF SWAY II MOMENTS: FIGURE 6.28(c)

|   | ↓ AC | AB → | BA ← |   | ↓ BD |   |
|---|---|---|---|---|---|---|
|   | 1/3 | 2/3 | 2/3 |   | 1/3 |   |
|   |   | + 340<br>− 113 | + 340<br>− 227 |   | − 113 | 1 |
| 2 | − 76 | − 151 | − 75 |   | + 14 |   |
|   | + 14 | + 20 | + 41 |   | + 20 | 5 |
| 6 | − 11 | − 23 | − 11 |   | − 1 |   |
|   | − 2 | + 4 | + 8 |   | + 4 | 9 |
|   |   | − 2 |   |   |   |   |
|   | − 75 | + 75 | + 76 |   | − 76 | Totals |

|   | ↓ CE | ↑ CA | CD → | DC ← | ↓ DF | ↑ DB |   |
|---|---|---|---|---|---|---|---|
|   | 2/7 | 2/7 | 3/7 | 3/7 | 2/7 | 2/7 |   |
|   | − 200 |   | + 138 | + 138 | − 200 |   |   |
|   |   | − 38 |   |   |   | − 56 |   |
| 3 | + 28 | + 29 | + 43 | + 21 |   |   |   |
|   |   | − 5 | + 20 | + 41 | + 28 | + 28 | 4 |
| 7 | − 4 | − 4 | − 7 | − 3 | + 10 |   |   |
|   |   |   |   | − 3 | − 2 | − 2 | 8 |
|   | − 176 | − 18 | + 194 | + 194 | − 174 | − 20 | Totals |

| EC |   |   |   | FD |   |   |
|---|---|---|---|---|---|---|
| − 200<br>+ 14<br>− 2 |   |   |   | − 200<br>+ 14<br>− 1 |   |   |
| − 188 |   |   |   | − 187 |   | Totals |

In TABLE 6.5(d) are set out the original distribution, the Sway I results multiplied by − 0·761 and the Sway II results multiplied by + 0·205 together with their sums which are the required end moments. The sway equations are found to be practically satisfied by these moments. Further examples of the application of this method will be found in a paper by Matheson[9].

## FRAMES WITH RIGID JOINTS

Although the simultaneous equations which must be solved in the final stage of this method are much less numerous than would have been the case if the slope-deflexion technique had been used the fact that they appear at all rather destroys the special advantage of moment distribution. In the next Section we show how the sway corrections can be incorporated in the distribution process as a series of successive approximations.

TABLE 6.5(d)

CALCULATION OF FINAL END MOMENTS

|  | AC | AB | BA |  | BD |  |
|---|---|---|---|---|---|---|
|  | + 133 | − 133 | + 247 |  | − 247 | Original Distribution |
|  | + 164 | − 164 | − 164 |  | + 164 | Sway I × − 0·761 |
|  | − 15 | + 15 | + 15 |  | − 15 | Sway II × 0·205 |
|  | + 282 | − 282 | + 98 |  | − 98 | Totals |

| CE | CA | CD | DC | DF | DB |  |
|---|---|---|---|---|---|---|
| − 29 | + 43 | − 14 | + 46 | + 44 | − 90 | Original Distribution |
| − 39 | + 128 | − 89 | − 89 | − 39 | + 128 | Sway I × − 0·761 |
| − 36 | − 4 | + 40 | + 40 | − 36 | − 4 | Sway II × 0·205 |
| − 104 | + 167 | − 63 | − 3 | − 31 | + 34 | Totals |

| EC |  |  |  | FD |  |  |
|---|---|---|---|---|---|---|
| − 14 |  |  |  | + 22 |  | Original Distribution |
| − 20 |  |  |  | − 20 |  | Sway I × 0·761 |
| − 39 |  |  |  | − 39 |  | Sway II × 0·205 |
| − 73 |  |  |  | − 37 |  | Totals |

### 6.4.3.2. *Correction for Sway Effects by Successive Approximations*

The procedure now is to test the moments for sway, by substitution in the sway equations, after each cycle of distribution. If the sway equations are not satisfied then a correction is introduced before the next cycle of distribution is begun. These pairs of operations are repeated until both the joint equations and the sway equations are sufficiently nearly satisfied. The physical interpretation of the procedure is that cycles of joint balance with sway prevented, and sway correction with joint rotation prevented, follow one another in succession until all external restraint, both of joint rotation and of joint translation, has been liquidated. The method was described by Grinter[12] and systematized by Clyde[7] and is now illustrated by reference to the problem of FIGURE 6.28.

It is convenient at this stage to introduce factors which show how the sway correction moments, found necessary after each substitution in the sway equations, are divided between the members participating in each joint translation or sway correction movement.

## 6.4. THE MOMENT DISTRIBUTION METHOD 6.4.3.2

Thus for Sway I (p. 279)    $M_{SCA} = M_{SAC} = -E\delta_1$
$$M_{SAB} = M_{SBA} = +1 \cdot 7 E\delta_1$$
$$M_{SBD} = M_{SDB} = -E\delta_1$$

Hence if substitution reveals that an amount $S_I$ must be added to the right-hand side of equation I in order to satisfy it, we have

$$(M_{SAC} + M_{SBD}) + \frac{10}{18 \cdot 5}(M_{SCA} + M_{SDB}) = S_I$$

TABLE 6.6(a)

CORRECTION FOR SWAY EFFECTS BY SUCCESSIVE APPROXIMATIONS: FIGURE 6.28

| | ↓AC | AB → | ← BA | | ↓BD | |
|---|---|---|---|---|---|---|
| | 1/3 | 2/3 | 2/3 | | 1/3 | Distribution factor |
| | 0·325 $S_I$ | −0·552 $S_I$ | −0·552 $S_I$ | | 0·325 $S_I$ | Sway I correction factor |
| | | −0·505 $S_{II}$ | −0·505 $S_{II}$ | | | Sway II correction factor |
| | | −160 | +640 | | | Fixed end moments |
| | | −213 | −427 | | −213 | 1 |
| 2 | +124 | +249 | +124 | | +15 | |
| | −12 | | | | | |
| | (+112) | | | | (−198) | Sum after 1st cycle |
| | +130 | −220 | −220 | | +130 | Sway I correction |
| | | +110 | +110 | | | Sway II correction |
| | | −53 | −106 | | −53 | 5 |
| 6 | +15 | +30 | +15 | | −9 | |
| | −15 | | | | | |
| | (+242) | | | | (−130) | Sum after 2nd cycle |
| | +31 | −54 | −54 | | +31 | Sway I correction |
| | | −32 | −32 | | | Sway II correction |
| | | +16 | +33 | | +16 | 9 |
| 10 | +18 | +36 | +18 | | −3 | |
| | −6 | | | | | |
| | (+285) | | | | (−86) | Sum after 3rd cycle |
| | −8 | +13 | +13 | | −8 | Sway I correction |
| | | −12 | −12 | | | Sway II correction |
| | | −2 | −5 | | −3 | 13 |
| 14 | +5 | +10 | +5 | | +2 | |
| | (+282) | | | | (−95) | Sum after 4th cycle |
| | −2 | +3 | +3 | | −2 | Sway I correction |
| | | +2 | +2 | | | Sway II correction |
| | | −3 | −7 | | −3 | 17 |
| 18 | — | — | | | | |
| | +281 | −281 | +100 | | −100 | TOTAL |

TABLE 6.6(a) (contd.)

| | ↓ CE | ↑ CA | CD → | DC ← | ↓ DF | ↑ DB | |
|---|---|---|---|---|---|---|---|
| | 2/7 <br> 0·297 $S_{II}$ | 2/7 <br> 0·325 $S_I$ | 3/7 <br> −0·204 $S_{II}$ | 3/7 <br> −0·204 $S_{II}$ | 2/7 <br> 0·297 $S_{II}$ | 2/7 <br> 0·325 $S_I$ | |
| | | +62 <br> +23 | +46 | +30 | −107 <br> +31 | 3 | |
| 4 | −24 | −24 | −37 | −18 | | | |
| | (−24) <br> −65 | (+38) <br> +130 <br> +7 | +45 <br> −14 | +45 <br> −28 | (+30) <br> −65 <br> −19 | (−76) <br> +130 <br> −26 <br> −19 | Sum after 1st cycle <br> Sway I correction <br> Sway II correction <br> 7 |
| 8 | −29 | −30 | −44 | −22 | | | |
| | (−118) <br> +19 | (+145) <br> +31 <br> +9 | −13 <br> −5 | −13 <br> −10 | (−54) <br> +19 <br> −6 | (+9) <br> +31 <br> +8 <br> −7 | Sum after 2nd cycle <br> Sway I correction <br> Sway II correction <br> 11 |
| 12 | −12 | −12 | −17 | −8 | | | |
| | (−111) <br> +7 | (+173) <br> −8 <br> +2 | −5 <br> +3 | −5 <br> +7 | (−41) <br> +7 <br> +4 | (+41) <br> −8 <br> −1 <br> +4 | Sum after 3rd cycle <br> Sway I correction <br> Sway II correction <br> 15 |
| 16 | | | +1 | | | | |
| | (−104) <br> −1 | (+167) <br> −2 | +1 | +1 | (−30) <br> −1 <br> +1 | (+36) <br> −2 <br> −1 <br> +1 | Sum after 4th cycle <br> Sway I correction <br> Sway II correction <br> 19 |
| 20 | | +1 | +1 | | | | |
| | −105 | +166 | −61 | −4 | −30 | +34 | TOTAL |

| EC | | | | FD | | |
|---|---|---|---|---|---|---|
| 0·297 $S_{II}$ | | | | 0·297 $S_{II}$ | | |
| −12 | | | | +15 | | Carry over from 1st cycle |
| −65 <br> −14 | | | | −65 <br> −9 | | Sway II correction |
| (−91) <br> +19 <br> −6 | | | | (−59) <br> +19 <br> −3 | | Sum after 2nd cycle <br> Sway II correction |
| (−78) <br> +7 <br> — | | | | (−43) <br> +7 <br> +2 | | Sum after 3rd cycle <br> Sway II correction |
| (−71) <br> −1 | | | | (−34) <br> −1 | | Sum after 4th cycle <br> Sway II correction |
| −72 | | | | −35 | | TOTAL |

## 6.4. THE MOMENT DISTRIBUTION METHOD 6.4.3.2

i.e. $\quad (-E\delta_1 - E\delta_1) + \dfrac{10}{18\cdot 5}(E\delta_1 - E\delta_1) = S_I$

i.e. $\quad E\delta_1 = -\dfrac{18\cdot 5}{57} S_I = -0\cdot 325 S_I$

Hence $\quad M_{SCA} = M_{SAC} = M_{SBD} = M_{SDB} = 0\cdot 325 S_I$

$\quad M_{SAB} = M_{SBA} = -0\cdot 552 S_I$

In a similar way, for Sway II (p. 282), we have

$$M_{SEC} = M_{SCE} = M_{SDF} = M_{SFD} = \frac{27}{91} S_{II} = 0\cdot 297 S_{II}$$

$$M_{SCD} = M_{SDC} = -0\cdot 689 \times 0\cdot 297 S_{II} = -0\cdot 204 S_{II}$$

$$M_{SAB} = M_{SBA} = -1\cdot 7 \times 0\cdot 297 S_{II} = -0\cdot 505 S_{II}$$

TABLE 6.6(a) gives the necessary working in tabular form.

Steps 1, 2, 3 and 4 are the first balance of joints $B$, $A$, $D$, and $C$ respectively. Following this first cycle of moment distribution the current values of the appropriate end moments are inserted in the sway equations, as in TABLE 6.6(b), and it appears that corrections $S_I = +399$ and $S_{II} = -219$ are required. These are inserted in TABLE 6.6(a) in the proportions indicated

TABLE 6.6(b)

CALCULATION OF SWAY CORRECTIONS

| Cycle | Substitution in Sway Equation I $M_{AC}+M_{BD}+0\cdot 541\,(M_{CA}+M_{DB})$ | R.H.S. | i.e. $S_I$ |
|---|---|---|---|
| After 1st | $112 - 198 + 0\cdot 541\,(\ 38 - 76) =$ | $-107$ | $+399$ |
| ,, 2nd | $242 - 130 + 0\cdot 541\,(145 + \ 9) =$ | $+195$ | $+97$ |
| ,, 3rd | $285 - \ 86 + 0\cdot 541\,(173 + 41) =$ | $+320$ | $-23$ |
| ,, 4th | $282 - \ 95 + 0\cdot 541\,(167 + 36) =$ | $+297$ | $-5$ |
| Finally | $281 - 100 + 0\cdot 541\,(166 + 34) =$ | $+289$ | $+3$ |

| Cycle | Substitution in Sway Equation II $M_{CE}+M_{DF}+0\cdot 685\,(M_{EC}+M_{FD})$ | R.H.S. | i.e. $S_{II}$ |
|---|---|---|---|
| After 1st | $-\ 24 + 30 + 0\cdot 685\,(-\ 12 + 15) = +\ \ 8$ | | $-219$ |
| ,, 2nd | $-118 - 54 + 0\cdot 685\,(-91 - 59) = -275$ | | $+64$ |
| ,, 3rd | $-111 - 41 + 0\cdot 685\,(-78 - 43) = -240$ | | $+24$ |
| ,, 4th | $-104 - 30 + 0\cdot 685\,(-71 - 34) = -205$ | | $-5$ |
| Finally | $-105 - 30 + 0\cdot 685\,(-72 - 35) = -208$ | | $-3$ |

by the sway correction factors evaluated above and recorded at the head of each column. A further cycle of distribution is followed by another check in the sway equations and the consequent sway corrections. After five such complete cycles the column moments are found to satisfy the sway equations sufficiently closely and the final moments, obtained by summing each column of figures, are seen to agree quite closely with those previously obtained in TABLE 6.5(d).

The method just described shows to particularly good advantage when the columns are vertical for the sway corrections do not then involve the beams and the calculations are often simple enough to be done mentally. In many cases the various sway corrections are quite independent and it is then possible to speed up convergence by judicious over-correction; this would not have been easy in the example just worked.

This completes the account of the principles of moment distribution; the methods described are applicable to rigid frames of every type including those having members of varying cross-section (See Appendix A). A great many authors have presented developments of the basic method but it is not possible now to describe more than two of these; references to others will be found in the bibliography at the end of this Chapter.

### 6.4.4. Developments of the Moment Distribution Method
#### 6.4.4.1. *Naylor's Method for Symmetrical Frames*[13]

Within its range of applicability this is a most advantageous system which enables joint and sway distributions to be made simultaneously. The

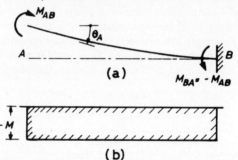

Figure 6.29. (a) A couple $M_{AB}$ applied at the free end $A$ of a cantilever $AB$ of length $L$ induces an equal and opposite couple at the fixed end; (b) The corresponding bending moment diagram

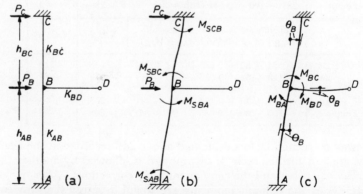

Figure 6.30. (a) In this frame joints $B$, $C$ and $D$ are supposed to be free to move horizontally while $B$ and $C$ are prevented from rotating; (b) The sway moments that are induced initially; (c) When joint $B$ is allowed to rotate the out-of-balance moment there is distributed between $BA$, $BC$ and $BD$ according to equation 6.16 if $B$, $C$ and $D$ are still free to translate

## 6.4. THE MOMENT DISTRIBUTION METHOD

necessary distribution factors are developed by reference to FIGURES 6.29 and 30. FIGURE 6.29(a) shows a cantilever $AB$ with a positive couple $M_{AB}$ applied at the free end $A$. The bending moment is evidently constant along the whole length of the cantilever, as indicated at (b), while a negative reactive couple $M_{BA} = -M_{AB}$ acts on the beam at $B$. The carry-over factor is thus $-1$ while the angle $\theta_A$ is given by equation 3.21(a):

$$i_B - i_A = \left[\frac{A_m}{EI}\right]_A^B$$

i.e. $\quad 0 - \theta_A = -\dfrac{ML}{EI} \quad$ i.e. $\quad M_{AB} = EK\theta_A \quad \ldots\ldots$ (6.15)

FIGURE 6.30(a) shows a pin-ended beam $BD$ attached to a column $ABC$; if the joints $B$, $C$ and $D$ are free to move sideways, under the lateral loads $P_C = P_B$, while $B$ and $C$ are prevented from rotating by imaginary clamps, sway moments $M_{SCB}$, $M_{SBC}$, etc., are induced which have the following values, obtained by cutting the column just above $B$ and just above $A$ respectively.

$$M_{SCB} = M_{SBC} = -\frac{1}{2} P_C . h_{BC}\,; \quad M_{SBA} = M_{SAB} = -\frac{1}{2}(P_C + P_B)h_{AB}$$

There is thus an out-of-balance moment at $B$ provided by the imaginary clamp and equal to the algebraic sum of $M_{SBA}$ and $M_{SBC}$. This moment is liquidated by releasing the clamp and distributing an equal and opposite

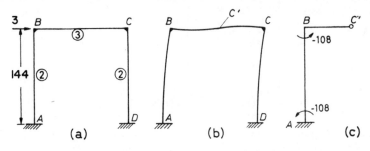

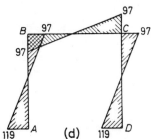

Figure 6.31. (a) Symmetrical portal frame under lateral load; (b) In the distorted position the beam has a point of contraflexure at its mid-point; (c) The simplified frame to be analysed; (d) The resulting bending moment diagram

moment between $BA$, $BC$ and $BD$ as indicated at (c). $M_{BA}$ and $M_{BC}$ are given by equation 6.15 above while $M_{BD}$ is given by equation 6.10. Thus

$$M_{BA} : M_{BC} : M_{BD} = EK_{AB}.\theta_B : EK_{BC}\,\theta_B : 3EK_{BD}\,\theta_B$$

$$= K_{AB} : K_{BC} : 3K_{BD} \quad \ldots\ldots \text{(6.16)}$$

## FRAMES WITH RIGID JOINTS

The above distribution process differs from the usual one in that the joints $B$, $C$ and $D$ are free to move sideways at all times although $C$ is prevented from rotating. In applying this result to symmetrical frames, FIGURE 6.31(a) for example, we first notice that there are points of contraflexure at the mid points of the beams, as at (b). For the purposes of analysis the frame can therefore be divided into two halves, as at (c); each half corresponds to FIGURE 6.30(a) (with the upper column omitted) and distribution can be performed according to equation 6.16. It is to be noted, however, that $K_{BC}' = 2K_{BC}$ since $L_{BC}' = \frac{1}{2}L_{BC}$.

The initial sway moments are

$$M_{SAB} = M_{SBA} = M_{SCD} = M_{SDC} = -\tfrac{1}{4} \times 3 \times 144 = -108$$

The distribution factors are

$$M_{BA} : M_{BC}' = K_{AB} : 3K_{BC}' = 2 : 3 \times 2 \times 3 = 1/10 : 9/10$$

and the actual distribution is

| BA | BC' | |
|---|---|---|
| 1/10 | 9/10 | Distribution Factor |
| − 108 | | Initial Sway Moment |
| + 11 | + 97 | Balance Joint $B$ |
| − 97 | + 97 | Total |
| AB | | |
| − 108 | | Initial Sway Moment |
| − 11 | | Carry over |
| − 119 | | Total |

The final bending moment diagram is shown at (d). In this simple example the result was reached after a single joint balance, at $B$, but multi-storey frames will normally require a series of such distributions. Unless the beams are abnormally slender convergence is rapid because the distribution factor is so much larger for the beams than for the columns; moreover there is no carry-over to the beams. It is noteworthy that the sway equations are automatically satisfied at all stages; the initial sway moments, of course, are effectively obtained from the sway equations while the carry-over factor of $-1$ means that the total of the column moments in each storey is never disturbed.

Several interesting and important examples are given in Naylor's paper[13]; we conclude by demonstrating his method for dealing with sway due to unsymmetrical vertical loads. Here he proceeds by replacing the actual loading by two other arrangements, one symmetrical and one anti-symmetrical. The latter is analysed as above while the former is easily dealt with by normal moment distribution since sway can be excluded by inspection. FIGURE 6.32(a) shows a frame carrying unsymmetrical vertical loads together with

## 6.4. THE MOMENT DISTRIBUTION METHOD 6.4.4.2

the anti-symmetrical arrangement, stage A, and the symmetrical arrangement, stage B, to which it is equivalent. The fixed end moments, calculated by means of equations 6.1(a) and 6.2, are shown at (b). The distributions for stage A, according to Naylor's method, and stage B, by normal distribution using the short method for symmetrical frames (see example (b), Section 6.4.2), are shown in TABLES 6.7(a) and (b) respectively, for one-half the

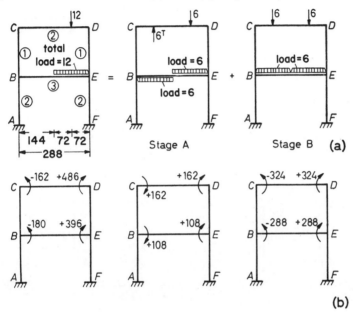

Figure 6.32. (a) Symmetrical frame with unsymmetrical loading together with the equivalent anti-symmetrical (Stage A) and symmetrical (Stage B) load systems; (b) The corresponding fixed end moments

frame in each case. The second half is written in by inspection and the final result obtained by addition as in TABLE 6.7(c). Naylor's method has been still further developed by other writers, notably Wood[14,15] and Lightfoot[16].

### 6.4.4.2. Bolton's Method

In part of his rather comprehensive paper[17] Bolton described a method of procedure which has some points of similarity with Naylor's method but which is not restricted to symmetrical frames. The idea is to build up distribution operations which can be used to liquidate out-of-balance joint moments, in the usual manner, but which do not disturb the sway relationships in the process.

The distribution operations are synthesized from the basic individual joint rotations and translations appropriate to the particular problem under examination, the relationships of equations 6.11 to 6.14 being used for the purpose.

Example (a):

In order to bring out the similarities to and differences from Naylor's method the frame FIGURE 6.32 is analysed by Bolton's method; it should

## FRAMES WITH RIGID JOINTS

be realized that no advantage is now taken of symmetry in order to show how an unsymmetrical frame would be tackled. FIGURE 6.33(a) shows the

TABLE 6.7(a)
STAGE A.  FIGURE 6.32

| ↓ CB | CD' | |
|---|---|---|
| 1/13 | 12/13 | } Distribution factor (Naylor) |
| −12 | +162<br>−150 | 1 |
| +6<br>0 | −6 | 3 |
| −6 | +6 | |

| ↓ BA | ↑ BC | BE' | |
|---|---|---|---|
| 2/21 | 1/21 | 18/21 | |
| −11 | +12<br>−6 | +108<br>−103 | 2 |
| −11 | +6 | +5 | |

| AB |
|---|
| +11 |
| +11 |

TABLE 6.7(b)
STAGE B.  FIGURE 6.32

| ↓ CB | CD' | |
|---|---|---|
| 1/2 | 1/2 | } Distribution factor (short method for symmetrical frames) |
| +162 | −324<br>+162 | 1 |
| +23<br>−12 | −11 | 3 |
| +173 | −173 | |

| ↓ BA | ↑ BC | BE' | |
|---|---|---|---|
| 4/9 | 2/9 | 3/9 | |
| +92 | +81<br>+46 | −288<br>+69 | 2 |
| +3 | −6<br>+1 | +2 | 4 |
| +95 | +122 | −217 | |

| AB |
|---|
| +46<br>+1 |
| +47 |

TABLE 6.7(c)
COMBINATION OF STAGES A & B:  FIGURE 6.32

| CD | CB | BC | BE | BA | AB | FE | EF | EB | ED | DE | DC |
|---|---|---|---|---|---|---|---|---|---|---|---|
| +6<br>−173 | −6<br>+173 | +6<br>+122 | +5<br>−217 | −11<br>+95 | +11<br>+47 | +11<br>−47 | −11<br>−95 | +5<br>+217 | +6<br>−122 | −6<br>−173 | +6<br>+173 |
| −167 | +167 | +128 | −212 | +84 | +58 | −36 | −106 | +222 | −116 | −179 | +179 |

moments produced at the ends of the members affected when the joint $C$ is rotated by an amount $\theta_C$ while all the other joints are clamped. At (b) these moments are shown in proportional form, while at (c), (d) and (e) are shown the corresponding moment patterns produced by rotations of joints $D$, $B$ and $E$ respectively. These moments are all proportional to the

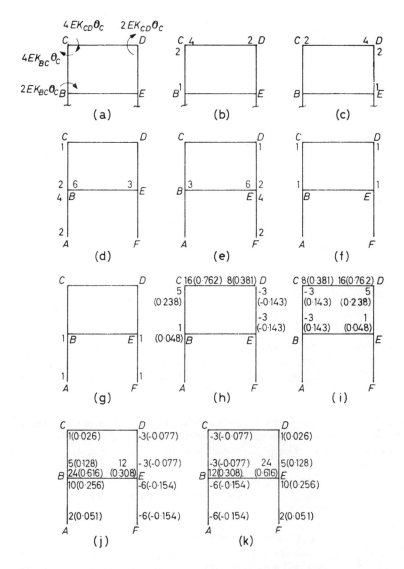

*Figure 6.33.* (a) Moments produced in the frame of FIGURE 6.32 when a rotation $\theta_C$ is imposed at $C$; (b) The above moments arranged in a proportional pattern; (c) Pattern of moments produced by a rotation of $D$; (d) Pattern of moments produced by a rotation of $B$; (e) Pattern of moments produced by a rotation of $E$; (f) Pattern of moments produced by sway of the upper storey; (g) Pattern of moments produced by sway of the lower storey; (h) Patterns (b) and (f) combined to give the effect of rotation of $C$ and sway of the upper storey so proportioned that sway equation I is not disturbed; this is the 'no shear' pattern for rotation of $C$; (i) No shear pattern for rotation of $D$; (j) No shear pattern for rotation of $B$; (k) No shear pattern for rotation of $E$

usual distribution and carry-over factors. At (f) and (g) are shown the moments produced by sway movements, without rotation, of the upper and lower storeys respectively.

We now seek to find the pattern of movements produced by rotations of the joints together with such sway movements as will leave the sway equations undisturbed. The sway equations are

$$M_{CB} + M_{BC} + M_{DE} + M_{ED} = 0 \quad \ldots (I)$$

$$M_{BA} + M_{AB} + M_{EF} + M_{FE} = 0 \quad \ldots (II)$$

Dealing first with joint $C$ (pattern (b)), we see that substitution in equation I, the only sway equation involved, gives

$$2 + 1 + 0 + 0 = 3$$

Substituting in I the pattern corresponding to sway of the upper storey (pattern (f)), we have

$$1 + 1 + 1 + 1 = 4$$

Hence if we combine $4 \times$ pattern (b) with $-3 \times$ pattern (f) we get pattern (h) which satisfies I, for

$$5 + 1 - 3 - 3 = 0$$

This pattern, then, gives an array of moments which corresponds to a rotation and a sway of $C$ and a sway, without rotation, of $D$, $B$ and $E$, combined in such proportions that the relevant sway equation I is satisfied. If unit moment is applied at $C$ then $\frac{16}{21} = 0.762$ goes into $CD$, $\frac{5}{21} = 0.238$ into $CB$ and so on. These figures, shown in brackets at (h), are the distribution and carry-over factors modified to represent this combined rotation and sway process. The process itself will be recognized as a development, applicable to unsymmetrical frames, of Naylor's process as embodied in equation 6.16. Similar reasoning leads to the pattern (i) and we have now developed what can be called the 'no shear' patterns for rotation of joints $C$ and $D$ respectively.

Rotation of $B$ (pattern (d)) involves sway of both storeys; in order to satisfy both I and II simultaneously we shall have to combine $4 \times$ pattern (d) with $-3 \times$ pattern (f) and $-6 \times$ pattern (g) to give pattern (j); pattern (k) is obtained similarly and these are the 'no shear' patterns for rotation of $B$ and $E$ respectively. Dividing by 39 gives the modified distribution and carry-over factors for unit moment applied at $B$ and $E$ respectively.

The distribution is given in TABLE 6.8. The first joint to be balanced, $D$, is dealt with in accordance with pattern (i) and requires $-486$ units divided between $DC$ and $DE$ in the proportion $0.762$ $(-370)$ to $0.238$ $(-116)$; the carry-over moments are:

$$CB: \; +70 \, (-0.143) \qquad CD: \; -185 \, (-0.381)$$

$$BC: \; +70 \, (-0.143) \qquad ED: \; -23 \, (0.048)$$

## TABLE 6.8
### ANALYSIS OF FRAME OF FIGURE 6.32 BY BOLTON'S METHOD

| | CB | CD | DC | | DE | |
|---|---|---|---|---|---|---|
| | | −162 | +486 | | | Fixed end moments |
| | | | −370 | | −116 | 1 Balance $D$ (i) |
| | +70 | −185 | | | | Carry over from 1 |
| | +29 | | | | −10 | C.O. from 2 |
| | +59 | +189 | | | | 3 (h) |
| | | | +95 | | −35 | C.O. from 3 |
| | +3 | | | | −10 | C.O. from 4 |
| | | | −30 | | −10 | 5 (i) |
| | +6 | −15 | | | | C.O. from 5 |
| | −2 | | | | +1 | C.O. from 6 |
| | +2 | +6 | | | | 7 (h) |
| | | | +3 | | −1 | C.O. from 7 |
| | 0 | | | | +1 | C.O. from 8 |
| | | | −3 | | −1 | 9 (i) |
| | +1 | −2 | | | | C.O. from 9 |
| | | +1 | | | | 11 (h) |
| | +168 | −168 | +181 | | −181 | TOTAL |

| BA | BC | BE | EB | EF | ED | |
|---|---|---|---|---|---|---|
| | | −180 | +396 | | | Fixed end moments |
| | +70 | | | | −23 | C.O. from 1 |
| | | | −230 | −95 | −48 | 2 (k) |
| +57 | +29 | −115 | | | | C.O. from 2 |
| | +12 | | | | −35 | C.O. from 3 |
| +33 | +16 | +78 | | | | 4 (j) |
| | | | +39 | −20 | −10 | C.O. from 4 |
| | +6 | | | | −2 | C.O. from 5 |
| | | | +17 | +7 | +4 | 6 (k) |
| −4 | −2 | +9 | | | | C.O. from 6 |
| | 0 | | | | −1 | C.O. from 7 |
| −2 | −1 | −6 | | | | 8 (j) |
| | | | −3 | +1 | +1 | C.O. from 8 |
| | +1 | | | | 0 | C.O. from 9 |
| | | +1 | +2 | | | 10 (k) (& C.O.) |
| | | −2 | | | | 12 |
| +84 | +131 | −215 | +221 | −107 | −114 | TOTAL |

| AB | | | | FE | | |
|---|---|---|---|---|---|---|
| +57 | | | | −19 | | C.O. from 2 |
| +6 | | | | −20 | | C.O. from 4 |
| −4 | | | | +1 | | C.O. from 6 |
| 0 | | | | +1 | | C.O. from 8 |
| +59 | | | | −37 | | TOTAL |

## FRAMES WITH RIGID JOINTS

These moments are entered in the table and attention turns to joint $E$ (pattern (k)) and so on. The totals at the foot of each column agree closely with the previous solution obtained by Naylor's method in TABLE 6.7.

It should be stated that Bolton's method does not show to particularly good advantage in this rather simple problem; its power is only apparent in more involved unsymmetrical cases, especially if the frame is to be analysed for several different load systems. Here the procedure is especially economical since the 'no shear' patterns are obtained once for all.

Example (b):

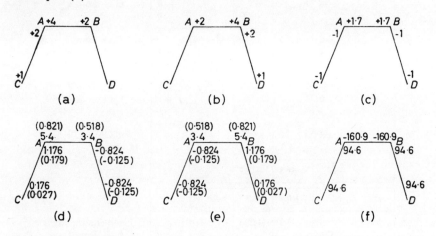

Figure 6.34. (a) Pattern of moments produced in the frame of FIGURE 6.27(a) by a rotation of $A$ only; (b) Pattern of moments produced in the frame of FIGURE 6.27(a) by a rotation of $B$ only; (c) Pattern of moments produced in the frame of FIGURE 6.27(a) by sway without joint rotation; (d) 'No shear' pattern for rotation of A; (e) 'No shear' pattern for rotation of $B$; (f) Fixed end moments produced by sway displacements

Consider again the frame of FIGURE 6.27(a). The pattern of moments corresponding to a rotation of $A$ only is shown in FIGURE 6.34(a), while rotation of $B$ produces pattern (b). The sway of FIGURE 6.27(b) is scaled down to give pattern (c). The sway equation, obtained on page 261, is

$$M_{AC} + M_{BD} + 0.541(M_{CA} + M_{DB}) = 292 \quad \ldots \text{(I)}$$

If this equation is to be undisturbed by the distribution operations the resulting moments, when substituted in the left-hand side, must give zero. Let this be achieved by an operation comprising the sum of pattern (a) and $x \times$ pattern (c).

Substituting pattern (a) gives:

$$2 + 0 + 0.541(1 + 0) = 2.541$$

Substituting $x \times$ pattern (c) gives:

$$-x - x - 0.541(x + x) = -3.082x$$

i.e. $x = 0.824$

The 'no shear' pattern for $A$ is therefore as shown at (d);

$M_{AB}$, for example, is given by

$$+ 4 + 0{\cdot}824 \times 1{\cdot}7 = 5{\cdot}4$$

The 'no shear' pattern for $B$ is similarly obtained and is given at (e). The fixed end moments for the beam, so far as the vertical load is concerned, are $-160$ and $+640$ but to these must be added the moments caused by the lateral force. These are shown at (f) and have been obtained by imagining the frame to be pushed sideways with the joints clamped. The resulting moments are in the proportion of pattern (c) and of sufficient magnitude to satisfy the sway equation I. The final array of fixed end moments, to be distributed, is therefore

$$CA = AC = BD = DB = + 94{\cdot}6$$
$$AB = -160 - 160{\cdot}9 = -320{\cdot}9$$
$$BA = +640 - 160{\cdot}9 = +479{\cdot}1$$

Distribution is then carried out according to Bolton's routine.

### 6.4.5. Comments on the Moment Distribution Method

The moment distribution method can now be usefully compared with the relaxation process described in Chapter 4. The procedure of balancing the joints corresponds essentially to the liquidation of the residuals while the distribution and carry-over factors take the place of the relaxation operators. The process is one of successive approximations, however, rather than one of successive corrections so that mistakes are perpetuated in moment distribution while they are eliminated in relaxation. The most important characteristic of moment distribution, as compared with relaxation, is that the calculations are conducted entirely in terms of moments and not in terms of displacements. Not only does this save the labour of constructing operators in terms of displacements but it eliminates the final step of converting displacements back into moments. This advantage is also apparent when moment distribution is compared with slope deflexion; it enables the designer to have in mind a clearer picture of his problem and the significance of the figures he is handling than if he were working in terms of displacements.

It is fortunate that the physical nature of the problems is usually such that the basic distribution process converges quite rapidly. There is thus no need to seek means of speeding up the rate of convergence by constructing special operators as must often be done with relaxation. When the fundamental methods of distribution and sway correction have been mastered the designer has at hand a routine which will dispose of a wide variety of problems, for even the simultaneous equations of Section 6.4.3.1.2 can easily be solved, especially if a desk calculator is available.

The developments due to Naylor and Bolton correspond with the construction of block and group operators. The former is so easily mastered that it should always be used for symmetrical frames but the latter involves quite formidable preparatory work which can only be justified if complicated frames have to be analysed for several loadings. The growth of the use of electronic computers may soon eliminate the need for such devices for speeding up convergence.

# FRAMES WITH RIGID JOINTS

Finally it is to be noted that moment distribution only deals with the simplified problem left when axial effects are eliminated; if these are to be included the more complex relaxation procedures described, for example, by Allen[18] would normally be used in the absence of a computer.

## 6.5. FURTHER METHODS FOR THE ANALYSIS OF RIGID FRAMES

The preceding Sections, which have dealt fairly completely with several important methods for analysing rigid frames, should serve as an adequate guide to the extensive body of literature on this subject that has grown up in recent years. No mention has yet been made, however, of certain other ideas which are still being used by writers on rigid frame analysis.

It is not possible to do more now than summarize these notions and to indicate where a full treatment can be found but it can be stated generally that, although these methods are often very convenient for a limited range of problems, attempts to extend them usually involve considerable complications especially where sway is involved. If a series of problems has to be solved, for which a certain method seems particularly suitable, then a special study of the nomemclature and technique can be made.

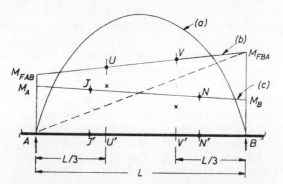

*Figure 6.35.* One span, *AB*, of a continuous beam with (a) The 'simply-supported' bending moment diagram; (b) The fixing moment diagram, and (c) The support moment diagram, showing the location of the characteristic points *U* and *V*, the fixed points *J'* and *N'* and the focal or conjugate points *J* and *N*

The following notes will perhaps help in a preliminary study and assessment of any method that may be under consideration. We begin by a discussion of certain features of the geometry of bending moment diagrams that have been extensively used in devising graphical methods of analysis.

### 6.5.1. THE GEOMETRY OF BENDING MOMENT DIAGRAMS

#### 6.5.1.1. *Definitions*

FIGURE 6.35 shows one span *AB* of a continuous beam together with the relevant bending moment diagrams: these are

(a) The diagram given by the applied loads acting on the simply-supported span *AB*.

## 6.5. FURTHER METHODS FOR ANALYSIS OF RIGID FRAMES 6.5.1.2

(b) The diagram given by the fixing moments $M_{FAB}$ and $M_{FBA}$ acting on $AB$.

(c) The diagram given by the actual support moments $M_A$ and $M_B$ acting on $AB$. The determination of $M_A$ and $M_B$ in any particular case is the central problem of rigid frame analysis.

Associated with these diagrams are certain points which are now defined.

*The Characteristic Point U* is the point at which the vertical through the centroid of the $M_{FAB}/EI$ diagram intersects the fixed end moment line. $V$ is the point where the vertical through the centroid of the $M_{FBA}/EI$ diagram intersects the fixed end moment line. If the beam is uniform between supports the verticals pass through the third points of the beam.

*The Fixed Point J'* is the point of contraflexure of the unloaded span $AB$ when a moment is applied at $B$; its position depends only on the degree of restraint afforded by the adjoining spans at $A$. The fixed point $N'$ is the point of contraflexure when a moment is applied at $A$.

*The Focal or Conjugate Points J and N* lie at the intersections of the verticals through $J'$ and $N'$ and the support moment line.

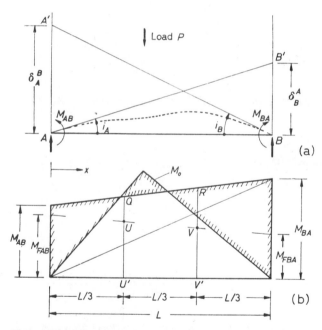

*Figure 6.36.* (a) One span, $AB$, of a continuous beam under loads $P$ and support moments $M_{AB}$ and $M_{BA}$; (b) The corresponding bending moments. The simply-supported moments are designated $M_0$

### 6.5.1.2. *Characteristic Points*

In FIGURE 6.36(a) $AB$ represents one span of a continuous beam uniform between supports and acted on by loads $P$ and support moments $M_{AB}$ and $M_{BA}$; the dotted line represents the deflected shape of the beam.

$$i_A = \frac{BB'}{L} = \frac{\delta_B^A}{L} = \frac{L}{2EI}\left\{\frac{2.M_{AB}}{3} + \frac{M_{BA}}{3} - \frac{2}{L^2}\int_0^L M_o(L-x)dx\right\}$$
$$i_B = \frac{AA'}{L} = \frac{\delta_A^B}{L} = \frac{L}{2EI}\left\{\frac{M_{AB}}{3} + \frac{2M_{BA}}{3} - \frac{2}{L^2}\int_0^L M_o x\,dx\right\} \quad \ldots\text{(i)}$$

If perpendiculars $U'Q$ and $V'R$ are erected at $U'$ and $V'$

$$U'Q = \frac{2M_{AB}}{3} + \frac{M_{BA}}{3} \text{ and } V'R = \frac{M_{AB}}{3} + \frac{2M_{BA}}{3}$$

Hence if $U'U$ and $V'V$ are made equal to $\dfrac{2}{L^2}\displaystyle\int_0^L M_o(L-x)dx$ and

$\dfrac{2}{L^2}\displaystyle\int_0^L M_o x\,dx$ respectively we shall have

$$i_A = \frac{L}{2EI}UQ \text{ and } i_B = \frac{L}{2EI}VR \quad \ldots\text{(ii)}$$

The points $U$ and $V$ obtained in this way are called characteristic points and were first described by Claxton Fidler[19]; their position depends only on the beam loading and so can be found at once. Their use in continuous beam analysis, which has been described in some detail by Salmon[20], depends on the fact that $UQ$ and $VR$ are proportional to the end slopes $i_A$ and $i_B$. Thus if another span $BC$ is continuous with $AB$ at $B$ there will

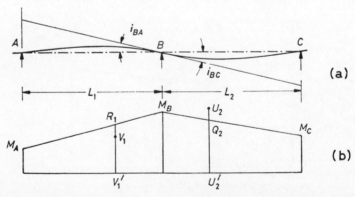

*Figure 6.37.* (a) Adjacent spans of a continuous beam having a common tangent at $B$; (b) The support moment diagram with the relevant characteristic points $V_1$ and $U_2$

be a common tangent at $B$ and the slopes of the two spans there will be numerically equal but opposite in sign, as indicated in FIGURE 6.37(a). Hence if the appropriate characteristic points $V_1$ and $U_2$ are plotted, as at (b), the support moment line $M_A - M_B - M_C$ will be in such a position that

$$\frac{R_1 V_1}{K_1} = -\frac{Q_2 U_2}{K_2} \quad \ldots\text{(iii)}$$

## 6.5. FURTHER METHODS FOR ANALYSIS OF RIGID FRAMES 6.5.1.3

Various graphical constructions are available which employ this relationship to find the support moment line.

It is to be noticed that if the end slopes at $A$ and $B$ are zero the intercepts $UQ$ and $VR$ are also zero; the characteristic points $U$ and $V$ thus lie on the fixing moment line and this is perhaps the easiest way of locating them. Also if the span carries no external loads the intercepts $U'U$ and $V'V$ are zero and the characteristic points lie on $AB$.

### 6.5.1.3. Fixed Points

FIGURE 6.38(a) shows a continuous beam with an indefinite number of spans of which only one single internal span is loaded. The bending moment diagram is of the general shape indicated at (b) from which it appears that a point of contraflexure $J'$ or $N'$ occurs in each unloaded span. The position

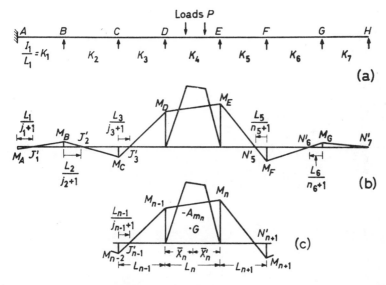

*Figure 6.38.* (a) A continuous beam of many spans of which only one interior span is loaded; (b) The bending moment diagram: points of contraflexure $J'$ and $N'$ in the unloaded spans are known as 'fixed points'; (c) Determination of the moments at the ends of the loaded span

of these points depends only on the $K$-values of successive spans and is independent of the loading; the points are therefore known as 'fixed points' and they can be located in various ways. The following method, described by Timoshenko and Young[10], depends on successive applications of Clapeyron's Equation. Two fixed points can be found at once: in the span $GH$ the fixed point $N'_7$ coincides with $H$ where the moment is zero; in $AB$, where $M_A = \frac{1}{2}M_B$, $J'_1$ occurs at the third point and coincides with $U'_1$. The remaining points are given by

$$\frac{M_n}{M_{n-1}} = -\left\{2 + \frac{K_n}{K_{n-1}}\left(2 - \frac{1}{j_{n-1}}\right)\right\} = -j_n \quad \ldots \ldots (6.17)$$

The bending moment diagram can be completed when the moments $M_D$ and $M_E$ are known; referring to the general diagram (c), it can be shown that

$$\left. \begin{array}{r} M_{n-1} = \dfrac{6 A_{mn} (\bar{x}_n' \, n_n - \bar{x}_n)}{L_n^2 \, (j_n \, n_n - 1)} \\[1em] M_n = \dfrac{6 A_{mn} (\bar{x}_n j_n - \bar{x}_n')}{L_n^2 \, (j_n \, n_n - 1)} \end{array} \right\} \quad \ldots \ (6.18)$$

Steinman's[21] construction for the fixed points, which is shown in FIGURE 6.39(a), is essentially based on the ideas of moment distribution. A moment $M'$ applied at $C$ produces the moment line $1-1$ which passes through the third point $U_2'$ if $B$ is clamped. When $B$ is released the carry-over moment there, $Bb'$, is divided in the ratio $K_1 : K_2$ and the moment line swings to the position $2-2$ crossing $1-1$ at $d$ vertically above the third point $V_2'$

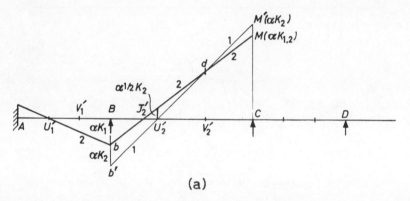

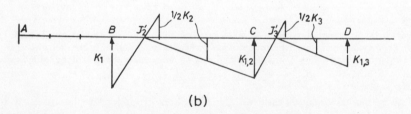

*Figure 6.39.* (a) Location of a fixed point in span $BC$; (b) Determination of the 'linked rigidity' of spans $A-B-C$

(since $C$ is clamped during this operation); the moment $M'$ is thus reduced to $M$. Both $M'$ and $M$ correspond to the same rotation of $C$, but the former was acting while $B$ was clamped and the latter after it had been released. $M'$ is therefore proportional to $K_2$ while $M$ is proportional to the 'linked rigidity' $K_{1,2}$ of the two spans $AC$. The intersection of $2-2$ with $BC$ locates the fixed point $J_2'$ while in $AB$, since $A$ is clamped, the fixed point $J_1'$ coincides with the third point $U_1'$.

## 6.5. FURTHER METHODS FOR ANALYSIS OF RIGID FRAMES  6.5.1.4

FIGURE 6.39(b) shows a simple geometrical construction which is equivalent to the above and which enables the 'linked rigidities' and the fixed points to be obtained in sequence.

### 6.5.1.4. *Focal or Conjugate Points*

If the focal points $J$ and $N$, of FIGURE 6.35, are known the support moment diagram can be drawn and the problem is solved. A construction due to Bowles and Cornish[22] exploits the relationship (iii), in Section 6.5.1.2 above, for this purpose and is quite convenient but Steinman's[21] method, described below, is interesting because of its relationship with moment distribution.

FIGURE 6.40 shows a three-span continuous beam $ABCD$ with fixed ends $A$ and $D$. The fixed end moments in $AB$ and $BC$ are drawn to scale at $Aa_1$, $Bb_1$, $Bb_2$ and $Cc_2$ respectively; $CD$ is unloaded. The fixed points $J_2'$ and $N_2'$

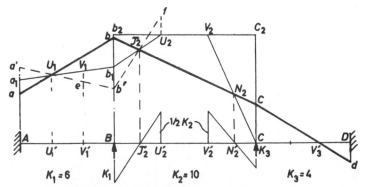

Figure 6.40. Location of the focal points $J$ and $N$

are located as above and, by definition, the focal points $J_2$ and $N_2$ are vertically above them; since $A$ is a fixed end the left-hand focal and conjugate points coincide at $U_1$ in $AB$. One way of locating $J_2$ is to draw a line $a'b'$ through $U_1$ (which must lie on the correct support moment line) and to join $b'$ to $f$ where $\dfrac{V_1 e}{K_1} = \dfrac{U_2 f}{K_2}$ in accordance with (iii) above; then, as demonstrated by Bowles and Cornish, $J_2$ lies on $b'f$ and can be located by drawing another pair of lines in the same way. But it will be seen that the lines $a_1 b_1 U_2$ satisfy (iii) so that $J_2$ can be found by projecting vertically upwards from $J_2'$ on to $b_1 U_2$. The right-hand focal point $N_2$ has been found in this way, $V_2$ being joined, in this instance, to $C$ because the fixed end moment $M_{FCD}$ happens to be zero. The final support moment line $abcd$ is obtained by extending $J_2 N_2$ to $b$ and $c$, and then extending $bU_1$ and $cV_3'$ to $a$ and $d$ respectively. A slight extension of this procedure leads to the solution of the beam $A - E$ shown in FIGURE 6.41. The fixed end moments are given by $Aa_1$, $Bb_1$, $Bb_2$, $Cc_2$, $Cc_3$, $Dd_3$, $Dd_4$ and $Ee_4$, while the fixed points are located by the usual construction which, it is to be noticed, makes the effective stiffness of the pin-ended member $DE$ equal to $\frac{3}{4}K_4$, as expected. The steps in the construction are:

(i) Join $EU_4$ and extend to give $d_4'$. This is equivalent to relaxing the clamp at $E$ and makes $Dd_4'$ equal to the effective fixed end moment in $DE$ at $D$.

(ii) Join $V_3 d_4'$ and project up from $N_3'$ to give the focal point $N_3$.

(iii) Join $b_1 U_2$ and project up from $J_2'$ to give the focal point $J_2$.

(iv) Join $J_2 V_2$ and extend to $b'$ and $C_2'$; the point $b'$ divides $b_2 b_1$ in the ratio $K_2 : K_1$ and gives the effect of balancing the joint $B$ while $A$ and $C$ are clamped. $Cc_2'$ thus represents the fixed end moment of the spans to the left of $C$ with the intermediate joints (in this case $B$ only) free to rotate.

(v) Join $c_2' U_3$ and project up from $J_3'$ to give the focal point $J_3$.

(vi) Join $J_3 N_3$ and extend to $c$ and $d$. This enables the final support moment line $a U_1 b J_2 c J_3 N_3 d E$ to be drawn.

The analyst who takes the trouble to master this construction, which is much harder to describe and justify than to perform, will find that it is often quicker than the arithmetical procedure to which it is equivalent.

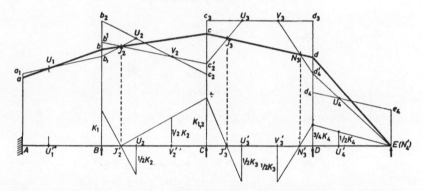

*Figure 6.41.* Graphical solution of a continuous beam problem by the use of fixed, focal and characteristic points

Sway problems can either be treated by combining individual solutions in the correct proportion, in the manner of Section 6.4.3.1 or by the modified construction given by Steinman. Finally it is of interest to note that Steinman, in deference to the many who prefer arithmetic to geometry, has produced an arithmetical equivalent of his construction which, by employing the conception of linked rigidities, has the effect of making moment distribution a direct rather than a convergent process.

### 6.5.2. The Concept of 'Degree of Fixity'

The fact that the support moment actually developed at an end of a member of a rigid frame depends on the resistance to rotation offered by the adjoining members has been used as the basis of several methods of analysis. Although different nomenclature and details of procedure are used by the various authors they all begin from the same fundamental idea of the 'degree of fixity' of the end of a member. Shepley[23], for example, describes a member $LR$ as having a degree of fixity $f_L^o$ at the end $L$ if the adjoining restraints rotate that end through the angle $f_L^o \times \theta$ from the position it would have occupied if $L$ had been pinned. Thus in FIGURE 6.42(a) the end $L$ is

## 6.5. FURTHER METHODS FOR ANALYSIS OF RIGID FRAMES  6.5.2

pinned and the slope there caused by the moment $M_R$ is $\theta$. Hence, by definition,

$$f_L^\circ = 0$$

and, from equations 6.9 and 6.10

$$M_R = 6EK\theta \qquad \ldots\ldots \text{(i)}$$

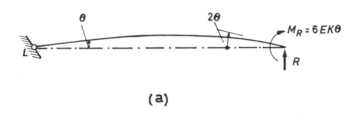

(a)

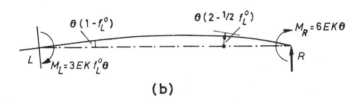

(b)

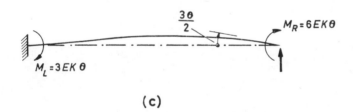

(c)

*Figure 6.42.* Relationships between the moment applied at the right-hand end of a beam and the angle and moment induced at the left-hand end in terms of the degree of fixity $f_L^\circ$; (a) $L$ pinned. $f_L^\circ = 0$, $\theta_L = \theta$, $M_L = 0$; (b) $L$ attached to other members. $\theta_L = \theta(1 - f_L^\circ)$, $M_L = 3EKf_L^\circ\theta$; (c) $L$ fixed. $f_L^\circ = 1$, $\theta_L = 0$, $M_L = EK\theta$

Diagram (b) shows the situation if the other members attached to $L$ now rotate $L$ through the angle $f_L^\circ\theta$; the moment required to do this is

$$M_L = 3EK(f_L^\circ\theta) \qquad \ldots\ldots \text{(ii)}$$

the slope at $L$ is now

$$\theta - f_L^\circ\theta = \theta(1 - f_L^\circ) \qquad \ldots\ldots \text{(iii)}$$

and the slope at $R$ becomes

$$2\theta - \tfrac{1}{2}f_L^\circ\theta = \theta(2 - \tfrac{1}{2}f_L^\circ) \qquad \ldots\ldots \text{(iv)}$$

as $M_R$ retains its original value of $6EK\theta$ during this process the carry-over factor from $R$ to $L$ is

$$\frac{M_L}{M_R} = \frac{3EK f_L^\circ \theta}{6EK\theta} = \frac{1}{2} f_L^\circ \qquad \ldots \text{(v)}$$

If $L$ is fixed, as at (c), the slope then is zero, so that from (iii)

$$f_L^\circ = 1$$

and, from (v), the carry-over factor is

$$\frac{M_L}{M_R} = \tfrac{1}{2}$$

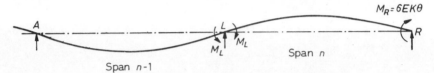

*Figure 6.43.* Adjacent spans of a continuous beam

These ideas can be used to obtain successive degrees of fixity in the spans of a continuous beam. Thus if the degree of fixity of the left-hand end $A$ of span $AL$ in FIGURE 6.43 is $f_{n-1}^\circ$ it can be shown by successive applications of (iii) and (iv) that

$$f_n^\circ = \frac{1}{1 + \dfrac{K_n}{K_{n-1}}(1 - \tfrac{1}{4} f_{n-1}^\circ)} \qquad \ldots \text{(vi)}$$

Where $f_n^\circ$ is the degree of fixity of $L$ in the span $LR$. Comparison with equation 6.17 reveals that

$$j = \frac{2}{f_L^\circ} \qquad \ldots \text{(vii)}$$

which provides a link between the ideas of degrees of fixity and of the fixed points of Section 6.5.1.3.

The degrees of fixity are determined once for all and it is then a simple matter to analyse a beam for any system of applied loads by means of equation 6.18 or its equivalent. Shepley presents a number of graphs which make the work very rapid in practice.

A disadvantage of this approach, which has perhaps prevented its being used more widely, is that the definition of the degree of fixity in terms of the angular deflexion produced by the restraints is an unfamiliar one. Wood[14,15] avoids this difficulty by working in terms of $K$-values.

Thus in FIGURE 6.44 a couple $M$ is applied at $L$ while a moment $M_R$ restrains the rotation of $R$ to a value $\theta_R$.

Then

$$M_R = 2EK\left(1 - \frac{K}{4\Sigma K}\right) 2\theta_R + M\frac{K}{2\Sigma K} \qquad \ldots \text{(viii)}$$

## 6.5. FURTHER METHODS FOR ANALYSIS OF RIGID FRAMES 6.5.3

In this equation, which is obtained by the use of equation 6.5, $\Sigma K$ is the sum of the $K$-values of the four members meeting at $L$. If $LR$ were fixed at the end $L$, a moment $M_R$ applied at $R$ would be connected with the rotation $\theta_R$ by the relation

$$M_R = 2EK'.2\theta_R \qquad \ldots \text{(ix)}$$

Comparing (viii) and (ix) it appears that the equivalent stiffness of $LR$ for rotation of $R$ is

$$K' = K\left(1 - \frac{K}{4\Sigma K}\right) \qquad \ldots \text{(x)}$$

The stiffness has thus been reduced by the amount $\tfrac{1}{4}K\dfrac{K}{\Sigma K}$ while the carry-over from $L$ to $R$ is seen from (viii) to be $\tfrac{1}{2}M\dfrac{K}{\Sigma K}$.

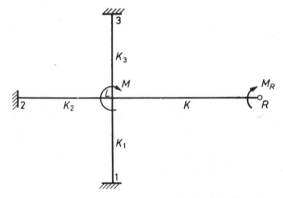

*Figure 6.44.* The moment $M_R$ developed at $R$ depends on the stiffnesses of the members meeting at $L$, where $M$ is applied, as well as on the rotation $\theta_R$ which is permitted at $R$

If the joints 1, 2 and 3 had been attached to other members, instead of being fixed, the stiffnesses $K_1$, $K_2$ and $K_3$ would have been modified to $K_1'$, $K_2'$ and $K_3'$ in accordance with (x); the carry-over from $L$ to $R$ would then have been $\tfrac{1}{2}M\dfrac{K}{\Sigma K'}$, and the stiffness reduction $\tfrac{1}{4}K\dfrac{K}{\Sigma K'}$.

Using these relationships it is possible to reduce a whole frame to a single equivalent member. Using this conception Wood[14] has developed a design method in which an extensive structure is built up from pre-analysed substitute frames while by means of a modification[15] of Bolton's method (Section 6.4.4.2) he has applied the idea to sway problems.

### 6.5.3. The Beam-Line Method

The methods of using the degree of fixity which were briefly introduced above can only be used for elastic structures having a linear moment-angle relationship. Batho's 'Beam-Line' method,[24] however, although applicable to—and often rather convenient for—such problems was originally devised to deal with the non-linear moment-angle curves given by cleated connexions between beams and stanchions. Although the graphical procedures it employs in such cases become very complicated when sway must be taken into account it is at least practicable in these inherently difficult problems.

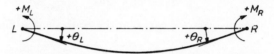

*Figure 6.45.* Sign conventions for the beam-line method

An introduction to the method is now given using the convention of signs indicated in FIGURE 6.45: the signs of terms in the slope-deflexion convention can be translated into the new beam-line convention as follows:

| *Slope-deflexion convention* | | *Beam-line convention* |
|---|---|---|
| $+ M_L$ | | $- M_L$ |
| $+ M_R$ | $=$ | $+ M_R$ |
| $+ \theta_L$ | | $+ \theta_L$ |
| $+ \theta_R$ | | $- \theta_R$ |

### 6.5.3.1. *Symmetrical Loading and Restraints*

The principle of the method is illustrated by reference to FIGURE 6.46(a) which shows a beam with symmetrical loading and equal end restraints. Using the slope-deflexion equation we have

$$M_L = 2EK(2\theta_L + \theta_R) + M_{FL} \qquad \ldots \text{(i)}$$

After introducing the fact that $|\theta_L| = |\theta_R|$ and translating into the new convention, this becomes

$$M_L = M_{FL} - 2EK\theta_{LR} \qquad \ldots \text{(6.19)}$$

This is the beam line shown on FIGURE 6.46(c). The restraint line, which may be either straight or curved, is the moment-angle relationship for the members to which $L$ is attached. In either case the intersection of the beam and restraint lines gives the required value of $M_L$.

The restraint line for a member which is fixed at the far end, as in FIGURE 6.47(a), is obtained from equation 6.7 and is

$$M = 4EK\theta \qquad \ldots \text{(6.20)}$$

while if the member is pinned at the far end, as at (b), equation 6.10 gives

$$M = 3EK\theta \qquad \ldots \text{(6.21)}$$

Thus the total restraint offered by the three members $L_1$, $L_2$ and $L_3$, in FIGURE 6.48(a), is given by

$$M = E\theta(4K_1 + 4K_2 + 3K_3) \qquad \ldots \text{(ii)}$$

The beam line is

$$M_L = PL/8 - 2EK\theta_L \qquad \ldots \text{(iii)}$$

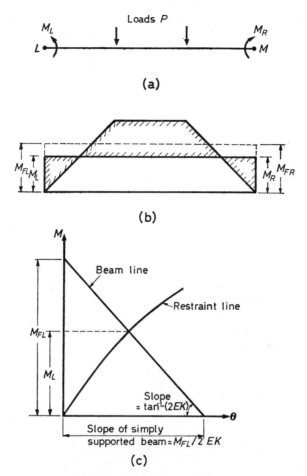

Figure 6.46. (a) Symmetrically loaded beam with symmetrical end restraining moments $M_L$ and $M_R$; (b) Corresponding bending moment diagram showing the fixed end moments $M_{FL}$ and $M_{FR}$; (c) Beam and restraint lines

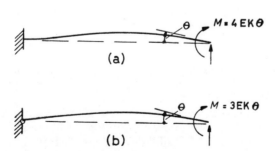

Figure 6.47. Restraints offered by a member when far end is (a) Fixed; (b) Pinned

# FRAMES WITH RIGID JOINTS

and the solution, obtained algebraically in this case, is

$$M_L = \frac{PL}{8}\left\{1 - \frac{2K}{(4K_1 + 4K_2 + 3K_3 + 2K)}\right\} \quad \ldots \text{(iv)}$$

and corresponds to the graphical solution shown at (b).

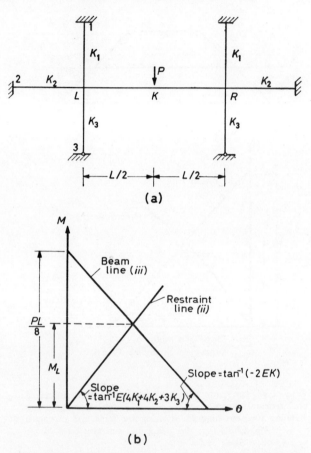

*Figure 6.48.* (a) Symmetrically loaded beam attached to identical restraining members at its ends; (b) Corresponding beam and restraint lines

It thus appears that when a beam is attached to a group of members 'in parallel' the restraint line is obtained by adding the effective stiffnesses of these members. But if the beam is attached to the members by cleats, which transmit the moment unchanged but deform in the process, the arrangement can be thought of as a 'series' one and the slope of the restraint line is the reciprocal of the sum of the 'flexibilities' of the cleat and the restraining members. Thus in FIGURE 6.49 the restraint line for the stanchions is

$$M = 4E(K_1 + K_2)\theta_S = s\theta_S\overline{K} \quad \text{(say)}$$

while that of the cleats is

$$M = K_C \theta_C$$

The total angle of rotation is

$$\theta_S + \theta_C = \frac{M}{K_S} + \frac{M}{K_C} = \theta$$

The combined restraint line is therefore

$$M = \frac{1}{\frac{1}{K_S} + \frac{1}{K_C}} \theta \qquad \ldots \ldots (6.22)$$

The graphical equivalent of the above is shown in FIGURE 6.50. At (a) the combined restraint line, obtained by adding the restraint due to the

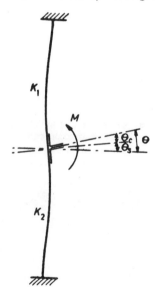

Figure 6.49. Distortion of a cleat and members to which it is attached 'in series'

stanchions to that due to the cleats, is shown intersecting the beam line to give the required value of $M$. An equivalent construction, much more convenient if the cleat restraint line happens to be curved, is shown at (b).

As an example of the calculation of the restraint line of a number of members connected by cleats consider the frame of FIGURE 6.51. The restraint line for the cleats is taken to be

$$M = 2EK\theta$$

The member $L2$ is in series with the cleats connecting it to the stanchion.

i.e.    Flexibility of $L2$ = flexibility of beam + flexibility of cleat

$$= \frac{1}{3EK} + \frac{1}{2EK} = \frac{5}{6EK}$$

$\therefore$ Stiffness of $L2 = \frac{6}{5} EK$

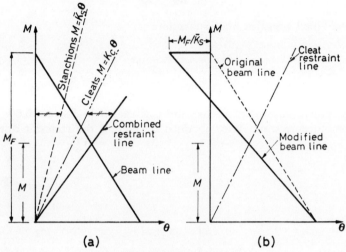

*Figure 6.50.* (a) Combined restraint line for cleat and stanchion in series; (b) Alternative construction

Stiffness of $1 - L - 3 = 4E\frac{K}{2} + 4E\frac{K}{2} = 4EK$

$\therefore$ Total stiffness of $L1$, $L2$ and $L3 = \frac{6}{5} EK + 4EK = \frac{26}{5} EK.$

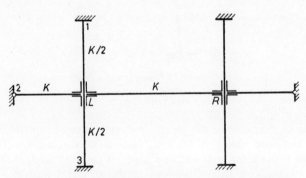

*Figure 6.51.* Symmetrical frame with semi-rigid connexions

Now $LR$ is in series with these three members and the cleats,

hence flexibility of total restraint at $L = \frac{5}{26EK} + \frac{1}{2EK} = \frac{9}{13EK}$

## 6.5. FURTHER METHODS FOR ANALYSIS OF RIGID FRAMES 6.5.3.2

Hence final restraint line at $L$ is $M = \dfrac{13}{9} EK\theta$.

### 6.5.3.2. *The General Beam Line for Unsymmetrical Loading and Restraints*

At the left-hand end of a beam $LR$ the slope-deflexion equation 6.5, converted to the beam-line sign convention, gives:

$$M_L = M_{FL} - 2EK(2\theta_L - \theta_R) \quad \ldots \text{(i)}$$

and at the right-hand end

$$M_R = M_{FR} - 2EK(2\theta_R - \theta_L) \quad \ldots \text{(ii)}$$

If the restraint line at $R$ is

$$M_R = nEK\theta_R \quad \ldots \text{(iii)}$$

We have, substituting from (ii) and (iii) in (i),

$$M_L = \left(M_{FL} + \dfrac{2}{n+4} M_{FR}\right) - 4\left(\dfrac{n+3}{n+4}\right) EK\theta_L \quad \ldots \text{(6.23)}$$

This is the General Beam Line.

If the restraint line at $L$ is

$$M_L = mEK\theta_L \quad \ldots \text{(iv)}$$

then, solving (iv) and 6.23 simultaneously, we have

$$M_L = \overline{M}_F \dfrac{mn + 4m}{mn + 4m + 4n + 12} \quad \ldots \text{(6.24)}$$

where

$$\overline{M}_F = M_{FL} + \dfrac{2}{n+4} M_{FR} \quad \ldots \text{(v)}$$

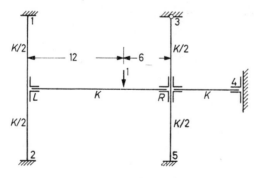

Figure 6.52. Unsymmetrical frame with semi-rigid connexions

These results are now applied to the frame shown in FIGURE 6.52 whose cleats have a restraint line $M = 2EK\theta$.
The fixed end moments are:

$$M_{FL} = +4/3 \quad \text{and} \quad M_{FR} = +8/3 \quad \ldots \text{(vi)}$$

315

FRAMES WITH RIGID JOINTS

The first step is to find the coefficients $m_L$ and $n_R$ at $L$ and $R$.

For $m_L$: Stiffness of stanchion $1 - L - 2 = 4EK/2 + 4EK/2 = 4EK$

$$\text{Flexibility of joint } L = \frac{1}{4EK} + \frac{1}{2EK} = \frac{3}{4EK}$$

$$\therefore \text{Stiffness of joint } L = \frac{4}{3} EK$$

i.e. $$m_L = 4/3 \qquad \ldots \text{(vii)}$$

For $n_R$: Applying the general beam-line equation 6.23 to the unloaded beam $R4$, and assuming for the time being that the joint $R$ is rigid, we have

$$M_R = -4\left(\frac{n+3}{n+4}\right) EK\theta_R$$

This applies generally to unloaded beams: the coefficient $n$ applies to the end remote from $M$ and $\theta$ and in this case refers to the cleats at 4.

i.e. $$n = 2 \text{ and } M_R = -4\left(\frac{2+3}{2+4}\right) EK\theta_R$$

$$= -\frac{10}{3} EK\theta_R$$

The restraint for the beam $R4$ to moments applied at $R$ is therefore

$$M = \frac{10}{3} EK\theta$$

Since this beam is in series with the cleats on the right-hand side of the column $3 - R - 5$, we have

$$\text{Flexibility of beam } R4 + \text{cleats} = \frac{1}{\frac{10}{3}EK} + \frac{1}{2EK} = \frac{4}{5EK}$$

$$\therefore \text{Stiffness of beam } R4 + \text{cleats} = \frac{5}{4}EK$$

The stanchions $R3$ and $R5$ and the cleated beam $R4$ are in parallel and therefore the total stiffness to the right of $3 - R - 5$ is

$$3EK/2 + 4EK/2 + 5EK/4 = \frac{19}{4}EK$$

This group of members is in series with $LR$ through a cleat and hence

$$\text{Flexibility of joint } R = \frac{1}{\frac{19}{4}EK} + \frac{1}{2EK} = \frac{27}{38EK}$$

## 6.5. FURTHER METHODS FOR ANALYSIS OF RIGID FRAMES 6.5.3.2

i.e. $$\text{Stiffness of joint } R = \frac{38}{27}EK$$

i.e. $$n_R = \frac{38}{27} \qquad \ldots \text{(viii)}$$

It is now possible to find $M_L$ from 6.23 or 6.24; first we require

$$\overline{M}_F = \frac{4}{3} + \frac{2}{\frac{38}{27} + 4} \times \frac{8}{3} = 2 \cdot 32 \quad \text{(from (v))}$$

Then $M_L = 2 \cdot 32 \dfrac{\frac{4}{3} \cdot \frac{38}{37} + 4 \cdot \frac{4}{3}}{\frac{4}{3} \cdot \frac{38}{27} + 4 \cdot \frac{4}{3} + 4 \cdot \frac{38}{27} + 12} = 0 \cdot 67 \quad \text{(from (6.24))}$

The simplest way of finding $M_R$ is to transpose $M_{FL}$ and $M_{FR}$ and $m$ and $n$ in (v) and 6.24. This gives $\overline{M}_F = 3 \cdot 17$ and $M_R = \cdot 95$.

In order to find the moments elsewhere in the frame we divide the moments at the joints in proportion to the members meeting there. Thus at $L$ it is obvious that there will be moments of $-\frac{1}{2} \times 0 \cdot 67 = -0 \cdot 33$ in $L1$ and $L2$. At $R$ the moment of $-0 \cdot 95$ is divided between $R3$, $R4$ and $R5$ in the ratio $3/2 : 5/4 : 2$. Hence $M_{R3} = -0 \cdot 3$; $M_{R4} = -0 \cdot 25$; $M_{R5} = -0 \cdot 4$. The moments at 3 and 5 are obviously 0 and $+0 \cdot 2$ respectively but the moment at 4 needs further consideration. FIGURE 6.53 shows a beam $AB$ which has a non-rigid connexion at $B$ having the restraint line $M_B = nEK\theta_B$.

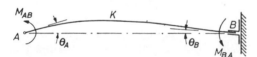

Figure 6.53. Carry-over of moment to the end of a member connected by cleats to adjacent members

Writing the slope-deflexion equations in the beam-line sign convention for the ends $A$ and $B$ we have

$$M_{AB} = -2EK(2\theta_A - \theta_B)$$

and $$M_{BA} = -2EK(2\theta_B - \theta_A)$$

But $$M_{BA} = +nEK\theta_B$$

i.e. $$M_{BA} = -M_{AB}\left(\frac{n}{2n+6}\right) \qquad \ldots \text{(6.25)}$$

[Note that if the joint $B$ is rigid $n = \infty$ and the carry-over factor $\dfrac{n}{2n+6}$ becomes $1/2$.]

It is to be noted that the restraint at $B$ depends on the stiffness of the cleat and on any members to the right of $B$. For the present purpose, however, the restraint must be expressed in terms of the $K$-value of $AB$.

In this case $n = 2$ and

$$M_{4R} = -M_{R4} \frac{2}{2 \times 2 + 6} = +0.05$$

Further details of this method will be found in the paper by Batho and Rowan[24] to which reference has already been made. The treatment of beams with unsymmetrical loading and restraints is described in a further paper by Batho[25] while an introduction to the treatment of sway problems is given in the discussion of a paper by Sourochnikoff[26].

## 6.6. THE ANALYSIS OF COLLAPSE LOADS

### 6.6.1. INTRODUCTION

The analytical methods that have been described so far permit the distribution of load between the members of a structure to be worked out with an accuracy that depends on the extent to which the assumptions underlying the calculations are realized in the completed structure. The precision with which the component parts are fabricated, for example, obviously affects the load distribution directly for, as we have seen, it is possible to pre-stress a hyperstatic structure deliberately by making the members of such a length that they have to be forced into position; as absolute dimensional accuracy is unattainable it is inevitable that some pre-stressing will occur. In large structures the self-weight of the members is a major component of the applied loads and it is by no means an easy task to ensure that the erection procedure is such that each member is carrying its share of the dead weight of the structure in a manner appropriate to the temperature conditions prevailing when the final joints are closed†.

The errors inherent in the assumption that rollers and pinned joints are frictionless will at once be apparent, as will the fact that the members will neither have the exact cross-sectional dimensions assumed during the analysis nor be ideally straight.

Nevertheless, in spite of these complications, the calculated force or bending moment in any member of a structure is usually close to that actually realized and, in view of the uncertainties that inevitably attend the estimation of live loads, can be accepted as being sufficiently accurate. The *stress* at a particular point, however, is a very different matter and many field measurements have revealed local stresses that depart considerably from the estimates[28,29]. It is obvious that this must be so for the details of the stress pattern round a joint, for example, depend on the fine details of the dimensions of the components and can be affected by minute changes in those dimensions. Photoelastic investigations show very vividly how stress concentrations occur particularly at load points and at reentrant angles and, indeed, it is apparent that a uniform stress condition is very difficult to achieve. These facts are of the greatest possible importance to the machine designer conscious, as he always must be, of the danger of fatigue but their significance to the structural designer is rather different. A design specification based on a maximum

---

† An interesting account of how this difficulty was met during the construction of the Sydney Harbour Bridge is given in a paper by Freeman and Ennis[27].

## 6.6. THE ANALYSIS OF COLLAPSE LOADS

allowable stress carries the implications of danger if that stress is exceeded and disaster if the metal should yield; but the fact that steel structures seldom fail suggests that stress concentrations are not so dangerous as might at first be thought. It is found, in fact, that the plastic distortion that occurs, at a more or less constant stress, when the yield point is reached in mild steel has the effect of transferring load from the more to the less highly stressed regions. For this reason yield can be regarded as a source of safety, rather than the reverse, if the metal is sufficiently ductile.

Many experiments have shown that a mild steel beam subjected to increasing bending moments has a considerable reserve of strength after the appearance of yield at the outer fibres and it is only when yield has spread right through the section that collapse finally occurs.

These considerations have led to the idea of regarding complete collapse, rather than first yield, as the limit of usefulness of a structure and of designing the structure in such a way that the actual loads are a suitable fraction of those required to cause collapse. The details of the distribution of stress are ignored in this process but the load factor—that is the ratio of the collapse load to the working load—is chosen so that yield, if it appears at all, occurs only at a few regions of high local stress where it can do no harm. The possibility of cumulative damage due to successive applications of unfavourable load combinations can usually be excluded. The conception of collapse as the limit of usefulness seems to have originated in the work of Kist, Kazinczy and Maier-Leibnitz. It probably first became at all widely known to English-speaking engineers through the writings of van den Broek[30,31] who described its application to highly redundant braced structures, such as transmission towers, as well as to rigidly jointed frames. The increasing attention now being paid to this approach to the design problem is undoubtedly due to the long series of researches carried out by Baker[32] and his associates. Concentrating on rigid frames and continuous beams they have carried our knowledge of the real—as opposed to the assumed—behaviour of these structures to a stage where it is possible to use their methods with confidence to design bare frames of the portal type; a great many structures so designed have already been erected. It therefore seems appropriate to conclude this Chapter on rigid frames with a brief introduction to the new ideas although, depending as they do on plastic behaviour, they may seem out of place in a book on elastic analysis. The reader may even entertain the thought that elastic analysis itself must be of diminishing importance as the conception of plastic or collapse design gains ground but, in fact, the reverse is the case. There are three reasons for this: the first is that although plastic design is a very convenient method of fixing the sizes of the members to withstand a certain set of working loads the resulting structure will, under normal conditions, be operating within the elastic range. Its deflexion, which may be critical and will often have to be checked, can only be found by elastic analysis. It is fortunate that deflexions, which are controlled by the average stresses prevailing rather than by local variations from the average, can be estimated with reasonable accuracy[33]. The second reason is that as designers become more confident they work to closer limits and, in consequence, there is a tendency for slenderer members to be used than formerly. The possibility of failure due to elastic instability therefore increases, and must be investigated; analytical methods that are closely allied to those described earlier in the present Chapter are being developed for this purpose, as indicated in Chapter 8. The third reason for the continuing importance of elastic analysis is to be found in the realization that

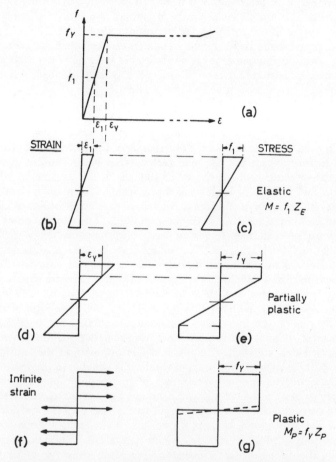

Figure 6.54. (a) Ideal stress-strain law used in the plastic theory; (b) and (c) Strain and corresponding elastic stress diagrams for a symmetrical section; (e) Yield has spread inwards from the outer fibres to the level where the strain is $\varepsilon_Y$, as given by the strain diagram (d); (f) Infinite strain is necessary for yield to spread completely across the section as at (g): in actual practice strain hardening will start at the outer fibres while the stress at the centre is still elastic as indicated by the dotted line

an adequate description of the behaviour of multi-storey buildings can only be made in terms of the composite action of the floors and walls with the frame; this is essentially a very complicated hyperstatic problem.

## 6.6.2. Basic Ideas of the Analysis of Collapse Loads

### 6.6.2.1. *The Plastic Hinge*

When a structural steel section is subjected to a gradually increasing bending moment the distribution of stress at first follows the familiar straight line rule of elementary beam theory. When a certain value of the bending moment is reached the most highly stressed fibres at the outsides of the beam reach the yield point and thereafter, it is assumed, remain at constant stress. This assumption implies that the stress-strain relationship for the material has the form shown in FIGURE 6.54(a), which is an idealized version† of that actually found in mild steel.

At still higher bending moments, if plane sections continue to remain plane, the strain diagram has the shape shown at (d) corresponding to the stress diagram shown at (e). When the bending moment reaches a certain value $M_P$ the stress diagram takes the form shown at (f) and the section deforms plastically under constant bending moment $M_P$ and fibre stress $f_Y$ until strain hardening begins at the outer fibres.

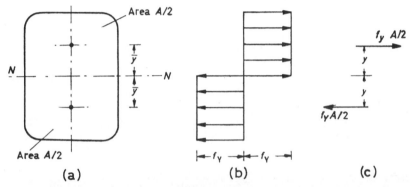

*Figure 6.55.* (a) The neutral plane of a fully plastic section intersects the section in the 'equal area axis' $NN$, which coincides with the centroidal axis if the section is symmetrical; (b) Fully plastic stress distribution; (c) Resultant forces on the section forming a couple $M_P$

The bending moment $M_P$ is known as the *plastic moment of resistance*. A section which is deforming plastically under a moment $M_P$ is said to form a *plastic hinge*.

### 6.6.2.2. *Plastic Moment of Resistance. Shape Factor*

The plastic moment can readily be calculated from the geometrical properties of the section. FIGURE 6.55(a) shows a symmetrical cross-section under the stress distribution (b). Since there is no resultant horizontal force we have

$$\int f_Y dA = f_Y \int dA = 0 \qquad \ldots \text{(i)}$$

---

† It will be observed that the distinction between the upper and lower yield points has been ignored and that yield is supposed to proceed at constant stress $f_Y$.

Hence the axis of zero stress must be the 'equal area axis' $NN$ rather than the centroidal axis familiar in elastic theory, although it happens that these coincide in the symmetrical case illustrated.

Equating the internal and external moments

$$M_p = f_Y \frac{A}{2} (2\bar{y}) = f_Y A \bar{y} \qquad \ldots \text{(ii)}$$

By analogy with the elastic case, it is convenient to define a plastic section modulus $Z_p$ such that

$$M_p = f_Y Z_p \qquad \ldots (6.26)$$

For a rectangular section, of width $b$ and depth $d$, we have

$$Z_p = A\bar{y} = bd\frac{d}{4} = \frac{bd^2}{4}$$

While for a circular cross section of radius $r$

$$Z_p = \pi r^2 \frac{4r}{3\pi} = \frac{4}{3} r^3$$

It is convenient to define a 'shape factor' $\nu$ such that

$$\nu = \frac{M_p}{M_Y} = \frac{f_Y Z_p}{f_Y Z_E} = \frac{Z_p}{Z_E} \qquad \ldots (6.27)$$

where $Z_E$ is the elastic section modulus.

Hence for the rectangular section $\quad \nu = \dfrac{bd^2/4}{bd^2/6} = 1 \cdot 5$

and for the circular section $\quad \nu = \dfrac{4/3 r^3}{\pi/4 r^3} = \dfrac{16}{3\pi} = 1 \cdot 7$

while for commercial I-sections $\quad \nu \simeq 1 \cdot 15$

### 6.6.2.3. Behaviour of Simply-supported Beam under Increasing Loads

If the load $P$ on the simply-supported uniform beam shown in FIGURE 6.56(a) is gradually increased yielding will eventually begin at the outside fibres of the most highly stressed section of the beam; this is obviously the point $B$ since the bending moment diagram (b), which depends only on the statics of the beam, has a sharp peak at that point. Yield will spread sideways, towards the ends of the beam, and inwards, towards the neutral axis, until eventually the whole section at $B$ has yielded and a plastic hinge has formed. The beam has then been transformed into a mechanism with real hinges at the ends and a plastic hinge at $B$; deformations will then increase rapidly and the beam can be said to be in a state of collapse†. The collapse load $P_c$ required to produce the plastic hinge is obtained by equating the maximum bending moment to the plastic moment of resistance $M_p$.

---

† In actual practice the onset of strain hardening and the change of geometry due to the large deformation may require the application of further load to bring about complete collapse, but the appearance of the plastic hinge is obviously crucial.

i.e.
$$P_c = \frac{4M_p}{L}$$

Alternatively, by considering the Virtual Work† involved in a small displacement from the original position, as at (c), we have

$$P_c \Delta = P_c \frac{L}{2} \theta = M_p 2\theta$$

i.e.
$$P_c = \frac{4M_p}{L} \quad , \text{ as above}$$

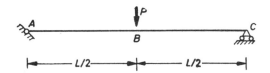

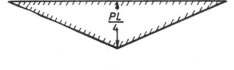

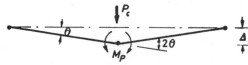

*Figure 6.56.* (a) Simply-supported beam subject to gradually increasing load $P$; (b) Bending moment diagram; (c) Condition when a plastic hinge has formed at $B$ under the collapse load $P_c$

### 6.6.2.4. *Behaviour of a Beam with Fixed ends under Increasing Loads*

The bending moment diagram for the fixed ended beam of FIGURE 6.57(a) is shown at (b). Yield will first appear at the ends, which carry the biggest bending moments, and plastic hinges will occur at the ends when the load is given by

$$\frac{w'L^2}{12} = M_p \qquad \ldots\ (i)$$

At this stage the central bending moment is only

$$\frac{w'L^2}{24} = \frac{M_p}{2}$$

so that from equation 6.27 we see that the beam is still behaving elastically at the centre provided the shape factor is less than 2.

The beam is evidently still capable of supporting load, being statically equivalent to a simply-supported beam with constant end moments $M_p$

---

† See Chapter 2, Section 2.6.2.1. Example (c).

## FRAMES WITH RIGID JOINTS

as shown at (c), and the load can be increased until a third plastic hinge appears at the centre; the beam has then become a mechanism and collapse occurs.

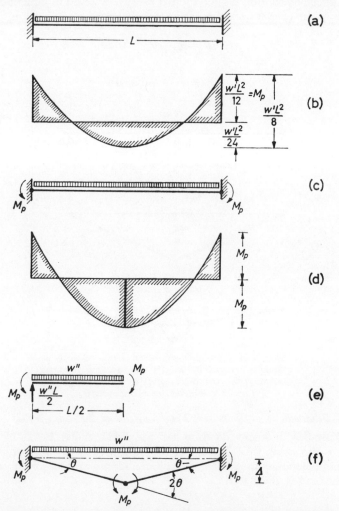

*Figure 6.57.* (a) Fixed ended beam under uniformly distributed load; (b) Bending moment diagram when plastic hinges have developed at the ends; (c) The beam is now equivalent to a simply-supported beam with constant couples $M_p$ acting on the ends; (d) Bending moment diagram when a plastic hinge has also developed at mid-span; (e) Equilibrium diagram for the situation at collapse; (f) Distortion diagram for the virtual work determination of the collapse load

The load $w''$ at this stage is found by considering the equilibrium of one half of the beam, as at (e) from which we have:—

$$\frac{w''L}{2} \cdot \frac{L}{2} - \frac{w''L}{2} \cdot \frac{L}{4} = 2 M_p$$

324

## 6.6. THE ANALYSIS OF COLLAPSE LOADS

i.e.
$$w'' = \frac{16 M_p}{L^2}$$

Alternatively, applying the principle of virtual work to the displacement diagram (f) we have:—

$$w''L \cdot \frac{L}{4} \cdot \theta = M_p \theta + M_p \theta + M_p 2\theta$$

i.e.
$$w'' = \frac{16 M_p}{L^2}$$

The designer is now faced with the problem of selecting a suitable beam to carry the actual working load $w''$: he must, in fact, choose an appropriate load factor. An extensive discussion of this difficult matter will be found in 'The Steel Skeleton', Vol. II[32]. So far as dead and superimposed vertical loads are concerned it is pointed out that the working stress permitted by B.S.449, which epitomises the accumulated experience of hundreds of steel-framed buildings, corresponds to a load factor of 1·75 with beams having a shape factor of 1·15.

Applying this to the result just obtained would give the working load $w = \frac{16 M_p}{1 \cdot 75 \, L^2}$. Under this load the fixed end moments would have the value $\frac{wL^2}{12} = \frac{16}{1 \cdot 75 \times 12} M_p = 0 \cdot 76 \, M_p$

From equation 6.27 we have

$$M_Y = \frac{1}{\nu} M_p = \frac{1}{1 \cdot 15} M_p = 0 \cdot 87 \, M_p$$

It thus appears that under this working load the beam behaves elastically even at the ends where the greatest bending moments occur.

### 6.6.3. General Methods for the Determination of Collapse Loads

It was possible to solve the above examples in a very simple and straightforward manner because the collapse mechanism investigated was in each case the only possible one. In more complicated cases there will be several possible collapse mechanisms and, although it is possible to investigate these in turn and then to select that which occurs under the lowest load, it is very convenient to have a more systematic method of procedure. It is first necessary, however, to introduce three theorems which have been shown to apply to the plastic collapse of all structures in which plastic hinges form in the manner described in Section 6.6.2.1. The theorems are here stated without proof; detailed discussions and proofs will be found in books by Neal[34] and by Baker, Horne and Heyman[32] together with references to papers by the originators of the theorems.

#### 6.6.3.1. *Theorems Governing Plastic Collapse*

(a) The uniqueness theorem:
   The bending moment distribution which fulfils the following conditions is unique:—

(i) Plastic hinges must be formed at a sufficient number of points to produce a mechanism.

(ii) Bending moments must be in equilibrium with the applied loads.

(iii) The allowable plastic moment must not be exceeded at any point in the structure.

It will be seen that these three conditions were in fact fulfilled in the examples worked above.

(b) The minimum principle, kinematic theorem[34] or 'unsafe' theorem[35]. This can be stated in several ways which can be seen to be equivalent.

    e.g. 'Of all the mechanisms formed by assuming plastic hinge positions, the correct mechanism for collapse is that which requires the minimum load'.

    or: 'Any arbitrary choice of hinge positions will lead to an estimate of the collapse load which is greater than, or equal to, the correct one'.

(c) The maximum principle, static theorem[34] or 'safe' theorem[35].

    'Of all the equilibrium states satisfying the condition that the plastic moment is nowhere exceeded, the equilibrium state at collapse gives the maximum estimate of load'.

    or: 'Any arbitrary equilibrium state satisfying the condition that the plastic moment is nowhere exceeded gives a load less than, or equal to, the collapse load'.

### 6.6.3.2. *Determination of Collapse Load by Trial*

The application of these theorems allows upper and lower bounds to be found for the collapse loads, as in the following example. FIGURE 6.58(a)

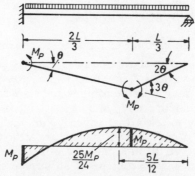

*Figure 6.58.* (a) Propped cantilever under uniformly distributed load; (b) Distortion diagram if plastic hinges are assumed at the fixed end and at a third point; (c) Corresponding bending moment diagram

shows a propped cantilever carrying a uniformly distributed load. It is obvious that one plastic hinge will occur at the fixed end but the position of the second must be determined. We begin by assuming, incorrectly as it turns out, that the second hinge forms at $x = L/3$ from the free end. The assumed conditions at collapse are then as shown at (b) and correspond to a load of $w' = 12\, M_p/L^2$, which gives the bending moment diagram (c). This has a maximum of $25\, M_p/24$ at a distance of $5L/12$ from the simply-supported end.

It thus appears that $w'$ is greater than the correct value $w_c$ in accordance with the minimum principle, and thus gives an upper bound on $w_c$.

A lower bound is obtained from the maximum principle by reducing the load $w'$ to $w'' = \dfrac{24}{25} w'$

i.e. $$w'' = \frac{24}{25} \cdot \frac{12\, M_p}{L^2} = 11\cdot 52\, M_p/L^2$$

Hence we have
$$12\, M_p/L^2 \geqslant w_c \geqslant 11\cdot 52\, M_p/L^2$$

It may well be thought that a close enough estimate of $w_c$ has now been found but, if necessary, the bounds can be narrowed by making new estimates based on the knowledge already gained. The exact value of $M_p$ can be found by assuming that the inner plastic hinge occurs at a distance $x$ from the simply-supported end. This gives a value of the load equal to $\dfrac{2M_p(L + x)}{Lx(L - x)}$. From the minimum principle above the value of $x$ corresponding to the smallest value of the load is the correct one, and this occurs when $x = 0\cdot 414L$, giving $w_c = 11\cdot 65 M_p/L^2$

Many examples of the above will be found in the works already cited, in a book by Heyman[35] and in the booklets published by the British Constructional Steelwork Association[36].

### 6.6.3.3. *Analysis by Combination of Elementary Mechanisms*

It will be apparent that if a frame has $n$ redundants it will become determinate if $n$ plastic hinges form in appropriate places and will collapse if $n + 1$ hinges form. The central problem of plastic analysis is to find the correct location of these hinges so that the theorems of Section 6.6.3.1 are satisfied. While the trial and error method described above is adequate in simple cases a more powerful technique is desirable for complicated structures. Two such techniques have been described: accounts of the method of plastic moment distribution which was originated independently by Horne[37] and by English[38], will be found in the books by Neal[34] and Baker, Horne and Heyman[32]; it will not be discussed here.

The method of combining elementary mechanisms, due to Neal and Symonds[39], depends on the realization that all possible mechanisms can be built up from three basic mechanisms:

(i) *beam mechanisms* can develop in any member carrying transverse loads;

(ii) *sway mechanisms* can develop whenever sway of members can occur;

(iii) *joint mechanisms* are possible at any joints connecting three or more members.

The number of such elementary mechanisms which must be considered in any particular problem can be shown to be equal to the number of independent equations of equilibrium relevant to the structure in question. If the structure has $n$ redundants then, by definition, there are $n$ more unknowns than there are equations of equilibrium. But if there are $p$ possible hinge locations and $m$ independent mechanisms (each corresponding to an equation of equilibrium) then $(p - m)$ is also the excess of unknowns over available equations.

Hence $$p - m = n$$

or $$m = p - n \qquad \ldots \ldots (6.28)$$

The portal frame of FIGURE 2.14, which was previously analysed in Chapter 2, Section 2.6.2, Example (c), by what amounted to the trial method, will now be examined from this point of view. In this case $n = 3$ and $p = 5$, so that there are $5 - 3 = 2$ independent mechanisms. These are the sway mechanism (b) and the beam mechanism (c). It is now necessary to see whether these mechanisms can be combined in such a way as to give a lower value of $W$ for a given value of $M_p$ (or a higher value of $M_p$ for a given value of $W$) in accordance with the minimum principle.

The application of the Virtual Work equation to the sway and beam mechanisms yielded the following equilibrium equations:

Sway mechanism $\qquad WL\,\theta = 4M_p\theta$

Beam mechanism $\qquad WL\,\theta = 5M_p\theta$

We now notice that if these two mechanisms are combined the plastic hinge at 2 is eliminated, for in the sway mechanism the movement at that hinge is such that the frame opens out there, while in the beam mechanism it closes up. The equilibrium equation corresponding to the combined mechanism is therefore obtained by adding the equations corresponding to the independent mechanisms and deducting the virtual work no longer performed at 2. A term $M_p\theta$ was previously included in each equation on account of the work done at 2, so that $2M_p\theta$ is deducted from the sum. Hence the new equilibrium is:—

$$2WL\theta = 9M_p\theta - 2M_p\theta = 7M_p\theta$$

i.e. $$W = \frac{3 \cdot 5\,M_p}{L}$$

This is the lowest value of $W$ obtained and so, as all possible combinations have been covered, it is the required collapse load.

The complete procedure can now be summarized:—

(i) Determine the number of independent mechanisms.

(ii) Sketch the possible beam, sway and joint mechanisms and check that their total number is correct.

(iii) Write down the equilibrium equation corresponding to each possible independent mechanism, preferably using the virtual work method for the purpose.

(iv) Investigate possible combinations of the independent mechanism, observing that progress towards the correct solution is only made if plastic hinges are eliminated by such combinations.

(v) Finally draw the bending moment diagram corresponding to the solution obtained and check that the plastic moment is nowhere exceeded.

## REFERENCES

[1] Reports of the Steel Structures Research Committee. H.M.S.O. 1931, 1934 and 1936

[2] CROSS, H. 'The Analysis of Continuous Frames by Distributing Fixed End Moments'. *Trans. Amer. Civ. Engrs.* Vol. 96. 1932

[3] CORNISH, R. J. 'The Magnitude of the Direct Stress in a Beam of Fixed Span'. *Phil. Mag.* Ser. 7. Vol. xxviii. Oct., 1939. p. 481

[4] CLAPEYRON, B. P. E. 'Calcul d'une poutre elastique'. *Comptes Rendus.* Vol. 45. 1857. p. 1076

[5] CORNISH, R. J. and JONES, E. 'Influence Lines for Beams Continuous over Three Spans'. *The Structural Engineer.* Jan., 1936. p. 2

[6] WILSON, W. M. and MANEY, G. A. 'Wind Stresses in Office Buildings'. Univ. of Illinois Engineering Experiment Station. Bull. No. 80. 1915

[7] CLYDE, D. H. 'The Problem of Sway in Moment Distribution'. *Engineering.* June 8, 1956. p. 457

[8] NEAL, B. G. 'Virtual Work and the Moment Distribution Method'. *Engineering.* Jan. 11, 1957. p. 47

[9] MATHESON, J. A. L. 'The Problem of Sway in Complicated Rigid Frames'. *J. Inst. Civ. Engrs.* Vol. 24. Feb., 1945. p. 99

[10] TIMOSHENKO, S. and YOUNG, D. H. *Theory of Structures.* McGraw-Hill. 1945

[11] BATEMAN, E. H. 'The Development of the Elastic Theory of Continuous Frames'. *The Structural Engineer.* March, 1955. p. 73

[12] GRINTER, L. E. *Theory of Modern Steel Structures.* Vol. II. Macmillan

[13] NAYLOR, N. 'Side Sway in Symmetrical Building Frames'. *The Structural Engineer.* April, 1950

[14] WOOD, R. H. *An Economical Design of Rigid Steel Frames for Multi-storey Buildings.* National Building Studies Research paper No. 10. H.M.S.O. 1951

[15] WOOD, R. H. 'Degree of Fixity Methods for Certain Sway Problems'. *The Structural Engineer.* July, 1952

[16] LIGHTFOOT, E. 'The Analysis for Wind Loading of Rigid-jointed Multi-storey Building Frames'. *Civil Engineering.* Vol. 51. No. 601. July, 1956

[17] BOLTON, A. 'A new approach to the elastic analysis of two-dimensional rigid frames'. *Structural Engineer.* 1952. p. 1

[18] ALLEN, D. N. de G. *Relaxation Methods.* McGraw-Hill. 1954

[19] CLAXTON FIDLER. 'Continuous Girder Bridges'. *Proc. Inst. Civ. Engrs.* Vol. 74. 1883. p. 196

[20] SALMON, E. H. *Materials and Structures*. Vol. I. Chap. VI. Longmans, Green and Co.
[21] STEINMAN, D. B. 'Moment Distribution by Linked Rigidities'. *Engineering News Record*. 1944. p. 802
[22] BOWLES, R. E. and CORNISH, R. J. 'Continuous Beam Analysis by Focal Points'. *The Structural Engineer*. July, 1944. p. 332
[23] SHEPLEY, E. *Continuous Beam Structures*. Concrete Publications
[24] BATHO, C. and ROWAN, H. C. 'The Analysis of the Moments in the Members of a Frame having Rigid or Semi-rigid Connexions, under Vertical Loads'. Second Report of the Steel Structures Research Committee. H.M.S.O. 1934. p. 177
[25] BATHO, C. 'The Analysis and Design of Beams under given end restraints'. Final Report of the Steel Structures Research Committee. H.M.S.O. 1936. p. 364
[26] SOUROCHNIKOFF, B. 'Wind Stresses in semi-rigid connexions of steel framework'. *Trans. Amer. Soc. Civ. Engrs*. Vol. 115. 1950. p. 382
[27] FREEMAN, R. and ENNIS, L. 'Sydney Harbour Bridge: Manufacture of the Structural Steelwork and Erection of the Bridge'. *Proc. Inst. Civ. Engrs*. Vol. 238 (1933-4). pp. 242-5
[28] BAKER, J. F. 'The Design of Steel Frames'. *The Structural Engineer*. 1949. p. 399
[29] Conference on the Correlation between calculated and observed Stresses and Displacements in Structures. *Inst. Civ. Engrs*. 1956
[30] VAN DEN BROEK, J. A. 'Theory of Limit Design'. *Trans. Amer. Soc. Civ. Engrs*. Vol. 105. 1940. p. 638
[31] VAN DEN BROEK, J. A. *Theory of Limit Design*. Wiley. 1948
[32] BAKER, J. F., HORNE, M. R. and HEYMAN, J. *The Steel Skeleton*. Vol. II. Cambridge, 1956
[33] CHARLTON, T. M. 'The analysis of structures with particular reference to the prediction of deflexions'. *Trans. N.E. Coast Inst. of Engineers and Shipbuilders*. Vol. 74. 1958. p. 163
[34] NEAL, B. G. *The plastic methods of structural analysis*. Chapman & Hall. 1956
[35] HEYMAN, J. *Plastic design of portal frames*. Cambridge. 1957
[36] *The Collapse Method of Design*. British Constructional Steelwork Association. Publications Nos. 5 and 11
[37] HORNE, M. R. 'A moment-distribution method for the analysis and design of structures by the plastic theory'. *Proc. Inst. Civ. Engrs*. Pt. III, Vol. 3. 1954
[38] ENGLISH, J. M. 'Design of frames by relaxation of yield hinges'. *Trans. Amer. Soc. Civ. Engrs*. Vol. 119. 1954. p. 1143
[39] NEAL, B. G. and SYMONDS, P. S. 'Rapid calculation of the plastic collapse load of a framed structure'. *Proc. Inst. Civ. Engrs*. Pt. III, Vol. 1. 1952. p. 58

## ADDITIONAL REFERENCES

ABELES, P. 'The Four Moment Theorem: Simplification in Calculation of Frame Structures'. *Concrete and Constructional Engineering*. 1942
AMERIKIAN, A. *Analysis of Rigid Frames*. U.S. Govt. Printing Office. 1942

# REFERENCES

ANGER. *Ten division influence lines for continuous beams.* Constable

BANERJEE, S. P. 'Influence lines for continuous girders'. *Civil Engineering and Public Works Review.* Vol. 48. Feb., 1953. p. 149

BEAUFOY, L. A. and DIWAN, A. F. S. 'Analysis of continuous structures by the stiffness factors method'. *Quart. J. Mech. and App. Maths.* Vol. II. Pt. 3. 1949. p. 263

BETTESS, F. 'Graphical Determination of Moments in rigid frameworks'. *Civil Engineering and Public Works Review.* Vol. 47. 1952. p. 100

BLEICH, F. *Die Berechnung Statisch Unbestimmte Tragwerke nach der Methode der Viermoment ensatzes.* Springer. 1918

BRISBY, M. D. 'The Four Moment Theorem applied to Multiple Bay Frames'. *Concrete and Constructional Engineering.* 1949

CORNISH, R. J. 'Moment Balance; a self-checking analysis of rigidly jointed frames'. *Trans. Amer. Soc. Civ. Engrs.* Vol. 108. 1943. p. 677

CSONKA, P. 'Analysis of frames with movable joints'. Publication of the University of Technical Sciences. Budapest. 1948

FRANCIS, A. J. 'Frames subjected to multiple sway'. *Concrete and Constructional Engineering.* Nov., 1949. p. 335

GRAY, C. S., KENT, L. E., MITCHELL, W. A. and GODFREY, G. B. *Steel Designers' Manual.* Crosby, Lockwood. 1955

GRIOT, G. *Four place influence line tables* . . . Ungar. 1952

HENDRY, A. W. 'An investigation of the stress distribution in steel portal frame knees'. *The Structural Engineer.* 1947

HENDRY, A. W. 'Test on Welded Portal Frame connexions under compressive structures'. *The Structural Engineer.* 1950

JOHANNES, H. and ANTONO, A. *Flexure Factors method for analysing structures.* Gadjah Mada State University. Jakarta. Indonesia

KLEINLOGEL, A. *Rahmenformeln.* W. Ernst u. Sohn. Berlin

KLEINLOGEL, A. *Mehrstielige Rahmen.* W. Ernst u. Sohn. Berlin

KLOUCEK, C. V. *Distribution of deformation.* Prague. 1950

LIGHTFOOT, E. 'Generalized slope-deflexion and moment distribution methods for the analysis of viaducts and multi-bay ridged portal frames'. *The Structural Engineer.* Vol. 36. Jan., 1958. p. 21

MANNING, G. P. *The displacement method of frame analysis.* Concrete Publications Ltd. 1952

MATHESON, J. A. L. 'Moment Distribution applied to rectangular rigid space frames'. *J. Inst. Civ. Engrs.* Vol. 29. 1948. p. 221

MAUGH, L. C. 'The Analysis of Vierendeel trusses by successive approximations'. Publications: International Association for Bridge and Structural Engineering. 1935

MICHALOS, J. P. and WILSON, E. H. 'Influence lines by corrections to an assumed shape'. *Trans. Amer. Soc. Civ. Engrs.* Vol. 118. 1953. p. 113

SKAYANNIS, A. P. *System of Tables for quick and accurate solving of any continuous beam.* Athens. 1949

STEWART, R. W. 'Analysis of frames with elastic joints'. *Trans. Amer. Soc. Civ. Engrs.* Vol. 114. 1949. p. 17

*Some notes on the Analysis of Rigid Frames.* British Constructional Steelwork Association. Publication No. 10. 1956

# Chapter 7
# ARCHES
## 7.1. INTRODUCTION

THE arch is a structural form of very great antiquity, isolated examples of vaulted construction having been found in Mesopotamia and elsewhere which show that the principle has been known for some six thousand years. The first extensive use of arches, however, was made by the Romans who carried the art to a high degree of perfection; some magnificent examples of their work still survive. Although the scale of construction reached by the Romans was hardly matched until quite recently the techniques they established for building masonry arches and vaults continued in use throughout the Middle Ages and, indeed, down to modern times.

Although the buttresses of the mediaeval Gothic cathedrals show that their builders had a sound intuitive understanding of the need to counteract the thrust of an arch no real progress in the theory of arch action was made

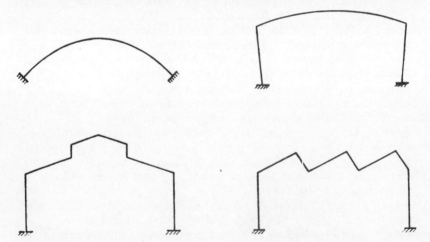

*Figure 7.1.* These structures can all be regarded as arches and analysed as such

until about the end of the 17th century. De la Hire, who died in 1718, originated the idea of investigating the stability of arch rings by the use of the line of pressure or linear arch. This led to much speculation on the correct method of drawing the linear arch to which Gerstner, Moseley and Villarceau made important contributions†.

Since the fixed arch is hyperstatic the correct solution could not be found by statics alone and it was not until Winkler and others introduced the

---
† Interesting accounts of the historical development of arches are given by Pippard and Baker[2], Timoshenko[3] and Straub[4].

## 7.1. INTRODUCTION

ideas of elasticity that a rational approach was possible. Indeed it was left to Pippard in 1936 to give a full explanation of the behaviour of arch rings constructed of discrete *voussoirs* or wedge-shaped stones.

In the 20th century interest has shifted away from this form of construction to steel and reinforced concrete arches with integral ribs. The elastic theory has now been extensively developed for structures of this type although it is perhaps insufficiently appreciated that, so far as its structural action is concerned, there is no need for an arch to have the familiar curved form. The frames illustrated in FIGURE 7.1, can all be regarded as arches and can often be treated more simply in this way than by other means.

### 7.1.1. THE PRINCIPLE OF ARCH ACTION

The basic principle of arch action is easily explained: FIGURE 7.2(a) shows a simply-supported beam while at (b) there is an arched rib with the same span and vertical loading but having, in addition, inward horizontal forces acting at the ends. The beam and the arch evidently have the same bending moments $M_o$ so far as vertical loads are concerned but the effect of the inward thrust is to reduce the bending moment in the arch; thus

$$M = M_o - Hy \qquad \ldots (7.1)$$

The central problem of arch analysis is to find the horizontal thrust acting in any given circumstances and so to find the effective bending moment to which the rib is subjected. This is clearly a hyperstatic problem, for in the absence of horizontal restraint at the supports the arch will tend to spread, as at (d). If the supports are arranged so as to prevent this spread completely then a horizontal thrust will be induced whose magnitude can be found, in principle, by direct comparison of displacements. Thus

$$\Delta_B = 0 = \Delta_{bo} - H\Delta_{bb}$$

or $\qquad H = \Delta_{bo}/\Delta_{bb} \qquad \ldots (7.2)$

If, however, the supports are not capable of suppressing the spread completely then we shall have

$$H = \frac{\Delta_{bo} - \Delta_B}{\Delta_{bb}} \qquad \ldots (7.3)$$

Where $\Delta_B$ is the final horizontal displacement of $B$. The effect of this is to reduce $H$ and so, from 7.1, to increase the nett bending moment acting on the rib. It thus appears that arches should only be used in situations where the necessary thrust can be relied upon.

No mention has yet been made of the effect of angular restraint at the ends of the arch; a full discussion of this matter is reserved until Section 7.4 but it can be said now that if the supports offer any angular restraint then fixing moments, akin to those at the ends of fixed ended beams, will be induced. These act in such a way as to reduce the effective bending moments everywhere in the rib except close to the ends.

The designer's problem is to proportion the rib so that the smallest possible nett stresses are produced at every section and this usually means that the bending moment must everywhere be minimized. If the arch is to carry a fixed load system, or if the dead loads outweigh the live loads, the designer is greatly helped by the conception of the linear arch; this is discussed in

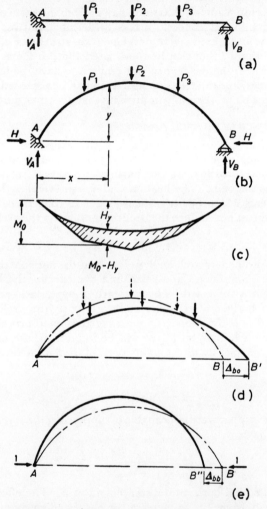

*Figure 7.2.* (a) Simply-supported beam carrying vertical loads $P$; (b) Arch rib carrying vertical loads $P$ and horizontal thrust $H$; (c) The bending moment diagram $M_o$ for the vertical loads is the same for beam and arch, but the thrust $H$ in the arch reduces the moment at all points by $Hy$; (d) If the arch is free to spread the loads will cause $B$ to move an amount $\Delta_{bo}$ relative to $A$; (e) Unit horizontal thrust moves $B$ to $B''$ by an amount $\Delta_{bb}$

Section 7.1.3 after a brief description, in Section 7.1.2, of the different types of arch.

## 7.1.2. Types of Arch

Arch bridges may be classified into different structural types according to the design of the rib, whether solid or braced, and to the arrangements made for supporting the deck. A few common types are shown in FIGURE 7.3.

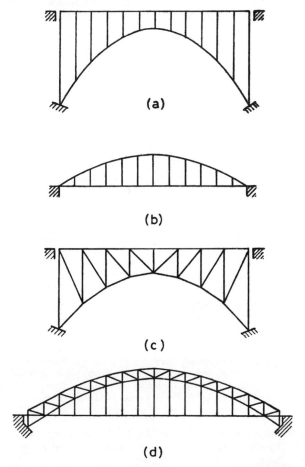

*Figure 7.3.* Schematic arrangement of arch bridges. (a) Many reinforced concrete arches are of this open spandrel type; (b) The bow-string arch is built either in steel or in reinforced concrete. The horizontal thrust can be provided by tension in the deck system giving a tied arch; (c) The spandrel braced arch lends itself to steel construction in steep sided gorges; (d) The braced rib is most frequently used for long span steel arches

They can also be classified according to the number of hinges incorporated in the structure; since this determines the degree of redundancy it is a very useful approach from the present point of view.

FIGURE 7.4(a) shows a three-hinged arch which, it will readily be seen, is statically determinate. It is thus possible to find the horizontal thrust from statical considerations without regard to the elastic distortions of the rib. Arches of this type have the advantage of being insensitive to foundation movements and to changes of temperature, and for this reason they have often been used to support wide-span roofs. They are less suitable in bridge construction since the discontinuity at the crown hinge is communicated to the deck system; this is particularly inconvenient in rail bridges.

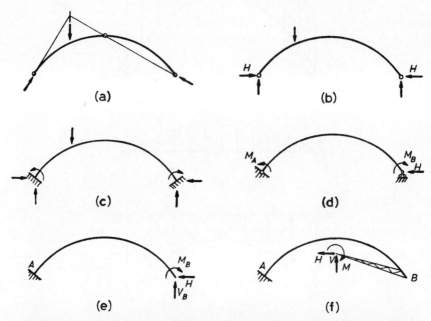

*Figure 7.4.* (a) Statically determinate three-hinged arch; (b) Two-hinged arch with one redundant, $H$; (c) Fixed arch with three redundants; (d) Redundants $M_A$, $M_B$ and $H$; (e) Redundants $M_B$, $H$ and $V_B$ acting on cantilever primary structure; (f) Redundants $M$, $H$ and $V$ acting at elastic centre

The two-hinged arch shown in FIGURE 7.4(b) has four reactive restraints of which one, usually taken as the horizontal component of reaction at one hinge, is redundant.

The fixed arch shown at (c) has six reactive restraints and there are therefore three redundants. These can be chosen in one of several ways; for example:

   (i) the fixing moments at the ends and the horizontal component of reaction at one end. The primary structure is then the simply-supported arch rib of FIGURE 7.2(b). (See FIGURE 7.4(d).)

  (ii) the fixing moment and the horizontal and vertical components of reaction at one end. The primary structure is then the curved cantilever springing from $A$. (See FIGURE 7.4(e).)

 (iii) a couple and horizontal and vertical forces acting at the 'elastic centre' of the arch and operating on the primary structure through an imaginary rigid arm. (See FIGURE 7.4(f).)

## 7.1. INTRODUCTION

It is also possible to take the bending moment, thrust and shear at some internal point, the crown for example, as the redundants and this is sometimes done in the case of symmetrical arches.

### 7.1.3. THE LINEAR ARCH

The linear arch, or line of thrust, is a development of the idea of the funicular polygon familiar from elementary graphic statics. It will be recalled

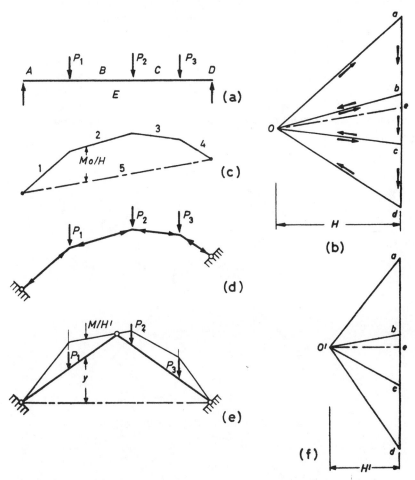

*Figure 7.5.* (a) Simply-supported beam under vertical loads; (b) Vector polygon for the loads $P$; (c) Corresponding funicular polygon; (d) A linear arch, constructed so that its sides coincide with those of a funicular polygon, is free from bending moment; (e) A triangular three-hinged arch under the same loads. The funicular polygon must now be adjusted so that it passes through the hinges where the bending moments are zero; (f) The vector polygon that gives the required funicular polygon

that the resultant of a series of coplanar forces can be obtained by drawing a force or vector polygon and a funicular polygon; the former gives the magnitude of the resultant and the latter its position. In the case of vertical loads acting on a beam the two vertical reactions are found as in FIGURE 7.5. In this case the vector polygon, at (b), degenerates into a straight line *abcd* and, as the position of the two reactions is known, a closing line *5* can be drawn at once by joining the points shown on the funicular polygon (c). The line *Oe* on the vector polygon, drawn through *O* parallel to *5* gives *de* and *ea*, the required reactions.

The proof of the validity of this construction turns on consideration of the series of force triangles obtained by joining the points, a, b, etc. to the pole *O* in the vector polygon. Thus $P_1$ represented by *ab*, can be balanced by forces parallel to and represented to scale by *bO* and *Oa* and intersecting, as do lines *1* and *2*, on the line of action of $P_1$. Similar considerations apply to the other forces $P_2$ and $P_3$ so that the whole load system can be supported by forces acting along the sides of the funicular polygon. If an arch were constructed of the same form as the funicular polygon, as at (d), it would not be subject to bending but would carry the loads by means of axial

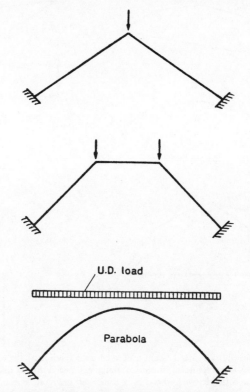

*Figure 7.6.* Linear arches for different load arrangements

forces only; moreover, the horizontal component of the thrust at every point of the arch would be represented, to the scale of force, by the polar distance *H*. Such an arch is called a *linear arch*; there is naturally an infinite number of such arches for every load pattern, corresponding to the infinite number of possible positions of the pole $O$. Some examples are shown in FIGURE 7.6.

It can also be shown from similar triangles that a vertical intercept on the funicular polygon, when multiplied by the polar distance $H$ on the vector diagram, gives the bending moment at the corresponding point on the beam. (Measurements on the funicular polygon are made to the scale of length while those on the vector polygon are made to the force scale.) This gives the clue to the construction of the correct linear arch in any particular case.

For example the triangular arch of FIGURE 7.5(e) must have zero bending moment at each of its three hinges; it can be seen from equation 7.1 that this implies that the linear arch must pass through the hinges†. It also appears that when this has been done intercepts between the linear arch and the actual arch represent the bending moments. It follows that if the arch is arranged to coincide with a funicular polygon for the loads it is free of bending moment even if it is fixed. This is important because, in practice, the dead load is often predominant especially if the span is large. The arch can thus be designed to be free from bending moment under the dead load apart from any change in its shape caused by strain; this is usually negligible unless the arch is very flat.

Finally it is to be noticed that if the load is uniformly distributed the linear arch will be parabolic.

## 7.2. THE THREE-HINGED ARCH

Certain features of arch theory, notably influence lines, can best be illustrated by reference to the three-hinged arch which would otherwise, as a statically determinate structure, fall outside the scope of this book. The following brief account of some of the properties of this type of arch is therefore included.

### 7.2.1. DETERMINATION OF THE HORIZONTAL THRUST

The horizontal thrust $H$ for a single load as in FIGURE 7.7(a), can be found very quickly, by drawing the linear arch passing through the crown hinge and then completing the triangle of forces (b). Alternatively, applying equation 7.1 to the crown hinge $C$, where the bending moment is zero, we have

$$M_C = 0 = (M_o)_C - Hh$$

i.e. $\qquad H = (M_o)_C / h \qquad \ldots . (7.4)$

where $(M_o)_C$ is the simply-supported bending moment at the crown. If there are several loads, and especially if the loading is continuous, the graphical method is inconvenient and it is better to use equation 7.4.

---

† This can be contrived by drawing any funicular polygon, as at (b), to obtain the point $e$: a new pole $O'$ is then chosen on the horizontal through $e$ so that the closing line $5$, which is parallel to $O'e$, is also horizontal. The polar distance $O'e = H''$ is made such a multiple of $H$ that the funicular polygon is the correct height to pass through the crown hinge. See FIGURE 7.5(f).

# ARCHES

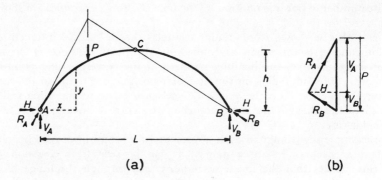

*Figure 7.7.* (a) Three-hinged arch; since $CB$ is unloaded $R_B$ must pass through $C$. This determines the line of action of $R_A$; (b) When the directions of action of $R_A$ and $R_B$ are known the triangle of forces can be drawn

Applying this result to the three-hinged parabolic arch shown in FIGURE 7.8 we have

$$H = (M_o)_C/h = wL^2/8h \qquad \ldots (7.5)$$

The bending moment at any other point $(x, y)$ is then

$$M = M_o - Hy = \left(\frac{wLx}{2} - \frac{wx^2}{2}\right) - \frac{wL^2}{8h} \cdot \frac{4hx}{L^2}(L-x) = 0$$

thus confirming the concluding statement of Section 7.1.3.

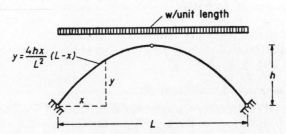

*Figure 7.8.* Three-hinged parabolic arch under uniformly distributed load

If the abutments are not at the same level it is not possible to proceed in quite such a simple manner since the vertical components of reaction at the hinges, being affected by the offset horizontal thrust, are no longer the same as in the corresponding simply-supported beam. Thus referring to FIGURE 7.9 we have

Moments about $C$:  $V_A.16 = H_A.12 + 1.8$ $\quad\quad H_A = 1$
                                                           i.e.
Moments about $B$:  $V_A.24 = H_A.6 + 1.16 + 2.4$ $\quad V_A = 5/4$

Resolve vertically:  $V_A + V_B = 3$ $\quad\quad$ i.e. $V_B = 7/4$

Resolve horizontally: $H_A = H_B$

340

## 7.2. THE THREE-HINGED ARCH

The bending moment can now be found at all points.

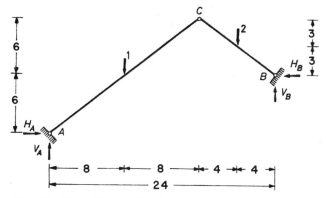

*Figure 7.9.* Three-hinged triangular arch with abutments at different levels

### 7.2.2. DETERMINATION OF THRUST AND SHEAR

When the horizontal component of thrust $H$ is known the axial thrust and normal shear at any section can be found by resolution. Thus in FIGURE 7.10 the vertical shear $V$ can readily be found while in the absence of horizontal loads on the arch $H$ is constant.
Then

$$\left. \begin{array}{l} \text{Shear } Q = V \cos \alpha - H \sin \alpha \\ \text{Thrust } T = V \sin \alpha + H \cos \alpha \end{array} \right\} \quad \ldots \ldots (7.6)$$

and

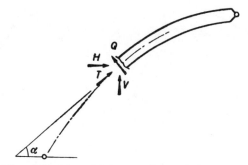

*Figure 7.10.* The axial thrust and normal shear at a section

### 7.2.3. DETERMINATION OF STRESS

When the thrust $T$ has been determined as above there remains the question of the stress it produces in combination with the bending moment $M$. Thus at the ' extrados ' of the rib shown in FIGURE 7.11(a) we shall have

$$f_e = \frac{T}{A} + \frac{My_e}{I}$$

and at the 'intrados'

$$f_i = \frac{T}{A} - \frac{My_i}{I}$$

There is some advantage, however, in using the 'core' theory† as follows:—in FIGURE 7.11(b) $I$ and $E$ are the interior and exterior kern points‡, and are respectively distant $r^2/y_i$ and $r^2/y_e$ from the neutral axis of the section, whose second moment of area $I = Ar^2$. We now observe that the combined effect of $T$ and $M$ would be given by a single thrust offset by an amount $a$, as at (b).

i.e.
$$f_e = \frac{T}{A} + \frac{Tay_e}{I}$$

$$= \frac{Tr^2}{Ar^2} + \frac{Tay_e}{I}$$

$$= T\left(\frac{r^2}{y_e} + a\right)\frac{y}{I}$$

Now $T\left(\dfrac{r^2}{y_e} + a\right)$ is the moment of $T$ about $E$, the exterior kern point, so that

the stresses are obtained directly if moments are calculated for the kern points instead of for the centroid of the section. Hence in equation 7.1 we should compute $M_o$ for and measure $y$ to the kern point.

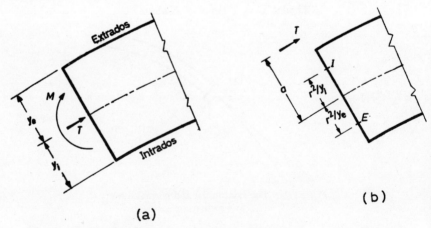

(a)     (b)

*Figure 7.11.* (a) Axial thrust and bending moment acting on a section; (b) Equivalent eccentric thrust. $I$ and $E$ are the internal and external kern points respectively

---

† The 'core' of a section is that region within which an eccentric thrust must act if the resulting stresses are to be wholly compressive. The extremes of this region are the kern points; if the section is rectangular they are the 'third' points.

‡ See for example Timoshenko, *Strength of Materials*, Part I, 1940, p. 235.

## 7.2.4. Influence Lines

### 7.2.4.1. *Influence Line for Horizontal Thrust*

Taking as an example the three-hinged circular arch of FIGURE 7.12(a) we begin by finding the influence line for the horizontal thrust. From equation 7.4 we see that this is given by drawing the bending moment influence line for that point on the corresponding simply-supported beam which corresponds to the crown hinge. (See Chapter 5, FIGURE 5.1(d)).

i.e.
$$H = (M_o)_C/h = \frac{bx}{Lh}$$

when unit load is between $A$ and $C$. Hence the influence for $H$ is as shown. at (b), the maximum ordinate, at $C$, being $\frac{ab}{Lh}$.

### 7.2.4.2. *Influence Line for Bending Moment*

From equation 7.1 we see that the influence line for bending moment at a particular point is obtained by compounding that for the same point on the corresponding beam with the influence line for $H$ which has just been obtained.

Thus for the point $D$ we have

$$M = (M_o)_D - Hy_D$$
$$= \frac{x(L-d)}{L} - \frac{bx}{Lh}y_D$$

when unit load is between $A$ and $D$. A similar expression can, if desired, be obtained for the situation when the unit load is to the right of $D$ by working from the other end but, as the influence diagram must consist of a pair of triangles, it can be constructed from the above expression alone. This has been done at (c).

### 7.2.4.3. *Influence Line for Axial Thrust*

Using the symbols of FIGURE 7.12(d) we have for the axial thrust at $D$

Load to left of $D$: $\quad T_D = H\cos\alpha - V_B\sin\alpha$

Load to right of $D$: $\quad T_D = H\cos\alpha + V_A\sin\alpha$

These expressions have been used to obtain the influence diagram of FIGURE 7.12(e).

### 7.2.4.4. *Influence Line for Normal Shear*

Again from FIGURE 7.12(d) we have

Load to left of $D$: $\quad Q_D = -H\sin\alpha - V_B\cos\alpha$

Load to right of $D$: $\quad Q_D = -H\sin\alpha + V_A\cos\alpha$

These expressions are plotted in FIGURE 7.12(f).

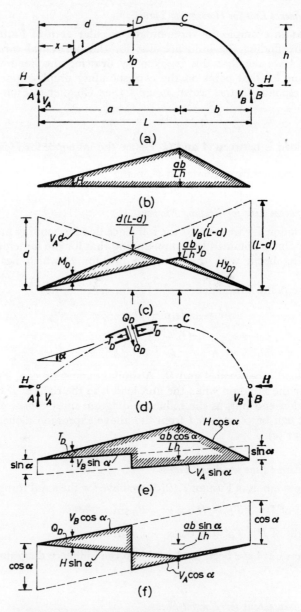

*Figure 7.12.* (a) Three-hinged unsymmetrical arch; (b) Influence line for horizontal thrust $H$; (c) Influence line for bending moment $M_D$ at $D$; (d) Axial thrust and normal shear at $D$; (e) Influence line for axial thrust $T_D$ at $D$; (f) Influence line for normal shear $Q_D$ at $D$

## 7.3. THE TWO-HINGED ARCH

### 7.3.1. Derivation of the Expression for Horizontal Thrust

It has already been indicated, in Section 7.1.1, that the two-hinged arch has one redundant quantity and that this is usually taken to be the horizontal component of the reaction at one hinge. The analysis therefore begins with the evaluation of this redundant thrust by solving equation 7.2. Any convenient means can be used for finding the necessary deflexions: a structural model can be used, or Hoadley's method, but it is usual to employ Castigliano's Theorem of Compatibility to obtain the necessary expressions in terms of bending moments and then to integrate either directly or numerically.

Using the arch of FIGURE 7.2(b) as an illustration we have from equation 7.1

$$M = M_o - Hy \qquad \ldots (7.1)$$

From the Theorem of Compatibility we have

$$\frac{\partial U}{\partial H} = 0 \text{ if the span } AB \text{ is constant}$$

i.e.
$$0 = \frac{\partial}{\partial H} \int \frac{M^2 ds}{2EI} = \int \frac{M}{EI} \cdot \frac{\partial M}{\partial H} \cdot ds$$

$$= \int \frac{(M_o - Hy)}{EI} (-y) \, ds$$

Hence
$$H = \frac{\int \frac{M_o y}{EI} \cdot ds}{\int \frac{y^2}{EI} \, ds} \qquad \ldots (7.7)$$

This equation coincides with equation 7.2 since Castigliano's Theorem, Part II shows that

$$\Delta_{bo} = \int \frac{M_o y}{EI} \, ds \text{ and } \Delta_{bb} = \int \frac{y^2}{EI} \, ds$$

if the deflexion is assumed to be entirely caused by bending action; this is the usual assumption but the effect of direct stress can be included quite readily as indicated in Section 7.3.4 below.

The integration of the terms of equation 7.7 can only be performed algebraically in the limited number of cases when $EI$ is a suitable function of $s$; otherwise Simpson's rule or some other technique of numerical integration must be employed. A few examples are given below.

#### 7.3.1.1. *Polygonal Arch of Uniform Cross-section*

FIGURE 7.13 shows a polygonal arch having $EI$ constant; equation 7.7 thus becomes

$$H = \frac{\int M_o y \, ds}{\int y^2 \, ds} \qquad \ldots (7.7(a))$$

ARCHES

and we can integrate the numerator and denominator directly, using $x, y$ or $s$ as the variable at choice; careful attention must be paid to the limits of integration which, by way of illustration is carried out below in terms of $x$ and $s$.

Integrating the numerator in terms of $x$:

$$\int_A^B M_0 y\, ds = 2 \int_A^{D_1} M_0 y\, ds = 2 \int_0^3 Px \cdot \frac{4x}{3} \cdot \frac{5dx}{3} \qquad = 40P$$

$$+ 2 \int_3^7 Px \cdot \left\{4 + \frac{3}{4}(x-3)\right\} \frac{5dx}{4} \qquad + 285P$$

$$+ 2 \int_7^{12} 7P \cdot 7 \cdot dx \qquad + 490P$$

$$\overline{\qquad\qquad 815P\dagger}$$

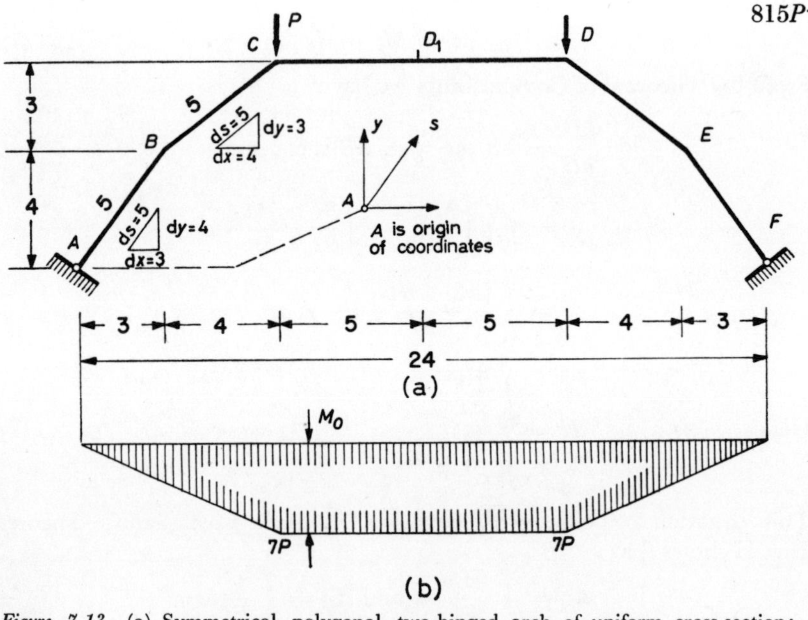

*Figure 7.13.* (a) Symmetrical polygonal two-hinged arch of uniform cross-section; (b) Simply-supported bending moment diagram

Integrating the denominator in terms of $s$

$$\int_A^B y^2 ds = 2 \int_A^{D_1} y^2 ds = 2 \int_0^5 \left(\frac{4s}{5}\right)^2 ds \qquad = 53\cdot 3$$

$$+ 2 \int_5^{10} \left\{4 + \frac{3}{5}(s-5)\right\}^2 ds \qquad + 310\cdot 0$$

$$+ 2 \int_5^{10} 7^2\, ds \qquad + 490\cdot 0$$

$$\overline{\qquad\qquad 853\cdot 3}$$

† This result was previously obtained by Hoadley's method. See Chapter 3, Section **3.7.5**, Example (b).

## 7.3. THE TWO-HINGED ARCH

Hence
$$H = \frac{815P}{853 \cdot 3} = 0 \cdot 954P$$

### 7.3.1.2. Circular Arch of Uniform Cross-section

FIGURE 7.14(a) shows a circular arch having $EI$ constant. Equation 7.7(a) is now best handled by changing the variable from $s$ to $\alpha$, so that we have

$$H = \frac{\int M_o . R(\cos \alpha - \cos \phi) . R\, d\alpha}{\int R^2 (\cos \alpha - \cos \phi)^2 . R\, d\alpha} \quad \ldots \text{(i)}$$

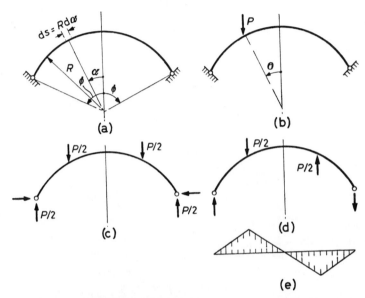

*Figure 7.14.* (a) Circular two-hinged arch of constant cross-section; (b) Point load $P$ is equivalent to (c) symmetrically placed loads $P/2$ together with (d) anti-symmetrically placed loads $P/2$; (e) Bending moment diagram for (d)

The unsymmetrical point load shown at (b) can best be dealt with by superimposing the symmetrical arrangement (c) on the anti-symmetrical arrangement (d). For (c) we have

$$M_o = PR(\sin \phi - \sin \alpha)/2 \text{ when } \phi > \alpha > \theta$$

and
$$M_o = PR(\sin \phi - \sin \theta)/2 \text{ when } \theta > \alpha > 0$$

Hence for this case the numerator of (i) is

$$\left. \begin{array}{l} 2 \int_0^\theta \frac{PR}{2} (\sin \phi - \sin \theta) R (\cos \alpha - \cos \phi) R\, d\alpha \\ + 2 \int_\theta^\phi \frac{PR}{2} (\sin \phi - \sin \alpha) R (\cos \alpha - \cos \phi) R\, d\alpha \end{array} \right\}$$

$$= PR^3 \{\cos \phi (-\phi \sin \phi + \theta \sin \theta + \cos \theta) + 1/2 (\cos^2 \theta - 3\cos^2 \phi)\} \quad \ldots \text{(ii)}$$

It will be seen from the form of the bending moment diagram (e) that the anti-symmetrical load system makes no contribution to the horizontal thrust; it remains only to integrate the denominator of (i) which gives

$$R^3 \{\phi(2 + \cos 2\phi) - (3 \sin 2\phi)/2\} \quad \ldots \text{(iii)}$$

The final value of $H$ is given by dividing (ii) by (iii); when $\theta = O$ and the load is located at the crown, we have

$$H = \frac{P\left\{\dfrac{1}{2} + \cos\phi - \phi\sin\phi\cos\phi - (3\cos^2\phi)/2\right\}}{2\phi + \phi\cos 2\phi - (3\sin 2\phi)/2} \quad \ldots \text{(iv)}$$

and if $\phi = \pi/2$ this reduces to

$$H = P/\pi \quad \ldots \text{(v)}$$

By putting $\phi = \tan^{-1} 4/3$ in (ii) and (iii) and substituting different values of $\theta$ we obtain an influence line for horizontal thrust for a circular arch with rise equal to one-quarter of the span. This is shown in FIGURE 7.16(a).

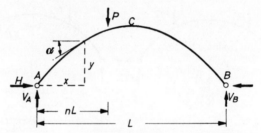

*Figure 7.15.* Parabolic arch with secant variation of flexural rigidity

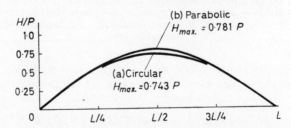

*Figure 7.16.* Influence lines for horizontal thrust for arch with rise equal to one quarter of the span: (a) Circular, of uniform cross-section; (b) Parabolic, with secant variation of flexural rigidity

### 7.3.1.3. *Parabolic Arch with Secant Variation of* EI

The equation to the parabolic rib shown in FIGURE 7.15 is

$$y = \frac{4hx}{L^2}(L - x) \quad \ldots \text{(i)}$$

In order to simplify the integration of equation 7.7 the assumption is often made that

$$EI_x = EI_C \sec \alpha = EI_C \frac{ds}{dx}$$

where $EI_C$ is the flexural rigidity of the crown. Although the values of $EI_x$ resulting from this assumption differ somewhat from those usually found in practice, the effect on $H$ of this error is not very great and is acceptable at least for a preliminary analysis. On this understanding the numerator of equation 7.7 becomes

$$\int_0^L \frac{M_o y\, ds}{EI} = \frac{1}{EI_C} \int_0^L M_o y\, dx = \frac{1}{EI_C}\left\{\int_0^{nL} V_A\, x\, y\, dx + \int_{nL}^L V_B (L-x) y\, dx\right\}$$

$$= \frac{PhL^2}{3EI_C}(n - 2n^3 + n^4) \qquad \ldots\ \text{(ii)}$$

The denominator is

$$\frac{1}{EI_C}\int_0^L y^2 dx = \frac{8h^2 L}{15} \qquad \ldots\ \text{(iii)}$$

The ratio of these two expressions is the influence line for $H$,

i.e. $$H = \frac{5PL}{8h}(n - 2n^3 + n^4) \qquad \ldots\ (7.8)$$

This is shown in FIGURE 7.16 at (b).

### 7.3.1.4. Numerical Evaluation of Thrust

In many practical cases equations 7.7 cannot be integrated algebraically and a numerical procedure must be adopted. The procedure is simply to

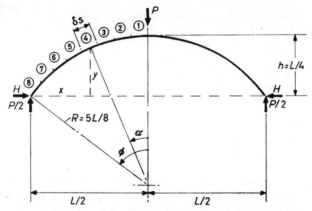

Figure 7.17. The arch equation can be integrated numerically by dividing the rib into a suitable number of segments

divide the arch into a suitable number of segments, for each of which $M_o$, $y$ and $EI$ can be considered constant, and to replace the integration by a summation; Simpson's rule can be used if the segments are of equal length. Timoshenko and Young[5] show that it is sufficiently accurate to divide the half arch into eight sections. In TABLE 7.1 the calculations are set out for the circular arch shown in FIGURE 7.17. Thus

$$\frac{H}{P} = \frac{\int \frac{M_o y\, ds}{EI}}{\int \frac{y^2 ds}{EI}} = \frac{\sum_1^8 \frac{M_o y\, \delta s}{EI}}{\sum_1^8 \frac{y^2 \delta s}{EI}} = \frac{0\cdot 9892 \times \frac{0\cdot 116}{2}}{0\cdot 6682 \times 0\cdot 116} = 0\cdot 740$$

This agrees very closely with the value $0\cdot 738$ obtained by direct integration. (See FIGURE 7.16).

### TABLE 7.1
#### CALCULATIONS FOR ARCH OF FIGURE 7.17

$L^2/4 = (2R - L/4)\, L/4$  i.e. $R = 5L/8$  i.e. $\phi = \tan^{-1} \dfrac{L/2}{3L/8} = \tan^{-1} 4/3$

$\phantom{L^2/4 = (2R - L/4)\, L/4 \quad} = 53°8' = 0\cdot 927$ radians

$\phantom{L^2/4 = (2R - L/4)\, L/4 \quad} \sin \phi = 0\cdot 8 \quad \cos \phi = 0\cdot 6$

$x = R(\sin \phi - \sin \alpha);\quad y = R(\cos \alpha - \cos \phi);\quad \delta s = R/8 \times 0\cdot 927 = 0\cdot 116 R.$

$M_o = Px/2$

| (1) | (2) | (3) | (4) | (5) | (6) | (7) | (8) |
|---|---|---|---|---|---|---|---|
| Section | $\alpha$ | $\sin \alpha$ | $\cos \alpha$ | $\dfrac{x}{xR}$ | $\dfrac{y}{xR}$ | $\dfrac{y\,ds/EI}{x\,\dfrac{0\cdot 116 PR^3}{2EI}}$ | $\dfrac{y^2\, \delta s/EI}{x\,\dfrac{0\cdot 116 R^3}{EI}}$ |
| 1 | 3°19¼′ | 0·0580 | 0·9983 | 0·7420 | 0·3983 | 0·2955 | 0·1586 |
| 2 | 9°57¾′ | 0·1730 | 0·9849 | 0·6270 | 0·3849 | 0·2413 | 0·1481 |
| 3 | 16°36¼′ | 0·2858 | 0·9583 | 0·5142 | 0·3583 | 0·1843 | 0·1284 |
| 4 | 23°14¾′ | 0·3946 | 0·9188 | 0·4054 | 0·3188 | 0·1292 | 0·1016 |
| 5 | 29°53¼′ | 0·4983 | 0·8670 | 0·3017 | 0·2670 | 0·0806 | 0·0713 |
| 6 | 36°31¾ | 0·5952 | 0·8036 | 0·2048 | 0·2036 | 0·0417 | 0·0415 |
| 7 | 43°10¼′ | 0·6841 | 0·7294 | 0·1159 | 0·1294 | 0·0150 | 0·0167 |
| 8 | 49°48¾′ | 0·7639 | 0·6453 | 0·0361 | 0·0453 | 0·0016 | 0·0020 |
|   |   |   |   |   |   | 0·9892 | 0·6682 |

NOTE: Line (7) = Line (5) × Line (6) ;   Line (8) = Line (6) × Line (6)

### 7.3.2. THE TIED ARCH

FIGURE 7.18 represents a two-hinged arch in which the horizontal thrust is provided by a tie connecting the hinges. The final displacement of $B$ relative to $A$ is now equal to the stretch of the tie so that equation 7.3 becomes

$$H = \frac{\Delta_{bo} - HL/A_t E_t}{\Delta_{bb}} = \frac{\int \dfrac{M_o y\, ds}{EI}}{\int \dfrac{y^2 ds}{EI} + \dfrac{L}{A_t E_t}} \quad \ldots (7.9)$$

where $A_t$ and $E_t$ are the cross-sectional area and modulus of elasticity of the tie respectively.

If the tie is very stiff then $A_t E_t$ is large and $\dfrac{L}{A_t E_t}$ correspondingly small; equation 7.9 then approximates to equation 7.7 and the arch acts as if it had

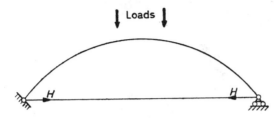

*Figure 7.18.* In the tied arch the horizontal thrust is provided by the tension in the tie

immovable abutments. On the other hand if the tie is comparatively slender then $A_t E_t$ is small and $\dfrac{L}{A_t E_t}$ is large: $H$ then tends to zero, as we should expect if the hinges experienced only a slight restraint against spreading.

### 7.3.3. The Effect of Temperature

It will be appreciated that if the arch experiences a rise of temperature which would, in the absence of the hinges, increase its length then an inward horizontal thrust is generated; this can be calculated by finding the horizontal thrust produced if the hinges are pushed towards one another by the horizontal component of the thermal expansion of the arch.

Increase of length of segment $\delta s \quad = \alpha_t t\, \delta s$

Horizontal component $\quad\quad\quad\quad\quad = \alpha_t t\, \delta s . \cos \alpha$

$\quad\quad\quad\quad\quad\quad\quad\quad\quad\quad\quad\quad = \alpha_t t . \delta x.$

∴ Horizontal component of increase of length of whole arch

$$= \int_0^L \alpha_t t\, dx = \alpha_t t L.$$

Hence in equation 7.3, considering temperature effects only,

$$H = \frac{-\Delta_B}{\Delta_{bb}} = \frac{-(-\alpha_t t L)}{\Delta_{bb}} = \frac{\alpha_t t L}{\int \dfrac{y^2 ds}{EI}} \quad \ldots .(7.10)$$

### 7.3.4. The Effect of Axial Thrust

In Section 7.3.1 the expression for the horizontal thrust was obtained by considering the strain energy of bending only. This was equivalent to assuming that the change of shape caused by the axial thrust was negligible. While this is often an acceptable simplification there are cases, notably when the rise of the arch is rather small compared with the span, where it is necessary to consider axial as well as bending deformation. The following elementary treatment will usually prove to be adequate; a more elaborate method of calculation, which also includes the effects of changes of curvature of the rib, is given by Johnson, Bryan and Turneaure[6].

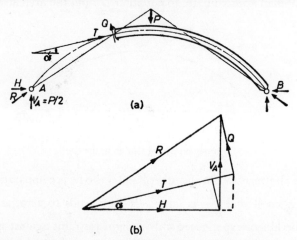

*Figure 7.19*. (a) Arch cut through at a point to find the forces acting there;   (b) Triangles of forces for this point

FIGURE 7.19(a) shows the arch of FIGURE 7.17 with a section at about one-third point; the triangles of forces for this point are shown at (d). We then have

arc $\delta s$, at this point, is shortened by $\dfrac{T\delta s}{AE}$

hence $\delta x$, at this point, is shortened by $\dfrac{T\delta s \cos \alpha}{AE} \simeq \dfrac{H\delta s}{AE}$

Hence total horizontal component of shortening of rib $\Big\} \simeq H \int \dfrac{ds}{AE}$

The effect of this rib shortening is exactly the same as an outward relative displacement of the hinges; hence in equation 7.3

$$H = \frac{\Delta_{bo} - H\int \dfrac{ds}{AE}}{\Delta_{bb}} = \frac{\int \dfrac{M_o y\, ds}{EI}}{\int \dfrac{y^2 ds}{EI} + \int \dfrac{ds}{AE}} \quad \ldots (7.11)$$

It can be seen from the form of equation 7.11 that $H$ is actually reduced by this effect so that the bending moment is increased.

Finally, the extent of the approximation involved in assuming that

$$H = T \cos \alpha$$

can be judged from FIGURE 7.19(b). It should be noted, however, that the more nearly the linear arch coincides with the arch axis the smaller is the discrepancy between $H$ and $T \cos \alpha$. It therefore appears that in a well-designed arch equation 7.11 will be satisfactorily accurate.

## 7.4. THE FIXED ARCH

Reference has already been made, in Section 1.2, to the fact that the three redundant reactions of the fixed arch can be chosen in several different ways. There is no great difference in principle involved in this choice but in most cases the arithmetic is simplified if the arrangement shown in FIGURE 7.4(f) is chosen. Since this scheme also leads on to interesting developments it is selected here for detailed discussion.

### 7.4.1. THE ELASTIC CENTRE

#### 7.4.1.1. *Cantilever Primary Structure*

FIGURE 7.20(a) shows an arch of arbitrary shape carrying loads which produce bending moments $M_0$ in the primary structure, here taken to be

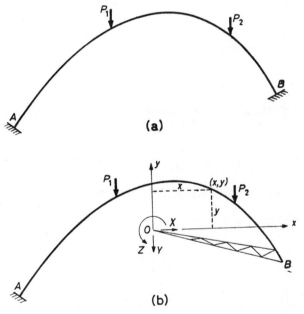

*Figure 7.20.* (a) Fixed arch of arbitrary shape carrying loads $P$; (b) The primary structure is the cantilever $AB$ fixed at $A$ and acted on by loads $P$ and redundants $X$, $Y$ and $Z$; the latter are considered to act at the end $O$ of a rigid arm $BO$ and are positive as shown

the cantilever shown at (b)†. The redundants have been taken to be the horizontal and vertical forces $X$ and $Y$ and the couple $Z$ acting at the end

---

† The convention of signs is now slightly modified so that a bending moment is considered to be positive if it produces tension at the outside of the arch.

of an imaginary rigid arm attached to $B$; the directions shown are taken as positive. It will be clear that this arm does not change the problem fundamentally since, being rigid, it makes no direct contribution to the deflexion of the rib. The magnitudes of $X$ and $Y$ are naturally affected by the position chosen for $O$, but the final value of the bending moments in the arch is quite independent of the position of $O$.

Compatibility requires that the horizontal, vertical and rotational movement of $B$ shall be zero (in the absence of abutment displacement) and, since $B$ and $O$ are connected by the rigid arm, this means that the three component displacements of $O$ are also zero. These displacements are conveniently found by means of Castigliano's Theorem, Part II, so that we have

$$\left.\begin{aligned} 0 = \Delta_x &= \frac{\partial U}{\partial X} = \frac{\partial}{\partial X}\int_A^B \frac{M^2 ds}{EI} = \int_A^B M\frac{\partial M}{\partial X}\frac{ds}{EI} \\ 0 = \Delta_y &= \frac{\partial U}{\partial Y} = \frac{\partial}{\partial Y}\int_A^B \frac{M^2 ds}{EI} = \int_A^B M\frac{\partial M}{\partial Y}\frac{ds}{EI} \\ 0 = \Delta_z &= \frac{\partial U}{\partial Z} = \frac{\partial}{\partial Z}\int_A^B \frac{M^2 ds}{EI} = \int_A^B M\frac{\partial M}{\partial Z}\frac{ds}{EI} \end{aligned}\right\} \quad \ldots \text{(i)}$$

where $\Delta_x$, $\Delta_y$ and $\Delta_z$ are the horizontal, vertical and rotational displacements of $O$ and therefore also of $B$.

But

$$M = M_o - Xy - Yx - Z \quad \ldots (7.12)$$

where $M$ and $M_o$ are positive according to the bending moment sign convention (FIGURE 3.19). Differentiating equation 7.12 we have

$$\frac{\partial M}{\partial X} = -y; \quad \frac{\partial M}{\partial Y} = -x; \quad \frac{\partial M}{\partial Z} = -1 \quad \ldots \text{(ii)}$$

Hence substituting equation 7.12 and (ii) in (i) we obtain

$$\left.\begin{aligned} 0 &= -\int \frac{M_o y\, ds}{EI} + X\int \frac{y^2\, ds}{EI} + Y\int \frac{xy\, ds}{EI} + Z\int \frac{y\, ds}{EI} \\ 0 &= -\int \frac{M_o x\, ds}{EI} + X\int \frac{xy\, ds}{EI} + Y\int \frac{x^2 ds}{EI} + Z\int \frac{x\, ds}{EI} \\ 0 &= -\int \frac{M_o\, ds}{EI} + X\int \frac{y\, ds}{EI} + Y\int \frac{x\, ds}{EI} + Z\int \frac{ds}{EI} \end{aligned}\right\} \quad \ldots (7.13)$$

integration being extended from $A$ to $B$ throughout. These are the general equations for the fixed arch; they apply whatever the position of $O$. Very similar equations are obtained if $M_A$, $M_B$ and $H$ are chosen as the redundants as in FIGURE 7.4(d). It is evidently quite a laborious business to find $X$, $Y$ and $Z$ from these equations which have to be solved simultaneously after nine different integrals have been evaluated.

We are therefore led to enquire whether the work can be simplified at all and attention turns to the possibility of choosing $O$ in such a way that some of the integrals disappear. One way of doing this is locate $O$ so that

$$\int_A^B \frac{x \, ds}{EI} = \int_A^B \frac{y \, ds}{EI} = 0 \qquad \ldots\ldots (7.14)$$

In so doing we are thinking of the rib as being composed of a series of 'elastic weights' $\frac{ds}{EI}$ and equations 7.14 then carry the implication that $O$ is the centroid of these elastic weights or 'elastic centre'.

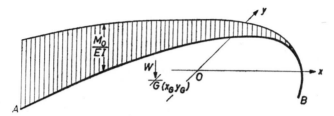

Figure 7.21. The $M_o/EI$ diagram, plotted perpendicularly to the rib, has its centroid at $G$

If we now imagine the $M_o/EI$ diagram to be constructed on the arch lying on its side, as in FIGURE 7.21, then the total area of this diagram is

$$W = \int \frac{M_o \, ds}{EI} \qquad \ldots\ldots (7.15)$$

which has a centroid at the point $G$ $(x_G, y_G)$. The moments of $W$ about the $x$- and $y$- axes are respectively

and
$$\left. \begin{array}{l} Wy_G = \int \dfrac{M_o \, y \, ds}{EI} = \mu_x \\[6pt] Wx_G = \int \dfrac{M_o \, x \, ds}{EI} = \mu_y \end{array} \right\} \qquad \ldots\ldots (7.16)$$

By analogy we can also think of 'moments of inertia of elastic weights' and write

$$\int \frac{x^2 \, ds}{EI} = I_y \,;\, \int \frac{y^2 \, ds}{EI} = I_x \,;\, \int \frac{x\, y \, ds}{EI} = I_{xy} \qquad \ldots\ldots (7.17)$$

It is also convenient to have a symbol for the total elastic weight of the rib; thus

$$A = \int \frac{ds}{EI} \qquad \ldots\ldots (7.18)$$

Introducing the equations 7.14 to 7.18 into 7.13 and solving simultaneously we obtain

$$X = \frac{\mu_x - \mu_y I_{xy}/I_y}{I_x - I_{xy}^2/I_y} = \frac{\mu_x'}{I_x'} \,;\, Y = \frac{\mu_y - \mu_x I_{xy}/I_x}{I_y - I_{xy}^2/I_x} = \frac{\mu_y'}{I_y'} \,;\, Z = \frac{W}{A}$$
$$\ldots\ldots (7.19)$$

It should be mentioned that if we had also chosen the directions of the $x$- and $y$- axes appropriately then $I_{xy}$ would have been zero† and equation 7.13 would have reduced to simple equations in $X$, $Y$ and $Z$. However if the arch happens to be symmetrical about the $y$-axis then every point $(x,y)$ on the rib corresponds to a point $(x, -y)$ on the other side of the axis and

$$\int_A^B \frac{x y\, ds}{EI} = I_{xy} = 0 \qquad \ldots \text{(iii)}$$

Equations 7.19 then reduce to

$$X = \frac{\mu_x}{I_x}; \qquad Y = \frac{\mu_y}{I_y}; \qquad Z = \frac{W}{A} \qquad \ldots \text{(7.19(a))}$$

Substituting in equation 7.12 we obtain the final expression for the bending moment at any point $(x, y)$ on the complete arch,

$$M = M_o - Z - Xy - Yx \qquad \ldots \text{(7.12)}$$

$$= M_o - \left( \frac{W}{A} + \frac{\mu_y y}{I_x} + \frac{\mu_y x}{I_y} \right) \qquad \ldots \text{(7.20)}$$

The simplification of equations 7.13 which followed when the redundants were made to act at the elastic centre is exactly the same in principle as the simplification of equations 5.2 previously discussed in Chapter 5, Section 5.3.4. It was shown there that if the redundants could be chosen in such a way that the application of each redundant to the primary structure produced zero displacement corresponding to the others then the equations for the redundants became independent rather than simultaneous. Applying this to the present problem (FIGURE 7.20(b)) we now seek to find how $O$ should be chosen so that it simply rotates, without translation, when $Z$ is applied alone. Equation 7.12 becomes

$$M = -Z \qquad \ldots \text{(iv)}$$

since $M_o$, $X$ and $Y$ are now zero, but equations (ii) are still valid. We thus have

$$\left. \begin{aligned} \Delta_x &= \int M \frac{\partial M}{\partial X} \frac{ds}{EI} = Z \int \frac{y\, ds}{EI} \\ \Delta_y &= \int M \frac{\partial M}{\partial Y} \frac{ds}{EI} = Z \int \frac{x\, ds}{EI} \end{aligned} \right\} \qquad \ldots \text{(v)}$$

The condition that $\Delta_x$ and $\Delta_y$ shall be zero is thus the same as before, namely

$$\int \frac{x\, ds}{EI} = \int \frac{y\, ds}{EI} = 0 \qquad \ldots \text{(7.14)}$$

#### 7.4.1.2. *Auxiliary Primary Structures*

Suppose that when dealing with the arch of FIGURE 7.20(a) we had chosen the cantilever of diagram (b), together with its imaginary rigid arm, as

---

† The axes would then have been ' principal axes.'

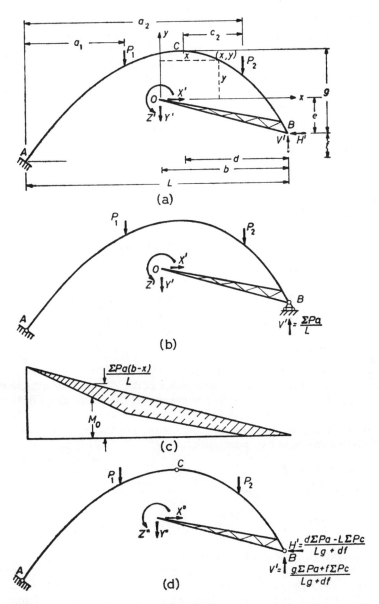

*Figure 7.22.* (a) The primary cantilever acted on by redundants $X'$, $Y'$ and $Z'$ together with *known* forces $V'$ and $H'$; (b) When the known force at $B$ is $V' = \Sigma Pa/L$ the moment at $A$ is zero and the primary structure becomes the simply-supported beam $AB$; (c) The bending moment diagram for the cantilever $AB$ under the loads $P$ (i.e. $M_o$) together with that for $V'$ (i.e. $\Sigma \dfrac{Pa}{L}(b-x)$); the shaded area shows the combined effect and is evidently the diagram for the simply-supported beam $AB$; (d) When the known forces $H'$ and $V'$ have the values shown the moments at $A$ and $C$ are zero and the primary structure becomes the three-hinged arch $ACB$

357

the primary structure, but that the redundants had been taken to be the forces $X'$ and $Y'$ and the couple $Z'$ acting in conjunction with known forces $H'$ and $V'$ at $B$. This is a valid procedure since the equilibrium conditions can readily be satisfied when the expression for the bending moment is written down while compatibility is assured, as before, if the translational and rotational movement of $O$ is zero. Referring to FIGURE 7.22(a) it will be seen that equation 7.12 has become

$$M = \{M_o + H'(e+y) - V'(b-x)\} - X'y - Y'x - Z' \quad \ldots(i)$$

**As before**

$$\frac{\partial M}{\partial X'} = -y; \quad \frac{\partial M}{\partial Y'} = -x; \quad \frac{\partial M}{\partial Z'} = -1 \quad \ldots(ii)$$

The compatibility conditions at $O$ give

$$0 = \Delta_x = \int_A^B M \frac{\partial M}{\partial X} \frac{ds}{EI}$$

$$= \int_A^B \{M_o + H'(e+y) - V'(b-x)\}(-y)\frac{ds}{EI}$$

$$+ X' \int_A^B \frac{y^2 ds}{EI} + Y' \int_A^B \frac{xy\,ds}{EI} + Z' \int_A^B \frac{y\,ds}{EI} \quad \ldots(iii)$$

and two other similar equations.

Comparison with equations 7.13 shows that only the first term has been materially altered and that the coefficients of $X'$, $Y'$ and $Z'$ are the same as those of $X$, $Y$ and $Z$. It therefore appears that the same simplification as before can be achieved by making $O$ the elastic centre of the arch. Now we are presumably quite free to assign any convenient values to $H'$ and $V'$ and we therefore examine two possibilities:

(a) $H' = 0$; $V' = \Sigma Pa/L$

Equation (i) then becomes

$$M = \left\{ M_o - \frac{\Sigma Pa}{L}(b-x) \right\} - X'y - Y'x - Z' \quad \ldots(i')$$

and equations (iii) are similarly altered.

The terms in brackets give the bending moment at a point on the primary cantilever when the loads $P$ and the vertical force $V' = \Sigma Pa/L$ are acting. But they also give the bending moment in a primary beam $AB$ having a hinge at $A$ and rollers at $B$, as in FIGURE 7.22(b); the identity of the two bending moment diagrams is shown clearly at (c).

Equation (i) can therefore be written

$$M = M'_o - X'y - Y'x - Z' \quad \ldots(i(a))$$

where $M'_o$ means the bending moments in the primary beam $AB$ and $X'$, $Y'$ and $Z'$ the corresponding redundants acting at $O$. With this modification all the equations 7.13—7.20 previously derived in Section 7.4.1.1 for the primary cantilever retain their validity.

(b) $H' = \dfrac{d\,\Sigma Pa - L\,\Sigma Pc}{Lg + df}$ ; $V' = \dfrac{g\,\Sigma Pa + f\,\Sigma Pc}{Lg + df}$

It is now found that the terms in brackets in (i) give the bending moments at points on the three-hinged arch $ACB$ shown in FIGURE 7.22(d). Equation (i) then becomes

$$M = M_o'' - X''y - Y''x - Z'' \qquad \ldots\ldots \text{(i'')}$$

where $M_o$ means the bending moments in the primary three-hinged arch $ACB$ and $X''$, $Y''$ and $Z''$ the corresponding redundants acting at $O$.

The discussion under (a) and (b) above has shown that the elastic centre method can be used with any convenient primary structure substituted for the original primary cantilever; the term $M_o$ in equation 7.12 then signifies the bending moment in this auxiliary primary structure under the applied loads. When the bending moments in the complete arch are finally calculated by means of equation 7.12 we are in effect abandoning the primary cantilever and retaining the auxiliary primary structure, and it will be observed that this procedure makes it unnecessary to evaluate the supplementary forces $H'$ and $V'$†. A worked example will be found in Section 7.4.3 below.

### 7.4.1.3. *The Elastic Centre of the two-hinged Arch*

It is now interesting to reconsider the two-hinged arch in the light of the above discussion of the significance of the elastic centre in fixed arch analysis.

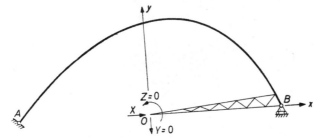

Figure 7.23. The two-hinged arch treated by the elastic centre method. The primary structure is the simply-supported beam $AB$ and the elastic centre $O$ is at the mid point of $AB$

A hinge can evidently be regarded as being equivalent to a short length of rib with a very small flexural rigidity and therefore with infinite elastic weight $ds/EI$. The elastic centre of a two-hinged arch will therefore lie at the centre of the line joining the hinges, which can conveniently be taken as the $x$- axis (FIGURE 7.23). Following the discussion of Section 7.4.1.2 we

---

† This important result is implicit in the writings of Hardy Cross[7,8] and is proved, in a similar manner to the above, by Parcel and Moorman[13] in their introduction to the Column Analogy. The suggestion that it was equally applicable to the elastic centre method was communicated to the Author, together with the ideas of Section 7.4.1.3 by Professor A. J. Francis.

ARCHES

can take the simply-supported beam $AB$ as the primary structure; we shall also have
$$I_y = \int \frac{x^2 ds}{EI} = \infty \; ; \; I_{xy} = \int \frac{xy\,ds}{EI} = 0 \; ; \; A = \int \frac{ds}{EI} = \infty$$

Hence equations 7.19 reduce to
$$X = \frac{\mu_x}{I_x} = \frac{\int \dfrac{M_0 y\, ds}{EI}}{\int \dfrac{y^2 ds}{EI}} \; ; \; Y = 0 \; ; \; Z = 0.$$

We have thus contrived to derive equation 7.7 again.

### 7.4.2. Formulae for Members of Polygonal Arch

The following formulae for straight uniform members are convenient and easily derived. See Figures 7.24(a) and (b).

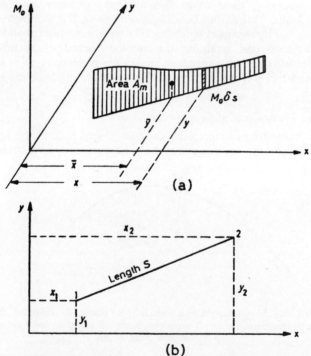

Figure 7.24. Derivation of formulae for uniformly straight members of a polygonal arch

$$W = \int \frac{M_0 ds}{EI} = \frac{A_m}{EI} \qquad \ldots\ldots (7.15(a))$$

$$\left. \begin{aligned} \mu_x &= \int \frac{M_0 y\, ds}{EI} = \frac{A_m \bar{y}}{EI} \\ \mu_y &= \int \frac{M_0 x\, ds}{EI} = \frac{A_m \bar{x}}{EI} \end{aligned} \right\} \qquad \ldots\ldots (7.16(a))$$

## 7.4. THE FIXED ARCH

$$I_x = \int \frac{y^2 ds}{EI} = \frac{S}{3EI}(y_1^2 + y_1 y_2 + y_2^2)$$

$$I_y = \int \frac{x^2 ds}{EI} = \frac{S}{3EI}(x_1^2 + x_1 x_2 + x_2^2) \qquad \ldots (7.17(a))$$

$$I_{xy} = \int \frac{xy\, ds}{EI} = \frac{S}{6EI}\{(x_1 y_2 + x_2 y_1) + 2(x_1 y_1 + x_2 y_2)\}$$

$$A = \int \frac{ds}{EI} = \frac{S}{EI} \qquad \ldots (7.18(a))$$

By means of these formulae and the methods described above fixed arch ribs composed of straight members of uniform cross-section can readily be analysed. In the following Section a simple example is worked in different ways in order to bring out the various points to which attention has been directed.

### 7.4.3. Example of Fixed Arch Analysis

The portal frame of FIGURE 7.25(a) is analysed below by three different methods.

This particular frame can, of course, be dealt with very quickly by means of moment distribution, or otherwise, but it is used here as an illustration in order that the principles shall not be obscured by arithmetical complexities.

(a) *Primary cantilever fixed at A.* FIGURE 7.25(b).

The position of the elastic centre, which must lie on the vertical axis of symmetry, is found by taking the moments of the elastic weights about $BC$; thus

$$OE = \frac{2 \times \frac{16}{2} \times 8}{2 \times \frac{16}{2} + \frac{16}{3}} = 6$$

The terms 7.15(a) — 7.18(a) are evaluated below in tabular form.

| Member | End | $x$ | $y$ | $EI$ | $I$ | $I_y$ | $I_{xy}$ | $A$ |
|---|---|---|---|---|---|---|---|---|
| AB | A | −8 | −10 | 2 | 608/3 | 512 | +128 | 8 |
|    | B | −8 | +6  |   |       |     |      |   |
| BC |   |    |     | 3 | 192   | 1024/9 | 0 | 16/3 |
|    | C | +8 | +6  |   |       |     |      |   |
| CD | D | +8 | −10 | 2 | 608/3 | 512 | −128 | 8 |
| Σ  |   |    |     |   | 1792/3 | 10240/9 | 0 | 64/3 |

| Member | $A_m$ | $\bar{x}$ | $\bar{y}$ | $EI$ | $W$ | $\mu_x$ | $\mu_y$ |
|---|---|---|---|---|---|---|---|
| AB | +448P | −8 | −2 | 2 | +224P | −448P | −1792P |
| BE | +112P | −16/3 | +6 | 3 | +112P/3 | +224P | −1792P/9 |
| Σ |  |  |  |  | +784P/3 | −224P | −1792P×10/9 |

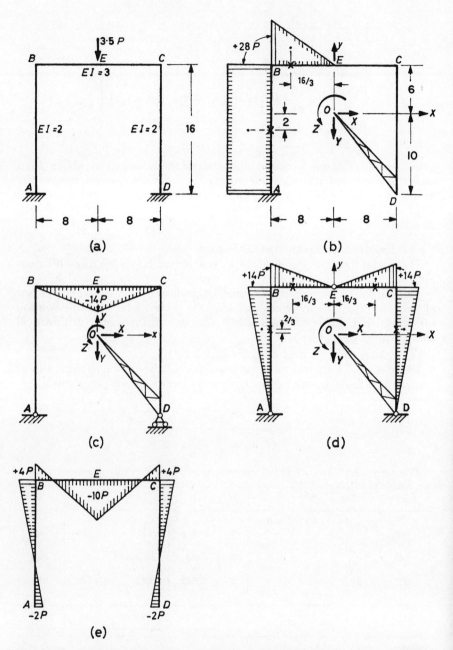

*Figure 7.25.* (a) Simple portal frame for solution as a fixed arch; (b) Primary cantilever fixed at $A$. The redundants $X$, $Y$ and $Z$ act at the end $O$ of a rigid arm $DO$, where $O$ is the elastic centre; (c) Auxiliary primary structure consisting of simply-supported beam $ABECD$; (d) Auxiliary primary structure consisting of three-hinged arch $AED$; (e) Final bending moment diagram

## 7.4. THE FIXED ARCH

Hence
$$X = \frac{\mu_x}{I_x} = -\frac{224P}{1792/3} = -3P/8$$

$$Y = \frac{\mu_y}{I_y} = -\frac{1792P \times 10}{10240} = -7P/4$$

$$Z = \frac{W}{A} = \frac{784P}{64} = +12\tfrac{1}{4}$$

Substituting in equation 7.12 for the final bending moments,

$$M = M_o - Xy - Yx - Z$$

$$M_A = +28P - \left(-\frac{3P}{8}\right)(-10) - \left(-\frac{7P}{4}\right)(-8) - 12\tfrac{1}{4}P = -2P$$

$$M_B = +28P - \left(-\frac{3P}{8}\right)(+6) - \left(-\frac{7P}{4}\right)(-8) - 12\tfrac{1}{4}P = +4P$$

$$M_E = 0 - \left(-\frac{3P}{8}\right)(+6) - \left(-\frac{7P}{4}\right)(0) - 12\tfrac{1}{4}P = -10P$$

$$M_C = 0 - \left(-\frac{3P}{8}\right)(+6) - \left(-\frac{7P}{4}\right)(+8) - 12\tfrac{1}{4}P = +4P$$

$$M_D = 0 - \left(-\frac{3P}{8}\right)(-10) - \left(-\frac{7P}{4}\right)(+8) - 12\tfrac{1}{4}P = -2P$$

These figures have been plotted at (e) to give the final bending moment diagram.

(b) *Auxiliary primary structure consisting of simply supported beam ABCD.* FIGURE 7.25(c).

The position of $O$ and the values of $I_x$, $I_y$, $I_{xy}$ and $A$ are unchanged, but for $W$, $\mu_x$ and $\mu_y$ we have

| Member | $A_m$ | $\bar{x}$ | $\bar{y}$ | $EI$ | $W$ | $\mu_x$ | $\mu_y$ |
|---|---|---|---|---|---|---|---|
| BC | $-112P$ | 0 | $+6$ | 3 | $-112P/3$ | $-224P$ | 0 |

Hence
$$X = \frac{\mu_x}{I_x} = -\frac{224P \times 3}{1792} = -3P/8$$

$$Y = \frac{\mu_y}{I_y} = 0$$

$$Z = \frac{W}{A} = -\frac{112P}{64} = -7P/4$$

ARCHES

Substituting in equation 7.12 we have, for example,

$$M_E = -14P - (-3P/8)(+6) - 0 - (-7P/4) = -10P.$$

The bending moments elsewhere will also be found to coincide with those found previously. Evidently the arithmetic is somewhat reduced by adopting this symmetrical auxiliary primary structure.

(c) *Auxiliary primary structure consisting of three-hinged arch AED.* FIGURE 7.25(d).

We now have

| Member | $A_m$ | $\bar{x}$ | $\bar{y}$ | EI | W | $\mu_x$ | $\mu_y$ |
|---|---|---|---|---|---|---|---|
| AB | +112P | −8 | +2/3 | 2 | +56P | +112P/3 | −448P |
| CD | +112P | +8 | +2/3 | 2 | +56P | +112P/3 | +448P |
| BE | +56P | −16/3 | +6 | 3 | +56P/3 | +112P | −896P/9 |
| EC | +56P | +16/3 | +6 | 3 | +56P/3 | +112P | +896P/9 |
| | | | | Σ | +448P/3 | +896P/3 | 0 |

Hence

$$X = \frac{896P}{1792} = +P/2$$

$$Y = 0$$

$$Z = \frac{448P}{64} = +7P$$

Then in equation 7.12

$$M_E = 0 - \left(\frac{P}{2}\right)(6) - 7P = -10P \text{ as before.}$$

### 7.4.4. INFLUENCE LINES FOR THE SYMMETRICAL FIXED ARCH

The elastic centre method, in combination with Betti's reciprocal theorem, can be used to develop a rather neat method of finding the influence lines for the redundants of a symmetrical fixed arch. Such an arch is shown in FIGURE 7.26(a) carrying a unit load at a movable point $m$; the redundants are shown acting at the elastic centre $O$. Let this arrangement of forces be designated Group 1; the displacements corresponding to $X$, $Y$ and $Z$ are zero, by definition, while that corresponding to the unit load at $m$ is $\Delta_m$.

The forces and displacements of Group 2 are shown at (b), the former being so arranged that a purely horizontal movement, $\Delta_{Ox}$, has been imposed on $O$; the associated vertical movement of $m$ is $\Delta_{mX}$, downwards (i.e. in the same direction as the unit load in diagram (a)). Now it was shown above, in Section 7.4.1.1, that when the redundants act at the elastic centre and the coordinate axes are principal axes the application of each redundant alone produces zero deflexion corresponding to the others. A purely horizontal movement of $O$ is therefore produced by $X'$ acting alone as at (c).

## 7.4. THE FIXED ARCH

Applying Betti's Theorem (Chapter 3, Section 3.5.2) to the forces and displacements of Groups 1 and 2 as defined above, we have

$$1 \times \Delta_{mX'} + X \Delta_{Ox} + Y.0 + Z.0 = X'.0 + Y'.0 + Z'.0 \quad \ldots \text{(i)}$$

i.e. 
$$X = -\frac{\Delta_{mX'}}{\Delta_{Ox}} \quad \ldots \text{(ii)}$$

The numerator and denominator of (ii) are respectively the vertical displacement of $m$ and the horizontal displacement of $O$ when the horizontal

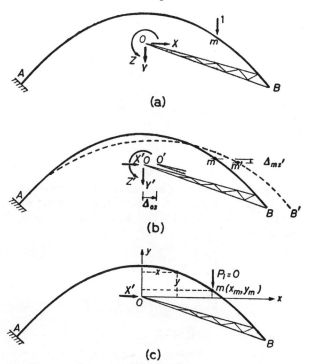

Figure 7.26. (a) Group 1: Fixed arch under the action of unit load at a movable point $m$ and redundants acting at the elastic centre; (b) Group 2: Forces are imposed at $O$ so that the movement there is exclusively in the $x$-direction. This means, in fact, that $Y' = Z' = 0$; (c) In order to find the deflexion of $m$, when $X'$ acts at $O$, an imaginary load $P_i = 0$ is imposed

force $X'$ acts at $O$. In order to find the former we impose an imaginary load $P_i = 0$ at $m$. Then

$$M_{(x,y)} = P_i(x_m - x) - X'y \quad \text{when } x < x_m$$

i.e. 
$$\frac{\partial M}{\partial P_i} = (x_m - x) \text{ and } \frac{\partial M}{\partial X'} = -y$$

Now 
$$\Delta_{mX'} = \int_A^m M \frac{\partial M}{\partial P_i} \frac{ds}{EI} = \int_A^m (-X'y)(x_m - x) \frac{ds}{EI}$$

and 
$$\Delta_{0x} = \int_A^B M \frac{\partial M}{\partial X'} \frac{ds}{EI} = \int_A^B (-X'y)(-y) \frac{ds}{EI}$$

Substituting in (ii) we have

$$X = \frac{x_m \int_A^m \frac{y\,ds}{EI} - \int_A^m \frac{xy\,ds}{EI}}{\int_A^B \frac{y^2 ds}{EI}} \quad \ldots\ldots (7.21)$$

Similar calculations with $Y'$ and $Z'$ acting alone in turn yield

$$Y = \frac{x_m \int_A^m \frac{x\,ds}{EI} - \int_A^m \frac{x^2 ds}{EI}}{\int_A^B \frac{x^2 ds}{EI}} \quad \ldots\ldots (7.22)$$

$$Z = \frac{x_m \int_A^m \frac{ds}{EI} - \int_A^m \frac{x\,ds}{EI}}{\int_A^B \frac{ds}{EI}} \quad \ldots\ldots (7.23)$$

These integrals can be obtained algebraically or numerically in specific cases; we now illustrate the calculations for the portal frame of FIGURE 7.25, using the formulae 7.17(a) and 7.18(a). In addition we note that for the member $1-2$ shown in FIGURE 7.24(b)

$$\left.\begin{array}{l} \int \dfrac{x\,ds}{EI} = \dfrac{S}{2EI}(x_1 + x_2) \\[6pt] \int \dfrac{y\,ds}{EI} = \dfrac{S}{2EI}(y_1 + y_2) \end{array}\right\} \quad \ldots\ldots (7.24)$$

When the point $m$ coincides with $E$, we have

| Member | End | $x$ | $y$ | $EI$ | $\int_A^m \frac{ds}{EI}$ | $\int_A^m \frac{x\,ds}{EI}$ | $\int_A^m \frac{y\,ds}{EI}$ | $\int_A^m \frac{x^2 ds}{EI}$ | $\int_A^m \frac{xy\,ds}{EI}$ |
|---|---|---|---|---|---|---|---|---|---|
| AB | A | −8 | −10 | 2 | 8 | −64 | −16 | +512 | +128 |
| BE | B | −8 | +6 | 3 | 8/3 | −32/3 | +16 | +512/9 | −64 |
|    | E | 0  | +6 |   |     |       |     |        |     |
|    |   |    |    |   | +32/3 | −224/3 | 0 | +5120/9 | +64 |

Substituting these results in the numerators of equations 7.21—7.23 and using the results obtained in the previous Section in the denominators, we have

$$X = \frac{-64}{1792/3} = -\frac{3}{28}; \quad Y = \frac{-5120/9}{10240/9} = -\frac{1}{2}; \quad Z = \frac{224/3}{64/3} = +\frac{7}{2}$$

## 7.5. THE COLUMN ANALOGY

These values coincide with those previously obtained. Values of $X$, $Y$ and $Z$ are similarly obtained for other values of the movable point; the influence lines thus obtained are shown in FIGURE 7.27.

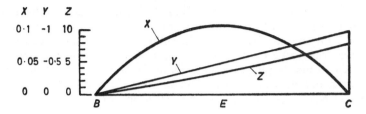

Figure 7.27. Influence lines for $X$, $Y$ and $Z$ for the portal frame of FIGURE 7.25(a) using as primary structure the cantilever shown at (b)

It is possible to use the above method of find influence lines for unsymmetrical arches but, in order to secure the independence of $X$, $Y$ and $Z$, it is necessary to use the principal axes as the coordinate axes. This involves the preliminary steps of using arbitrary axes to locate the elastic centre and the principal axes and then recalculating all the integrals in terms of the latter. Rather than embark on this complicated procedure it is probably simpler to use equations 7.19 as they stand to find $X$, $Y$ and $Z$ for a series of positions of the unit load.

## 7.5. THE COLUMN ANALOGY

### 7.5.1. PRINCIPLE

In 1930 Hardy Cross[7] suggested an interesting analogy between the bending moment in a fixed arch and the stress in an eccentrically loaded short column. Restricting the discussion, for the present, to symmetrical arches the bending moment at any point $(x, y)$ is given by equation 7.20 which can be written

$$M = M_o - \left(\frac{W}{A} + \frac{Wy_G}{I_x}y + \frac{Wx_G}{I_y}x\right) \quad \ldots (7.20(a))$$

using the symbols as defined in equations 7.15, 7.17 and 7.18. Now in FIGURE 7.28 is shown a thin short column under the action of an eccentric load $W$; the compressive stress at any point $(x, y)$ is given by

$$f = \frac{W}{A} + \frac{Wy_G}{I_x}y + \frac{Wx_G}{I_y}x \quad \ldots (7.25)$$

The right-hand side of this equation is identical with the terms in brackets in equation 7.20(a) but the symbols $A$, $I_x$ and $I_y$ have slightly different meanings in the two equations. So far as the column is concerned

$$A = \Sigma\, bds, \quad I_x = \Sigma bds.y^2 \quad \text{and} \quad I_y = \Sigma bds.x^2.$$

while equations 7.17 defined the 'moments of inertia of the elastic weights' as

$$I_x = \int \frac{1}{EI}ds.y^2 \quad \text{and} \quad I_y = \int \frac{1}{EI}ds.x^2$$

and from equation 7.18

$$A = \frac{1}{EI} ds$$

It is evident then that if the column is made the same shape as the arch but with its thickness everywhere inversely proportional to the flexural rigidity of the arch the compressive stress in the column will be equal to the moments in the arch caused by the redundants $X$, $Y$ and $Z$. It also appears from equation 7.15 that the load on the column, $W$, must be the $M_o/EI$ diagram for the arch and we conclude from the discussion of Section 7.4.1.2 that any convenient primary structure can be used to find $M_o$.

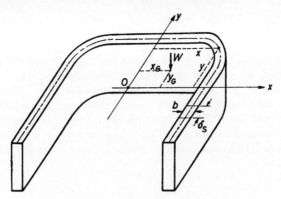

*Figure 7.28.* The analogous column

If the column of Figure 7.28 had been unsymmetrical then equation 7.25 for the compressive stress would have become

$$f = \frac{W}{A} + \frac{Wy_G - Wx_G I_{xy}/I_y}{I_x - I_{xy}^2/I_y} y + \frac{Wx_G - Wy_G I_{xy}/I_x}{I_y - I_{xy}^2/I_x} x \quad \ldots (7.25(a))$$

unless the axes had happened to be the principal axes. Comparison with equations 7.19 shows that again there is an analogy between the column stress and the moments due to the redundants in the arch. In both the symmetrical and unsymmetrical cases we can therefore write

$$M = M_o - f \qquad \ldots (7.26)$$

where $M$ and $M_o$ refer to the arch proper while $f$ is the compressive stress in the analogous column given by equations 7.25 or 7.25(a).

Corresponding terms in the analogy are:

|  | *Elastic centre analysis* | *Analogous column* |
|---|---|---|
| $A = \int \frac{ds}{EI}$ | Total elastic weight | Cross-sectional area |
| $I_x = \int \frac{y^2 ds}{EI}$ | Moment of inertia of elastic weight about $x$-axis | Second moment of area of section about $x$-axis |

## 7.5. THE COLUMN ANALOGY

| | | |
|---|---|---|
| $I_y = \int \dfrac{x^2 ds}{EI}$ | Moment of inertia of elastic weight about $y$-axis | Second moment of area of section about $x$-axis |
| $I_{xy} = \int \dfrac{xy\,ds}{EI}$ | Product of inertia of elastic weight | Product moment of area of section |
| $W = \int \dfrac{M_o\,ds}{EI}$ | Total area of $\dfrac{M_o}{EI}$ diagram | Resultant load on column |
| $\mu_x = W y_G = \int \dfrac{M_o y\,ds}{EI}$ | Moment of $\dfrac{M_o}{EI}$ diagram about $x$-axis | Moment of this load about $x$-axis |
| $\mu_y = W x_G = \int \dfrac{M_o x\,ds}{EI}$ | ,, about $y$-axis | ,, about $y$-axis |

### 7.5.2. Examples

(a) The first example is the symmetrical portal frame of FIGURE 7.25 that has already been analysed by the elastic centre method. The frame is shown

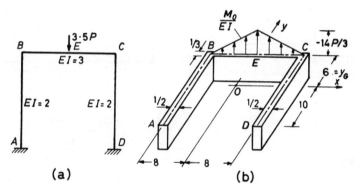

Figure 7.29. (a) Symmetrical portal frame; (b) Analogous column loaded with the $M_o/EI$ diagram

again in FIGURE 7.29 together with the analogous column loaded with the $M_o/EI$ diagram for the primary structure. Here we are using the auxiliary primary beam of FIGURE 7.25(c) and, as $M_o$ is negative in this case, the load on the analogous column will be upwards.

The required values of $I_x$, $I_y$ and $A$ have already been worked out in Section 7.4.3 and are

$$I_x = 1792/3 \quad I_y = 10240/9 \quad A = 64/3$$

and we also have

$$W \times \frac{1}{2} \times \frac{14P \times 16}{3} = \frac{112P}{3}\,;\quad W y_G = \frac{112P \times 6}{3} = 224P\,;\quad W x_G = 0$$

The compressive stress at $A$ is therefore

$$f_A = -\frac{112P}{64} + \frac{224P \times 10}{1792/3} = -\frac{7P}{4} + \frac{30P}{8} = 2P$$

Similarly the compressive stress at $B$ is

$$f_B = -\frac{112P}{64} - \frac{224P \times 6}{1792/3} = -\frac{7P}{4} - \frac{9P}{4} = -4P$$

and this is also the compressive stress at $E$.

Returning to the original arch we obtain from equation 7.26

$$M_A = 0 - 2P = -2P \; ; \quad M_B = 0 - (-4P) = +4P$$
$$M_E = -14P - (-4P) = -10P$$

These values are the same as those previously obtained.

In the remaining two examples the ordinary elastic centre arch analysis is given first in order to show the similarities to and differences from the column analogy.

(b) The non-uniform fixed beam shown in FIGURE 7.30 is rectangular in cross-section so that the second moments of area of the two sections are

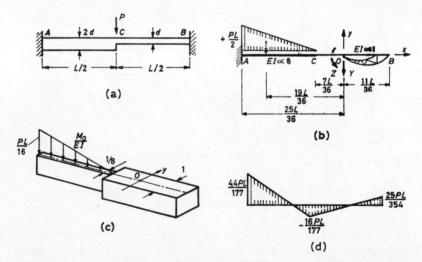

Figure 7.30. (a) Non-uniform beam fixed at the ends; (b) Primary cantilever for elastic centre analysis; (c) Analogous column loaded with the $M_0/EI$ diagram; (d) Final bending moment diagram

proportional to $8I$ and $I$ respectively. The elastic centre is obtained by taking moments about $B$, thus

$$BO = \frac{L/2 \times L/4 + 1/8 \times L/2 \times 3L/4}{L/2 + 1/8 \times L/2} = \frac{11L}{36}$$

## 7.5. THE COLUMN ANALOGY

Since for the present purpose the depth of the beam is supposed to be negligible in comparison with its length the $x$-axis, coinciding with $OB$, is an axis of symmetry and equations 7.19(a) apply. The cantilever fixed at $A$ is taken as the primary structure which is shown at (b) with the redundants $Y$ and $Z$ acting at the elastic centre on the end of the rigid arm. (We can be sure that $X = 0$ since $I_x = 0$). The calculations follow the same course as previously.

| Member | End | $x$ | $EI$ | $I_y$ | $A$ |
|---|---|---|---|---|---|
| AC | A | $-\dfrac{25L}{36}$ | 8 | $\dfrac{849\,L^3}{48\times 36^2}$ | $\dfrac{L}{16}$ |
|  | C | $-\dfrac{7L}{36}$ |  |  |  |
| CB |  |  | 1 | $\dfrac{744\cdot L^3}{48\times 36^2}$ | $\dfrac{L}{2}$ |
|  | B | $+\dfrac{11L}{36}$ |  |  |  |
|  |  |  | $\sum$ | $\dfrac{1593\cdot L^3}{48\cdot 36^2}$ | $\dfrac{9L}{16}$ |

| Member | $A_m$ | $\bar{x}$ | $EI$ | $W$ | $\mu_y$ |
|---|---|---|---|---|---|
| AC | $\dfrac{PL^2}{8}$ | $-\dfrac{19L}{36}$ | 8 | $\dfrac{PL^2}{64}$ | $-\dfrac{19PL^3}{64\times 36}$ |

Hence
$$Y = \frac{\mu_y}{I_y} = -\frac{19P}{59}$$

$$Z = \frac{W}{A} = +\frac{PL}{36}$$

Hence
$$M_A = \frac{PL}{2} - \left(-\frac{19P}{59}\right)\left(\frac{25L}{36}\right) - \frac{PL}{36} = \frac{44PL}{177}$$

$$M_B = 0 - \left(-\frac{19P}{59}\right)\left(\frac{11L}{36}\right) - \frac{PL}{36} = \frac{25PL}{354}$$

$$M_C = 0 - \left(-\frac{19P}{59}\right)\left(\frac{-7L}{36}\right) - \frac{PL}{36} = -\frac{16PL}{177}$$

The analogous column is shown at (c). The position of the centroid and the values of $I_y$ and $A$ have already been calculated but it is of interest to recalculate $I_y$ using the formulae familiar from elementary beam theory†.

$$I_y = \frac{1}{12}\times\frac{1}{8}\times\frac{L^3}{8} + \frac{1}{8}\times\frac{L}{2}\left(\frac{16L}{36}\right)^2 + \frac{1}{12}\times\frac{L^3}{8} + \frac{L}{2}\left(\frac{2L}{36}\right)^2 \quad \frac{1593\,L^3}{48\times 36^2}$$

† $I = bd^3/12 + A\bar{x}^2$

The analogous load to be applied is

$$W = \frac{1}{2} \times \frac{PL}{16} \times \frac{L}{2} = \frac{PL^2}{64}$$

whose moment about $O$ is

$$Wx_G = \frac{PL^2}{64} \times \frac{19L}{36} = \frac{19PL^3}{64 \times 36}$$

Hence the compressive stress at $A$, $B$ and $C$ is

$$f_A = \frac{PL^2/64}{9L/16} + \frac{19PL^3}{64 \times 36} \times \frac{25L}{36} \times \frac{48 \times 36^2}{1593} = \frac{PL}{36} + \frac{475PL}{2124} = \frac{267PL}{1062}$$

$$f_B = \frac{PL^2/64}{9L/16} - \frac{19PL^3}{64 \times 36} \times \frac{11L}{36} \times \frac{48 \times 36^2}{1593} = \frac{PL}{36} - \frac{209PL}{2124} = -\frac{25PL}{354}$$

$$f_C = \frac{PL^2/64}{9L/16} + \frac{19PL^3}{64 \times 36} \times \frac{7L}{36} \times \frac{48 \times 36^2}{1593} = \frac{PL}{36} + \frac{133PL}{2124} = \frac{16PL}{177}$$

Hence from equation 7.26

$$M_A = \frac{PL}{2} - f_A = \frac{PL}{2} - \frac{267PL}{1062} = \frac{44PL}{177}$$

$$M_B = 0 - f_B = \frac{25PL}{354}$$

$$M_C = 0 - f_C = \frac{16PL}{177}$$

These results are identical with those obtained above and are shown on the bending moment diagram at (d). Once again the close similarity between the two sets of calculations will be observed.

(c) As the abutments of an arch are connected by the rock upon which they are constructed no less than by the rib itself the complete structure can be regarded as a ring having infinite stiffness over part of its perimeter. It would therefore seem likely that the methods of arch analysis can be used to study rings and closed frames of finite stiffness. Such a frame is shown in FIGURE 7.31(a); it typifies a culvert carrying a traffic load on its upper slab and subject to a reactive ground pressure on its lower slab. Simple numerical values are assumed and it can be seen from symmetry that the elastic and geometrical centres coincide at $O$.

A solution can be reached quite easily using the primary cantilever shown at (b), but considerations of symmetry suggest that there will be some advantages in using the auxiliary primary structure (c). We now discard the imaginary rigid arm which has been used hitherto and put our trust in equation 7.12 with the extended meaning attached to it at the end of Section 7.4.1.2.

The convention of signs for arches (that positive bending moments produce tension at the outside of the arch), which has so far been quite consistent with that used for beams, now implies negative bending moments in both $BC$ and $AD$. This may at first sight seem anomalous but a little consideration of FIGURE 7.31(c) will show that the loads produce 'coil up' the primary cantilever and so must be given the same sign.

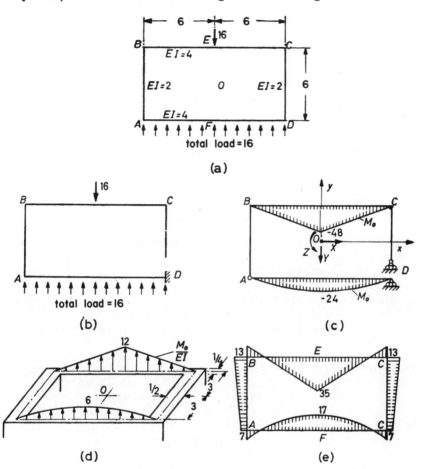

*Figure 7.31.* (a) Closed frame representing culvert with traffic load and uniform ground reaction; (b) The primary cantilever; (c) Auxiliary primary structure with redundants acting at the elastic centre; (d) Analogous column loaded with the $M_o/EI$ diagram; (e) Final bending moment diagram with moments plotted on the tension side of members

As the auxiliary primary structure (c) is symmetrical about the $y$-axis the term $\mu_y$ will be zero. But we have

$$I_x = 2\left(12 \times \frac{1}{4} \times 3^2\right) + 2\left(\frac{1}{12} \times \frac{1}{2} \times 6^3\right) = 72$$

$$A = 2\left(12 \times \frac{1}{4}\right) + 2\left(6 \times \frac{1}{2}\right) = 12$$

Calculating the remaining terms in tabular form:—

| Member | $A_m$ | $\bar{x}$ | $\bar{y}$ | $EI$ | $W$ | $\mu_x$ | $\mu_y$ |
|---|---|---|---|---|---|---|---|
| BC | −288 | 0 | +3 | 4 | −72 | −216 | 0 |
| AD | −192 | 0 | −3 | 4 | −48 | +144 | 0 |
| | | | | $\Sigma$ | −120 | −72 | 0 |

Hence $X = \dfrac{\mu_x}{I_x} = -1$ ; $Y = \dfrac{\mu_y}{I_y} = 0$ ; $Z = \dfrac{W}{A} = -10$.

Substituting in equation 7.12 to find the final bending moments at various points, we have

$$M = M_o - Z - Xy - Yx \qquad \ldots\ldots(7.12)$$

At $B$ and $C$: $M = 0 - (-10) - (-1)(+3) = +13$

At $E$: $M = -48 - (-10) - (-1)(+3) = -35$

At $A$ and $D$: $M = 0 - (-10) - (-1)(-3) = +7$

At $F$: $M = -24 - (-10) - (-1)(-3) = -17$

The final bending moment diagram is shown at (e) with the moments platted on the tension side of the members.

The analogous column is shown at (d) loaded with the $M_o/EI$ diagrams drawn according to the arch convention of signs. $I_x$, $A$ and $W$ have the values calculated above while the nett bending moment on the section is

$$\frac{1}{2} \times 12 \times 12 \times 3 - \frac{2}{3} \times 12 \times 6 \times 3 = 72 \text{ (also as before)}$$

The compressive stress at $B$ is

$$f_B = -\frac{120}{12} - \frac{72 \times 3}{72} = -13$$

While at $A$ we have

$$f_A = -\frac{120}{12} + \frac{72 \times 3}{72} = -7$$

From equation 7.26 we have

$$M_E = -48 - (-13) = -35 \text{ and}$$
$$M_F = -24 - (-7) = -17$$

Evidently the bending moment diagram (e) previously obtained is confirmed by these figures.

### 7.5.3. Comments on the Column Analogy

In the preceding paragraphs the attempt has been made to emphasize the close similarity that exists between the elastic centre method of fixed arch analysis and the column analogy; it is now pertinent to enquire which is the best to use in given circumstances.

From the pedagogical point of view there can be no doubt that the column analogy is dangerous since it is purely mathematical in origin and gives a false picture of the inherent structural action.  The same can be said of the soap bubble analogy to the torsion of non-circular shafts but in this case the equations are often insoluble; when it was first introduced by Griffith and Taylor[9] the analogy provided the only practicable way of finding the torsional rigidity and the shearing stress in irregular sections such as those of propellers.

But no such advantage accrues to the use of the column analogy since, as we have seen, the figures that actually emerge are identical with those of the elastic centre method.  It is true that the calculation of the second moments of area of the analogous column come readily to the minds of most engineers and it is perhaps true that mistakes of sign are easy to avoid in this method.  On the other hand the use of the column analogy in less obvious cases, such as those suggested by Grinter[10], requires very great care if the analogy is to be correctly interpreted.  In such cases there seems to be little point in departing from orthodox methods.

It is very probable that the popularity of the column analogy stems from its tacit employment of auxiliary primary structures.  Hardy Cross himself dealt very cryptically with this device and seemed to regard it as being obviously valid.  Perhaps it is to an intuitive genius of his calibre but more pedestrian students of the subject will perhaps welcome the justification given above.  When in the course of this justification it transpired that the device was applicable to elastic centre analysis no less than to the column analogy the latter was deprived of what had seemed to be a unique advantage.  There now seems to be little point in retaining the column analogy.

## 7.6. INTERCONNECTED ARCHES

### 7.6.1. Introduction

The viaduct (FIGURE 7.32(a)) consisting of a series of interconnected arches supported on columns is a familiar type of bridge which is often used in circumstances when intermediate supports are permissible, while welded steel structures similar to those shown at (b) and (c) are being increasingly used in factory construction.  All three structures illustrated have nine redundant reactions so that the labour of carrying out an analysis by classical methods is quite considerable.  Several attempts have therefore been made to adopt the ideas of slope deflexion and moment distribution to this type of structure.  Amerikian[11], Fowler[12], Lightfoot[13], and Parcel and Moorman[14] have described variants of the slope deflexion equations while Maugh[15], and Lightfoot[13] have developed distribution procedures which, although much more complicated than the corresponding methods for rectangular frames, seem to offer possible methods of solution.  A different line of attack has been proposed by Beaufoy and Diwan[17] whose 'stiffness factors' method is the analytical equivalent to structural models.

The Author considers that the obvious way to deal with such structures is to use models to obtain influence lines but if, for some reason, it is imperative to use an analytical procedure then Fowler's is as convenient and straightforward a method as any.  The following account is based on his paper[12].

## ARCHES

### 7.6.2. GENERALIZED SLOPE DEFLEXION PROCEDURE

It will be realized that the slope deflexion equations 6.5 give the moment acting at the end of a straight member in terms of the rotations of the joints to which it is attached, the relative displacement of those joints (perpendicular to the axis of the member) and the fixed end moment. In using these

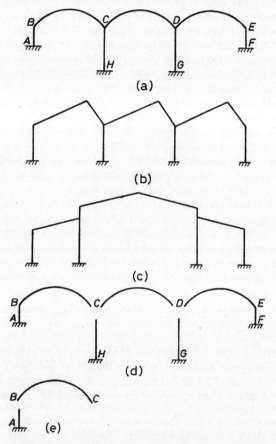

*Figure 7.32.* (a) Viaduct with curved arches on columns; (b) A north light roof can be treated as series of interconnected polygonal arches; (c) A factory building frame that is structurally identical to (a) and (b); (d) In applying the slope deflexion method the elastic constants for the component members shown must be computed first; (e) Instead of treating *ABC* as a single member it can be subdivided at *B*

equations to analyse a rigid frame the rotations and displacements of those joints which are free to move are regarded as the unknowns, equations 6.5 being used to set up equilibrium equations in terms of those unknowns. Compatibility is assured since the members are rigidly connected at the joints.

376

## 7.6. INTERCONNECTED ARCHES         7.6.2

Exactly the same procedure can be used for curved members but it is now necessary to have three equations corresponding to 6.5 for each end of the member which give the moment and the horizontal and vertical forces there in terms of the relative movement of the ends. The horizontal and vertical displacements and the rotation of each joint now become the unknowns so that the total number of simultaneous equations to be solved is equal to three times the number of joints that are free to move. The viaduct of FIGURE 7.32(a), for example, can be regarded as comprising the members $ABC, CD, DEF, CH$ and $DG$ shown at (d). The joints $C$ and $D$ are free to move horizontally (but not vertically if compression of the columns is neglected) and to rotate so that four simultaneous equations must be solved. The slope deflexion equations for each member must be derived, conveniently by means of the elastic centre method, so that a good deal of preliminary calculation is involved.

This work is reduced, in the example shown, by separating the members $ABC$ and $DEF$ at $B$ and $E$ respectively, as shown at (e). In this symmetrical case the three curved members $BC, CD$ and $DE$ are then identical while equations 6.5 apply directly to the straight vertical members. This procedure reduces the preliminary work but the number of simultaneous equations to be solved rises to eight.

We begin by finding the changes in the forces at the elastic centre produced by horizontal, vertical and rotational movements of the ends of a curved member 1–2. The conventions of signs must be carefully observed: clockwise rotations and clockwise couples acting on the ends of the member are taken as positive while forces and translations are positive when they are in the positive directions of the coordinate axes. This is shown in FIGURE 7.33(a). Two rigid arms are now imagined extending from the ends 1 and 2 to the elastic centre and the member is supposed to be in equilibrium under the forces $X$ and $Y$ and the couple $Z$ acting at $O_2$ and the corresponding reactions acting at $O_1$. The convention of signs used in Section 7.4.1 has now served its purpose and can be discarded; the positive directions are those shown in FIGURE 7.33(b). The relative movements of the ends of the imaginary rigid arms corresponding to movements of the ends 1 and 2 are obtained from diagrams (c), (d) and (e).

When joints 1 and 2 move horizontally, without vertical or rotational movement, the horizontal movement of $O_2$ relative to $O_1$ is

$$\Delta_{Ox} = \Delta_{2x} - \Delta_{1x} \qquad \ldots\ldots \text{(i)}$$

Similarly for vertical movements only:

$$\Delta_{Oy} = \Delta_{2y} - \Delta_{1y} \qquad \ldots\ldots \text{(ii)}$$

When ends 1 and 2 rotate through $+\theta_1$ and $+\theta_2$, respectively, without translating as at (e), we have

$$\left. \begin{aligned} \theta_O &= \theta_2 - \theta_1 \\ \Delta_{Ox} &= -y_2\theta_2 + y_1\theta_1 \\ \Delta_{Oy} &= x_2\theta_2 - x_1\theta_1 \end{aligned} \right\} \qquad \ldots\ldots \text{(iii)}$$

# ARCHES

Combining (i), (ii) and (iii), we have

$$\left.\begin{array}{l} \Delta_{Ox} = \Delta_{2x} - \Delta_{1x} - y_2\theta_2 + y_1\theta_1 \\ \Delta_{Oy} = \Delta_{2y} - \Delta_{1y} + x_2\theta_2 - x_1\theta_1 \\ \theta_O = \theta_2 - \theta_1 \end{array}\right\} \quad \ldots \text{(iv)}$$

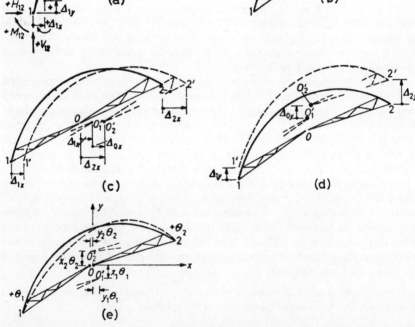

*Figure 7.33.* (a) Positive forces and couples acting on and translations and rotations of the ends of a curved member 1 — 2; (b) Positive pairs of couples and forces acting at the elastic centre on the ends of rigid arms. (The points $O_1$ & $O_2$ really coincide but have been separated for clarity); (c) The effect of horizontal displacement of the ends; (d) The effect of vertical displacement of the ends; (e) The effect of rotational displacement of the ends

In these equations the signs of $x_1$, $x_2$, $y_1$ and $y_2$ have been taken as positive; they must be given their proper signs in a numerical problem.

In order to find the values of $X$, $Y$ and $Z$ corresponding to these movements we note that the bending moment at the point $(x, y)$ in FIGURE 7.33(b) is

$$M = Z + Yx - Xy \quad \ldots \text{(v)}$$

Hence, considering bending only,

$$U = \int_1^2 \frac{M^2 ds}{2EI} = \int_1^2 \frac{(Z + Yx - Xy)^2 ds}{2EI} \quad \ldots \text{(vi)}$$

i.e.
$$\Delta_{Ox} = \frac{\partial U}{\partial X} = \int_1^2 \frac{(Z + Yx - Xy)(-y)\, ds}{EI} \quad \ldots \text{(vii)}$$

Recalling that $O$ is the elastic centre and therefore that

$$\int_1^2 \frac{x\, ds}{EI} = \int_1^2 \frac{y\, ds}{EI} = 0 \quad \ldots \text{(viii)}$$

equation (viii) becomes

$$\Delta_{Ox} = -Y \int_1^2 \frac{xy\, ds}{EI} + X \int_1^2 \frac{y^2\, ds}{EI}$$
$$= -Y I_{xy} + X I_x \quad \ldots \text{(ix)}$$

In a similar manner
$$\Delta_{Oy} = Y I_y - X I_{xy} \quad \ldots \text{(x)}$$
$$\theta_O = Z \cdot A \quad \ldots \text{(xi)}$$

Solving (ix) (x) and (xi) simultaneously for $X$, $Y$ and $Z$ we have

$$\left. \begin{array}{l} X = \dfrac{\Delta_{Ox} I_y + \Delta_{Oy} I_{xy}}{I_x I_y - I_{xy}^2} \\[1em] Y = \dfrac{\Delta_{Ox} I_{xy} + \Delta_{Oy} I_x}{I_x I_y - I_{xy}^2} \\[1em] Z = \dfrac{\theta_O}{A} \end{array} \right\} \quad \ldots (7.27)$$

If the curved member is symmetrical one of the axes of symmetry can conveniently be taken as a reference axis and $O_x$ and $O_y$ become principal axes: $I_{xy}$ is then zero.

i.e.
$$X = \frac{\Delta_{Ox}}{I_x} \; ; \; Y = \frac{\Delta_{Oy}}{I_y} \; ; \; Z = \frac{\theta_O}{A} \quad \ldots (7.27(a))$$

Combining 7.27 and (iv) we have

$$\left. \begin{array}{l} X = \dfrac{(\Delta_{2x} - \Delta_{1x} - y_2\theta_2 + y_1\theta_1) I_y + (\Delta_{2y} - \Delta_{1y} + x_2\theta_2 - x_1\theta_1) I_{xy}}{I_x I_y - I^2_{xy}} \\[1em] Y = \dfrac{(\Delta_{2x} - \Delta_{1x} - y_2\theta_2 + y_1\theta_1) I_{xy} + (\Delta_{2y} - \Delta_{1y} + x_2\theta_2 - x_1\theta_1) I_x}{I_x I_y - I^2_{xy}} \\[1em] Z = \dfrac{\theta_2 - \theta_1}{A} \end{array} \right\} \quad (7.28)$$

Equations 7.28 give the forces at the elastic centre in terms of the movements of the ends. It now remains to write down the corresponding equations for the joints so that equilibrium equations can be written for each joint.

Thus at 1 we have, from FIGURE 7.33(b)

$$\left.\begin{array}{l} M_{12} = M_{F12} - Z - Yx_1 + Xy_1 \\ \text{also} \quad V_{12} = V_{F12} - Y \\ \text{and} \quad H_{12} = H_{F12} - X \\ \text{while at end 2 we have} \\ M_{21} = M_{F21} + Z + Yx_2 - Xy_2 \\ V_{21} = V_{F21} + Y \\ H_{21} = H_{F21} + X \end{array}\right\} \quad \ldots (7.29)$$

Here $M_{F12}$, $V_{F12}$, $H_{F12}$ are the ' fixed-end ' moment, vertical reaction and horizontal reaction at the end 1 of the curved member 1–2 caused by the applied loading; they are worked out for all loaded members by the methods of Section 7.4.1. The above equations are now used in the following sequence:

1. Assume convenient directions for the reference axes.
2. Compute $A$, $I_x$, $I_y$ and $I_{xy}$ for each of the members into which the complete structure has been divided, using axes parallel to the reference axes originating at the elastic centre of each member.
3. Compute the terms $M_{F12}$, $V_{F12}$, $H_{F12}$ etc. for each loaded member.
4. Write down equations 7.29 for each end of each member; equations 7.28 are used to give $X$, $Y$ and $Z$ for each member in terms of such end translations and rotations as can in fact occur.
5. Equilibrium equation $\Sigma M = 0$, $\Sigma V = 0$, $\Sigma H = 0$ are written for each joint in terms of the equations resulting from step 4. This gives enough equations to find the translations and rotations of each joint.
6. These translations and rotations are inserted in the equations of step 4 to give the end reactions for each member from which the final bending moments, thrusts and shears throughout the frame can be obtained. A worked example of this procedure will be found in Volume II.

## REFERENCES

[1] PIPPARD, A. J. S., TRANTER, E. and CHITTY, L. ' The Mechanics of the Voussoir Arch '. *J. Inst. Civ. Engrs.* Vol. 4. 1936. p. 281

[2] PIPPARD, A. J. S. and BAKER, J. F. *The Analysis of Engineering Structures*. Arnold. 3rd. edn. 1957

[3] TIMOSHENKO, S. *History of Strength of Materials*. McGraw-Hill. 1953

[4] STRAUB, H. *A History of Civil Engineering*. Leonard Hill, Ltd. 1952

# REFERENCES

[5] TIMOSHENKO, S. and YOUNG, D. H. *Theory of Structures.* McGraw-Hill. 1945

[6] JOHNSON, J. B., BRYAN, C. W. and TURNEAURE, F. E. *Modern Framed Structures.* Vol. 2. Wiley

[7] HARDY CROSS. 'The Column Analogy'. Bull. No. 215. Univ. of Illinois

[8] CROSS, H. and MORGAN, N. B. *Continuous Frames of Reinforced Concrete.* Wiley. 1932

[9] GRIFFITH, A. A. and TAYLOR, G. J. 'The Use of Soap Films in Solving Torsion Problems'. *Proc. Inst. Mech. Engrs.* 1917. p. 753

[10] GRINTER, L. E. *Theory of Modern Steel Structures.* Vol. 2. Macmillan

[11] AMERIKIAN, A. *Analysis of Rigid Frames.* U.S. Govt. Printing Office. 1942

[12] FOWLER, K. T. 'Slope Deflexion Equations for Curved Members'. *Proc. Amer. Soc. Civ. Engrs.* Vol. 76. Separate No. 6. 1950

[13] LIGHTFOOT, E. 'Generalized Slope deflexion and Moment Distribution Methods for the Analysis of Viaducts and Multi-bay, Ridged Portal Frames'. *The Structural Engineer.* Jan. 1958. p. 21

[14] PARCEL, J. I. and MOORMAN, R. B. B. *Analysis of Statically Indeterminate Structures.* Wiley. 1955

[15] MAUGH, L. C. *Statically Indeterminate Structures.* Wiley. 1946

[16] FRANCIS, A. J. 'The Analysis of Single-storey, Multi-bay, Gabled Rigid Frames'. *The Structural Engineer.* July, 1951. p. 189

[17] BEAUFOY, L. A. and DIWAN, A. F. S. 'Analysis of Continuous Structures by the Stiffness Factors Method'. *Quart. J. Mech. & App. Maths.* Vol. 2. Pt. 3. 1949. p. 263

# Chapter 8

# STABILITY OF STRUTS AND FRAMEWORKS†

## 8.1. STABILITY OF ISOLATED STRUTS

The discussion in Section 6.6.1 of Chapter 6 included a reference to the collapse of structures because of the instability of one or more of its members. In this Chapter methods are described which enable the stability of complete structures to be investigated in terms of the properties of its component members. The phenomenon of elastic instability can be demonstrated very easily: suppose that a long, slender bar that is reasonably straight is subjected to a gradually increasing axial compressive load. It will be found that at a certain load the bar will suddenly bend laterally and that if the load is removed the bar will return to its original straight condition. For example, if a round mild steel bar whose diameter is 2% of its length is loaded axially as a pin ended strut, then it will buckle when the axial stress is only about 30% of the yield stress. For loads less than this the strut is said to be stable and for loads greater than this some external form of restraint must be applied to prevent failure. This type of failure can also be demonstrated on models of portal frames and braced frameworks by constructing them of slender members. At definite loads one or more of the struts of these models will buckle and no further load can be applied to the structure.

The behaviour of individual members is dominated by the fact that it is not possible to make a perfectly straight strut; some lack of straightness, no matter how small, occurs in all members. The initial shape of the pin-ended strut (Figure 8.1) can be expressed approximately as

$$y_0 = a_0 \sin \frac{\pi x}{L}$$

where $a_0$ is the lack of straightness at the centre of the strut, i.e. at $x = \frac{L}{2}$.

When an axial force $F$ is applied to the end of this strut, the additional deflexion at point $x$ from the origin is $y$. The bending equation can be applied in the following way.

$$EI \frac{d^2y}{dx^2} = -Fy - Fa_0 \sin \frac{\pi x}{L} \qquad \ldots \ldots (8.1)$$

where $I$ is the second moment of area of the strut section about the axis of bending.

---

† This chapter was contributed by Prof. N. W. Murray of Monash Univerity, Australia.

## 8.1. STABILITY OF ISOLATED STRUTS

Substitution will show that the solution of this equation is†

$$y = C_1 \sin x \sqrt{\frac{F}{EI}} + C_2 \cos x \sqrt{\frac{F}{EI}} - \frac{Fa_0}{\left(F - \frac{\pi^2 EI}{L^2}\right)} \sin \frac{\pi x}{L}$$

where $C_1$ and $C_2$ are arbitrary constants to be determined from the boundary conditions. The boundary condition that $y = 0$ when $x = 0$ gives $C_2 = 0$

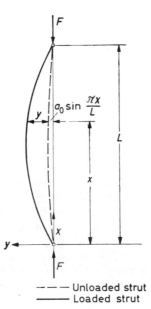

Figure 8.1. Deflexion of an initially bent pin-ended strut

– – – – Unloaded strut
———— Loaded strut

and the other boundary condition, viz. $y = 0$ and when $x = L$ gives $C_1 = 0$. Thus the deflexion of an initially curved strut is given by

$$y = -\frac{Fa_0}{\left(F - \frac{\pi^2 EI}{L^2}\right)} \sin \frac{\pi x}{L}$$

It is seen that the deflexion is theoretically infinite when the denominator is zero,

i.e. $$F = \frac{\pi^2 EI}{L^2} \qquad \ldots (8.2)$$

This load is known as the Euler load of the strut and it is the buckling load of the pin-ended column. It is usually given the symbol $P_E$ and it should be noted that its value depends upon the geometry of the strut and its elastic modulus.

The reader should carefully notice the assumptions on which the above theory is based. It has been assumed that the material is elastic so that the

---

† The first two terms are the complementary function of 8.1 and the last term is the particular integral which can be obtained by assuming a solution $y = k \sin \frac{\pi x}{L}$.

yield stress has not been exceeded. This enables the modulus $E$ to be treated as a constant. Thus the above theory applies to long slender struts because they behave elastically. A short stocky strut tends to yield long before its Euler load is reached and its failure load is determined largely by the value of the yield stress. Strictly speaking the bending moment at a point is given by $\frac{EI}{R}$ but if the deflexions are small it is permissible to write $EI\frac{d^2y}{dx^2}$ as the bending moment. Thus, for the above theory to be valid the material must remain elastic and the deflexions must be small.

It is interesting to compare the behaviour of the simple beam with that of a pin-ended strut. We saw in Chapter 3, Section 3.7 that the bending moment in a simple beam, $EI\frac{d^2y}{dx^2}$, is a function of $x$ only. However, the bending moment in the strut is a function of both $x$ and $y$, i.e. of the shape of the strut, because of the term $-Fy$ (equation 8.1). This point will be mentioned later when the application of the Principle of Superposition to struts is discussed.

The behaviour of other struts can be studied in a similar way. By developing equations similar to 8.1 it can be shown that a strut which is initially bent and built-in at each end will buckle at a load which is four times the load at which it would buckle if the ends were pinned, i.e. four times the Euler load, $P_E$. This can be demonstrated in another way. In FIGURE 8.2 the buckling modes of some simple struts are illustrated. The simple pin-ended strut assumes the shape of a sine curve when loaded axially. Consideration of the symmetry of the strut built-in at both ends (FIGURE 8.2) shows that points of inflexion occur at the quarter points $A$ and $B$ and that the part between these points is a pin-ended strut of half the length of the original. Equation 8.2 shows that the buckling load of the pin-ended strut $AB$ of length $\frac{L}{2}$ and, therefore, of the built-in strut of length $L$, is four times as great as that of a pin-ended strut of length $L$. Similar reasoning shows that the strut which is fixed at its base and pinned at its upper end which is also allowed to sway laterally (FIGURE 8.2(c)) has a buckling load of $\frac{P_E}{4}$. The distance between the points of inflexion is known as the effective length of the strut.

In later sections the stability of frameworks is treated. The struts in a framework are partially restrained against buckling by the other members of the structure. If the joints of the framework are effectively held fixed in space and if the members adjacent to the strut in question are flimsy, the strut will tend to behave as a pin-ended Euler column because its effective length is nearly equal to its axial length, i.e. the points of inflexion are near the ends of the strut. However, when the surrounding members are very rigid, the strut will behave as if its ends were fixed and its effective length is halved. In practice a strut will behave in a manner which is between these two limits. If the joints of the framework are allowed to move in space, the strut will behave in a manner which is between the limits shown in FIGURES 8.2(c) and (d).

## 8.1. STABILITY OF ISOLATED STRUTS

The methods of slope deflexion and moment distribution which are used to analyse the behaviour of frameworks have been described in Chapter 6. These methods are based upon the moment-rotation relationship of a simple

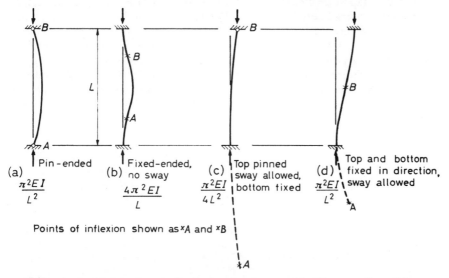

(a) Pin-ended
$\frac{\pi^2 EI}{L^2}$

(b) Fixed-ended, no sway
$\frac{4\pi^2 EI}{L}$

(c) Top pinned sway allowed, bottom fixed
$\frac{\pi^2 EI}{4L^2}$

(d) Top and bottom fixed in direction, sway allowed
$\frac{\pi^2 EI}{L^2}$

Points of inflexion shown as ×A and ×B

*Figure 8.2.* Buckling loads and effective lengths of struts with different end conditions

beam. When a simple cantilever is bent by a couple, $M_{AB}$, at its free end $A$ (FIGURE 8.3), the relationship between moment $M_{AB}$ and angle of rotation $\theta_A$ is given by equation 6.7

$$M_{AB} = \frac{4EI}{L} \theta_A \qquad \ldots (6.7)$$

However, suppose that the cantilever is also subjected to an axial load $F$ (FIGURE 8.4) and that it is desired to determine the moment-rotation relationship. It is seen that as soon as the cantilever deflects from its originally

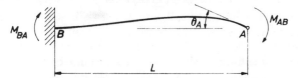

*Figure 8.3.* Moment applied at end of simple cantilever

straight position a moment is introduced whose value is the product of axial load, $F$, and deflexion, $y$. It is this moment which causes the cantilever to become more flexible than the simple cantilever shown in FIGURE 8.3 when the load $F$ is compressive and more rigid when $F$ is tensile.

In the treatment of rigidly-jointed frameworks in Chapter 6 it was assumed that the axial loads in the members were zero. This assumption is satisfactory for frameworks whose members carry only small axial loads but when these loads become large the struts become more flexible and the ties more rigid.

As the ratios of the axial loads in the members to the Euler loads of the corresponding members increase it is found that the effects of stability become more important.

In this Chapter the behaviour of frameworks in which some of the members carry large axial loads is studied. Before it is possible to understand the behaviour of actual frameworks, it is necessary to study their purely elastic behaviour by assuming that the yield stress is infinite. The behaviour of an actual framework is introduced in Section 8.5 in a simple way by combining its elastic behaviour with its plastic behaviour. Thus in Sections 8.2, 8.3 and 8.4, elastic conditions are assumed to prevail. It is also assumed that the deflexions are not large enough to change appreciably the geometrical configuration of the frameworks.

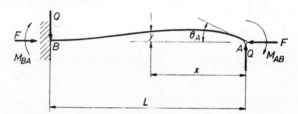

*Figure 8.4.* Moment applied at end of axially loaded cantilevers

As the external loads applied to a framework are increased, the struts rapidly become more flexible. The ties become more rigid but at a slower rate. Finally, one of the struts becomes so flexible that it can no longer be supported by the adjacent members. Assuming that only elastic conditions exist, one of the struts buckles at this load and the frame collapses. This load is known as the 'critical load' of the framework. Methods of determining the critical loads of frameworks are briefly treated in Section 8.4.

## 8.2. EFFECT OF AXIAL LOAD ON THE BEHAVIOUR OF A SIMPLE CANTILEVER

To study the effect of axial loading on framework members it is necessary for the reader to become familiar with certain dimensionless stability functions. The simplest of these are the $s$- and $c$-functions which are stiffness and carry-over functions modified to take account of the effects of axial load.

### 8.2.1. $s$- AND $c$-FUNCTIONS FOR AN AXIALLY LOADED CANTILEVER

When a cantilever (FIGURE 8.4), which is straight when unloaded is subjected to an axial load $F$ and a moment $M_{AB}$ is applied to its end $A$, which is free to rotate but not to change in position, a rotation $\theta_A$ occurs at $A$. At the fixed end $B$ a moment $M_{BA}$ is induced. $\theta_A$ and $M_{BA}$ may be evaluated from the equilibrium equation

$$EI\frac{d^2y}{dx^2} = -Fy - M_{AB} + Qx \qquad \ldots . (8.3)$$

## 8.2. EFFECT OF AXIAL LOAD ON CANTILEVER

This equation is conveniently solved by using the dimensionless variables $s$ and $c$ which are defined by the following equations.

$$M_{AB} = sk\theta_A = s\frac{EI}{L}\theta_A \qquad \ldots\ldots(8.4)$$

$$\text{and } M_{BA} = cM_{AB} \qquad \ldots\ldots(8.5)$$

These equations are similar to equations 6.7 and 6.8 which were used in Chapter 6 as a basis for Hardy Cross moment distribution. In that method the effect of the axial forces in the members was neglected and for all members $s = 4$ and $c = 0\cdot5$. Here it will be shown that $s$ and $c$ are functions of the axial loads in the members.

By writing $Q$ in terms of $M_{AB}$ and $M_{BA}$ and using equations 8.4 and 8.5, the equilibrium equation becomes

$$\frac{d^2y}{dx^2} + \frac{Fy}{EI} = -\frac{s}{L}\theta_A + s(1+c)\frac{\theta_A}{L^2}x$$

The solution of this equation is

$$y = -\frac{sk}{FL}[L - (1+c)x]\,\theta_A + C_3\cos x\sqrt{\frac{F}{EI}} + C_4\sin x\sqrt{\frac{F}{EI}}$$

where the first term is the particular integral and the last two terms are the complementary function. $C_3$ and $C_4$ are constants which are determined by considering the conditions at ends $A$ and $B$. Four end conditions are known, viz.,

$$\text{At } x = 0, \quad y = 0 \text{ and } \frac{dy}{dx} = \theta_A$$

$$\text{and at } x = L, \quad y = 0 \text{ and } \frac{dy}{dx} = 0$$

Not only do these conditions allow $C_3$ and $C_4$ to be determined, but they also allow $s$ and $c$ to be evaluated in terms of

$$\mu = \frac{L}{2}\sqrt{\frac{F}{EI}} = \frac{\pi}{2}\sqrt{\frac{F}{P_E}}$$

where $P_E$ is the Euler load of the member, i.e. $\frac{\pi^2 EI}{L^2}$ (by definition)

Thus
$$s = \left(\frac{1 - 2\mu\cot 2\mu}{\tan\mu - \mu}\right)\mu$$

$$\text{and } c = \frac{2\mu - \sin 2\mu}{\sin 2\mu - 2\mu\cos 2\mu}$$

$s$ and $c$ have been tabulated for $\frac{F}{P_E}$ values on the Manchester University Electronic Digital Computer and the complete tables are published[1]. A short table is included here (TABLE 8.1) and graphs of $s$ and $c$ against $\dfrac{F}{P_E}$ are shown in FIGURE 8.5. They show that $s$ decreases as the axial compressive load $F$ increases and equation 8.4 then shows that the stiffness of the member at $A$, i.e. the moment at $A$ required to produce unit rotation at $A$, decreases with increasing $F$. A negative value of $s$ means that the

TABLE 8.1

VALUES OF $s$, $c$ AND $m$ (N.B. $F$ is positive when *compressive*)

| Tension | | | | Compression | | | |
|---|---|---|---|---|---|---|---|
| $F/P_E$ | $s$ | $c$ | $m$ | $F/P_E$ | $s$ | $c$ | $m$ |
| 0 | + 4·000 | + 0·500 | + 1·000 | + 0 | + 4·000 | + 0·500 | + 1·000 |
| − 0·5 | + 4·619 | + 0·402 | + 0·724 | + 0·1 | + 3·867 | + 0·526 | + 1·091 |
| − 1·0 | + 5·175 | + 0·338 | + 0·584 | + 0·2 | + 3·730 | + 0·550 | + 1·205 |
| − 1·5 | + 5·681 | + 0·293 | + 0·498 | + 0·3 | + 3·589 | + 0·587 | + 1·351 |
| − 2·0 | + 6·147 | + 0·260 | + 0·440 | + 0·4 | + 3·444 | + 0·624 | + 1·545 |
| − 2·5 | + 6·581 | + 0·235 | + 0·397 | + 0·5 | + 3·294 | + 0·666 | + 1·817 |
| − 3·0 | + 6·988 | + 0·214 | + 0·364 | + 0·6 | + 3·140 | + 0·714 | + 2·223 |
| − 3·5 | + 7·372 | + 0·198 | + 0·338 | + 0·7 | + 2·981 | + 0·769 | + 2·900 |
| − 4·0 | + 7·737 | + 0·185 | + 0·317 | + 0·8 | + 2·816 | + 0·833 | + 4·253 |
| | | | | + 0·9 | + 2·645 | + 0·909 | + 8·307 |
| | | | | + 1·0 | + 2·467 | + 1·000 | ∞ |
| | | | | + 1·1 | + 2·283 | + 1·111 | − 7·902 |
| Compression (cont.) | | | | + 1·2 | + 2·090 | + 1·249 | − 3·847 |
| $F/P_E$ | $s$ | $c$ | | + 1·3 | + 1·889 | + 1·424 | − 2·495 |
| | | | | + 1·4 | + 1·678 | + 1·656 | − 1·818 |
| | | | | + 1·5 | + 1·457 | + 1·973 | − 1·411 |
| + 2·6 | − 2·249 | − 2·231 | | + 1·6 | + 1·224 | + 2·435 | |
| + 2·7 | − 2·809 | − 1·928 | | + 1·7 | + 0·978 | + 3·166 | |
| + 2·8 | − 3·445 | − 1·708 | | + 1·8 | + 0·717 | + 4·497 | |
| + 2·9 | − 4·176 | − 1·543 | | + 1·9 | + 0·439 | + 7·661 | |
| + 3·0 | − 5·032 | − 1·416 | | + 2·0 | + 0·143 | + 26·684 | |
| + 3·1 | − 6·052 | − 1·316 | | + 2·1 | − 0·176 | − 21·072 | |
| + 3·2 | − 7·297 | − 1·236 | | + 2·2 | − 0·519 | − 7·511 | |
| + 3·3 | − 8·863 | − 1·173 | | + 2·3 | − 0·893 | − 4·623 | |
| + 3·4 | − 10·908 | − 1·122 | | + 2·4 | − 1·301 | − 3·370 | |
| + 3·5 | − 13·719 | − 1·082 | | + 2·5 | − 1·750 | − 2·673 | |

TABLE 8.1 is extracted from *Stability Functions for Structural Frameworks* by courtesy of Manchester University Press.

cantilever has to be supported to prevent buckling. In other words, a positive rotation $\theta_A$ can be maintained by applying a negative moment $M_A$ given by equation 8.4 when $s$ is negative. The graph also shows that the stiffness of the member when $F$ is tensile increases as the axial load. A similar study of the graph of $c$ against $\dfrac{F}{P_E}$ and equations 8.4 and 8.5 shows that the moment of $B$ increases with the compressive load $F$ and decreases when $F$ is tensile.

The graph (FIGURE 8.5) shows that when $\dfrac{F}{P_E} = 2·04$, $s = 0$ and $c$ is theoretically infinite. At this load equation 8.4 shows that the axially loaded cantilever (FIGURE 8.4) has zero stiffness where the stiffness is defined

## 8.2. EFFECT OF AXIAL LOAD ON CANTILEVER 8.2.1

as $\dfrac{M_{AB}}{\theta_A}$. This means that a very small disturbance will cause elastic buckling of the cantilever. Although $c$ is infinite at this load and its graph is discontinuous, the product $sc$ is finite and has the value 3·59. Equations 8.4 and 8.5 demonstrate that at the buckling load $M_{BA}$ is finite and continuous. It is also seen that the stiffness of a strut, $\dfrac{M_{AB}}{\theta_A}$, is not directly proportional to the axial load $F$; i.e. for a given moment $M_{AB}$ the end rotation $\theta$ does not become $2\theta_A$ when $F$ is increased to $2F$. This is because

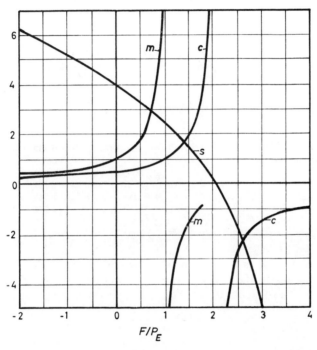

Figure 8.5. Graphs showing variation of stability functions with axial load

the bending moments depend upon the deflexions of the strut (see equations 8.1 and 8.3). The Principle of Superposition does not apply to the load on the strut. However, if the axial load $F$ is fixed $s$ is constant and, from equation 8.4, it is seen that $\theta_A$ is proportional to $M_{AB}$; i.e. for a given axial load the Principle of Superposition can be used. It is this property which allows the $s$- and $c$-functions to be used in a moment-distribution process for analysing frameworks. The loads must remain constant while the distribution process. is performed.

In 1916 Berry[2] published functions which enabled the rotations at the ends of the members to be written down as functions of the moments. The Berry functions are defined as

$$f(\mu) = \frac{3}{2} \frac{2\mu \ \text{cosec} \ 2\mu - 1}{\mu^2} \quad \ldots \text{(i)}$$

$$\phi(\mu) = \frac{3(1 - 2\mu \cot 2\mu)}{4\mu^2} \qquad \ldots \text{(ii)}$$

$$\psi(\mu) = 3\,\frac{\tan \mu - \mu}{\mu^3} \qquad \ldots \text{(iii)}$$

Thus the rotation-moment relation of a uniform strut $AB$ which is rotated through $\theta_A$ and $\theta_B$ at its ends by moments $M_A$ and $M_B$ at its ends and a distributed load $w$ per unit length can be written down in the form

$$k\theta_A = \frac{M_A}{3}\phi(\mu) - \frac{M_B}{6}f(\mu) + \frac{wL^2}{24}\psi(\mu) \qquad \ldots \text{(iv)}$$

$$k\theta_B = \frac{M_B}{3}\phi(\mu) - \frac{M_A}{6}f(\mu) - \frac{wL^2}{24}\psi(\mu) \qquad \ldots \text{(v)}$$

where $k = \dfrac{EI}{L}$

To convert from $s$- and $c$-functions to Berry functions the following equations can be used.

$$s = \frac{4\,\phi(\mu)}{\mu\,\psi(\mu)} \qquad \ldots \text{(vi)}$$

$$c = \frac{f(\mu)}{2\,\phi(\mu)} \qquad \ldots \text{(vii)}$$

The function $\psi(\mu)$ is a function which magnifies the initial slope due to $w$ alone (i.e. when $M_A = M_B = F = 0$) and can be very closely approximated as

$$\psi(\mu) = \frac{1}{1 - \dfrac{F}{P_E}} \qquad \ldots \text{(viii)}$$

It is seen from (iv) and (v) that when a structure is analysed using Berry functions the number of simultaneous equations to be solved is equal to the number of unknown moments. $s$- and $c$-functions will, in general, allow a considerable reduction in the number of simultaneous equations since the unknowns are joint rotations. However, the main advantage of $s$- and $c$-functions is that they are more convenient for moment distribution and the analyst can readily visualise the mathematical processes.

The $s$- and $c$-functions form the basis of the methods recently developed to solve stability problems. They may be used directly in a modified form of moment distribution or slope-deflexion and examples of their use in this way are given in Section 8.3. Certain types of stability problems require them to be used in other forms and it is convenient to introduce other functions which have been tabulated[1]. Some of these functions are now briefly described.

## 8.2.2. $s''$ Function for Pin-ended Members which are Axially Loaded

When $B$ is pinned instead of fixed, the dimensionless coefficient $s$ in equation 8.4 becomes $s''$,

i.e. 
$$M_{AB} = s''k\theta_A \qquad \ldots\ldots (8.6)$$

| Operation | $M_{BA}$ | $M_{AB}$ |
|---|---|---|
| B held fixed A rotated | $sck\theta_A$ ← | $sk\theta_A$ |
| B released | $-sck\theta_A$ → | $-sc^2k\theta_A$ |
| Total | 0 | $s(1-c^2)k\theta_A$ |

*Figure 8.6. Moment distribution applied to a pin-ended strut*

The relation between $s$ and $s''$ is found by the moment-distribution operations shown in Figure 8.6. Thus the moment-rotation relationship at the end of a pin-ended strut is given by equation 8.6 where

$$s'' = s(1 - c^2) \qquad \ldots\ldots (8.7)$$

## 8.2.3. Functions for Members Subjected to Sway

In this case $A$ is allowed to move through a distance which is perpendicular to the original position of the member but rotation of $A$ and $B$ is prevented (Figure 8.7). By symmetry $M_{BA}$ and $M_{AB}$ are equal and it can be shown by moment distribution that

$$M_{AB} = M_{BA} = -s(1+c)\frac{k\Delta}{L} \qquad \ldots\ldots (8.8)$$

Another function which is useful for sway problems is the $m$-function defined by

$$M_{AB} = M_{BA} = -\frac{mQL}{2} \qquad \ldots\ldots (8.9)$$

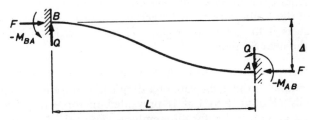

*Figure 8.7. Sidesway of strut*

On taking moments about $A$ in Figure 8.7 the equilibrium equation becomes

$$M_{AB} + M_{BA} + F\Delta + QL = 0$$

whence $$m = \cfrac{1}{1 - \cfrac{\pi^2 \dfrac{F}{P_E}}{2s(1+c)}}$$ ....(8.9(a))

Some values of $m$ are included in TABLE 8.1 and $m$ is plotted against values of $\dfrac{F}{P_E}$ in FIGURE 8.5. From these values and equations 8.8 and 8.9 it is seen that the strut which sways sideways (FIGURE 8.7) becomes more flexible as the axial load increases. When $F = \dfrac{\pi^2 EI}{L^2} = P_E$ the strut becomes theoretically infinitely flexible. It should be noted that this is a lower load than the load at which the cantilever (FIGURE 8.4) becomes infinitely flexible. Later it will be shown that this causes some structures to fail by swaying sideways.

8.2.4. OTHER FUNCTIONS WHICH ARE TABULATED

Livesley and Chandler[1] have tabulated other functions which are very useful for the analysis of frameworks. The $n$- and $o$-functions are used to speed up the analysis of frameworks which are subjected to sway loads.

When the members of the framework have gusset plates, the $A$- and $B$-functions[1] should be used to calculate modified $s$ and $c$ values.

## 8.3. ELASTIC BEHAVIOUR OF IDEAL FRAMEWORKS

In this section the behaviour of ideal frameworks is considered. An ideal framework is one which has no eccentricities at the joints, no initial curvature of the members and which is fabricated from a uniform elastic material. When studying the behaviour of frameworks, the reader should notice particularly how the struts are restrained from buckling by the adjacent members. Rectangular type frameworks, e.g. portal frames, derive their rigidity from the fixity of their joints and from the ability of their members to resist transverse bending moments. The loads are usually applied to the members between the joints and so give rise to transverse bending. A braced framework is usually loaded at its joints and apparently no transverse bending will arise. However, due to the large axial forces in the members, axial deformations will develop. The subsequent change in the geometry of the framework must be accommodated by transverse bending of its members if the joints are rigid. These two types of behaviour are treated in the next Sections.

8.3.1. BEHAVIOUR OF AN IDEAL RECTANGULAR TYPE FRAMEWORK

As an example consider the case of a simple portal frame which is loaded symmetrically (FIGURE 8.8) but is prevented in some way from moving laterally. $EI = 39 \cdot 6 \times 10^6$ for all members. The axial load in each of the stanchions is $F = W = 20,000$ but in the beam it is very small and can be neglected. On taking the ratio of the axial loads in the stanchions to their Euler load, values of $s$ and $c$ can be found from TABLE 8.1 or FIGURE 8.5 and then used in a simple moment-distribution method. The following values of $s$ and $c$ are found from TABLE 8.1.

## 8.3. ELASTIC BEHAVIOUR OF FRAMEWORKS

TABLE 8.2

|        | Stanchions | Beam   |
|--------|------------|--------|
| $P_E$  | 18,850     | 75,400 |
| $F/P_E$| 1·06       | 0      |
| $s$    | 2·358      | 4·000  |
| $c$    | 1·064      | 0·500  |

The fixed end moments are distributed in the following manner.

TABLE 8.3

| Moments | AB | BA | BE | EB | EF | FE |
|---|---|---|---|---|---|---|
| Distribution Factors | | 0·228 | 0·772 | 0·772 | 0·228 | |
| F.E.M. | 0 | 0 | − 130,000 | + 130,000 | 0 | 0 |
| Release B | | + 29,600 | + 100,400 | | | |
| Release E | | | − 100,400 | −29,600 | | |
| Carry-over | + 31,600 | | − 50,200 | + 50,200 | | − 31,600 |
| Release B | | + 11,500 | + 38,700 | | | |
| Release E | | | | − 38,700 | − 11,500 | |
| Carry-over | + 12,200 | | − 19,400 | + 19,400 | | − 12,200 |
| Release B | | + 4,400 | + 15,000 | | | |
| Release E | | | | − 15,000 | − 4,400 | |
| Carry-over | + 4,700 | | − 7,500 | + 7,500 | | − 4,700 |
| Release B | | + 1,700 | + 5,800 | | | |
| Release E | | | | − 5,800 | − 1,700 | |
| Carry-over | + 1,800 | | − 2,900 | + 2,900 | | − 1,800 |
| Release B | | + 700 | + 2,200 | | | |
| Release E | | | | − 2,200 | − 700 | |
| Carry-over | + 700 | | − 1,100 | + 1,100 | | − 700 |
| Total | 51,000 | 47,900 | − 49,000 | 49,000 | − 47,900 | − 51,000 |
| Moments from simple theory | 32,500 | 65,000 | − 65,000 | 65,000 | − 65,000 | − 32,500 |

The distribution factors are simply the ratios of $sk$ of the member to the sum of $sk$ values at the joint. The method is the same as that described in Chapter 6 except for the use of $s$- and $c$-functions. If the effects of stability are neglected (i.e. $s = 4$, $c = \frac{1}{2}$ for all three members) the moments in the last line are obtained. It is seen that the simple analysis can give rise to large errors when the axial load is large.

Convergence of the above method becomes increasingly slow as the critical load is approached and it is found expedient to use other methods. The same frame as that of FIGURE 8.8 is now analysed for a general value of $W$. In the following method the stiffnesses of joints $B$ and $E$ against equal and opposite moments are first evaluated. This enables the rotations of joints $B$ and $E$ to be found after they are released, i.e. after moments equal and opposite to the fixed end moments are applied.

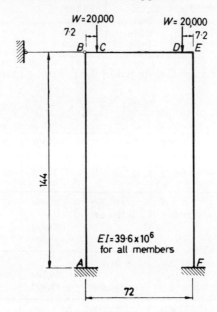

*Figure 8.8.* Portal frame prevented from failure by sidesway

Let
$$k_{BE} = \frac{EI_{BE}}{L_{BE}} \text{ and } k_{AB} = k_{EF} = \frac{EI_{AB}}{L_{AB}}$$

The fixed end moments when $B$ and $E$ are held against rotation are

$$M_{FBE} = -0.09\, WL_{BE} = -M_{EFB}$$

Give equal and opposite rotations $\theta$ to $B$ and $-\theta$ to $E$

TABLE 8.4

| Operation | AB | BA | BE | EB | EF | FE |
|---|---|---|---|---|---|---|
| $B$ rotated through $\theta$ | $sck_{AB}\theta$ | $sk_{AB}\theta$ | $4k_{BE}\theta$ | $2k_{BE}\theta$ | 0 | 0 |
| $C$ rotated through $-\theta$ | 0 | 0 | $-2k_{BE}\theta$ | $-4k_{BE}\theta$ | $-sk_{EF}\theta$ | $-sck_{EF}\theta$ |
| Total | $sck_{AB}\theta$ | $sk_{AB}\theta$ | $2k_{BE}\theta$ | $-2k_{BE}\theta$ | $-sk_{EF}\theta$ | $-sck_{EF}\theta$ |

i.e. moment required to rotate $B$ through angle $\theta$ is $(sk_{AB}\theta + 2k_{BE}\theta)$. Equating this to the moment required to hold $B$ against rotation

$$sk_{AB}\theta + 2k_{BE}\theta - 0.09\, WL_{BE} = 0$$

i.e.
$$\theta = \frac{0.09\ WL_{BE}}{sk_{AB} + 2k_{BE}}$$

## 8.3. ELASTIC BEHAVIOUR OF FRAMEWORKS 8.3.1

Since $k_{BE} = 2k_{AB}$, the values of moments in TABLE 8.5 are obtained as the applied loads $W$ increase.

TABLE 8.5

| $W$ | 0 | 20,000 | 40,000 | 50,000 | 54,200 |
|---|---|---|---|---|---|
| $\theta$ radian | 0 | 0·0742 | 0·251 | 0·797 | $\infty$ |
| $M_{BA} = -M_{BE}$ | 0 | 48,100 | $-$ 16,700 | $-$552,000 | $\infty$ |
| $M_{AB} = -M_{FE}$ | 0 | 51,200 | 258,000 | 1,140,000 | $\infty$ |

In FIGURE 8.9 values of the moments at $A$ and $B$ are plotted against the applied load $W$. Also shown are the results of the analysis which neglects stability effects. The discrepancy between the two analyses becomes increasingly large as $W$ increases.

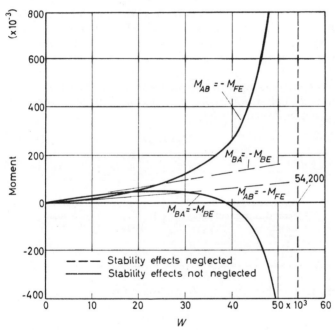

Figure 8.9. Variation of moments with load in a portal frame (FIGURE 8.8)

It is interesting to notice in FIGURE 8.9 that the moments $M_{BA}$ and $M_{BE}$ change sign after reaching maximum values and are zero at $W = 38,500$. This is the load at which the stanchion would buckle if joints $B$ and $E$ were pinned and the frame was restrained against lateral movement. It can be seen then that it is the rigidity at the joints $B$ and $E$ and the stiffness of the beam which prevent buckling of the stanchions until $W$ reaches 54,200. When this load is reached, the structure is elastically unstable since the stanchions are then so flexible that they cannot be supported by the beam against end rotations at $B$ and $E$. This load is the critical load of the frame.

## 8.3.2. Behaviour of an Ideal Braced Framework

Braced frameworks derive their rigidity from the power of their members to resist axial loads. The behaviour of braced frameworks which consist of members rigidly connected to one another can be studied using $s$- and $c$-functions. When loads are applied to a braced framework, the members are subjected to axial loads which cause them to deform axially. These deformations constitute a change in the geometry of the framework and the members must bend to accommodate this change because the joints are rigid. The stresses due to the axial loads are known as 'primary stresses' and the stresses due to the bending of the members are called 'secondary stresses'. Primary stresses and axial deformations are readily evaluated by using one of the methods of analysing pin-jointed trusses. Secondary stresses assumed great importance at the turn of the century possibly because the failure of the first Quebec Bridge in 1907 demonstrated the need for an accurate method of truss analysis. It was at first thought that secondary stresses induced collapse of a truss because the basis of design was that the stresses should be limited to a predetermined maximum. It will be shown that the moments which cause the secondary stresses also prevent the struts from failing at a lower load, i.e. their failure load if they were pin-jointed.

In FIGURE 8.10 a simple triangular frame is illustrated and demonstrates the method of analysis for secondary stresses. The properties of the members are given in TABLE 8.6.

TABLE 8.6

|  | AC & BC | AB |
|---|---|---|
| Area | 3·22 | 2·58 |
| Section modulus | 0·819 | 0·655 |
| Second moment of area | 0·415 | 0·461 |
| Length | 54·0 | 76·2 |
| $\dfrac{EI}{L}$ | 16,300 | 12,850 |

Values of $s$ and $c$ when $P = 4{,}550$ are found from TABLE 8.1.

TABLE 8.7

|  | AC & BC | AB |
|---|---|---|
| Axial load $F$ | − 3,215 | 2,280 |
| $P_E$ | 2,990 | 1,670 |
| $\dfrac{F}{P_E}$ | − 1·077 | 1·367 |
| $s$ | 5·255 | 1·748 |
| $c$ | 0·330 | 1·571 |

On loading the frame, apex $C$ is assumed to be fixed in space and $A$ and $B$ move to $A'$ and $B'$ respectively (FIGURE 8.10(b)) because of changes in member lengths. $A'$ and $B'$ are assumed to be in the positions which

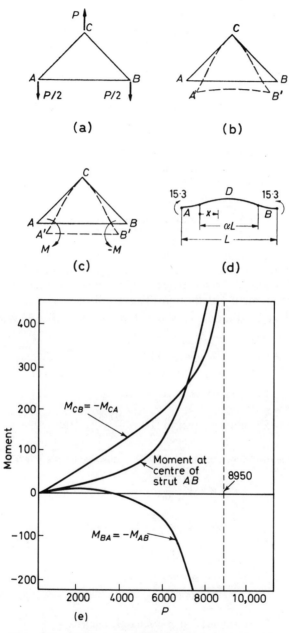

Figure 8.10. Moments in a single braced framework with rigid joints

would be occupied by the joints of the corresponding pin-jointed framework and their locations are found by use of the Williot diagram (see Chapter 3, Section 3.6.2).

$$\text{Extension of } AC \text{ and } CB = \frac{3{,}215 \times 54 \cdot 0}{3 \cdot 22 \times 2 \cdot 11 \times 10^6} = 0 \cdot 0254$$

$$\text{Compression of } AB = \frac{2{,}280 \times 76 \cdot 2}{2 \cdot 58 \times 2 \cdot 11 \times 10^6} = 0 \cdot 032$$

∴ lateral displacement of $A$ in the direction at right angles to $AC =$

$$\varDelta_A = 0 \cdot 16\sqrt{2} + 0 \cdot 0254 = 0 \cdot 048$$

To find the moments in the members at $A'$ and $B'$, use is made of moment distribution modified for stability effects. The analysis is started by applying external moments $M$ at $A'$ and $B'$ so that $A'B'$ is made straight (FIGURE 8.10(c)). Thus from equation 8.8

$$M = \left(\frac{s(1+c)k\varDelta_A}{L}\right)_{AC} = \frac{5 \cdot 255 \times 1 \cdot 330 \times 16{,}300 \times 0 \cdot 048}{54 \cdot 0} = 101$$

The moments at the ends of $A'B'$ are zero and, therefore, the externally applied moments are equal and opposite to the moments at the ends $A'$ and $B'$ of $A'C$ and $B'C$ respectively. The applied moments are liquidated by use of the moment distribution (TABLE 8.8).

Only 3 cycles of distribution are shown but the result after 8 cycles is given as

$$M_{AB} = -M_{BA} = -M_{BC} = 17 \cdot 6$$
$$-M_{CA} = M_{CB} = -140 \cdot 1$$
$$M_D = \text{Moment at centre of } AB = 67 \cdot 4\dagger$$

Primary stress in $AB = 882$

$$\text{Secondary stress in } AB = \frac{67 \cdot 4}{0 \cdot 655} = 103$$

---

† Consider the strut shown in Figure 8.10(d). Effective length $= \alpha L$

$$\alpha = \sqrt{\frac{1}{1 \cdot 367}}$$

$$M_x = A \sin \frac{\pi x}{\alpha L} \quad (A = \text{const.})$$

$$\text{when } x = \frac{L(1-\alpha)}{2} \quad M_x = 17 \cdot 65$$

$$A = 17 \cdot 65 \operatorname{cosec} \frac{\pi(L - L\alpha)}{2L\alpha}$$

Substitute in expression for $M_x$ and let $x = \frac{\alpha L}{2}$

i.e. $M_D = 17 \cdot 65 \operatorname{cosec} \frac{\pi}{2} (\sqrt{1 \cdot 367} - 1)$

$$= 17 \cdot 65 \times 3 \cdot 81 = 67 \cdot 4$$

8.4. CRITICAL LOADS OF FRAMEWORKS 8.4.1

Primary stress in $AC = 1,000$

$$\text{Secondary stress in } AC = \frac{140 \cdot 1}{0 \cdot 819} = 171$$

TABLE 8.8

|  | CA | AC | AB | BA | BC | CB |
|---|---|---|---|---|---|---|
| Distribution Factors |  | 0·793 | 0·207 | 0·207 | 0·793 |  |
| A and B held | − 101 | 101 | 0 | 0 | − 101 | − 101 |
| A and B released |  | − 80·3 | − 20·7 | 20·7 | 80·3 |  |
| Carry-over | − 26·4 |  | 32·8 | −32·8 |  | 26·4 |
| A and B released |  | − 26·0 | − 6·8 | 6·8 | 26·0 |  |
| Carry-over | − 8·5 |  |  |  |  | 8·5 |
| A and B released |  | − 8·4 | − 2·2 | 2·2 | 8·4 |  |
| Carry-over | − 2·8 |  | 3·5 | − 3·5 |  | 2·8 |

It is seen that in this framework the secondary stresses are about 17 per cent. of the primary stresses. If the base angle of the framework is decreased to about 30°, this figure may rise to about 40 per cent. It was values of this order which caused such concern to early designers. Instead of using moment distribution this frame could have been analysed by the general method used to analyse the portal frame (FIGURE 8.8). In FIGURE 8.10(e) the moments at joints in the frame have been plotted for values of P. These graphs show that changes in the lengths of the members induce moments which cause the strut AB to bend inwards. As the load P increases the strut has a tendency to buckle in this direction but the moments at its ends change their sign and restrain it. Collapse of the strut is prevented until P reaches 8,950 when it becomes so flexible that it cannot be supported by the moments from the ties. This load is then the critical load of the frame.

The moment distribution method just described is quite general and may be applied to more extensive braced frameworks.

## 8.4. METHODS OF EVALUATING THE CRITICAL LOADS OF FRAMEWORKS

The critical load of a framework is the load at which the stiffness of the framework is zero. In other words, it is the minimum load at which the framework offers no resistance to a small disturbance. In this Section methods of determining the critical loads of frameworks are described.

### 8.4.1. MOMENT DISTRIBUTION METHOD

Consider the pin-ended strut (FIGURE 8.11) which is rotated through equal angles $\theta$ at its ends. The simple moment distribution shown demonstrates that the moment required to rotate each end through angle $\theta$ is

$$M = sk(1 - c)\theta \quad \ldots \text{(8.10)}$$

## STABILITY OF STRUTS AND FRAMEWORKS

The stiffness of the strut to a moment disturbance at its ends is defined as $\dfrac{M}{\theta}$ and equation 8.10 shows that this is $sk(1 - c)$. Stability tables (TABLE 8.1) have been used to develop FIGURE 8.12 which shows that the stiffness decreases as the axial load $F$ increases, and that the stiffness is zero when $\dfrac{F}{P_E}$ is unity. The strut then has no rigidity and therefore collapse is imminent.

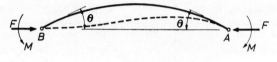

| Operation | $M_{BA}$ | $M_{AB}$ |
|---|---|---|
| B held fixed<br>A rotated through $\theta$ | $sck\,\theta$ ← | $sk\,\theta$ |
| A held at above position<br>B rotated through $-\theta$ | $-sk\,\theta$ | $-sck\,\theta$ → |
| Total | $-s(1-c)k\,\theta$ | $s(1-c)k\,\theta$ |

*Figure 8.11.* Determination of stiffness of pin-ended strut

This load is the critical load of the strut. The same principle is used for determining the critical loads of frameworks. The stiffness of the framework to a disturbing moment or force is determined for values of the applied

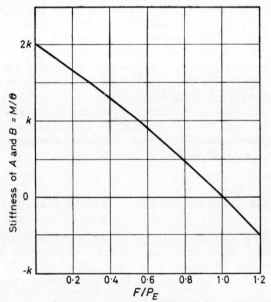

*Figure 8.12.* Variation of stiffness of a pin-ended strut with axial load

loads. As the loads increase it will be found that the stiffness of the frame decreases until it reaches zero when the critical load is attained. The stiffness of simple frameworks can often be expressed in terms of the stability

## 8.4. CRITICAL LOADS OF FRAMEWORKS

functions and solved by trial and graphical interpolation. Frameworks consisting of many members are more difficult to treat and it is more expedient to resort to approximate methods.

One example of a simple framework is that shown in FIGURE 8.10(a), the properties of whose members are given in TABLE 8.6. If small

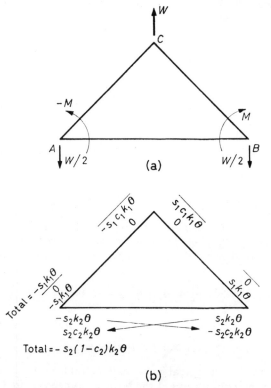

*Figure 8.13.* Determination of stiffness of a simple braced framework with rigid joints

disturbing moments $M$ are applied at $A$ and $B$ to rotate each joint through angle $\theta$ (FIGURE 8.13(a)), the stiffness of these two joints can be obtained. The simple moment distribution (FIGURE 8.13(b)) shows that

$$M = s_1 k_1 \theta + s_2 k_2 (1 - c_2) \theta$$

where the suffix 1 refers to members $AC$ and $BC$ and suffix 2 refers to member $AB$. The stiffness is zero when

$$s_2(1 - c_2) = -s_1 \frac{I_1}{I_2} \frac{L_2}{L_1}$$
$$= -1 \cdot 273 \, s_1 \qquad \ldots (8.11)$$

Equation 8.11 is solved by assuming values of $W$ and plotting the left- and right-hand sides of the equation against $W$. The point of intersection of the two curves gives the value of $W$ at the critical load. In this case $W_C = 8,950$ which is the value obtained in Section 8.3.2.

## STABILITY OF STRUTS AND FRAMEWORKS

The next example illustrates the method again but also shows that a framework may have a number of critical loads. Consider the portal frame shown in FIGURE 8.8. If the pinned member at $B$ which prevents sway of the frame is now removed, disturbing moments and forces can be applied in a number of ways. FIGURE 8.14(a) shows one method of applying the disturbance; $B$ and $E$ are given equal and opposite rotations $\theta$. In this case moment distribution (FIGURE 8.14(b)) shows that the stiffness of joints $B$ and $E$ is

$$\frac{M}{\theta} = s_1 k_1 + s_2 k_2 (1 - c_2)$$

Remembering that the axial load in $BE$ is small and can be neglected (i.e. $s_2 = 4$, $c_2 = 0.5$) the stiffness becomes zero when

$$s_1 = -2 \frac{I_2}{I_1} \frac{L_1}{L_2}$$

$$= -4$$

Stability tables show that this is so when $\frac{F_1}{P_{E_1}}$ is 2·87, i.e. $W_C = 2\cdot 87 \times 18,850$ = 54,200 which is the value obtained in Section 8.3.1.

Another way of applying the disturbances is shown in FIGURE 8.14(c). From the moment distribution (FIGURE 8.14(d)) the stiffness of joints $B$ and $E$ is

$$\frac{M}{\theta} = s_1 k_1 + s_2 k_2 (1 + c_2)$$

The stiffness is zero when $s_1 = -12$ and stability tables show that $\frac{F_1}{P_{E_1}}$ is then 3·44, i.e. the critical load is 64,900.

The critical load of the portal for a small lateral force $P$ can be determined by considering the equilibrium of the frame. When a load $P$ acts laterally at $B$, the frame will sway sideways through a small distance $\Delta$ at $B$ and joints $B$ and $E$ will each rotate through a small angle $\theta$. Thus there are two unknowns, viz., $\theta$ and $\Delta$, but it is possible to write two equilibrium equations. The first equilibrium equation is based on the condition that joints $B$ and $E$ are in equilibrium and the second from the horizontal equilibrium of the beam $BE$. The latter equation is developed from an extension of the equilibrium condition that in a stanchion such as $AB$ the shear in the stanchion at $B$ is equal to the force it exerts on the beam $BE$. This force, $Q$, is given by the expression

$$\frac{M_{AB} + M_{BA}}{m_1 L_1}$$

The full analysis is demonstrated in TABLE 8.9, in which the column headed $QL_1$ takes account of the sway or horizontal equilibrium of beam $BE$. Equation 8.9 explains the way in which the term $\frac{4(1 + c_1) s_1 k_1 \Delta}{m_1 L_1}$ arises.

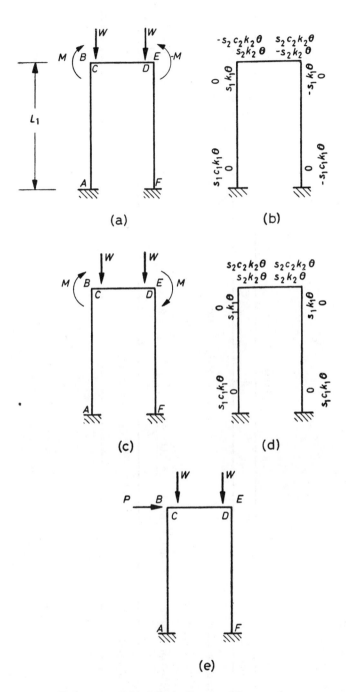

Figure 8.14. Critical loads of a portal framework

TABLE 8.9

| Operation | A | B | E | F | $QL_1$ |
|---|---|---|---|---|---|
| Rotate $B$ through $\theta$ | $s_1c_1k_1\theta$ | $4k_2\theta$ | 0 | 0 | $-s_1(1+c_1)k_1\theta$ |
| Rotate $E$ through $\theta$ | 0 | $2k_2\theta$ | $4k_2\theta$ | $s_1c_1k_1\theta$ | $-s_1(1+c_1)k_1\theta$ |
| Sway through distance $\Delta$ | $-\dfrac{s_1(1+c_1)k_1\Delta}{L_1}$ | $-\dfrac{s_1(1+c_1)k_1\Delta}{L_1}$ | $-\dfrac{s_1(1+c_1)k_1\Delta}{L_1}$ | $-\dfrac{s_1(1+c_1)k_1\Delta}{L_1}$ | $\dfrac{4(1+c_1)s_1k_1\Delta}{m_1L_1}$ |
| Totals at Joint $B$ | | $-\dfrac{s_1(1+c_1)k_1\Delta}{L_1} + s_1k_1\theta + 6k_2\theta$ | | | |

## 8.4. CRITICAL LOADS OF FRAMEWORKS

Since joints $B$ and $E$ are in equilibrium

$$(s_1 k_1 + 6k_2)\theta - s_1(1 + c_1)k\frac{\Delta}{L_1} = 0$$

For sway equilibrium

$$PL_1 = -2s_1(1 + c_1)k_1\theta + \frac{4(1 + c_1)\,s_1 k_1 \Delta}{m_1 L_1}$$

On eliminating $\theta$ from these two equations

$$PL_1 = \frac{2\{s_1(1+c_1)k_1\}^2 \Delta}{(s_1 k_1 + 6k_2)\,L_1} + \frac{4s_1(1+c_1)k_2 \Delta}{m_1 L_1}$$

The stiffness $\dfrac{P}{\Delta}$ is zero when

$$m_1 s_1 (1 + c_1) = 2\left(s_1 + \frac{6k_2}{k_1}\right)$$

i.e.

$$\frac{m_1 s_1 (1 + c_1)}{2} - s_1 = 6\frac{k_2}{k_1}$$

$$= 12$$

Stability tables show that $\dfrac{F_1}{P_{E_1}}$ is then 0·855, i.e. the critical load is 16,100.

Just as a beam, string or any vibrating body may have many modes of vibration, so a structure may have many modes of failure. Each mode of failure has a corresponding critical load. Since a designer is interested in the lowest critical load of his structure, he has to be careful to ensure that the correct disturbance is applied. The example above demonstrates that the structure may have several critical loads each of which corresponds to a particular mode of failure. If lateral movement of $B$ and $E$ is prevented in some way, say by light cross ties, the critical load is raised from 16,100 to 54,200.

The stiffness of frameworks which consist of several members cannot, in general, be expressed as a simple function. Lundquist[3] developed an early method of determining the critical load of a framework using stability tables produced by James[4]. In this method a unit disturbing moment is applied to the least rigid joint of the loaded framework. Each joint in turn is released in a moment-distribution process until the original moment is absorbed into the framework. If the distribution converges the framework is stable, but should it not converge the critical load of the structure has been exceeded. Near the critical load the method becomes indefinite and laborious due to poor convergence.

Allen[5] overcame this difficulty for braced frameworks by replacing in turn each triangle of members of the framework by an equivalent member until one member remained. A simple stability criterion was then applied to determine whether the applied loads exceeded the critical load. The method gives an accurate value of the critical load but it becomes laborious if the framework has a large number of members.

### 8.4.2. Bolton's Approximate Method

An approximate method due to Bolton[6] enables a rapid evaluation of the critical load of a braced framework to be made. The method is based upon the idea of simplifying the structure by assuming that the members remote from the least rigid joint have little effect upon the behaviour of the framework. The remote joints are assumed to be fixed. The least rigid or critical joint is determined by selecting the member whose $\dfrac{F}{P_E}$ value is greatest and the critical joint is taken as the one at the end of this member with the lesser value of stiffness ($= \Sigma sk$). The structure is simplified by assuming that the

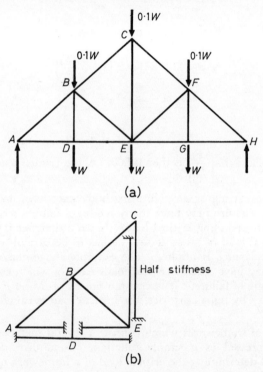

*Figure 8.15.* Actual and simplified framework for determination of critical load

remote ends of the members radiating from the joints adjacent to the critical joint are fixed. A typical example is shown in FIGURE 8.15(a) and FIGURE 8.15(b), the former being a diagram of an actual framework and the latter that of the corresponding simplified framework. In this case it is only necessary to treat one half of the framework because of symmetry. $B$ and $F$ are the critical joints and a joint on the centreline such as $C$ takes half of its moment from the left-hand half of the truss and the other half of its moment from the right. Thus when treating only one half of the framework it is necessary to halve the total stiffness at $C$ and $E$. The members of the simplified framework are allowed to carry the same axial loads as the actual framework. The

## 8.4. CRITICAL LOADS OF FRAMEWORKS 8.4.2

disturbance to be applied to give the lowest critical load is a moment at the critical joint. A sway type failure does not in general give the lowest critical load for this type of structure. It is now found that the stiffness of the framework to a disturbing moment applied at the critical joint is determined in one cycle of moment distribution. On applying a rotation $\theta$ to joint $B$, which is assumed to be the critical joint, it is seen that the moment carried back to $B$ after releasing the surrounding joints is $-\theta_B \sum \frac{(sck)^2{}_{AB}}{T_A}$ where the summation is made for all members such as $AB$ radiating from the critical joint and $T_A$ is the total stiffness of joint $A$, i.e. $T_A = \sum_{\text{joint } A} sk$. Thus the total moment at $B$ is $\left[ T_B \theta_B - \theta_B \sum \frac{(sck)^2}{T_A} \right]$ and the stiffness of the critical joint is $T_B - \sum \frac{(sck)^2}{T_A}$ where $T_B = \sum_{\text{joint } B} sk$. At the critical load of the simplified framework the stiffness is zero and the method is applied in the following steps:

1. Choose a value of the load parameter $W$.

2. Determine the ratio of $\dfrac{F}{P_E}$ for the members and select the one whose ratio is greatest.

3. Evaluate $\Sigma sk$ at each joint of the two joints at the ends of this member and select the joint which has the lesser value of $\Sigma sk$. This joint is the critical joint, say joint $B$. Draw the simplified framework.

4. Evaluate $\Sigma sk$ for each joint of the simplified framework and evaluate
$$T_B - \sum \frac{(sck)^2}{T}$$

5. Repeat the above steps for different values of $W$ until this expression is reduced to zero.

TABLE 8.10

| Member | L | $P_E$ | Axial Load | |
|---|---|---|---|---|
| AB | 129·24 | 87,100 | strut | 4·443 $W$ |
| BC | 129·24 | 46,600 | strut | 2·962 $W$ |
| AD | 120 | 38,000 | tie | 4·125 $W$ |
| DE | 120 | 38,000 | tie | 4·125 $W$ |
| BD | 48 | 3,360 | tie | 1·000 $W$ |
| BE | 129·24 | 17,900 | strut | 1·481 $W$ |
| CE | 96 | 7,830 | tie | 2·100 $W$ |

The method of computation of the critical load of the framework shown in FIGURE 8.15(a) using Bolton's approximate method[6] is carried out in the following way. The properties of the members are given in TABLE 8.10. This is the frame analysed by Allen's[5] exact method by which the critical load was found to be $W = 22{,}200$.

It is only necessary to treat the left-hand half of the framework because of symmetry.

From the values of $\frac{F}{P_E}$ (TABLE 8.1) it is seen that $BE$ and $EF$ are the members with the highest $\frac{F}{P_E}$ ratio. Of the joints $B$, $F$ and $E$, the values of the total stiffness show that joints $B$ and $F$ are the critical ones to which the disturbing moments are applied in the same sense (i.e. say both clockwise moments). The simplified framework is shown in FIGURE 8.15(b).

TABLE 8.11

| Member | $k = \frac{EI}{L}$ | $W = 22,400$ | | | $W = 21,300$ | | |
|---|---|---|---|---|---|---|---|
| | | $\frac{F}{P_E}$ | $(sk)$ | $\frac{(sck)^2}{T_{FAR}}$ | $\frac{F}{P_E}$ | $(sk)$ | $\frac{(sck)^2}{T_{FAR}}$ |
| $AB$ | 1,137,000 | 1·14 | 2,500,000 | 1,540,000 | 1·09 | 2,620,000 | 1,485,000 |
| $BC$ | 609,000 | 1·42 | 993,000 | 2,372,000 | 1·35 | 1,082,000 | 1,930,000 |
| $BD$ | 16,400 | −6·62 | 154,000 | 1,000 | −6·29 | 151,000 | 1,000 |
| $BE$ | 234,000 | 1·86 | 131,000 | 168,000 | 1·76 | 193,000 | 158,000 |
| $AD$ | 461,000 | −2·43 | 3,010,000 | | −2·31 | 2,960,000 | |
| $ED$ | 461,000 | −2·43 | 3,010,000 | | −2·31 | 2,960,000 | |
| $EC$ | 76,100 | −6·00 | 689,000 | | −5·70 | 674,000 | |
| $CF$ | 609,000 | 1·42 | 993,000 | | 1·35 | 1,082,000 | |
| | | | Total = | 4,081,000 | | Total = | 3,574,000 |
| | | $T_A = 5,510,000$ | | 4,081,000 | $T_A = 5,580,000$ | | 3,574,000 |
| | | $T_B = 3,778,000$ | | 3,778,000 | $T_B = 4,046,000$ | | 4,046,000 |
| | | $\tfrac{1}{2}T_C = 1,338,000$ | | = 1·08 | $\tfrac{1}{2}T_C = 1,419,000$ | | = 0·884 |
| | | $T_D = 6,174,000$ | | > 1 | $T_D = 6,071,000$ | | < 1 |
| | | $\tfrac{1}{2}T_E = 3,486,000$ | | ∴ Unstable | $\tfrac{1}{2}T_E = 3,490,000$ | | ∴ Stable |

In this example the simplified framework included all of the members of the actual frame and the approximate value of the critical load was close to the true value. However, it is not always possible to obtain such good agreement for a number of reasons. In frameworks which lack symmetry two disturbing moments cannot be applied and then the simplified framework departs more from the actual frame. The framework could have been simplified in a number of ways other than that used by Bolton. Close agreement to the critical load of the actual framework is obtained only if the behaviour of the simplified framework approximates to that of the actual framework. A number of ways of simplifying the actual framework of FIGURE 8.15(a) are demonstrated in FIGURE 8.16. In FIGURE 8.16(a) and (b) all joints surrounding the critical joint are considered to be fixed and pinned respectively. In FIGURE 8.16(c) and (d) the same procedure is applied to joint $C$ which is the least rigid joint of the frame. In FIGURE 8.16(e) and (f) a disturbing moment is applied at only one critical joint, $B$, and the remote joints are considered to be fixed and pinned respectively. The former would be the normal way of using Bolton's method in a frame which lacked symmetry. It will be seen from the values of the critical loads of these simplified frameworks that the simplified framework obtained by assuming the joints remote from the critical joint to be fixed is more rigid than the actual frame; an upper bound to the critical load is thus obtained. Conversely, by assuming that the joints remote from the critical joint are pinned a lower bound to the critical load is found. Results given here and experience show that in those frames which lack symmetry the average of the upper and lower bounds is a good approximation to the true value.

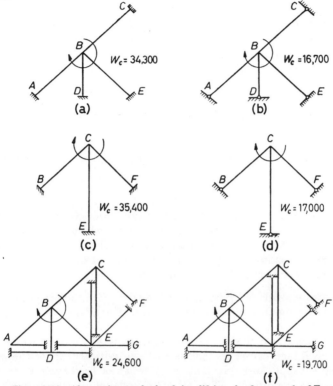

Figure 8.16. Alternative methods of simplifying the framework of FIGURE 8.15(a)

Bolton's method enables a rapid estimate to be made of the critical load of a framework. So far the theory has been concerned with buckling of the members in the plane of the frame. However, the members may have a tendency to buckle out of this plane if they are deep in the plane of the frame but narrow normal to it. A method similar to that of Bolton has been used to determine the approximate load at which lateral buckling of the members takes place in an elastic frame[7]. Merchant[8] has also derived analytical expressions which allow the critical loads of tall building frames to be evaluated.

### 8.4.3. EXPERIMENTAL METHODS

The critical loads of models can be obtained experimentally by using a Southwell Plot. Southwell[9] examined the equation of a hyperbola which passes through the origin and is asymptotic to $y = u$. This hyperbola can be expressed as

$$xy - ux + vy = 0$$

where $x$ and $y$ are variables and $u$ and $v$ are constants. On dividing by $y$ the equation becomes

$$x - u\frac{x}{y} + v = 0 \qquad \ldots (8.12)$$

This is a linear equation in variables $x$ and $\frac{x}{y}$ of slope $u$. Southwell applied equation 8.12 to determine the load at which a pin-ended strut fails. A pin-ended strut which is not quite straight before loading will bend as the axial load is applied. If the initial shape can be expressed as $y_0 = a_0 \sin\frac{\pi x}{L}$ then the deflexion at the centre under load $P$ has been shown to be (see Section 1)

$$y_c = \frac{a_0}{1 - \dfrac{F}{P_E}} - a_0 \qquad \ldots (8.13)$$

This is the equation of a hyperbola which passes through the origin and is asymptotic to $F = P_E$. The experimental value of the Euler load of a strut can be found by plotting experimental values of the deflexion, $y_c$, at the centre of the strut against the ratio of $y_c$ to the axial load. The slope of the graph enables the Euler load to be evaluated.

An elastic framework will behave in a similar way to the pin-ended strut, the graph of load against deflexion being asymptotic to the critical load. It is found that the most accurate values of the critical load are obtained by measuring the central deflexion, relative to its ends, of the strut which fails. The great advantage of the method is that the critical load of the model can be found without causing its failure and it can be used again to examine different loading conditions. The mode of collapse of structures is also conveniently studied by the use of models.

It is well known that the period of oscillation of a body undergoing simple harmonic motion is inversely proportional to the square root of the restoring force per unit distance. Chandler[10] has used this principle to determine the critical loads of structural models. As loads on the models are increased, the period of oscillation increases and by extrapolation it is possible to find the critical load when the period of oscillation is theoretically infinite since the stiffness of the structure is then zero. If the dimensions of a prototype framework are known, then its critical load can be determined experimentally by constructing and testing a model. In what follows the suffix $m$ will refer to the model, $p$ to the prototype, 1 to member 1, 2 to member 2, and so on.

$W_c$ is the critical load of the framework. The $\frac{L}{r}$ values of the members of the model can be chosen to be high (say about 200) so that plastic conditions do not affect its behaviour. The dimensions of the members of the model are calculated so that

$$G = \frac{(P_E)_{m1}}{(P_E)_{p1}} = \frac{(P_E)_{m2}}{(P_E)_{p2}} = \frac{(P_E)_{m3}}{(P_E)_{p3}} = \text{ etc. for all members}$$

where $G$ is some convenient constant which will be quite small and will be chosen to give a suitable size of model. After constructing the model its critical load is found by experiment using either the Southwell Plot or Chandler's oscillation method. The critical load of the prototype can then be found since at any load

$$\frac{(F)_{m1}}{(P_E)_{m1}} = \frac{(F)_{p1}}{(P_E)_{p1}} \; ; \; \frac{(F)_{m2}}{(P_E)_{m2}} = \frac{(F)_{p2}}{(P_E)_{p2}} \; ; \text{ etc. for all members}$$

and, in particular, these equations hold at the critical load. Hence

$$(W_c)_p = \frac{1}{G}(W_c)_m$$

This experimental method can be applied to design by building a model of the prototype. If it is found that the critical load is too small, stronger members can be substituted for those which are found to be very flexible and the procedure repeated. The advantages are that tedious computation is eliminated, the designer can see the actual behaviour of the elastic structure and the model can be subjected to a variety of loading conditions with little extra work.

## 8.5. ACTUAL BEHAVIOUR OF FRAMEWORKS WHEN THE AXIAL FORCES IN THEIR MEMBERS ARE LARGE

So far in this Chapter it has been assumed that the yield stress of the steel has not been exceeded. When some of the steel begins to yield, the behaviour of the framework differs from that given by the elastic analysis in the previous Sections. It has also been assumed that there was no initial imperfection in the framework. It is not possible to construct an ideal frame in practice because the members are always bent to a greater or lesser degree and their centre lines and the load lines never intersect at a common point at a given joint. Also the steel is never homogeneous and isotropic and it usually contains initial stresses. In this Section it will be shown that the elastic analysis already dealt with can be modified to account for some of these effects. It has not been found possible, up to the present, to take account of the initial stresses and material properties which differ from those of an ideal material.

Attempts[11] have been made to analyse model test results using a simplified stress-strain relationship (FIGURE 8.17). The analysis of stanchion tests has been developed by systematically tracing out the plastic zones which formed in the stanchions as the axial load was increased. As each new plastic zone appeared, a new set of equations was derived and solved. Good agreement between experimental and theoretical results were obtained but the analysis had the disadvantage that it was long. It did not seem practicable to extend the method to frameworks consisting of many members.

Another way of treating the problem is to divide the stress-strain curve of FIGURE 8.17 into two approximately equivalent parts, FIGURES 8.18(a) and (b), and to study the behaviour of the framework as if it obeyed each part

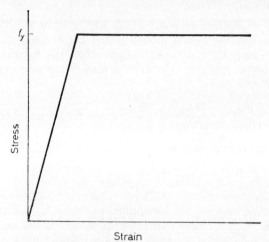

*Figure 8.17.* Simplified stress-strain relationship of mild steel

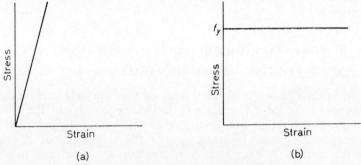

*Figure 8.18.* (a) Purely elastic stress-strain relationships; (b) Purely plastic stress-strain relationships

separately. If the material of the framework obeys the stress-strain curve of FIGURE 8.18(a), a purely elastic analysis will apply, while a purely plastic analysis will apply if the material obeys the stress-strain curve of FIGURE 8.18(b).

### 8.5.1. ELASTIC ANALYSIS

Before considering the behaviour of frameworks, the elastic behaviour of a pin-ended strut which has initial imperfections is studied. The shape of an unloaded strut with pin ends can be expressed in the form of a Fourier Series

$$y_0 = \sum_{n=1}^{\infty} a_n \sin \frac{n\pi x}{L} \quad \ldots (8.14)$$

## 8.5. BEHAVIOUR OF FRAMEWORKS WHEN FORCES ARE LARGE 8.5.1

which simply means that its shape can be made up of a number of sine curves added together. If the load is applied with eccentricity $\epsilon$ then it is found that as the load on the strut is increased the deflexions at the centre of the strut depend almost entirely upon $a_1$ and $\epsilon$ and are given very closely by

$$y_c = \frac{\epsilon + a_1}{1 - \dfrac{F}{P_E}} \qquad \ldots(8.15)$$

Thus the graph of $F$ against the deflexion at the centre of the strut approximates to a hyperbola which has asymptotes at $F = P_E$ and $y_c = 0$ and which passes through the point $y_c = (\epsilon + a_1)$, $F = 0$.

The deflexions at the centre of the strut which fails in a given framework will depend not only upon the dimensions of the framework and its members but also upon their shape when the frame is unloaded. Attempts to determine the exact deflexions are tedious and complicated[12] and it is found more expedient to resort to an approximate method.

In the case of a framework the load-deflexion curve is asymptotic to the critical load and the relation between deflexion at the centre of the strut which fails and applied load $W$ is given by

$$y_c = \frac{a}{1 - \dfrac{W}{W_c}} \qquad \ldots(8.16)$$

where $a$ is a measure of the imperfections and effects due to axial shortening of the members, i.e. secondary stresses. The parameter $a$ is the sum of three terms[14]. The first term is the initial deflexion at the centre of the strut which fails with respect to the straight line joining its ends. It is the first term of the Fourier Series (equation 8.14) of all practical struts. At the joints at the ends of the strut the eccentricities of the members give rise to moments which can be distributed to the members in proportion to their stiffnesses. Thus the moment taken by the strut $AB$ at end $A$ is $\dfrac{k_{AB}}{T_A} \Sigma F\epsilon$ where the summation of the moments $F\epsilon$ is taken for all members meeting at $A$. The equivalent eccentricity at the end of $AB$ is therefore

$$\epsilon_A = \frac{k_{AB}}{F_{AB} T_A} \Sigma F\epsilon \qquad \ldots(8.17)$$

A similar value $\epsilon_B$ can be obtained at joint $B$ and the second term of the parameter $a$ is $\dfrac{\epsilon_A + \epsilon_B}{2}$. It is also found that in an ideal framework the deflexions at the centre of the strut due to changes in axial length of the members are given by the third term $\dfrac{\gamma W}{1 - \dfrac{W}{W_C}}$ where $\gamma$ is the deflexion per unit applied load when the axial loads in the members are small ($s$ is then 4 and $c$ is 0·5 for all members). $\gamma$ is determined from a moment distribution analysis similar to that of TABLE 8.8 but with these values of $s$ and $c$.

## STABILITY OF STRUTS AND FRAMEWORKS

After $a$ and $W_c$ are evaluated the elastic behaviour of the framework can be plotted in the form of a graph of load $W$ against the deflexion at the centre of the strut by using equation 8.16.

### 8.5.2. Plastic Analysis

To illustrate the plastic analysis of a structure consider a pin-ended strut of rectangular cross-section $b \times d$ ($b > d$). A plastic hinge will form in the

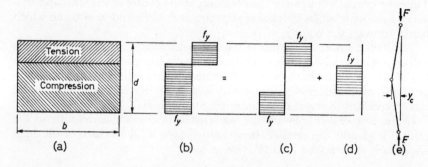

*Figure 8.19.* (a) Cross-section of strut at plastic hinge; (b) Stress distribution; (c) Stress due to bending moment $M'_p$; (d) Stress due to axial load $F$; (e) Pin-ended strut with plastic hinge at centre

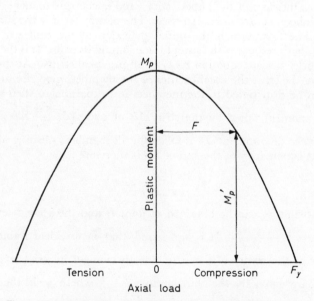

*Figure 8.20.* Reduction of plastic moment with axial load for a member of rectangular cross-section

strut when the stress distribution of FIGURE 8.19(b) exists. By considering the geometry of these stress distributions it can be shown that the plastic

## 8.5. BEHAVIOUR OF FRAMEWORKS WHEN FORCES ARE LARGE  8.5.2

moment $M_P'$ when an axial force $F$ is acting on the section is given by

$$M_P' = M_P \left(1 - \frac{F^2}{F_y^2}\right) \qquad \ldots (8.18)$$

where $M_P$ denotes the fully plastic moment when no axial force acts $\left(= f_y \dfrac{bd^2}{4}\right)$ and $F_y$ denotes the plastic load when no moments acts $(= f_y bd)$. The relation between $M_P'$ and $F$ is graphed in FIGURE 8.20. When the

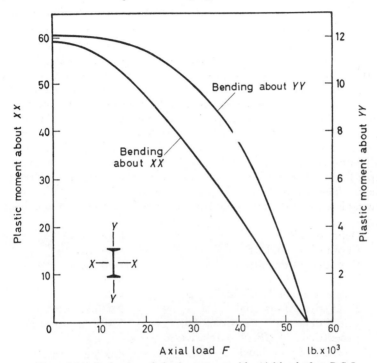

Figure 8.21. Reduction of plastic moment with axial load of an R.S.J.

plastic hinge has formed at the centre of the strut the equilibrium condition gives the relation between the load $F$, $M_P'$ and the deflexion at the centre (FIGURE 8.19(e)).

$$M_P' = F y_c \qquad \ldots (8.19)$$

From 8.18 and 8.19

$$\frac{F^2 d}{4 F_y} + y_c F - f_y \frac{bd^2}{4} = 0$$

Whence

$$F = f_y bd \left[\sqrt{\frac{4 y_c^2}{d^2} + 1} - \frac{2 y_c}{d}\right] \qquad \ldots (8.20)$$

This relation gives the condition of the strut once the plastic hinge has formed.

The plastic analysis of a framework is similar to that of the pin-ended strut. The reduced plastic moment $M'_p$ cannot be expressed as a simple function of the axial load in the member if the member is a standard structural section but a graph similar to FIGURE 8.20 may be obtained by an analysis of the shape of the section. FIGURE 8.21 shows the corresponding curves for an $I$-section when bending takes place about the major and minor axes. The equilibrium condition of the framework is obtained by considering the collapse mechanism of the structure.

### 8.5.3. COMBINED ELASTIC AND PLASTIC BEHAVIOUR OF THE FRAMEWORK

The elastic and plastic load-deflexion graphs may now be superimposed on one another. During the early part of the loading programme the load-deflexion graph of the frame will follow the curve given by the elastic analysis. Once the yield stress of the steel has been reached at some point then the experimental points are found to depart from the elastic curve. This departure is not, however, appreciable until one of the plastic zones has penetrated a considerable depth into the member. As the plastic hinges form, it is found that the applied load reaches a maximum after which it must be reduced to maintain equilibrium. Further increase in the deflexion shows that the experimental points begin to approach closer to the plastic collapse line. The following is an analysis and test results of a triangular frame[13]. The shape of the frame is shown in FIGURE 8.10(a) and the following is a list of the measured properties of the members of the frame.

| | |
|---|---|
| Depth of $AB$ | = 1·27 cm |
| Depth of $AC$ and $CB$ | = 1·90 cm |
| Width of all members | = 1·90 cm |
| Length of $AB$ | = 53·3 cm |
| Base angles $CAB$ and $CBA = \beta$ | = 48° |
| Yield stress of material | = 3,940 kgf/cm$^2$ |
| Initial deflexion at centre of $AB$ | = 0·109 cm inwards |
| Eccentricity of $AB$ at $A$ relative to centre of joint block | = 0·008 cm outwards |
| Eccentricity of $AC$ at $A$ relative to centre of joint block | = 0·005 cm inwards |
| Eccentricity of $AB$ at $B$ relative to centre of joint block | = 0·033 cm outwards |
| Eccentricity of $BC$ at $B$ relative to centre of joint block | = 0·025 cm inwards |

*Elastic Behaviour*

† Critical load (see equation 8.11) = 20,300 kgf
Determination of $a$ (see equation 8.14 et seq.):
  First term = + 0·109 cm
  Second term. Put unit load in $AB$

$$\epsilon_A = \left(\frac{I_{AB}}{I_{AB} + I_{AC}}\right) \cdot \frac{\Sigma F\epsilon}{F_{AB}}$$

---

† Allowance has been made for the size of the joints in calculating $W_c$.

## 8.5. BEHAVIOUR OF FRAMEWORKS WHEN FORCES ARE LARGE  8.5.3

$$= 0.228 \times \left(\frac{-0.008 + 1.5 \times 0.005}{1}\right)$$

$$= 0$$

$$\epsilon_B = 0.228 \times \left(\frac{-0.033 + 1.5 \times 0.025}{1}\right)$$

$$= 0$$

i.e. second term $= 0$.

The third term is found as indicated at the foot of p. 413 and for this framework $= 1.280 \times 10^{-6}$ cm

Hence $a = (0.109 + 1.280 \, W \times 10^{-6})$ cm and the deflexion at the centre

of the strut $AB = \dfrac{0.109 + 1.280 W \times 10^{-6}}{1 - \dfrac{W}{20{,}300}}$

This curve has been plotted in FIGURE 8.22.

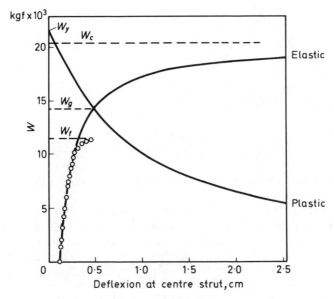

*Figure 8.22.* Theoretical and experimental load-deflexion curves of a simple braced framework with rigid joints

*Plastic Analysis*

The collapse mechanism of the frame is shown in FIGURE 8.23(a). In this frame $W = 2 \tan \beta \times$ axial load in member $AB$, and from the equilibrium of $AB$ (FIGURE 8.23 (b))

$$F_{AB}\, y_c = 2 M'_P$$

On eliminating $M_P'$ from this equation by use of equation 8.18, it is seen that

$$F_{AB} = f_y bd \left[ \sqrt{\left(\frac{y_c}{d}\right)^2 + 1} - \frac{y_c}{d} \right]$$

whence $W = 2 f_y bd \tan \beta \left[ \sqrt{\left(\frac{y_c}{d}\right)^2 + 1} - \frac{y_c}{d} \right]$

$$= 2 \times 3{,}940 \times 1\cdot 27 \times 1\cdot 90 \times 1\cdot 117 \left[ \sqrt{\left(\frac{y_c}{1\cdot 27}\right)^2 + 1} - \frac{y_c}{1\cdot 27} \right]$$

$$= 21{,}200 [\sqrt{0\cdot 620 y_c^2 + 1} - 0\cdot 788 y_c]$$

This equation has been plotted in FIGURE 8.22 where the experimental points are compared with this curve and the elastic curve.

Several simple frameworks have been analysed and it appears that the method can be used to determine the collapse loads of frameworks. It is

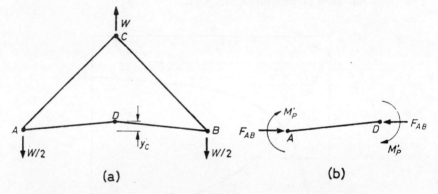

*Figure 8.23.* Collapse mechanism of a simple braced framework with rigid joints

found that the failure load $W_f$ of the framework is approximately equal to 80 per cent. of the load $W_g$ at the point of intersection of the elastic and plastic lines. This is a useful rule since $W_g$ can be obtained theoretically. For frameworks which have short stocky members and those which have long slender members this ratio can be raised. In general, the failure loads are conservative for structures which have members of rectangular cross-section.

The third term of the parameter $a$ in the above analysis has only a small effect upon the behaviour of the framework treated in the above example. Its calculation can be tedious for a frame consisting of many members. It is found that its inclusion has little effect upon the value of the collapse load.

## 8.6. DISCUSSION AND CONCLUSION

A general method has been described for studying the behaviour to collapse of frameworks in which the members carry large axial forces. The analysis of a framework is separated into a purely elastic analysis and a

## 8.6. DISCUSSION AND CONCLUSION

purely plastic analysis which are superimposed on one another in graphical form.

In the elastic analysis account is taken of the geometry of the framework and of stability effects in its members in calculating the critical load. Analytical, approximate and experimental methods are available for determining the critical load of the framework. It was shown that in the case of the analytical methods care must be taken to ensure that the correct (i.e. lowest) critical load is determined. In the case of a simple portal frame the lowest critical load is the one associated with sway failure. It is interesting to notice that if light cross ties are inserted to prevent this mode of failure, the lowest critical load of the framework is raised to a considerable extent. Initial imperfections in the framework are taken into account in drawing the elastic load-deflexion curve. The effects of axial shortening can also be included but their omission in practical analysis is not serious when the failure load of the structure is being evaluated.

The plastic collapse load-deflexion graph is determined by analysing the collapse mechanism which corresponds to the mode of collapse associated with the lowest critical load. Account is taken of the reduction in plastic moment of the members with axial load and the equilibrium condition of the mechanism.

It is seen that experimental load-deflexion results lie very close to the elastic curve when the deflexions and applied loads on the framework are small and approach the plastic curve as the deflexions increase. Experimental evidence available to date indicates that the collapse load of a framework is approximately 80 per cent. of the load at the intersection point of the theoretical curves ($W_g$, FIGURE 8.22). The collapse load applied to a framework should be divided by a load factor to obtain the working load.

The failure load of a framework depends to some extent upon initial imperfections but not to any marked degree. If the imperfections of the triangular frame analysed (see Example, SECTION 8.5.3.) had been doubled, the failure load would have been reduced by 11 per cent. It thus appears that the inclusion of elaborate secondary stress calculations in the analysis of braced frameworks for collapse load is of academic interest only. The failure load of a framework depends upon its critical load and it has been shown that the critical load of a framework can be raised considerably by use of ties which prevent the lower modes of failure. This method is not, of course, always practicable or desirable where an uninterrupted span is required.

In designing a framework special attention may have to be paid to some points. Once a plastic hinge has been formed in a strut it has no rigidity at that point, i.e. $\dfrac{dM}{d\theta} = 0$. This means that the critical load is immediately reduced when a plastic hinge is formed. Thus if a plastic hinge is formed in an elastically restrained strut, rapid changes of curvature occur in its vicinity and the effective length of the strut increases rapidly. This results in a decrease in the critical load or a 'deterioration in stability' and, in some cases, in sudden collapse[15].

The lowest critical load of a given framework may be the one which allows twisting of the member out of the plane of the frame. As a result the collapse mode associated with the lowest critical load may be that in which the members fail by torsional instability or lateral instability. A designer must provide adequate restraint against these modes of buckling.

## REFERENCES

[1] LIVESLEY, R. K. and CHANDLER, D. B. *Stability Functions for Structural Frameworks*. Manchester University Press

[2] BERRY, A. 'The Calculation of Stresses in Aeroplane Wing Spars'. *Trans. Royal Aero. Soc. G.B.* No. 1. 1916

[3] LUNDQUIST, E. 'Stability of Structural Members under Axial Load'. *N.A.C.A.* T.N. 617. 1937

[4] JAMES, B. W. 'Principal effects of Axial Loads on Moment-distribution Analysis of Rigid Structures'. *Tech. Note No.* 543 *N.A.C.A.* 1935

[5] ALLEN, H. G. 'The Estimation of the Critical Load of a Braced Framework'. *Proc. Royal Soc. Series A.* Vol. 231. 1955

[6] BOLTON, A. 'A Quick Approximation to the Critical Load of Rigidly Jointed Trusses'. *The Structural Engineer.* March, 1955

[7] MURRAY, N. W. 'A Method of Determining an Approximate Value of the Critical Loads at which Lateral Buckling occurs in Rigidly Jointed Trusses'. *Proc. Inst. Civ. Engrs.* Vol. 7. 1957

[8] MERCHANT, W. 'Critical Loads of Tall Building Frames'. *The Structural Engineer.* March, 1955

[9] SOUTHWELL, R. V. 'On the Analysis of Experimental Observations in Problems of Elastic Stability'. *Proc. Royal Soc. Series A.* Vol. 135

[10] CHANDLER, D. B. *The Prediction of Critical Loads of Elastic Structures.* Ph.D. Thesis, University of Manchester, 1955

[11] RODERICK, J. W. and HORNE, M. R. *The Behaviour of a Ductile Stanchion Length when Loaded to Collapse.* Brit. Welding Research Assn. Report July, 1948

[12] MURRAY, N. W. 'Behaviour in the Elastic Range of Triangular Frameworks with Rigid Joints'. *Civil Engineering and Public Works Review.* Vol. 50. 1955

[13] MURRAY, N. W. 'The Determination of the Collapse Loads of Rigidly Jointed Frameworks with Members in which the Axial Forces are Large'. *Proc. Inst. Civ. Engrs.* Pt. III. Vol. 5. April, 1956

[14] MERCHANT, W. and RASHID, C. A. Correspondence on Reference (13)

[15] WOOD, R. H. 'The Stability of Tall Buildings'. *Proc. Inst. Civ. Engrs.* Vol. 11. 1958

# CHAPTER 9

# MATRIX METHODS†

## 9.1. INTRODUCTION

DURING the last few years a number of methods of analysing structures have been suggested which use matrix notation. Most of these involve rather more numerical work than is required in more elementary methods, but the computing is entirely systematic and can easily be mechanised. They appear somewhat cumbersome when applied to simple structures, but form a very powerful tool for dealing with complex highly-redundant systems. They have already been used extensively in the aircraft industry.

Matrix notation is simply a useful shorthand invented by mathematicians for discussing problems of linear algebra. Almost any method of analysis which treats a structure as a linear elastic system can be written in matrix terms, but the notation appears to its best advantage when it is used to set up the load-displacement equations in explicit form. Such an approach leads, of course, to the computational problem of solving sets of linear simultaneous equations, and for this reason it has in the past been restricted to simple structures with only a few degrees of freedom. However, the development of the desk calculating machine and, more recently, the automatic digital computer has made it easy to solve large sets of equations, so that the main objection to direct methods of this type has disappeared.

As with other techniques of structural analysis, we can divide matrix methods into two categories‡. In the 'compatibility method' redundant forces or moments are taken as the basic unknowns, and these are obtained by solving the displacement compatibility equations. In the 'equilibrium method' all internal forces and moments are expressed in terms of joint displacements and rotations, and the latter are found by solving the equations of joint equilibrium. This approach does not involve the concept of redundancy. In either case, the appropriate set of equations can be obtained by direct means or by the use of strain energy. The two approaches are complementary, and in most cases the 'better' method is the one which gives rise to the smaller number of simultaneous equations. The second group of methods, however, possess certain advantages when an automatic computer is available, since the actual process of setting up the equations is usually easier to mechanise.

The techniques described in the present Chapter are equilibrium methods. They are only directly applicable to structures acted on by concentrated forces and moments applied at their joints, but this involves no real loss of generality. If some of the members of a structure are subject to distributed

---

† This Chapter was contributed by Dr. R. K. Livesley of Cambridge University.
‡ See Chapter 4, Section 4.1.

loads, or concentrated loads at points along their length, the loading on the structure can be treated as the sum of

(a) The actual applied loads, together with a system of forces and moments which act at the joints and are such as to prevent any joint displacement or rotation.
(b) A system of forces and moments which are equal in magnitude to those introduced in (a) but opposite in sign. These are usually termed the 'equivalent fixed end forces and moments'.

Analysis of the structure under load system (a) is trivial, since each member is regarded as being fixed at its ends and can be treated separately from the rest. This analysis gives the forces and moments which make up load system (b), which is the type of loading dealt with in this Chapter. When the second analysis is complete, a simple superposition gives the behaviour of the structure under the sum of the two loadings which is, of course, the actual loading applied.

Much of the material in this Chapter has only previously been available in published papers, and unfortunately there seems to be no agreement among authors on the matter of notation. The notation and sign-convention used here have been chosen to preserve continuity, as far as possible, with previous Chapters†, and are not in general the same as in the original papers. A short introduction to matrix algebra now follows.

## 9.2. MATRIX ALGEBRA

### 9.2.1. GENERAL IDEAS AND NOTATION

We say that two variables $x$ and $y$ are linearly related if they obey an equation of the form

$$x = a y \qquad \ldots\ldots \text{(i)}$$

where $a$ is a constant. More generally, we say that a set of $n$ variables $x_1, x_2, \ldots x_n$, is linearly related to a set of $m$ variables $y_1, y_2, \ldots y_m$, if there exists a relationship between the two sets of the form

$$\left.\begin{aligned} x_1 &= a_{11} y_1 + a_{12} y_2 - - - - - a_{1m} y_m \\ x_2 &= a_{21} y_1 + a_{22} y_2 - - - - - a_{2m} y_m \\ &\phantom{=}\vdots \\ x_n &= a_{n1} y_1 + a_{n2} y_2 - - - - - a_{nm} y_m \end{aligned}\right\} \quad \ldots\text{(ii)}$$

where the coefficients $a_{11}$, $a_{12}$, etc. are constant. (It should be noted that in this set of equations the first suffix of each coefficient defines the row in which it stands, while the second defines the column.) The study of relationships such as (ii) forms the branch of mathematics known as linear algebra.

In discussing the general properties of linear systems it is extremely tedious to write out equations such as (ii) in full at each step. Apart from the

---

† However, the symbol $F$ now signifies an end-load vector comprising axial and transverse forces and a couple; the symbol $S$ refers to the axial component.

actual labour involved, general ideas tend to become obscured by a mass of symbols. One means of overcoming this difficulty is suffix notation. In suffix notation the symbol $x_i$ is used to represent any one of the variables $x_1, x_2, \ldots x_n$, so that the suffix $i$ really means any one of the numbers $1, 2 \ldots n$. In the same way $y_j$ represents any one of the variables $y_1, y_2, \ldots y_m$, while $a_{ij}$ represents the coefficient of $y_j$ in the equation for $x_i$. Thus equations (ii) can be written

$$x_i = \sum_{j=1}^{m} a_{ij} y_j \qquad \ldots \text{(iii)}$$

It is common practice to omit the summation symbol and write (iii) merely as

$$x_i = a_{ij} y_j \qquad \ldots \text{(iv)}$$

the repetition of the suffix $j$ being used to indicate summation over all its possible values.

A more powerful tool for dealing with linear systems is provided by introducing the concepts of vectors and matrices. These ideas can be developed from the suffix notation described above by giving the symbols appearing in (iv) a slightly different interpretation. Instead of the symbol $x_i$ representing *one particular member* of a set of $n$ variables, we regard it as denoting the *whole* set $x_1, x_2, \ldots x_n$. Similarly $y_j$ is taken to represent the complete set of variables $y_1, y_2, \ldots y_m$, the order of the variables being preserved in each case. Such sets are commonly referred to as 'vectors'. This use of the word is merely an extension of its use in defining points in ordinary space. Just as an ordered set of three quantities may be regarded as a point in three-dimensional space, so a set of $n$ quantities can be thought of as the coordinates of a point in an $n$-dimensional space. We shall write vectors as columns of variables enclosed by square brackets.

Almost all the rules of three-dimensional vector analysis apply in the more general case. The statement that two $n$-dimensional vectors are equal implies equality between all $n$ components, while the sum of two vectors is formed by adding corresponding components. Multiplication of a vector by a scalar merely increases each component in the same proportion, or, in geometrical terms, changes the 'length' of the vector without altering its 'direction'. The *scalar* product of two vectors $a_1, a_2 \ldots a_n$ and $b_1, b_2, \ldots b_n$ is defined as $a_1 b_1 + a_2 b_2 \ldots a_n b_n$, but the idea of a *vector* product cannot usefully be carried over from three-dimensional vector analysis in the same way.

The symbol $a_{ij}$ appearing in (iv) is now given a rather different interpretation. We use it to represent the *complete set* of $n \times m$ constants appearing in (ii), writing these coefficients as an array of $n$ rows and $m$ columns. Such an array we call a 'matrix', and we write it in square brackets in a similar manner to a vector.

We now interpret (iv) as stating that the *vector* $x_i$ is equal to the *vector* formed by the multiplication of the *matrix* $a_{ij}$ and the *vector* $y_j$. Since these symbols do not now represent single algebraic quantities, it is necessary to define the meaning of the word 'multiplication' as used in this context.

We do this by saying that the matrix equation

$$\begin{bmatrix} x_1 \\ x_2 \\ | \\ | \\ x_n \end{bmatrix} = \begin{bmatrix} a_{11} & a_{12} & ---- & a_{1m} \\ a_{21} & a_{22} & ---- & a_{2m} \\ | & | & & | \\ | & | & & | \\ a_{n1} & a_{n2} & ---- & a_{nm} \end{bmatrix} \begin{bmatrix} y_1 \\ y_2 \\ | \\ | \\ y_m \end{bmatrix} \quad \ldots \text{(v)}$$

must have exactly the same meaning as the set of equations (ii). In other words we multiply a matrix and a vector by taking the scalar product of the vector with each row of the matrix in turn. For such a multiplication to be possible the number of *columns* in the matrix must be equal to the number of components in the vector, the result of the multiplication being a vector with as many components as there are *rows* in the matrix. For example, the equation

$$\begin{bmatrix} x_1 \\ x_2 \end{bmatrix} = \begin{bmatrix} 3 & -2 \\ 4 & 1 \end{bmatrix} \begin{bmatrix} y_1 \\ y_2 \end{bmatrix}$$

is identical in meaning to the equations

$$x_1 = 3y_1 - 2y_2$$
$$x_2 = 4y_1 + y_2$$

It is common practice to denote vectors and matrices by capital letters without suffixes. Thus (v) might be written simply as

$$X = AY$$

in which form the analogy with (i) is complete.

Apart from its conciseness, matrix notation is useful in that it clearly separates the constants appearing in a set of equations such as (ii) from the particular sets of variables which happen to be related. In any physical system whose behaviour is governed by linear algebraic equations, the matrix $A$ is always an invariant function and can indeed be regarded as forming a complete mathematical statement of the properties of the system. The vectors $X$ and $Y$, on the other hand, are merely related to one particular set of conditions.

Other properties of matrices can be derived by a further consideration of linear algebraic equations. In this way we can build up an algebra of matrices similar in many respects to the ordinary algebra of scalars. For instance, if

$$\left. \begin{array}{ll} x_1 = 3y_1 + 2y_2 & x'_1 = 5y_1 + 3y_2 \\ x_2 = 4y_1 + 3y_2 & x'_2 = 2y_1 + 4y_2 \end{array} \right\} \quad \ldots \text{(vi)}$$

then it follows that

$$\left. \begin{array}{l} x_1 + x'_1 = (3+5)\,y_1 + (2+3)\,y_2 \\ x_2 + x'_2 = (4+2)\,y_1 + (3+4)\,y_2 \end{array} \right\} \quad \ldots \text{(vii)}$$

## 9.2. MATRIX ALGEBRA

In matrix notation we might write equations (vi) as
$$X = AY \qquad X' = BY$$
and (vii) as
$$X + X' = (A + B)Y = CY$$
where $C$, the 'sum' of the matrices $A$ and $B$, is formed by adding corresponding elements of the two matrices. It follows that one cannot add two matrices unless they have the same number of rows and the same number of columns.

The multiplication of two matrices can be dealt with in a similar manner.

If
$$\left.\begin{array}{l} x_1 = a_1 y_1 + b_1 y_2 + c_1 y_3 \\ x_2 = a_2 y_1 + b_2 y_2 + c_2 y_3 \end{array}\quad \begin{array}{l} y_1 = d_1 z_1 + d_2 z_2 \\ y_2 = e_1 z_1 + e_2 z_2 \\ y_3 = f_1 z_1 + f_2 z_2 \end{array}\right\} \dots\text{(viii)}$$

eliminating $y_1, y_2, y_3$ gives
$$\left.\begin{array}{l} x_1 = (a_1 d_1 + b_1 e_1 + c_1 f_1) z_1 + (a_1 d_2 + b_1 e_2 + c_1 f_2) z_2 \\ x_2 = (a_2 d_1 + b_2 e_1 + c_2 f_1) z_1 + (a_2 d_2 + b_1 e_2 + c_2 f_2) z_2 \end{array}\right\} \dots\text{(ix)}$$

In matrix notation we write equations (viii) as
$$X = AY \qquad Y = BZ$$
and (ix) as
$$X = ABZ = CZ$$

Comparison of these equations with (viii) and (ix) will show that $C$, the 'product' of matrices $A$ and $B$ is formed by taking the scalar product of each *row* of $A$ with each *column* of $B$. We can state this rule for multiplying matrices as follows: If $C = AB$, then the element in the i'th row and the j'th column of $C$ is formed by taking the scalar product of the i'th row of $A$ and the j'th column of $B$. It follows that, for such a multiplication to be possible, the number of *columns* in $A$ must equal the number of *rows* in $B$. It also follows that the product $BA$ is not necessarily equal to $AB$, and may in fact not exist at all.

In the same way it is possible to show that matrices obey the ordinary associative laws,
$$A(B+C) = AB + AC, \quad (AB)C = A(BC)$$
as long as the order of the multiplications is maintained. It should be noted that the division of two matrices has no meaning except in a rather restricted sense discussed in the next Section.

### 9.2.2. SOME PROPERTIES OF SQUARE MATRICES

In practical problems one is normally concerned with matrices having the same number of rows and columns. Such matrices appear when two vectors containing equal numbers of components are linearly related, and while they obey the general rules discussed in the previous Section, they also have certain important special properties.

Returning to equations (ii), we now consider the problem of 'solving' such a set of equations. In other words, given a set of $n$ numbers $x_1, x_2, ---- x_n$, under what conditions can we find a unique set of $m$ numbers $y_1, y_2, ---- y_m$ which satisfies (ii)? If $m > n$ we have too few equations to determine all the $m$ unknowns, while if $m < n$ some of the equations are redundant, and it is impossible to specify all the numbers $x_i$ independently. If $m = n$, however, we have as many equations as unknowns, and a unique solution exists for any arbitrary set of numbers $x_i$ provided that the equations are 'linearly independent'. By this we mean that each equation must give us some new information about the unknowns and must not be merely a linear combination of some of the other equations. If we have a set of six equations in ten unknowns, for instance, it is quite easy to manufacture four new equations out of the original six, but this provides no new information about the unknowns, and does not make the equations any more solvable.

If $m = n$ and the $n$ equations are linearly independent, the matrix $A$ of the coefficients in (ii) is said to be non-singular. From the definition of linear independence given above, it is clear that in a non-singular matrix no row of coefficients is a linear combination of the other rows. (This means that the determinant of a non-singular matrix is always non-zero.) If

$$X = AY \qquad \ldots (\text{x})$$

where $A$ is non-singular, it follows that for every given vector $X$ there exists a unique vector $Y$ and vice versa. We can therefore regard $Y$ as a linear function of $X$, connected to it by a relationship of the form

$$Y = BX \qquad \ldots (\text{xi})$$

where $B$ is also a non-singular square matrix of $n$ rows and columns. The matrix $B$ is said to be the 'inverse' of the matrix $A$, and is usually written $A^{-1}$. [This notation is preferable to writing $B = 1/A$, since with matrices the order of a multiplication is important. If we write $C/A$, for instance, we may mean $CA^{-1}$ or $A^{-1}C$, and these will in general be quite different things.] Combining (x) and (xi) it follows that

$$X = AA^{-1}X$$

Now the only matrix which leaves all vectors unchanged on multiplication is the unit matrix

$$I = \begin{bmatrix} 1 & 0 & 0 & ----- & 0 \\ 0 & 1 & 0 & & | \\ 0 & 0 & \ddots & & | \\ | & & & \ddots & | \\ | & & & & | \\ | & & & & 0 \\ 0 & ----- & & 0 & 1 \end{bmatrix}$$

which has unity for all 'leading diagonal' elements and zeros everywhere else. It follows that $A A^{-1} = I$

just as in ordinary scalar algebra $a \times a^{-1} = 1$. Since (x) and (xi) can also be combined to give
$$Y = A^{-1} A Y$$
it follows that
$$A A^{-1} = A^{-1} A = I$$
the order of multiplication being immaterial in this case.

The solution of (x) for a given value of the vector $X$ forms the central computational problem in what may be termed 'linear equilibrium problems'. In structural analysis, for instance, (x) may represent the equations connecting the known loads $X$ with the unknown displacements $Y$, while in electric network theory the vectors $X$ and $Y$ may represent voltages and currents. If a solution is required for only one set of values $X_1$, it is simplest to solve the equations without determining the inverse matrix. If solutions are required for several different vectors $X_1, X_2, ---$, however, (as, for example, in a structure under various loading conditions), it is better to calculate $A^{-1}$ once, and then use (xi) to find each solution in turn. Methods of solving sets of simultaneous equations and inverting matrices are discussed briefly in the following Section.

While 'equilibrium' problems require the solution of simultaneous equations such as (x), in which the vector $X$ is known, the determination of natural frequencies and normal modes of vibration leads to equations of the type
$$AX = \zeta X \qquad \ldots \ldots (\text{xii})$$
where $A$ is a square matrix and $\zeta$ is a scalar. The problem here is to find the vector $X$ which remains unchanged (apart from the factor $\zeta$ which also has to be determined) when multiplied by the given matrix.

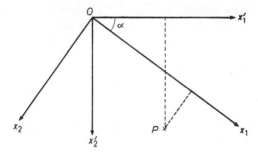

*Figure 9.1.* Change of coordinate axes in a plane

If the matrix $A$ has $n$ rows and columns it can be shown that there are in fact $n$ vectors which possess this property. These are known as the 'latent vectors' or 'eigenvectors' of $A$, and with each vector is associated a particular multiplying factor $\zeta$ called the 'latent root' or 'eigenvalue'. Since the vector $X$ appears on both sides of equation (xii) it follows that if $X$ is a latent vector of $A$, then so is any scalar multiple of it. In geometrical terms this means that the *directions* of the latent vectors are fixed but not their *magnitudes*. One component of each vector may therefore be given an arbitrary magnitude, and it is common to 'normalize' the vectors by making them of 'unit length'. That is to say, the magnitudes of the components $x_i$ are chosen so that $x_1^2 + x_2^2 + \ldots x_n^2 \doteq 1$. The problem of determining the latent roots and vectors of a matrix is more difficult than that of solving the associated set of equations or inverting the matrix.

Matrix notation is also useful in describing changes in coordinate systems. In FIGURE 9.1, for instance, if the point $P$ has coordinate $x_1, x_2$ in one reference frame and $x_1', x_2'$ in the other, it is clear that

$$x_1 = x_1' \cos \alpha + x_2' \sin \alpha$$
$$x_2 = -x_1' \sin \alpha + x_2' \cos \alpha$$

We may write this in the form

$$\begin{bmatrix} x_1 \\ x_2 \end{bmatrix} = \begin{bmatrix} \cos \alpha & \sin \alpha \\ -\sin \alpha & \cos \alpha \end{bmatrix} \begin{bmatrix} x_1' \\ x_2' \end{bmatrix}$$

or, more briefly, as

$$X = TX'$$

where the matrix

$$T = \begin{bmatrix} \cos \alpha & \sin \alpha \\ -\sin \alpha & \cos \alpha \end{bmatrix}$$

defines the transformation. The inverse of $T$

$$T^{-1} = \begin{bmatrix} \cos \alpha & -\sin \alpha \\ \sin \alpha & \cos \alpha \end{bmatrix}$$

enables us to carry out the reverse transformation

$$X' = T^{-1} X$$

and it will be noticed that the inverse of $T$ is equal to its 'transpose',—the matrix formed by interchanging rows and columns. This relationship always hold when the transformation of axes is equivalent to a pure rotation.

### 9.2.3. SOME NOTES ON NUMERICAL TECHNIQUES

We have seen that three important computational problems arise in matrix work. They are,

(1) The solution of sets of linear algebraic equations
(2) The inversion of matrices
(3) The extraction of latent roots and vectors

There are many methods of tackling each of these problems and we shall not attempt even to mention all of them here. For a full account of the relevant numerical techniques the reader is referred to textbooks on numerical analysis, such as those by Hartree[1] or Crandall[2].

Methods of solving sets of simultaneous equations may be divided into those which obtain an exact solution (except for numerical rounding-off errors), and those which derive a sequence of approximations to the true solution.

The simplest form of approximate method is the Gauss-Seidel. If, for example, we have to solve the equations

$$10x + 6y - z = 19$$
$$3x + 9y + 2z = 27$$
$$x - 4y + 8z = 17$$

we can start from an arbitrary set of numbers such as $x_0 = y_0 = z_0 = 2$, and use the first equation to re-calculate a value for $x$,

$$x_1 = (19 - 6 \times 2 + 2) / 10 = 0{\cdot}9$$

We now use this value, and the original value of $z$, to calculate a new value of $y$ from the second equation, writing it as

$$y_1 = (27 - 3 \times 0{\cdot}9 - 2 \times 2)/9 = 2{\cdot}255$$

In the same way the third equation gives a new approximation for $z$,

$$z_1 = (17 - 0{\cdot}9 + 4 \times 2.255)/8 = 3{\cdot}14$$

The process can now be repeated, taking the new values $x_1$, $y_1$, $z_1$ as the starting point. The reader will find that in this example a few cycles give a good approximation to the true solution, which is $x = 1$, $y = 2$, $z = 3$.

The convergence of the process depends on the coefficients in the equations. An analytical condition for convergence may be stated as follows: if the equations can be derived by equating to zero the partial derivatives with respect to the variables of some function $U$, then the process will converge if the solution corresponds to a minimum of $U$. For example, if the equations are the load-displacement equations of a structure, the solution always corresponds to a minimum value of the total potential energy, provided that the equilibrium is a stable one. The iterative process will therefore converge, although there are many cases in which convergence is so slow that the method becomes valueless.

Better results may be obtained by the more informal iterative process known as 'relaxation', in which the variables are not necessarily adjusted in any set order. By exercising judgement and intuition an experienced computer can often speed up convergence considerably. As far as structural analysis is concerned, however, it seems preferable to apply relaxation techniques in a semi-physical manner, as in moment-distribution, where the actual load displacement equations are never formally written down.

With 'exact' processes of solution the question of convergence does not arise, so that such methods can be used where an iterative approach is unsatisfactory. They are also the most convenient methods to use where an automatic digital computer is available. The best-known exact method is that of successive elimination of variables. If we have a set of $n$ equations

$$\begin{bmatrix} a_{11} & \cdots & a_{1n} \\ \vdots & & \vdots \\ a_{n1} & \cdots & a_{nn} \end{bmatrix} \begin{bmatrix} x_1 \\ \vdots \\ x_n \end{bmatrix} = \begin{bmatrix} b_1 \\ \vdots \\ b_n \end{bmatrix}$$

the variable $x_n$ may be eliminated from the first $(n-1)$ equations by subtracting suitable multiples of the last equation from the others. This leaves

a modified set of equations

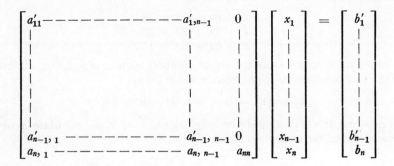

where primes indicate altered coefficients. In the same way we may subtract multiples of the $(n-1)$'th equation from the first $(n-2)$ equations in such a way as to make the coefficients $a'_{1,\,n-1}\text{------}a'_{n-2,\,n-1}$ all zero, and the process may be continued until the equations are in the form

$$\begin{bmatrix} a'_{11} & 0 & \text{------} & 0 \\ a'_{21} & a'_{22} & 0 & \text{------} 0 \\ & & & \ddots & \\ & & & & 0 \\ & & & & \\ a_{n-1} & a_{n2} & \text{------} & a_{nn} \end{bmatrix} \begin{bmatrix} x_1 \\ x_2 \\ \vdots \\ \\ \\ x_n \end{bmatrix} = \begin{bmatrix} b'_1 \\ b'_2 \\ \vdots \\ \\ \\ b_n \end{bmatrix}$$

with all coefficients above the leading diagonal equal to zero. It is clear that in this form, the first equation immediately gives $x_1$. This value can be substituted in the second equation to obtain $x_2$, and the process continued right through to $x_n$. The total number of operations is of the order of $\tfrac{1}{3}n^3$.

A similar technique can be used for inverting matrices. Consider, for instance, the equations

$$6x_1 - 4x_2 = y_1$$
$$-3x_1 + 10x_2 = y_2$$

These may be written

$$\begin{bmatrix} 6 & -4 \\ -3 & 10 \end{bmatrix} \begin{bmatrix} x_1 \\ x_2 \end{bmatrix} = \begin{bmatrix} 1 & 0 \\ 0 & 1 \end{bmatrix} \begin{bmatrix} y_1 \\ y_2 \end{bmatrix}$$

We can eliminate $x_1$ from the second equation by adding half the first equation to it. This involves adding half the first row to the second row in each matrix, to give

$$\begin{bmatrix} 6 & -4 \\ 0 & 8 \end{bmatrix} \begin{bmatrix} x_1 \\ x_2 \end{bmatrix} = \begin{bmatrix} 1 & 0 \\ \tfrac{1}{2} & 1 \end{bmatrix} \begin{bmatrix} y_1 \\ y_2 \end{bmatrix} \quad \ldots \text{(xiii)}$$

In the same way we eliminate $x_2$ from the first equation of (xiii) by adding half the second equation to it, giving

$$\begin{bmatrix} 6 & 0 \\ 0 & 8 \end{bmatrix} \begin{bmatrix} x_1 \\ x_2 \end{bmatrix} = \begin{bmatrix} 1\frac{1}{4} & \frac{1}{2} \\ \frac{1}{2} & 1 \end{bmatrix} \begin{bmatrix} y_1 \\ y_2 \end{bmatrix} \qquad \ldots\ldots \text{(xiv)}$$

All we have to do now is to divide the first equation of (xiv) by 6 and the second by 8,

$$\begin{bmatrix} 1 & 0 \\ 0 & 1 \end{bmatrix} \begin{bmatrix} x_1 \\ x_2 \end{bmatrix} = \begin{bmatrix} \frac{5}{24} & \frac{1}{12} \\ \frac{1}{16} & \frac{1}{8} \end{bmatrix} \begin{bmatrix} y_1 \\ y_2 \end{bmatrix} \qquad \ldots\ldots \text{(xv)}$$

The inverse matrix is the right-hand matrix in (xv), and as a check we have

$$\begin{bmatrix} 6 & -4 \\ -3 & 10 \end{bmatrix} \begin{bmatrix} \frac{5}{24} & \frac{1}{12} \\ \frac{1}{16} & \frac{1}{8} \end{bmatrix} = \begin{bmatrix} 1 & 0 \\ 0 & 1 \end{bmatrix}$$

It is clear that the process does not really need the presence of the vectors

$$\begin{bmatrix} x_1 \\ x_2 \end{bmatrix}, \begin{bmatrix} y_1 \\ y_2 \end{bmatrix}$$

and that it can easily be laid out in tabular form for matrices of any size.

Although general techniques for extracting the latent roots and vectors of matrices are too complicated to describe here, a simple numerical example will show the nature of the problem. Suppose we have to find the values of the scalar $\zeta$ and the vector $\begin{bmatrix} x_1 \\ x_2 \end{bmatrix}$ which satisfy the matrix equation

$$\begin{bmatrix} 3 & 1 \\ 2 & 2 \end{bmatrix} \begin{bmatrix} x_1 \\ x_2 \end{bmatrix} = \begin{bmatrix} \zeta x_1 \\ \zeta x_2 \end{bmatrix}$$

We may write this as

$$\begin{bmatrix} 3-\zeta & 1 \\ 2 & 2-\zeta \end{bmatrix} \begin{bmatrix} x_1 \\ x_2 \end{bmatrix} = \begin{bmatrix} 0 \\ 0 \end{bmatrix} \qquad \ldots\ldots \text{(xvi)}$$

and the theory of equations tells us that a non-zero solution of (xvi) cannot exist unless the determinant

$$\begin{bmatrix} 3-\zeta & 1 \\ 2 & 2-\zeta \end{bmatrix}$$

is zero. This condition gives the quadratic equation

$$(3-\zeta)(2-\zeta) - 2 = 0$$

whence we obtain $\zeta = 1$ or $4$. If $\zeta = 1$, the first equation of (xvi) gives $2x_1 + x_2 = 0$, so that the normalized latent vector is $\begin{bmatrix} 1/\sqrt{5} \\ -2/\sqrt{5} \end{bmatrix}$. In the same way if $\zeta = 4$, we obtain $-x_1 + x_2 = 0$, so that the normalized latent vector associated with this root is $\begin{bmatrix} 1/\sqrt{2} \\ 1/\sqrt{2} \end{bmatrix}$. With larger matrices this approach becomes somewhat tedious, and other methods are available which do not involve determining the roots of a high-order polynomial.

## 9.3. THE MATRIX ANALYSIS OF CONTINUOUS BEAMS

The method described in this Section is in essence the one suggested by S. U. Benscoter[3]. The sign convention we shall use is shown in FIGURE 9.2—

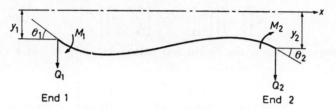

*Figure 9.2.* Conventions for the matrix analysis of continuous beams

apart from the convention for shear forces normal slope-deflexion notation is used. We shall only consider uniform beams, although much of the analysis is applicable to the non-uniform case.

The slope-deflexion equations for the beam shown in the figure are

$$M_1 = (6EI/L^2)\, y_1 + (4EI/L)\, \theta_1 - (6EI/L^2)\, y_2 + (2EI/L)\, \theta_2$$
$$M_2 = (6EI/L^2)\, y_1 + (2EI/L)\, \theta_1 - (6EI/L^2)\, y_2 + (4EI/L)\, \theta_2$$

which give by addition

$$(M_1 + M_2)/L = Q_1 = -Q_2$$
$$= (12EI/L^3)y_1 + (6EI/L^2)\theta_1 - (12EI/L^3)y_2 + (6EI/L^2)\theta_2$$

These three equations may be written in matrix notation as

$$\left.\begin{aligned}\begin{bmatrix} Q_1 \\ M_1 \end{bmatrix} &= \begin{bmatrix} 12EI/L^3 & 6EI/L^2 \\ 6EI/L^2 & 4EI/L \end{bmatrix} \begin{bmatrix} y_1 \\ \theta_1 \end{bmatrix} + \begin{bmatrix} -12EI/L^3 & 6EI/L^2 \\ -6EI/L^2 & 2EI/L \end{bmatrix} \begin{bmatrix} y_2 \\ \theta_2 \end{bmatrix} \\ \begin{bmatrix} Q_2 \\ M_2 \end{bmatrix} &= \begin{bmatrix} -12EI/L^3 & -6EI/L^2 \\ 6EI/L^2 & 2EI/L \end{bmatrix} \begin{bmatrix} y_1 \\ \theta_1 \end{bmatrix} + \begin{bmatrix} 12EI/L^3 & -6EI/L^2 \\ -6EI/L^2 & 4EI/L \end{bmatrix} \begin{bmatrix} y_2 \\ \theta_2 \end{bmatrix}\end{aligned}\right\} \ldots(9.1)$$

We now call the column vectors $\begin{bmatrix} y_1 \\ \theta_1 \end{bmatrix}$, $\begin{bmatrix} y_2 \\ \theta_2 \end{bmatrix}$, the 'displacements' of ends 1 and 2 of the beam and write them $D_1$ and $D_2$. In the same way we use symbols $F_1$, $F_2$ to denote the column vectors $\begin{bmatrix} Q_1 \\ M_1 \end{bmatrix}$, $\begin{bmatrix} Q_2 \\ M_2 \end{bmatrix}$ and term these quantities the 'end-loads' of the beam. [In the remainder of this Chapter the words 'displacements' and 'end-loads' will be used with this extended meaning, the number of components in the vectors depending on the type of structural element considered.] We also introduce symbols $Y_{11}$, $Y_{12}$, $Y_{21}$, $Y_{22}$ to denote the square matrices

## 9.3. MATRIX ANALYSIS OF CONTINUOUS BEAMS

appearing in equations 9.1, so that these equations can be written

$$F_1 = Y_{11} D_1 + Y_{12} D_2 \\ F_2 = Y_{21} D_1 + Y_{22} D_2 \qquad \ldots (9.2)$$

The matrices $Y_{11}$, etc., which depend only on the physical characteristics of the member, we shall term 'stiffness matrices'.

Equations 9.2 describe in shorthand form the relationship between displacements and end-loads for a single member. We cannot, at this stage, solve these equations to obtain the displacements in terms of the loads, since it is obviously possible to give the member an arbitrary rigid-body movement. Mathematically this corresponds to saying that equations 9.2 are singular. We shall see, however, that this apparent difficulty disappears when the equations for the individual elements are combined to give the equations for the whole structure. The latter set of equations is always non-singular, provided of course that the structure is not in fact a mechanism, and is sufficiently anchored to prevent overall rigid-body motion.

The next step is to use equations similar to 9.2 to build up the complete set of load-displacement equations mentioned above. As an example, we shall consider the analysis of the two-span beam on elastic supports shown in FIGURE 9.3. In this figure the applied joint loads (or the equivalent

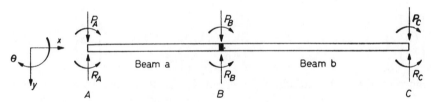

*Figure 9.3.* Example of a continuous beam

fixed end loads) each comprise a force and a moment, and are represented by the symbols $P_A, P_B, P_C$. In the same way the symbols $R_A, R_B, R_C$ represent the forces and moments applied by the elastic supports.

We shall assume that at each support the vertical reaction is $k_y$ times the vertical displacement of the joint, while the reactive moment is $k_\theta$ times the angular rotation.

For each beam we can write down equations 9.2.

$$F_{1a} = (Y_{11})_a D_{1a} + (Y_{12})_a D_{2a} \\ F_{2a} = (Y_{21})_a D_{1a} + (Y_{22})_a D_{2a} \qquad \ldots (9.3(a))$$

$$F_{1b} = (Y_{11})_b D_{1b} + (Y_{12})_b D_{2b} \\ F_{2b} = (Y_{21})_b D_{1b} + (Y_{22})_b D_{2b} \qquad \ldots (9.3(b))$$

where the letter suffix denotes the member referred to.

If we denote the displacements of the joints $A, B, C$, by $\Delta_A, \Delta_B, \Delta_C$, we have the following compatibility equations

$$D_{1a} = \Delta_A, \quad D_{2a} = D_{1b} = \Delta_B, \quad D_{2b} = \Delta_C \qquad \ldots (9.4)$$

We also have the following equations for equilibrium of the joints

$$F_{1a} = P_A - R_A, \quad F_{2a} + F_{1b} = P_B - R_B, \quad F_{2b} = P_C - R_C \quad \ldots (9.5)$$

the reactions $R_A$, $R_B$, $R_C$, being given by

$$R_A = Z\Delta_A, \quad R_B = Z\Delta_B, \quad R_C = Z\Delta_C \quad \ldots (9.6)$$

where $Z$ is the stiffness matrix of the supports, defined by

$$Z = \begin{bmatrix} k_y & 0 \\ 0 & k_\theta \end{bmatrix}$$

We now replace the $D$'s in equations 9.3 by $\Delta$'s in accordance with 9.4, and use these equations, together with equations 9.6 to substitute for the $F$'s and $R$'s in equations 9.5. This gives,

$$\left. \begin{aligned} P_A &= [Z + (Y_{11})_a]\Delta_A + (Y_{12})_a \Delta_B \\ P_B &= (Y_{21})_a \Delta_A + [Z + (Y_{22})_a + (Y_{11})_b] \Delta_B + (Y_{12})_b \Delta_C \\ P_C &= (Y_{21})_b \Delta_B + [Z + (Y_{22})_b]\Delta_C \end{aligned} \right\} \quad \ldots (9.7)$$

or, in matrix form,

$$\begin{bmatrix} P_A \\ P_B \\ P_C \end{bmatrix} = \begin{bmatrix} Z + (Y_{11})_a & (Y_{12})_a & 0 \\ (Y_{21})_a & Z + (Y_{22})_a + (Y_{11})_b & (Y_{12})_b \\ 0 & (Y_{21})_b & Z + (Y_{22})_b \end{bmatrix} \begin{bmatrix} \Delta_A \\ \Delta_B \\ \Delta_C \end{bmatrix} \quad \ldots (9.8)$$

Since each $P$ and $\Delta$ stands for a column vector of two quantities, equation 9.8 corresponds to six scalar equations, relating the applied forces and moments to the joint displacements and rotations. The equations are non-singular and can be solved for the unknowns $\Delta_A, \Delta_B, \Delta_C$. Equations 9.3 can then be used to find the internal moments and shearing forces. The arrangement of the physical structure is clearly reflected in the form of the matrix in 9.8. The elements $Z + (Y_{11})_a$, $Z + (Y_{22})_a + (Y_{11})_b$, $Z + (Y_{22})_b$ represent the 'direct' stiffnesses of the joints to applied loads, while the other terms represent cross-coupling between loads and displacements at different joints. The direct stiffness of the system at $B$, for instance, is the sum of the direct stiffness $Z$ of the support, the stiffness $(Y_{22})_a$ of end 2 of beam $a$, and the stiffness $(Y_{11})_b$ of end 1 of beam $b$. The off-diagonal term $(Y_{21})_b$, on the other hand, relates the load $P_C$ to the displacement $\Delta_B$. The matrix appearing in 9.8 may be termed the 'stiffness matrix' of the complete structure.

If a joint is known to have either or both components of its displacement vector identically zero, the appropriate variables can be dropped from the final equations and one or more of the equations omitted. If, for instance, joint $A$ in FIGURE 9.3 were fixed so that $\Delta_A = 0$, the equations 9.8 would become

$$\begin{bmatrix} P_B \\ P_C \end{bmatrix} = \begin{bmatrix} Z + (Y_{22})_a + (Y_{11})_b & (Y_{12})_b \\ (Y_{21})_b & Z + (Y_{22})_b \end{bmatrix} \begin{bmatrix} \Delta_B \\ \Delta_C \end{bmatrix}$$

## 9.3. MATRIX ANALYSIS OF CONTINUOUS BEAMS

The equation omitted becomes

$$P_A = (Y_{12})_a \Delta_B + R_A$$

and this can be used to determine the unknown fixing reaction $R_A$, once $\Delta_B$ has been found.

As a second example, consider the problem of analysing the three-span continuous beam shown in FIGURE 9.4, in which each beam has the same

Figure 9.4. Second example of a continuous beam

flexural rigidity $EI$. Each support $A, B,$ & $C$, has stiffness $k$ against vertical deflexion but offers no restraint against rotation. Using the notation of the previous example we have

$$Z = \begin{bmatrix} k & 0 \\ 0 & 0 \end{bmatrix}$$

The load-displacement equations can be written symbolically in the same manner as before

$$\begin{bmatrix} P_A \\ P_B \\ P_C \end{bmatrix} = \begin{bmatrix} Y_{22} + Y_{11} + Z & Y_{12} & 0 \\ Y_{21} & Y_{22} + Y_{11} + Z & Y_{12} \\ 0 & Y_{21} & Y_{22} + Z \end{bmatrix} \begin{bmatrix} \Delta_A \\ \Delta_B \\ \Delta_C \end{bmatrix} \quad \ldots (9.9)$$

where the letter suffix has been dropped from the $Y$-matrices since all the beams are equal. The loading condition shown in the figure can be written

$$P_A = \begin{bmatrix} 0 \\ M \end{bmatrix} \qquad P_B = P_C = \begin{bmatrix} 0 \\ 0 \end{bmatrix}$$

Written out in full, equations 9.9 are

$$\begin{bmatrix} 0 \\ M \\ 0 \\ 0 \\ 0 \\ 0 \end{bmatrix} = \begin{bmatrix} (24EI/L^3+k) & 0 & -12EI/L^3 & 6EI/L^2 & 0 & 0 \\ 0 & 8EI/L & -6EI/L^2 & 2EI/L & 0 & 0 \\ -12EI/L^3 & -6EI/L^2 & (24EI/L^3+k) & 0 & -12EI/L^3 & 6EI/L^2 \\ 6EI/L^2 & 2EI/L & 0 & 8EI/L & -6EI/L^2 & 2EI/L \\ 0 & 0 & -12EI/L^3 & -6EI/L^2 & (12EI/L^3+k) & -6EI/L^2 \\ 0 & 0 & 6EI/L^2 & 2EI/L & -6EI/L^2 & 4EI/L \end{bmatrix} \begin{bmatrix} y_A \\ \theta_A \\ y_B \\ \theta_B \\ y_C \\ \theta_C \end{bmatrix} \quad (9.10)$$

It will be noticed that these equations, like 9.1, form a symmetric set. This is of course a direct consequence of the reciprocal theorem. The equations can now be solved for the displacements.

If we consider the case where the supports are very stiff ($k \to \infty$) in which $y_A, y_B, y_C$ all tend to zero, we may drop all terms involving these variables from the second, fourth and sixth equations. These equations then become

$$\begin{bmatrix} M \\ 0 \\ 0 \end{bmatrix} = \begin{bmatrix} 8EI/L & 2EI/L & 0 \\ 2EI/L & 8EI/L & 2EI/L \\ 0 & 2EI/L & 4EI/L \end{bmatrix} \begin{bmatrix} \theta_A \\ \theta_B \\ \theta_C \end{bmatrix} \quad \ldots (9.11)$$

and their solution is

$$\theta_A = \frac{7}{52} \frac{ML}{EI} \qquad \theta_B = \frac{-2}{52} \frac{ML}{EI} \qquad \theta_C = \frac{1}{52} \frac{ML}{EI}$$

The internal forces and moments follow from equations 9.1.

This example shows the power and also the defects of a matrix method. Equation 9.9 states the load-displacement equations of the system in their most general form, from which 9.11 is derived as a special case. The latter equations could have been written down directly from the slope-deflexion equations without any difficulty, so that no great advantage can be claimed for matrix methods in such a simple case.

## 9.4. PROBLEMS INVOLVING COORDINATE TRANSFORMATIONS

When applying the same approach to frameworks a new problem arises. In the case of continuous beams the most suitable reference frame for analysing each component part is the natural reference frame for the complete system. In two or three dimensional structures, however, it is simplest to derive the load-displacement equations for each member in its own reference frame, and then transform into the common reference frame for the whole structure†. We illustrate this process with several examples.

### 9.4.1 PLANE PIN-JOINTED FRAMES

FIGURE 9.5(a) shows a member of a pin-jointed frame which has end-displacements $D_1 = \begin{bmatrix} x_1 \\ y_1 \end{bmatrix}$, $D_2 = \begin{bmatrix} x_2 \\ y_2 \end{bmatrix}$, under end-loads $F_1 = \begin{bmatrix} S_1 \\ Q_1 \end{bmatrix}$, $F_2 = \begin{bmatrix} S_2 \\ Q_2 \end{bmatrix}$, in the reference frame shown. Provided that the displacements are small (a normal requirement of linear theory), the relation between end-loads and end-displacements is simply

$$\left.\begin{aligned} \begin{bmatrix} S_1 \\ Q_1 \end{bmatrix} &= \begin{bmatrix} EA/L & 0 \\ 0 & 0 \end{bmatrix} \begin{bmatrix} x_1 \\ y_1 \end{bmatrix} - \begin{bmatrix} EA/L & 0 \\ 0 & 0 \end{bmatrix} \begin{bmatrix} x_2 \\ y_2 \end{bmatrix} \\ \begin{bmatrix} S_2 \\ Q_2 \end{bmatrix} &= -\begin{bmatrix} EA/L & 0 \\ 0 & 0 \end{bmatrix} \begin{bmatrix} x_1 \\ y_1 \end{bmatrix} + \begin{bmatrix} EA/L & 0 \\ 0 & 0 \end{bmatrix} \begin{bmatrix} x_2 \\ y_2 \end{bmatrix} \end{aligned}\right\} \ldots (9.12)$$

or

$$\begin{aligned} F_1 &= YD_1 - YD_2 \\ F_2 &= -YD_1 + YD_2 \end{aligned} \quad \ldots (9.13)$$

---

† In this Section we have taken the axis of positive $y$-displacement to be downwards, in order to maintain continuity with Section 9.3. All the equations are unchanged, however, if the $y$-axis is reversed in direction, provided that the sign conventions for moments, angles and shear forces are also reversed.

## 9.4. PROBLEMS INVOLVING COORDINATE TRANSFORMATIONS   9.4.1

where
$$Y = \begin{bmatrix} EA/L & 0 \\ 0 & 0 \end{bmatrix}$$

Although 9.13 is formally similar to 9.2 we cannot write down compatibility and equilibrium equations for the joints of the structure immediately, since the end-loads and end-displacements of different members are in different reference frames. We now consider the way in which equation 9.13 behaves under a change of coordinate axes.

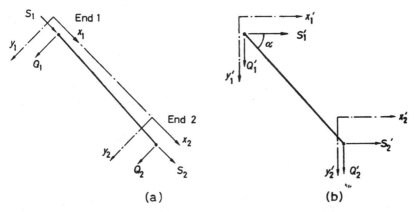

Figure 9.5. Change from 'member' to 'frame' coordinates for a pin-ended member

We choose arbitrary directions $x'$ and $y'$ for the whole structure (normally we take axes which are horizontal and vertical) and denote by $D_1'$, $D_2'$, $F_1'$, $F_2'$ the displacements and end-loads of a member referred to these directions, as shown in FIGURE 9.5(b). As mentioned in Section 9.2.2 the change in coordinates can most simply be expressed by writing

$$D_1 = TD_1', \quad D_2 = TD_2'; \quad F_1 = TF_1', \quad F_2 = TF_2' \quad \ldots \ldots (9.14)$$

where $T$ is the orthogonal transformation matrix
$$\begin{bmatrix} \cos\alpha & \sin\alpha \\ -\sin\alpha & \cos\alpha \end{bmatrix}$$

Substituting for $D_1$, $D_2$, $F_1$, $F_2$, equation 9.13 becomes

$$TF_1' = YTD_1' - YTD_2'$$
$$TF_2' = -YTD_1' + YTD_2'$$

or, multiplying by the inverse matrix $T^{-1}$

$$\left. \begin{array}{l} F_1' = (T^{-1}YT)D_1' - (T^{-1}YT)D_2' \\ F_2' = -(T^{-1}YT)D_1' + (T^{-1}YT)D_2' \end{array} \right\} \quad \ldots \ldots (9.15)$$

If we now write
$$Y' = T^{-1}YT$$

equation 9.15 becomes

$$F'_1 = Y'D'_1 - Y'D'_2 \\ F'_2 = -Y'D'_1 + Y'D'_2 \biggr\} \quad \ldots (9.16)$$

It is easy to show that

$$Y' = T^{-1} Y T = (EA/L) \begin{bmatrix} \cos^2\alpha & \cos\alpha \sin\alpha \\ \cos\alpha \sin\alpha & \sin^2\alpha \end{bmatrix}$$

a result which can also be deduced by direct geometrical argument.

With all displacements and end-loads expressed in a single coordinate system, the assembly of the complete equations for a structure proceeds in exactly the same way as before. Consider, for instance, the frame shown in FIGURE 9.6, acted on by arbitrary loads $P_B$ and $P_E$ at joints $B$ and $E$.

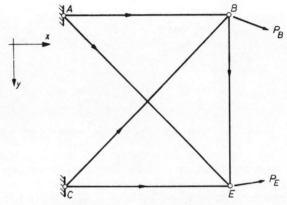

Figure 9.6. Example of a pin-jointed frame

The arrows on the members in that figure are inserted to show their positive $x$-directions. This is a convenient way of distinguishing ends 1 and 2 of a member in a drawing. If individual members are referred to by the joints at their ends, the compatibility equations are

$$(D'_1)_{AB} = (D'_1)_{CB} = (D'_1)_{AE} = (D'_1)_{CE} = 0$$
$$(D'_2)_{AB} = (D'_2)_{CB} = (D'_1)_{BE} = \Delta_B, \quad (D'_2)_{AE} = (D'_2)_{CE} = (D'_2)_{BE} = \Delta_E$$

where primes are omitted from the quantities $\Delta_B$, $\Delta_E$, since these must obviously be measured in the overall reference frame. Similarly, the equilibrium equations are

$$(F'_2)_{AB} + (F'_2)_{CB} + (F'_1)_{BE} = P_B$$
$$(F'_2)_{AE} + (F'_2)_{CE} + (F'_2)_{BE} = P_E$$

Combining these equations with equations 9.16 for the individual members it follows that

$$\begin{bmatrix} P_B \\ P_E \end{bmatrix} = \begin{bmatrix} Y'_{AB} + Y'_{CB} + Y'_{BE} & -Y'_{BE} \\ -Y'_{BE} & Y'_{AE} + Y'_{CE} + Y'_{BE} \end{bmatrix} \begin{bmatrix} \Delta_B \\ \Delta_E \end{bmatrix} \quad \ldots (9.17)$$

## 9.4. PROBLEMS INVOLVING COORDINATE TRANSFORMATIONS 9.4.1

Here again, the physical structure of the frame is reflected in the arrangement of the elements of the matrix in 9.17. The four scalar equations can now be solved for the displacements, and the internal forces obtained from 9.16.

Consider now the case in which all the members of the frame in FIGURE 9.6 have the same cross-sectional area, the members $AE$ and $CB$ being at 45° to the horizontal. The stiffness matrices for the various members are:

$$Y'_{AB} = Y'_{CE} = (EA/L) \begin{bmatrix} 1 & 0 \\ 0 & 0 \end{bmatrix}$$

$$Y'_{AE} = (EA/L\sqrt{2}) \begin{bmatrix} \frac{1}{2} & \frac{1}{2} \\ \frac{1}{2} & \frac{1}{2} \end{bmatrix} = (EA/L) \begin{bmatrix} \frac{1}{2\sqrt{2}} & \frac{1}{2\sqrt{2}} \\ \frac{1}{2\sqrt{2}} & \frac{1}{2\sqrt{2}} \end{bmatrix}$$

$$Y'_{CB} = (EA/L\sqrt{2}) \begin{bmatrix} \frac{1}{2} & -\frac{1}{2} \\ -\frac{1}{2} & \frac{1}{2} \end{bmatrix} = (EA/L) \begin{bmatrix} \frac{1}{2\sqrt{2}} & \frac{-1}{2\sqrt{2}} \\ \frac{-1}{2\sqrt{2}} & \frac{1}{2\sqrt{2}} \end{bmatrix}$$

$$Y'_{BE} = (EA/L) \begin{bmatrix} 0 & 0 \\ 0 & 1 \end{bmatrix}$$

If $P_B = \begin{bmatrix} 0 \\ 0 \end{bmatrix}$ and $P_E = \begin{bmatrix} 0 \\ 1 \end{bmatrix}$ (i.e. a unit vertical downward load at $E$), the load-displacement equations can be written down from 9.17 as

$$\begin{bmatrix} 0 \\ 0 \\ 0 \\ 1 \end{bmatrix} = \left(\frac{EA}{L}\right) \begin{bmatrix} 1 + \frac{1}{2\sqrt{2}} & \frac{-1}{2\sqrt{2}} & 0 & 0 \\ \frac{-1}{2\sqrt{2}} & 1 + \frac{1}{2\sqrt{2}} & 0 & -1 \\ 0 & 0 & 1 + \frac{1}{2\sqrt{2}} & \frac{1}{2\sqrt{2}} \\ 0 & -1 & \frac{1}{2\sqrt{2}} & 1 + \frac{1}{2\sqrt{2}} \end{bmatrix} \begin{bmatrix} x_B \\ y_B \\ x_E \\ y_E \end{bmatrix}$$

with the solution:—

$$x_B = 0.442\ L/EA \qquad y_B = 1.693\ L/EA$$
$$x_E = -0.558\ L/EA \qquad y_E = 2.135\ L/EA$$

from which values the forces in the members can be determined.

The method described above was first suggested by Chen[4]; it is most useful in dealing with highly redundant trusses, where the number of members is large and the number of joints small. It is obviously not suitable for analysing statically determinate trusses.

## 9.4.2. Plane Rigidly Jointed Frames

A member of a plane rigidly jointed frame can sustain both direct forces and bending moments at its ends, and may be regarded as a combination of the two cases already considered. Thus the displacement and end-load vectors for the member shown in FIGURE 9.7 now have three components, and can be written

$$D_1 = \begin{bmatrix} x_1 \\ y_1 \\ \theta_1 \end{bmatrix} \quad D_2 = \begin{bmatrix} x_2 \\ y_2 \\ \theta_2 \end{bmatrix} \quad F_1 = \begin{bmatrix} S_1 \\ Q_1 \\ M_1 \end{bmatrix} \quad F_2 = \begin{bmatrix} S_2 \\ Q_2 \\ M_2 \end{bmatrix}$$

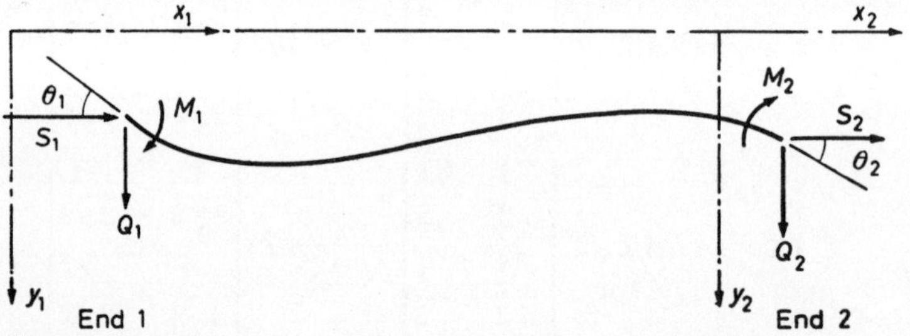

*Figure 9.7.* Notation for a beam in a rigidly jointed plane framework

Combining equations 9.1 and 9.12 it follows immediately that

$$\left. \begin{aligned} \begin{bmatrix} S_1 \\ Q_1 \\ M_1 \end{bmatrix} &= \begin{bmatrix} EA/L & 0 & 0 \\ 0 & 12EI/L^3 & 6EI/L^2 \\ 0 & 6EI/L^2 & 4EI/L \end{bmatrix} \begin{bmatrix} x_1 \\ y_1 \\ \theta_1 \end{bmatrix} + \begin{bmatrix} -EA/L & 0 & 0 \\ 0 & -12EI/L^3 & 6EI/L^2 \\ 0 & -6EI/L^2 & 2EI/L \end{bmatrix} \begin{bmatrix} x_2 \\ y_2 \\ \theta_2 \end{bmatrix} \\ \begin{bmatrix} S_2 \\ Q_2 \\ M_2 \end{bmatrix} &= \begin{bmatrix} -EA/L & 0 & 0 \\ 0 & -12EI/L^3 & -6EI/L^2 \\ 0 & 6EI/L^2 & 2EI/L \end{bmatrix} \begin{bmatrix} x_1 \\ y_1 \\ \theta_1 \end{bmatrix} + \begin{bmatrix} EA/L & 0 & 0 \\ 0 & 12EI/L^3 & -6EI/L^2 \\ 0 & -6EI/L^2 & 4EI/L \end{bmatrix} \begin{bmatrix} x_2 \\ y_2 \\ \theta_2 \end{bmatrix} \end{aligned} \right\} \quad (9.18)$$

and these equations can now be written in the usual way as

$$\left. \begin{aligned} F_1 &= Y_{11} D_1 + Y_{12} D_2 \\ F_2 &= Y_{21} D_1 + Y_{22} D_2 \end{aligned} \right\} \quad \ldots (9.19)$$

The transformation of these equations is carried out in a similar manner to the transformation described in the previous Section. The transformation matrix is

$$T = \begin{bmatrix} \cos\alpha & \sin\alpha & 0 \\ -\sin\alpha & \cos\alpha & 0 \\ 0 & 0 & 1 \end{bmatrix}$$

since the variables $M$ and $\theta$ are not affected by the rotation of axes. The transformed equations may be written

$$\left. \begin{aligned} F_1' &= Y_{11}' D_1' + Y_{12}' D_2' \\ F_2' &= Y_{21}' D_1' + Y_{22}' D_2' \end{aligned} \right\} \quad \ldots (9.20)$$

## 9.4. PROBLEMS INVOLVING COORDINATE TRANSFORMATIONS 9.4.3

where $\quad Y'_{ij} = T^{-1} Y_{ij} T$, $(i, j = 1, 2)$

The matrix $Y'_{11}$ has components

$$\begin{bmatrix} C^2.EA/L + S^2.12\,EI/L^3 & SC\,(EA/L - 12EI/L^3) & -S.6EI/L^2 \\ SC\,(EA/L - 12EI/L^3) & S^2.\,EA/L + C^2.12\,EI/L^3 & C.\,6EI/L^2 \\ -S.6\,EI/L^2 & C.6EI/L^2 & 4\,EI/L \end{bmatrix}$$

where $\quad S = \sin\alpha$, $C = \cos\alpha$

and the other $Y'$ matrices are similar. Details of these matrices have been given by Livesley[5].

The assembly of the complete equations for a frame follows a pattern similar to that already described for pin-jointed frames. If the frame in FIGURE 9.6 had rigid joints, for instance, the compatibility and equilibrium equations written down for the pin-jointed frame would still hold, the symbols $D$ and $F$ now being interpreted as including angular displacements and moments. Equations 9.17 become

$$\begin{bmatrix} P_B \\ P_E \end{bmatrix} = \begin{bmatrix} (Y'_{22})_{AB} + (Y'_{22})_{CB} + (Y'_{11})_{BE} & (Y'_{12})_{BE} \\ (Y'_{21})_{BE} & (Y'_{22})_{AE} + (Y'_{22})_{CE} + (Y'_{22})_{BE} \end{bmatrix} \begin{bmatrix} \Delta_B \\ \Delta_E \end{bmatrix} \quad (9.21)$$

where the symbols $\Delta$ and $P$ now represent vectors which include rotations and moments respectively. Equations 9.21 correspond to six scalar equations, and when these have been solved equations 9.20 give the internal forces and moments.

It is obvious that a direct application of the method always gives three equations for each joint, corresponding to two degrees of freedom in displacement and one in rotation. The strains produced by axial forces are automatically included whether they are important or not. This may be desirable in analysing rigidly jointed trusses, while in frameworks of rectangular form it may be an unnecessary refinement. In the latter case it may be possible to reduce the number of equations considerably by the application of a little commonsense. In a portal-type frame, for instance, one would normally neglect vertical movements of the joints, setting the appropriate variables to zero and deleting the relevant equations and coefficients from the full set of equilibrium equations. It is appropriate here to add a word of warning. Many designers automatically assume that strains due to axial forces produce negligible effects in frames which have no cross-bracing. This is often true, but in structures with stiff beams and relatively weak stanchions the vertical loads on the stanchions may cause vertical deflexions which produce significant moments in the beams. In such a case a matrix method provides a valuable check on more elementary methods. It also allows one to tackle problems where axial forces must obviously be considered, such as the calculation of the stresses induced by the settlement of a stanchion foundation. The analysis of plane rigidly jointed frames by matrix methods similar to the one described above has been discussed in papers by Kron[6] and Livesley[5].

### 9.4.3. SPACE FRAMES

In rigidly jointed space frames each joint has six degrees of freedom—three components of displacement and three of rotation. The $D$ and $F$

vectors thus have six components each, and the matrices $Y_{ij}$ in equation 9.19 each have six rows and six columns. The transformation matrix $T$ is also more complicated, being equal to

$$\begin{bmatrix} l & m & n & 0 & 0 & 0 \\ l_1 & m_1 & n_1 & 0 & 0 & 0 \\ l_2 & m_2 & n_2 & 0 & 0 & 0 \\ 0 & 0 & 0 & l & m & n \\ 0 & 0 & 0 & l_1 & m_1 & n_1 \\ 0 & 0 & 0 & l_2 & m_2 & n_2 \end{bmatrix}$$

where $l$, $m$, $n$, are the direction cosines of the axis of the beam relative to the overall system coordinates, and $l_1$, $m_1$, $n_1$; $l_2$, $m_2$, $n_2$ are the direction cosines of the principal axes of bending. However, the assembly of the complete set of equations from the individual member equations and the joint equilibrium and compatibility conditions follows exactly the same pattern as before. A similar treatment can be developed for space frames with pin-joints.

Matrix methods provide a general approach to the analysis of complex space frames, although it must be admitted that the computational work is extremely laborious when carried out by hand. They have so far been applied mainly in aircraft stress analysis. An aircraft wing, besides being a complex three-dimensional structure, contains stress-carrying webs and cover-plates as well as beams. These contribute to the stiffness of the structure, and their stiffness matrices must be included in the final set of load-displacement equations. A review of matrix methods in aircraft structural analysis has been given by Turner[7]. In practice the large scale matrix operations involved are normally carried out on automatic digital computers.

## 9.5. NON-LINEAR EFFECTS

The matrix method developed for plane rigidly jointed frames in Section 9.4.2 takes account of the direct strains produced by axial forces. If these forces are sufficiently large, however, they may also considerably influence the stiffness of the members in bending, and, in the limit, may cause the structure to become unstable.

An extension of the normal relaxation method for analysing rigid frames which takes account of these effects has been given by Merchant[8] and Bolton[9]. (see Chapter 8). A similar extension to the matrix method described in Section 9.4.2 of this Chapter can easily be made. If the member shown in FIGURE 9.7 carries a known axial force $S$, whose ratio to the Euler buckling load $\frac{\pi^2 EI}{L^2}$ is $\frac{S}{P_E}$ then the stiffness coefficients $6EI/L^2$, $4EI/L$, $2EI/L$, $12EI/L^3$ appearing in the matrices in equation 9.18 must be multiplied by certain factors. These factors are functions of $\frac{S}{P_E}$ only, and are closely related to the $s$- and $c$-functions used by Merchant, as well as to the Berry functions described in many standard textbooks. Details of these functions have been given by Livesley[10]. The important point as far as matrix methods are concerned is that if the axial forces in a structure are non-zero but known, then the equations 9.19 remain linear, only the numerical values of the coefficients being changed. Transformation from 9.19 to 9.20 and the assembly of the complete set of equations follows the procedure outlined in Section 9.4.3, and the equations may be solved in the usual way.

In most treatments of instability effects it is tacitly assumed that the values of the axial forces in the members are known as part of the data of the problem. In fact, of course, the axial forces are themselves part of the internal force system, and are related to the external loads by the equations of joint equilibrium. It follows that the coefficients in these equations are really functions of the external loading system, so that the equations are non-linear. A practical solution to this problem is to use a method of successive approximation. First an analysis is carried out which does not include the instability effects at all. This gives approximate values for the axial forces, and these are used to modify the various coefficients of the member stiffness matrices in the manner outlined above. The solution of the load-displacement equations derived from these stiffness matrices gives a new set of values for the axial forces, which can be compared with the previous set to see whether any significant change has occurred. If not, then the analysis is complete. If a change has occurred, a third analysis based on the second axial force system gives a third set of internal forces, which can be compared with the second in the same way. In practice a third analysis is rarely required. It should be noted that this technique is not an approximation in the sense of, say, the method of moment distribution. It consists of a sequence of *complete* linear analyses tending to a non-linear analysis, while moment distribution is merely a sequence of steps tending to a single *linear* analysis. The technique is well suited to automatic computing machines, and is described in more detail in Livesley's paper[10].

A similar approach can be adopted in dealing with other causes of non-linearity. In problems involving gross distortion, for instance, the non-linearity is often due to the large changes which occur in the angles of inclinations of the members. In such a case, an initial linear analysis can be used to find approximate values for these angles in the deformed structure, and the member stiffness matrices re-calculated. A second analysis based on these matrices leads to new values for the angles, and the process can be repeated until changes are insignificant. By this means a solution is obtained which satisfies the equilibrium equations for the distorted structure, however large the distortions may be.

The problem of calculating the loading at which a structure becomes elastically unstable is a more difficult one. In most cases the only approach is to carry out a series of analyses for gradually increasing loads, plotting a representative deflexion or rotation to determine the critical load. At this load, of course, the deflexion of the structure becomes indeterminate, and the load-displacement equations cease to have a unique solution. In matrix terminology, the critical load is the load which makes the stiffness matrix singular, or alternatively the load which makes the determinant of the stiffness matrix zero. Since the coefficients which make up the stiffness matrix are complex transcendental functions of the applied loading system, a direct solution of the determinantal equation is usually impracticable.

## 9.6. PARTITIONING TECHNIQUES

The use of matrix techniques leads to the computational problem of solving sets of linear equations, and if a structure has a large number of joints the number of equations will also be large. The maximum number of equations which can be solved as a single entity on an automatic computer is at present of the order of 100, and although this is an improvement on the dozen or so that a human being might regard as his limit, it is still small when one considers the analysis of a large and complicated structure.

Fortunately the equations which relate loads and displacements in a structure contain a large number of zero coefficients, and the proportion of such coefficients tends to increase with the number of joints. However large the structure, any one joint is not likely to be directly connected to more than three or four others, and its equilibrium equations will only involve those joints. If a joint $A$, for instance, is joined by members to joints $P$, $Q$, $R$, $S$, then the only non-zero coefficients in the equilibrium equations for joint $A$ will be those associated with the displacements of joints $A$, $P$, $Q$, $R$, $S$. If the equations are solved by the process of successive elimination of variables, mentioned in Section 9.2.3, the existence of a large number of zero coefficients considerably reduces the work of solution.

An even more important feature of most practical structures is that joints are only connected to their nearer neighbours. In many cases this makes it possible to treat a structure as a number of smaller structures connected at certain joints, each sub-structure being dealt with in turn. This does not affect the total amount of computing which has to be done, but it does break it down into stages of more manageable size. As a simple example we shall consider the structure shown in FIGURE 9.8(a). As in other parts of this Chapter we refer to member stiffness matrices as though they were single coefficients, and treat joint load and displacement vectors as single variables.

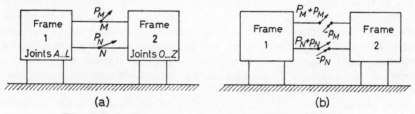

*Figure 9.8.* Division of a structure into two simpler sub-structures

We consider the structure as two simpler structures with common joints $M$ and $N$, the external loads at these joints being $P_M$ and $P_N$. It is clear that apart from $M$ and $N$ no joints in either frame are connected to joints in the other, so that by introducing the unknown internal loads $p_M$, $p_N$ we can treat the structure as two entirely independent smaller frames, as shown in FIGURE 9.8(b).

## 9.6. PARTITIONING TECHNIQUES

For frame 2 in that figure we can use the methods described earlier in this Chapter to write down the load-displacement equations,

$$\begin{bmatrix} -p_M \\ -p_N \\ P_O \\ | \\ | \\ | \\ P_Z \end{bmatrix} = \begin{bmatrix} Y_2 \end{bmatrix} \begin{bmatrix} \Delta_M \\ \Delta_N \\ \Delta_O \\ | \\ | \\ | \\ \Delta_Z \end{bmatrix} \quad \ldots (9.22)$$

where $Y_2$ is the stiffness matrix of the frame. Since we do not know $p_M$, $p_N$ we cannot solve the equations completely, but we can eliminate all coefficients of $\Delta_O$, $\Delta_P \ldots \Delta_Z$ from the first two equations by adding to them suitable multiples of the others. By this means we obtain

$$\begin{bmatrix} -p_M + \phi_M (P_O \ldots P_Z) \\ -p_N + \phi_N (P_O \ldots P_Z) \end{bmatrix} = \begin{bmatrix} Y^*_2 \end{bmatrix} \begin{bmatrix} \Delta_M \\ \Delta_N \end{bmatrix} \quad \ldots (9.23)$$

where $Y^*_2$ indicates the altered coefficients produced by the elimination process, and $\phi_M$, $\phi_N$ are linear combinations of the loads, $P_O \ldots P_Z$. One may regard $Y^*_2$ as defining the effective stiffness of frame 2 at the joints $M$ and $N$, and $\phi_M$, $\phi_N$ as the effective loads which the forces on frame 2 induce at these joints.

In the same way the equations for frame 1 are

$$\begin{bmatrix} P_A \\ | \\ | \\ | \\ P_L \\ P_M + p_M \\ P_N + p_N \end{bmatrix} = \begin{bmatrix} Y_1 \end{bmatrix} \begin{bmatrix} \Delta_A \\ | \\ | \\ | \\ \Delta_L \\ \Delta_M \\ \Delta_N \end{bmatrix} \quad \ldots (9.24)$$

and we can now eliminate $p_M$, $p_N$ by combining equations 9.23 and 9.24 to give,

$$\begin{bmatrix} P_A \\ | \\ | \\ | \\ P_L \\ P_M + \phi_M \\ P_N + \phi_N \end{bmatrix} = \begin{bmatrix} Y^*_1 \end{bmatrix} \begin{bmatrix} \Delta_A \\ | \\ | \\ | \\ \Delta_L \\ \Delta_M \\ \Delta_N \end{bmatrix} \quad \ldots (9.25)$$

where $Y^*_1$ indicates the stiffness matrix $Y_1$ of frame 1 with the coefficients of $Y^*_2$ from 9.23 added into the appropriate places. The equations 9.25 can now be solved by normal methods; once $\Delta_M$, $\Delta_N$ are known the first two equations of 9.22 can be discarded and the remainder solved to give the displacements of the joints in frame 2.

This method is in essence the same as the 'method of tearing', advocated by Kron[11]. It can be extended to the division of a structure into any number of smaller parts.

## 9.7. DYNAMIC PROBLEMS

We have seen in this Chapter that matrix methods are particularly suitable for analysing complex highly-redundant structures, and it is not surprising that they have been used extensively in the stress analysis of aircraft. The aeronautical designer, however, is not merely interested in static loading. He also has to deal with a large range of dynamic problems, most of them of considerably greater complexity than the static load problems encountered by ordinary structural engineers. For dealing with dynamic problems some form of matrix method is almost essential.

Although problems of structural vibration normally involve systems of distributed mass, it is usually possible to replace a continuous mass distribution by a series of concentrated masses without introducing much change in the dynamic behaviour of the system. If these masses are concentrated at the joints, the matrix techniques outlined in this Chapter can be used to construct equations of motion just as easily as equations of static equilibrium. For instance, if a structure has displacements $x_j$ under a set of static loads $P_i$, the methods outlined in this Chapter provide a systematic means of setting up the equilibrium equations

$$P_i = a_{ij} x_j$$

where $a_{ij}$ is the stiffness matrix of the structure. If now the $P_i$'s are time-dependent forces, and the mass of the structure can be treated as a series of concentrated masses $M_i$ at the joints, elementary mechanics tells us that the equations of motion of these masses are

$$P_i(t) - a_{ij} x_j = M_i \ddot{x}_i \qquad \ldots. (9.26)$$

(where the repetition of the suffix $i$ on the right-hand side does not imply summation). Rotational inertia and angular displacements can of course be included in these equations where appropriate.

The simplest special case of these equations is where the $P_i$'s are all zero. This arises in the important problem of finding the normal modes and natural frequencies of a structure in free undamped vibration. Putting $x_i = \bar{x}_i \sin \omega t$ in 9.26 leads immediately to the standard matrix problem of finding latent roots and vectors.

A more complex problem in which the $P_i$'s are non-zero is that of wing flutter. Here the applied forces depend on the velocities of the points on the wing relative to the air, so that the $P_i$'s in equation 9.26 are functions of both the aircraft velocity $V$ and the velocities $\dot{x}_i$ of the mass points relative to the aircraft. The problem is to find the aircraft velocity, or 'flutter speed', at which sustained vibration can occur under the action of these forces. A matrix approach to the problem involves finding the values of aircraft velocity and flutter frequency which make a determinant zero, the coefficients of the determinant being complex functions of these unknowns, as well as involving the stiffness of the wing structure.

There is a considerable body of literature on the matrix analysis of aircraft structures for both static and dynamic loading conditions, but unfortunately much of it is written in language unfamiliar to the average civil engineer. Apart from the paper by Turner[7] mentioned earlier a good account of the application of matrices to dynamic problems has been given by Williams[12]. For a general account of the use of matrices in aircraft structures the reader is referred to a series of papers by Argyris[13].

## 9.8. CONCLUSION

This Chapter has attempted to show that matrix notation has two aspects which are useful in structural analysis. First it provides a flexible but precise shorthand for discussing general problems of structural behaviour, and on this account alone it deserves the attention of engineers. Secondly it leads to the development of methods of structural analysis in which the computational work is completely systematic. This feature makes matrix methods especially suitable for an automatic computer, or for a human computer with a general understanding of numerical techniques but no specialist structural knowledge.

The methods described in this Chapter are representative of the family of equilibrium methods which take the displacements of the joints of a structure to be the basic unknowns. For a treatment of the approach by way of redundant forces (compatibility method) the reader is referred to the papers by Argyris[13] mentioned earlier and to a recent book by Morice[14].

Although the large high speed digital computer may always be a rarity, confined on the whole to Universities, research establishments, and the largest firms, it seems likely that smaller machines, designed in the first place for clerical and accounting work, will soon become fairly common. These machines, although slow by automatic computer standards, are yet many times faster than a human computer. Although their relatively small storage capacity is something of a drawback in structural work, partitioning techniques can nearly always be used to break down large problems into a series of smaller ones. It may not be long before structural designers regard automatic computers as much a part of their analytical equipment as the traditional slide-rule.

## REFERENCES

[1] HARTREE, D. R. *Numerical Analysis.* Oxford University Press (1952)

[2] CRANDALL, S. H. *Engineering Analysis.* McGraw-Hill (1956)

[3] BENSCOTER, S. U. ' The Matrix Analysis of Continuous Beams '. *Trans. Amer. Soc. Civ. Engrs.* Vol. 112. 1947. p. 1109

[4] CHEN, P. P. ' The Matrix Analysis of pin-connected structures '. *Trans. Amer. Soc. Civ. Engrs.* Vol. 114. 1949. p. 181

[5] LIVESLEY, R. K. ' The Analysis of Rigid Frames by an Electronic Computer '. *Engineering.* Vol. 176. pp. 230 and 277 (August 1953)

## REFERENCES

[6] KRON, G. 'Tensorial Analysis of Elastic Structures'. *J. Franklin Inst.* Vol. 238. No. 6. Dec. 1944

[7] TURNER, M. J., CLOUGH, R. W., MARTIN, H. C., and TOPP, L. J. 'Stiffness and Deflexion Analysis of Complex Structures'. *J. Aero. Sciences.* Vol. 23. No. 9 (September 1956)

[8] MERCHANT, W. 'Critical Loads of Tall Building Frames'. *The Structural Engineer.* Vol. 33. 1955. p. 84

[9] BOLTON, A. 'A Quick Approximation to the Critical Load of Rigidly Jointed Trusses'. *The Structural Engineer.* Vol. 33. 1955. p. 90

[10] LIVESLEY, R. K. 'The Application of an Electronic Digital Computer to some problems of Structural Analysis'. *The Structural Engineer.* Vol. 34. 1956. p. 1

[11] KRON, G. 'Solving Highly complex Elastic Structures in Easy Stages'. *J. Appl. Mech.* Vol. 22. 1955. p. 235

[12] WILLIAMS, D. 'Recent Developments in the Structural Approach to Aero-Elasticity Problems'. *J. Roy. Aero. Soc.* Vol. 58. 1954. p. 403

[13] ARGYRIS, J. H. 'Energy Theorems and Structural Analysis'. *Aircraft Engineering.* Vol. 27. 1955. p. 347, and subsequent issues

[14] MORICE, P. B. *Linear Structural Analysis.* Thames & Hudson (1959)

## Appendix A
# BEAMS OF NON-UNIFORM CROSS-SECTION
## A.1. DEFLEXION

In Chapter 3 various methods for computing deflexions were illustrated by examples where the flexural rigidity, $EI$, was constant; in one case, shown in Figure 3.37(a), a cantilever with a single change of section was studied. In practice, however, non-uniform beams are very common, especially in reinforced concrete construction where haunched beams are often used.

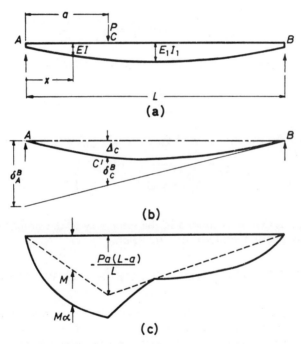

*Figure A.1.* (a) Simply-supported beam of non-uniform section; (b) The beam in its deflected position; (c) The bending moment diagram modified by multiplying the ordinate at each point by the value there of $\alpha = E_1 I_1 / EI$

Methods for computing deflexions based on the integration of equation 3.14

$$\frac{d^2 y}{dx^2} = + \frac{M}{EI}$$

are evidently no longer feasible unless the flexural rigidity is an integrable function of $x$, and even then the various limits that may have to be used complicate the final evaluation. In such cases it may be preferable to resort to numerical or graphical integration and this is essential when $EI$ varies irregularly.

The moment-area method, which was described in Chapter 3, Section 3.7.4, can be modified in the following manner to make it suitable for non-uniform beams:—

FIGURE A.1(a) shows a simply-supported beam $AB$, of variable cross-section, carrying a point load at $C$ whose deflexion is required. The beam is shown in its deflected position at (b), and the final position of $C$ is given by

$$\Delta_C = \delta_A^B \frac{L-a}{L} - \delta_C^B$$

The displacements of $A$ and $C$ from the tangents at $B$ are given by the moment-area equation 3.22

$$\delta_A^B = \int_0^L \frac{Mxdx}{EI} \quad \text{and} \quad \delta_C^B = \int_a^L \frac{M(x-a)\,dx}{EI}$$

Numerical or graphical integration of these expressions is facilitated by writing

$$\frac{E_1 I_1}{EI} = \alpha$$

where $E_1 I_1$ is the specific value of the flexural rigidity at a convenient point. We then have

$$\delta_A^B = \frac{1}{E_1 I_1} \int_0^L \alpha\, Mxdx \quad \text{and} \quad \delta_A^B = \frac{1}{E_1 I_1} \int_a^L \alpha\, M(x-a)dx$$

These integrals can be evaluated in the same way as the similar ones which appear throughout the following discussion.

## A.2. THE FIXING MOMENTS OF BEAMS WITH FIXED ENDS

Calculations of the values of the fixing moments acting at the ends of non-uniform beams with fixed ends follow similar lines to those for uniform beams (Chapter 6, Section 6.2.2), with suitable modifications to allow for the variation of the flexural rigidity. FIGURE A.2(a) shows such a beam; the corresponding simply-supported beam has the bending moment diagram shown at (b), while the fixing moments $M_{FAB}$ and $M_{FBA}$ give the diagram (c).

As there is no change of slope from $A$ to $B$, equation 3.21 gives

$$0 = \int_A^B \frac{Mdx}{EI} = \frac{1}{E_1 I_1} \left\{ \int_A^B \alpha\, M_o dx + \frac{M_{FAB}}{L} \int_A^B \alpha\, (L-x)dx + \frac{M_{FBA}}{L} \int_A^B \alpha\, xdx \right\}$$

$$= \frac{1}{E_1 I_1} \left\{ \int_A^B \alpha\, M_o dx + \frac{M_{FAB}}{L} G_1 + \frac{M_{FBA}}{L} G_2 \right\} \quad \ldots\ldots(\text{A.1})$$

## A.3. SLOPE-DEFLEXION EQUATIONS

As there is no deflexion of $A$ from the tangent at $B$, equation 3.22 gives

$$0 = \int_A^B \frac{Mxdx}{EI} = \frac{1}{E_1I_1}\left\{\int_A^B \alpha\, M_o\, xdx + \frac{M_{FAB}}{L}\int_A^B \alpha\, x(L-x)dx + \frac{M_{FBA}}{L}\int_A^B \alpha\, x^2 dx\right\}$$

$$= \frac{1}{E_1I_1}\left\{\int_A^B \alpha\, M_o\, xdx + \frac{M_{FAB}}{L}G_3 + \frac{M_{FBA}}{L}G_4\right\} \quad \ldots (A.2)$$

The first term in each equation must be evaluated for the particular

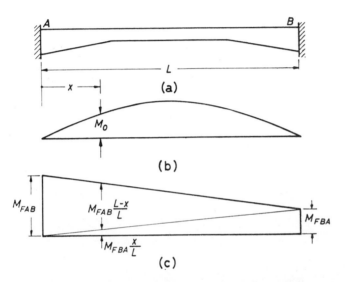

*Figure A.2.* (a) Non-uniform beam with fixed ends; (b) Bending moment diagram for the primary simply-supported beam under the applied loads. The correct sign must be given to $M_o$ in an actual problem; (c) The fixing moment diagram

loading in question, but the remaining terms $G_1 - G_4$ depend only on the elastic properties of the beam and can be computed once for all. Simultaneous solution of these equations yields the required fixing moments $M_{FAB}$ and $M_{FBA}$.

### A.3. THE SLOPE-DEFLEXION EQUATIONS

A similar modification of the discussion in Chapter 6, Section 6.3 leads to the slope-deflexion equations for non-uniform beams. FIGURE A.3(a) shows a beam $AB$ bent by the action of the terminal couples $m_{AB}$ and $m_{BA}$ and with $B$ displaced an amount $\delta$ (clockwise) relative to $A$. According to the slope-deflexion convention of signs all clockwise external couples and all clockwise rotations are positive. As far as internal bending moments are concerned, however, clockwise terminal couples make the beam sag or hog if they act at the left or right-hand end respectively, and are therefore of opposite sign in the bending-moment convention, as shown at (b).

## BEAMS OF NON-UNIFORM CROSS-SECTION

At the end $B$, we have

$$-\delta_B^A = L\theta_{AB} - \delta = \int_A^B \frac{M(L-x)\,dx}{EI} = -\frac{m_{AB}}{E_1 I_1 L}\int_0^L \alpha(L-x)^2 dx$$

$$+ \frac{m_{BA}}{E_1 I_1 L}\int_0^L \alpha x(L-x)\,dx$$

Hence $\quad E_1 I_1 L(L\theta_{AB} - \delta) = m_{AB} G_5 - m_{BA} G_3 \quad\quad\quad\ldots\ldots$ (i)

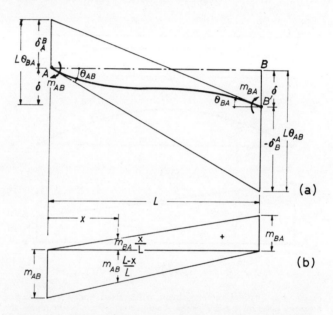

*Figure A.3.* (a) Beam with one end displaced an amount $\delta$ (clockwise) relative to the other and under the action of clockwise end moments; (b) Bending moment diagram for the end couples

Similarly, at end $A$,

$$\delta_A^B = L\theta_{BA} - \delta = \frac{m_{BA}}{E_1 I_1 L}\int_0^L \alpha x^2 dx - \frac{m_{AB}}{E_1 I_1 L}\int \alpha x(L-x)\,dx$$

$$E_1 I_1 L(L\theta_{AB} - \delta) = m_{BA} G_4 - m_{AB} G_3 \quad\quad\quad\ldots\ldots$ (ii)

Solving (i) and (ii) simultaneously for $m_{AB}$ and $m_{BA}$ and adding to each expression the appropriate fixing moment gives

$$M_{AB} = \frac{E_1 I_1 L^2}{G_4 G_5 - G_3{}^2}\{G_4 \theta_{AB} + G_3 \theta_{BA} - (G_3 + G_4)\delta/L\} + M_{FAB} \quad\ldots\text{(A.3)}$$

$$M_{BA} = \frac{E_1 I_1 L^2}{G_4 G_5 - G_3{}^2}\{G_5 \theta_{BA} + G_3 \theta_{AB} - (G_3 + G_5)\delta/L\} + M_{FBA} \quad\ldots\text{(A.4)}$$

## A.4. MOMENT DISTRIBUTION FACTORS

The slope-deflexion equations can conveniently be used to derive the distribution, carry-over and sway correction factors required for moment distribution.

### A.4.1. Stiffness; Beam Fixed at Far End

The stiffness at the end $A$ of the beam $AB$ (Figure A.4(a)) is the terminal couple required to produce unit rotation at $A$.

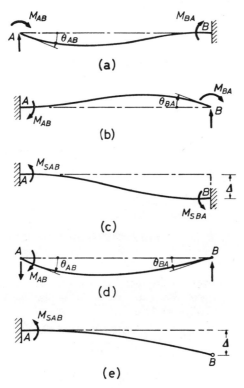

Figure A.4. Determination of moment distribution coefficients for non-uniform beams; (a) and (b) Stiffness and carry-over factors; (c) Sway moments; (d) Stiffness of beam pinned at far end; (e) Sway moments of beam pinned at one end

From equation A.3,

$$M_{AB} = \frac{E_1 I_1 L^2}{G_4 G_5 - G_3^2} G_4 \theta_{AB} \qquad \ldots\ldots \text{(i)}$$

i.e.

$$\frac{M_{AB}}{\theta_{AB}} = \frac{G_4}{G_4 G_5 - G_3^2} E_1 I_1 L^2$$

$$= \text{stiffness for end } A \qquad \ldots\ldots \text{(A.5)}$$

## BEAMS OF NON-UNIFORM CROSS-SECTION

Similarly, if $M_{BA}$ is applied at $B$ while $A$ is fixed, as at (b)

$$\frac{M_{BA}}{\theta_{BA}} = \frac{G_5}{G_4 G_5 - G_3^2} E_1 I_1 L^2$$

$$= \text{stiffness for end } B \qquad \ldots (A.6)$$

### A.4.2. Carry-over Factor

The carry-over factor from $A$ to $B$ is the ratio $M_{BA}/M_{AB}$ when the couple $M_{AB}$ is applied at $A$ as in Figure A.4(a).

From equation A.4 putting $\theta_{BA} = 0$ and $\delta = 0$

$$M_{BA} = \frac{E_1 I_1 L^2}{G_4 G_5 - G_3^2} G_3 \theta_{AB}$$

Combining this with (i), we have

$$\frac{M_{BA}}{M_{AB}} = \frac{G_3}{G_4} \qquad \ldots (A.7)$$

$$= \text{carry-over factor from } A \text{ to } B.$$

Similarly, when $M_{BA}$ is applied at $B$ as in (b)

$$\frac{M_{AB}}{M_{BA}} = \frac{G_3}{G_5} \qquad \ldots (A.8)$$

$$= \text{carry-over factor from } B \text{ to } A.$$

### A.4.3. Sway Moment; Both Ends Fixed

If one end of the beam is displaced, as in (c), without rotation of either end, moments $M_{SAB}$ and $M_{SBA}$ are called into play which can be found from the expressions A.3 and A.4.

$$M_{SAB} = -\frac{E_1 I_1 L \Delta (G_3 + G_4)}{G_4 G_5 - G_3^2} \qquad \ldots (A.9)$$

$$M_{SBA} = -\frac{E_1 I_1 L \Delta (G_3 + G_5)}{G_4 G_5 - G_3^2} \qquad \ldots (A.10)$$

### A.4.4. Stiffness; Beam Pinned at Far End. Figure A.4(d)

From equation A.4

$$\theta_{BA} = -\frac{G_3}{G_5} \theta_{AB}$$

In equation A.3

$$M_{AB} = \frac{E_1 I_1 L^2}{G_4 G_5 - G_3^2} \left\{ G_4 \theta_{AB} - G_3 \cdot \frac{G_3}{G_5} \theta_{AB} \right\}$$

$$\frac{M_{AB}}{\theta_{AB}} = \frac{E_1 I_1 L^2}{G_5} \qquad \ldots (A.11)$$

$$= \text{stiffness for end } A.$$

## A.4.5. Sway Moment; Beam Pinned at Far End. Figure A.4 (e)

From equation A.4

$$\theta_{BA} = \frac{G_3 + G_5}{G_5} \frac{\Delta}{L}$$

In equation A.3

$$M_{SAB} = \frac{E_1 I_1 L^2}{G_4 G_5 - G_3^2} \left\{ \frac{G_3(G_3 + G_5)}{G_5} \frac{\Delta}{L} - (G_3 + G_4) \frac{\Delta}{L} \right\}$$

$$= -\frac{E_1 I_1 L \Delta}{G_5} \quad \ldots \ldots (A.12)$$

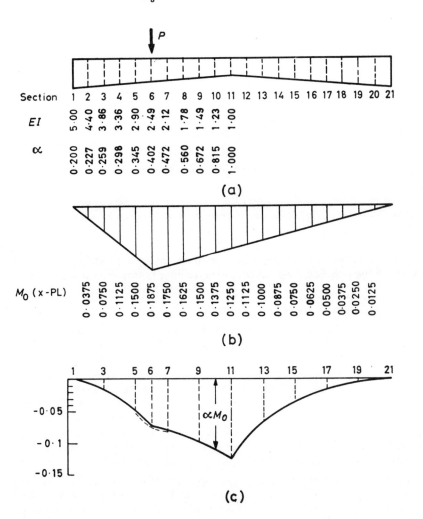

Figure A.5. (a) Data for non-uniform beam; (b) Bending moment diagram ($M_o$); (c) $\alpha\, M_o$ diagram

## A.5. EVALUATION OF COEFFICIENTS

It now remains to demonstrate the evaluation of the coefficients

$$G_1 = \int_0^L \alpha \, (L - x) \, dx$$

$$G_2 = \int_0^L \alpha x \, dx$$

$$G_3 = \int_0^L \alpha x \, (L - x) \, dx$$

$$G_4 = \int_0^L \alpha x^2 \, dx$$

$$G_5 = \int_0^L \alpha \, (L - x)^2 \, dx,$$

and of the bending moment integrals

$$\int \alpha \, M_o \, dx \quad \text{and} \quad \int \alpha \, M_o \, x \, dx$$

As an example, calculations will be made for the beam shown in FIGURE A.5(a). Flexural rigidities are referred to the central value $E_1 I_1$; the width is constant and the depth varies linearly being $d$ at the centre and $1.71 \, d$ at the ends. Proportional values of $EI$ and hence of $\alpha$ can now be evaluated and are shown for eleven sections along one half of the beam. Values of the bending moment at corresponding points are shown at (b).

### A.5.1. INTEGRATION BY SIMPSON'S RULE

Numerical integration can be performed by the mid-ordinate method, by the trapezoidal rule or by Simpson's rule†. There is little to choose between these so far as the amount of arithmetic is concerned, but as the last usually gives the closest approximation it is illustrated here. Sufficient accuracy is achieved by using eleven sections provided that an abrupt discontinuity does not coincide with one of the 'even' sections. In FIGURE A.5(c) for example, the dotted line shows the shape which the approximating curve takes at sections 5—6—7; the full line is the correct shape at this point. A much more serious error would have occurred at the central discontinuity had this not coincided with the 'odd' section 11.

These discontinuities evidently correspond to the peak of the bending-moment diagram at section 6 and to the break in the continuity of the soffit of the beam at section 11. It is advisable to subdivide the beam so that 'odd' sections fall at such points. If this is not possible then the areas between such discontinuities should be computed separately.

## A.5. EVALUATION OF COEFFICIENTS

The calculations for the beam of FIGURE A.5(a) are set out, using twenty-one sections, in Table A.1 and are almost self-explanatory. Columns 1—5 depend only on the spacing chosen for the sections, and can be set down at once. Column 6 records the values of $\alpha$, which depend on the variation of the cross-section, and column 7 gives the values of the bending moment which depend on the loading. The remaining columns 8—12 are obtained by multiplying together appropriate figures from the earlier columns and are arranged for easy application of Simpson's rule. The symmetry of this particular beam makes $C_1 = C_2$ and $C_4 = C_5$.

The fixing moments are given by solution of equations A.1 and A.2.

(A.1) $\quad - 0.0502\, PL^2 + 0.232\, L.M_{FAB} + 0.232\, L.M_{FBA} = 0$

(A.2) $\quad - 0.0221\, PL^3 + 0.0913\, L^2\, M_{FAB} + 0.1404\, L^2\, M_{FBA} = 0$

i.e. $M_{FAB} = + 0.169\, PL \qquad M_{FBA} = + 0.047\, PL$

These should be compared with the fixing moments of a uniform beam with a point load at quarter-span, namely

$$M_{FAB} = + 0.141\, PL \quad M_{FBA} = + 0.047\, PL$$

The slope-deflexion equations involve the term

$$G_4 G_5 - G_3^2 = (0.1404^2 - 0.0913^2)\, L^6 = 0.01137\, L^6$$

---

† Numerical integration methods.

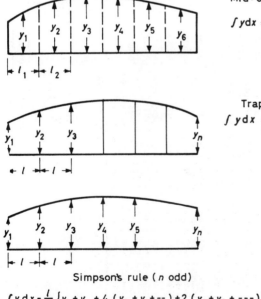

Mid-ordinate rule:
$\int y\,dx = \Sigma\, y\, l$
$= y_1 l_1 + y_2 l_2 + \cdots$

Trapezoidal rule:
$\int y\,dx = l \Sigma \{y_1 + y_n + 2(y_2 + y_3 \cdots)\}$
$= l \Sigma \{\text{end ordinates} + 2(\text{intermediate ordinates})\}$

Simpson's rule ($n$ odd)
$\int y.dx = \dfrac{l}{3}\{y_1 + y_n + 4(y_2 + y_4 + \cdots) + 2(y_3 + y_5 + \cdots)\}$
$= \dfrac{l}{3}\{\text{end ordinates} + 4(\text{even ordinates}) + 2(\text{odd ordinates})\}$

TABLE A.1.

| Section | 1 | 2 | 3 | 4 | 5 | 6 | 7 | 8 | | 9 | | 10 | | 11 | | 12 | |
|---|---|---|---|---|---|---|---|---|---|---|---|---|---|---|---|---|---|
| | $x$ | $x^3$ | $(L-x)$ | $(L-x)^3$ | $x(L-x)$ | $a$ | $M_0$ | $a M_0 = 6 \times 7$ | | $a M_0 x = 1 \times 8$ | | $a(L-x) = 3 \times 6$ | | $ax(L-x) = 5 \times 6$ | | $ax^2 = 2 \times 6$ | |
| | $\times L$ | $\times L^3$ | $\times L$ | $\times L^3$ | $\times L^2$ | | $\times PL$ | $\times PL$ | | $\times PL^2$ | | $\times L$ | | $\times L^2$ | | $\times L^2$ | |
| 1 | 0.00 | 0.0 | 1.00 | 1.000 | 0.000 | 0.200 | 0.0000 | 0.0000 | 0 | 0.0000 | 0 | 0.200 | 0.200 | 0.0000 | 0 | 0.0000 | 0 |
| 2 | 0.05 | 0.0025 | 0.95 | 0.902 | 0.047 | 0.227 | 0.0375 | 0.0085 | | 0.0004 | | 0.216 | | 0.0108 | | 0.0006 | |
| 3 | 0.10 | 0.01 | 0.90 | 0.810 | 0.090 | 0.259 | 0.075 | 0.0194 | | 0.0019 | | 0.233 | | 0.0233 | | 0.0026 | |
| 4 | 0.15 | 0.0225 | 0.85 | 0.722 | 0.127 | 0.298 | 0.1125 | 0.0336 | | 0.005 | | 0.253 | | 0.0380 | | 0.0067 | |
| 5 | 0.20 | 0.04 | 0.80 | 0.640 | 0.160 | 0.345 | 0.15 | 0.0517 | | 0.0103 | | 0.276 | | 0.0552 | | 0.0138 | |
| 6 | 0.25 | 0.0625 | 0.75 | 0.562 | 0.187 | 0.402 | 0.1875 | 0.0755 | | 0.0189 | | 0.302 | | 0.0753 | | 0.0251 | |
| 7 | 0.30 | 0.09 | 0.70 | 0.490 | 0.210 | 0.472 | 0.175 | 0.0826 | | 0.0248 | | 0.330 | | 0.0991 | | 0.0425 | |
| 8 | 0.35 | 0.1225 | 0.65 | 0.422 | 0.228 | 0.560 | 0.1625 | 0.0910 | | 0.0319 | | 0.364 | | 0.1273 | | 0.0686 | |
| 9 | 0.40 | 0.16 | 0.60 | 0.360 | 0.240 | 0.672 | 0.15 | 0.1008 | | 0.0403 | | 0.403 | | 0.1612 | | 0.1075 | |
| 10 | 0.45 | 0.2025 | 0.55 | 0.302 | 0.248 | 0.815 | 0.1375 | 0.1120 | | 0.0504 | | 0.448 | | 0.2015 | | 0.1652 | |
| 11 | 0.50 | 0.25 | 0.50 | 0.250 | 0.250 | 1.000 | 0.125 | 0.1250 | | 0.0625 | | 0.500 | | 0.2500 | | 0.2500 | |
| 12 | 0.55 | 0.3025 | 0.45 | 0.202 | 0.248 | 0.815 | 0.1125 | 0.0916 | | 0.0504 | | 0.367 | | 0.2015 | | 0.2465 | |
| 13 | 0.60 | 0.36 | 0.40 | 0.160 | 0.240 | 0.672 | 0.1 | 0.0672 | | 0.0403 | | 0.269 | | 0.1612 | | 0.2419 | |
| 14 | 0.65 | 0.4225 | 0.35 | 0.122 | 0.228 | 0.560 | 0.0875 | 0.0490 | | 0.0318 | | 0.196 | | 0.1273 | | 0.2366 | |
| 15 | 0.70 | 0.49 | 0.30 | 0.090 | 0.210 | 0.472 | 0.075 | 0.0354 | | 0.0248 | | 0.142 | | 0.0991 | | 0.2313 | |
| 16 | 0.75 | 0.5625 | 0.25 | 0.062 | 0.187 | 0.402 | 0.0625 | 0.0251 | | 0.0188 | | 0.101 | | 0.0753 | | 0.2261 | |
| 17 | 0.80 | 0.64 | 0.20 | 0.040 | 0.160 | 0.345 | 0.05 | 0.0172 | | 0.0138 | | 0.069 | | 0.0552 | | 0.2208 | |
| 18 | 0.85 | 0.7225 | 0.15 | 0.022 | 0.127 | 0.298 | 0.0375 | 0.0112 | | 0.0095 | | 0.045 | | 0.0380 | | 0.2153 | |
| 19 | 0.90 | 0.81 | 0.10 | 0.010 | 0.090 | 0.259 | 0.025 | 0.0065 | | 0.0058 | | 0.026 | | 0.0233 | | 0.2098 | |
| 20 | 0.95 | 0.9025 | 0.05 | 0.002 | 0.047 | 0.227 | 0.0125 | 0.0028 | | 0.0027 | | 0.011 | | 0.0108 | | 0.2048 | |
| 21 | 1.00 | 1.0 | 0.00 | 0.000 | 0.000 | 0.200 | 0.0000 | 0 | 0 | 0 | 0 | 0 | 0 | 0 | 0 | 0.2000 | 0.2000 |
| $\Sigma$ Even | | | | | | | | 0.5003 | | 0.2198 | | 2.303 | | 0.9058 | | 1.3955 | |
| $\Sigma$ Odd | | | | | | | | 0.5058 | | 0.2245 | | 2.248 | | 0.9276 | | 1.3202 | |
| 4 $\Sigma$ End | | | | | | | | | 0.0000 | | 0.0000 | | 0.200 | | 0.0000 | | 0.2000 |
| 2 $\Sigma$ Even | | | | | | | | | 2.0012 | | 0.8792 | | 9.212 | | 3.6232 | | 5.5820 |
| $\Sigma$ Odd | | | | | | | | | 1.0116 | | 0.4490 | | 4.496 | | 1.8552 | | 2.6404 |
| Sum | | | | | | | | | 3.0128 | | 1.3282 | | 13.908 | | 5.4784 | | 8.4224 |
| $\times L/60$ | | | | | | | | | $-0.0502\,PL^2$ | | $-0.0221\,PL^3$ | | $0.232\,L^2$ | | $0.0913\,L^3$ | | $0.1404\,L^3$ |
| | | | | | | | | | $\int a M_0\,dx$ | | $\int a M_0 x\,dx$ | | $G_1 = G_2$ | | $G_3$ | | $G_4 = G_5$ |

Equations A.3 and A.4 then become

$$M_{AB} = \frac{E_1 I_1}{L}\left\{12\cdot 35\,\theta_{AB} + 8\cdot 05\,\theta_{BA} - 20\cdot 4\,\delta/L\right\} - 0\cdot 141\,PL$$

$$M_{BA} = \frac{E_1 I_1}{L}\left\{12\cdot 35\,\theta_{BA} + 8\cdot 05\,\theta_{AB} - 20\cdot 4\,\delta/L\right\} + 0\cdot 047\,PL$$

The carry-over factor from $A$ to $B$ is

$$\frac{G_3}{G_4} = \frac{0\cdot 0913}{0\cdot 1404} = 0\cdot 64$$

### A.5.2. Graphical Integration

Quite acceptable accuracy can be obtained by measuring the derived areas given by the graphical construction shown in Figure A.6. The upper curve, of $\alpha$, is simply a plot, to a suitable scale, of the figures of column 6 of Table A.1. From a point $C$ on this curve the horizontal line $CD$ meets the vertical through $B$ at $D$. $AD$ meets the vertical $CG$ at $C'$, which is a point on the curve of $\alpha x/L$; the symmetry of this particular beam shows that the area under this curve is equal to that under the curve of $\alpha\,(L-x)/L$.

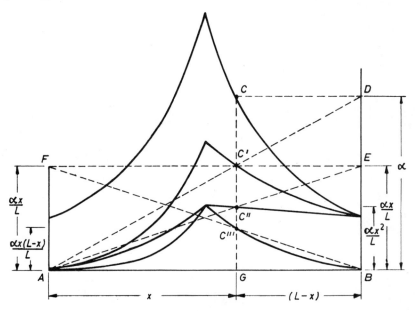

Figure A.6. Graphical construction for constants

A repetition of this construction gives $C''$, a point on the curve of $\alpha x^2/L$. To obtain $C'''$ on the curve of $\alpha x\,(L-x)/L$ we project from $C'$ to $F$, vertically above $A$, and then join $BF$ to cut $CG$ in $C'''$.

The areas under these curves are measured by planimeter and the values of $G_1 - G_5$ obtained by multiplying by the scale factor.

The integrals $\int \alpha M_o \, dx$ and $\int \alpha M_o \, x \, dx$ are obtained in a similar manner after plotting the figures of column 8 in TABLE A.1.

### A.5.3. TABULATED VALUES

While one or other of the above methods must be used in special cases, tabulated values of the constants for many standard cases of haunched beams are given in several publications[1,2]. The most useful of these is probably the *Handbook of Frame Constants*[3] which covers the great majority of the cases likely to be encountered in practice.

## REFERENCES

[1] RUPPEL, W. ' Moments in Restrained and Continuous Beams '. *Trans. Amer. Soc. Civ. Engrs*, Vol. 90. 1927. p. 152

[2] MARKLAND, E. ' Moment Distribution applied to non-prismatic members '. *Concrete and Constructional Engineering*. August and September, 1948

[3] *Handbook of Frame Constants*. Portland Cement Association, Chicago, Ill.

## Appendix B
# NOTES ON THE DESIGN OF HYPERSTATIC STRUCTURES

Although it has attracted much attention, as an interesting and important branch of the science of applied mechanics, the analysis of structures is only one of the series of processes that is involved in the production of the finished bridge or building. Many of these processes are quite outside the scope of this book: the preliminary investigation of the requirements, the formulation of the ruling dimensions of the structure, the choice of the critical live loading systems, the choice of a suitable technique of erection and so on: these are all preliminaries which are obviously of great importance. Then again, at a later stage, there are the organizational, financial and technical problems that accompany the actual construction. But between the preliminary studies and the manufacture and assembly of its components lie the analysis and design of the structure and the production of the detailed drawings from which manufacture can proceed. This is the responsibility of the design engineer, whose work clearly goes well beyond the principles and techniques of analysis with which we have so far been concerned.

It is hardly possible, within a reasonable compass, to deal adequately with the basic principles of analysis (which was our primary aim) and also with the details of design; but it is perhaps not satisfactory, especially when hyperstatic analysis is involved, to ignore the design problem completely for, as we have seen, the sizes of the members affect the distribution of load within a hyperstatic structure. The implication has been that the methods we have described are applied in the first instance to the analysis of a *trial design*; if the resulting stress distribution is reasonably satisfactory, from the point of view both of safety and of economy, then the design is complete and detailed drawings can be made; but if any members prove to be seriously over- or under-stressed then safety or economy require a revision. Surprisingly little has been written about the principles which should guide the designer in making such a revision but he can, at least, proceed on a trial and error basis until a reasonably satisfactory design has been achieved. This is a laborious business, however, and, since the consequences of failure are both more disastrous and more obvious than the possibility that a more economical design could have been obtained, it is not surprising that designers, of static structures at least, are more on their guard against over- than against under-stressing. Moreover the cost of the design process itself must not be overlooked and it is really only when the penalties for over-design are obvious and costly, notably in aircraft structures, that great refinement of design is justifiable. The selection of a suitable trial design is thus of crucial importance for, unless analysis shows it to be grossly extravagant, it is likely to be adopted if no members are over-stressed.

In making his trial design the engineer will draw on such existing information as is relevant to the problem in hand; where this is a comparatively simple structure, a continuous girder with a small number of spans, for example, complete solutions may already exist in the form of influence lines[1]. In the case of portal frames very effective guidance may be found in the works of Kleinlogel[2,3], who has published the solutions to a great many such frames under different loading systems. (Some of these results have recently been published in English[4].) But when there is no previous experience to draw on the designer must perforce fall back on one of several possibilities. In the case of rigidly jointed structures he will most probably adopt approximate methods (e.g. those suggested in C.P. 114[5] or B.S. 449[6]) and these may, in the case of minor structures at least, not require to be checked by an exact analysis. Or he may use the plastic method of design supplemented if thought necessary, by an elastic analysis to ensure that the deflexions are acceptable[7]. But the situation would be more satisfactory if a direct method of design of rigidly-jointed frames could be devised; some progress has already been made in this direction by Grinter[8] whose book *Automatic Design of Continuous Frames* describes a technique for balancing section moduli in the same sort of way as joint moments are balanced in moment distribution.

The position is perhaps not so satisfactory so far as hyperstatic braced frames are concerned. Certainly there is less in the way of published solutions of standard cases. One or two approximate methods of attack were briefly mentioned in Chapter I and there is scope for the development of these. Again, a method for obtaining a first approximation to the thrust of a two-hinged braced arch has been given by Haertlein[9]. But it is rather surprising that the direct design method devised by Pippard[10] has received so little attention. This method was rediscovered, independently, by Francis[11] who has shown how to incorporate prestressing so that all the members carry the desired stress under any given load system; he has also extended the method to non-linear frameworks[12] and so shown how to compute the collapse or limit loads of braced structures. As these ideas seem to be both important and capable of further development a brief introduction to Francis' technique is now given.

As we have seen, the solution to the analysis of a hyperstatic structure involves the satisfaction of two sets of conditions:—

(a) the equations of equilibrium, and

(b) the conditions of compatibility.

Any system of member forces satisfying condition (a) is acceptable but, in a hyperstatic structure, condition (b) requires that the changes of length of the redundant members shall conform to the pattern of deformation of the primary structure. In the method now under discussion a trial pattern of deformation is obtained by assuming that all the members are fully stressed; if condition (b) is not then satisfied (as will probably be the case) then the stresses in some members are suitably adjusted. A pattern of forces satisfying condition (a) is then chosen and the required cross-sectional areas can then be obtained immediately. A simple example will perhaps make the method clear.

# NOTES ON THE DESIGN OF HYPERSTATIC STRUCTURES

FIGURE B.1(a) shows an elementary framework carrying a single load $P$; it is desired to design this structure so that as many members as possible carry the working stress $f_w$ on the assumption that all the members are in compression. (It is not necessary that all the members should carry the

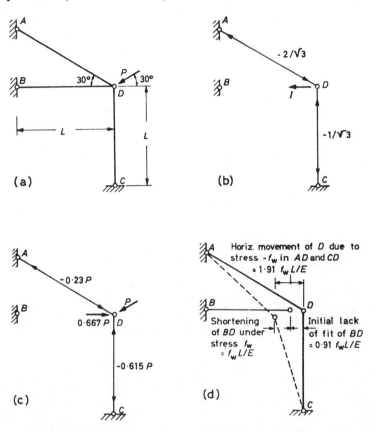

*Figure B.1.* (a) Simple hyperstatic framework under a fixed load system to be designed so that as many members as possible carry the working stress $f_w$; (b) Unit load applied at $D$ to find the horizontal displacement there; (c) Arbitrary load pattern defined initially by the force in $BD$; (d) If $BD$ is made $0 \cdot 91 f_w L/E$ too short before assembly all the members can be made to carry the desired stress $f_w$

same stress but, if different stresses are assigned to tension and compression members, it will be necessary to check that the force pattern which is finally chosen produces the assumed sign of stress in each member). Let $BD$ be regarded as the redundant member and, to begin with, assume that the stress in $AD$ and $CD$ is $f_w$. This assumption effectively fixes the position of $D$. We are particularly concerned with the horizontal displacement of $D$ (i.e. in the direction of $BD$) which is given by

$$\Delta_{BD} = \sum_{AD, DC} F'FL/AE = \Sigma F' f_w L/E$$

463

NOTES ON THE DESIGN OF HYPERSTATIC STRUCTURES

Here $F'$ is the force in members $AD$ and $DC$ due to unit horizontal load applied at $D$ to the primary structure $ADC$. (See diagram (b)). The calculations for $\Delta_{BD}$ are set out in TABLE B.1 in tabular form.

TABLE B.1

| Member | $L$ | $f$ | $F'$ | $F'fL/E$ |
|---|---|---|---|---|
| AD | $2L/\sqrt{3}$ | $-f_w$ | $-2/\sqrt{3}$ | $+4f_wL/3E$ |
| CD | $L$ | $-f_w$ | $-1/\sqrt{3}$ | $+f_wL/\sqrt{3}E$ |
| | | | $\Sigma$ | $=+1\cdot91 f_wL/E = \Delta_{BD}$ |

This displacement of $D$ corresponds to a stress in $BD$ of $1\cdot91 f_w$, which is unacceptable, so the stresses in $AD$ and $CD$ must be reduced; one possibility is to make them both equal to $-f_w/1\cdot91 = -0\cdot523f_w$. If this is done the stress in $BD$ will be $-f_w$ when the compatibility condition (b) is satisfied.

A suitable force pattern satisfying the equilibrium condition (a) and the assumption that all the members are compression can now be chosen. If the force in $BD$ is numerically greater than $-\sqrt{3}P/2$ the force in $AD$ is tensile but any value less than this acceptable; take $F_{BD} = -0\cdot667P$. Diagram (c) shows the corresponding force pattern from which the required areas are deduced to be

$$A_{AD} = \frac{0\cdot23P}{0\cdot523 f_w} = 0\cdot44\ P/f_w$$

$$A_{BD} = \frac{0\cdot67P}{f_w} = 0\cdot67\ P/f_w$$

$$A_{CD} = \frac{0\cdot615P}{0\cdot523 f_w} = 1\cdot18\ P/f_w$$

This is a design which satisfies conditions (a) and (b) and in which one member, $BD$, is fully stressed. It is now interesting to see how variations in the force in $BD$ affect the economy and this is readily done by computing the volumes of the members for different values of $F_{BD}$. The areas just computed correspond to a total volume of $2\cdot35\ PL/f_w$. If the determinate frame $ADC$ (omitting the redundant $BD$) is designed for a working stress of $f_w$ its volume is found to be $2\cdot15\ PL/f_w$ so that with this value of $F_{BD}$ the hyperstatic arrangement is some 9 per cent. the heavier. When different values are assigned to $F_{BD}$ the volume is found to vary as shown in FIGURE B.2 from which it appears that the hyperstatic arrangement is lighter than the determinate frame $ADC$ when $F_{BD}$ lies between $-0\cdot74P$ and $-0\cdot866P$.

It will be noticed, however, that there is no force in $AD$ when $F_{BD}$ is $-0\cdot866\ P$ and the frame then reduces to the alternative determinate form $BDC$. There is then no need to reduce the working stress in $CD$ to $0\cdot523f_w$ in order to ensure compatibility at $D$ and the working stress in both $BD$ and $DC$ can be $f_w$. This gives a total volume of $1\cdot366\ PL/f_w$. This calculation suggests that the stress of $-0\cdot523f_w$ previously chosen was not the most favourable and another possibility is given in TABLE B.2.

NOTES ON THE DESIGN OF HYPERSTATIC STRUCTURES

This shows that if the stress in $AD$ is limited to $-0.317f_w$ the other two members can be fully stressed. The areas corresponding to different load patterns, and the corresponding volumes, can be worked out as before and are also shown in FIGURE B.2.

TABLE B.2

| Member | $L$ | $f$ | $F'$ | $F'fL/E$ |
|---|---|---|---|---|
| AD | $2L/\sqrt{3}$ | $-0.317 f_w$ | $-2/\sqrt{3}$ | $+0.423 f_wL/E$ |
| CD | $L$ | $-f_w$ | $-1/\sqrt{3}$ | $+0.577 f_wL/E$ |
| | | | $\Sigma$ | $= +1.00 f_wL/E = \Delta_{BD}$ |

This very elementary example illustrates the complexity of the task of designing a hyperstatic structure for maximum economy and it also demonstrates the principle that for a *single load system* redundant internal members have the effect of increasing the volume of metal in a pin-jointed structure. This principle follows from the consideration that compatibility generally restricts the stress in at least one member to a value lower than is permissible[13].

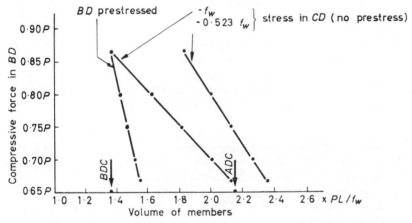

*Figure B.2.* The relation between the assumed load pattern, as defined by the compression force in *BD*, and the total volume of the members. The arrows give the volumes of the determinate frames *BDC* (i.e. *AD* omitted) and *ADC* (i.e. *BD* omitted)

We are thus led to see whether prestressing can improve the situation. Reverting to TABLE B.1 and the succeeding remarks we recall that the required reduction of the length of member $BD$ is $1.91f_wL/E$ if the other two members are to be fully stressed. If $BD$ also carries a stress $f_w$ it shortens by an amount $f_wL/E$ so that if it is to fit into the deformed frame $ACD$ it must be $0.91f_wL/E$ too short when inserted. This arrangement is sketched in FIGURE B.1(d). In the unloaded structure, therefore, $BD$ is in a state of tension but when the load $BD$ is applied all these members carry the compression stress $f_w$. The forces can again be assigned arbitrarily so as to satisfy equilibrium and the corresponding areas and volumes worked out. The results, which are also plotted on FIGURE B.2, show that the volume of the frame is now more nearly independent of the force pattern but the minimum possible weight is again when the member $AD$ is omitted.†

The next stage in the investigation would be to remove the restriction that the members are in compression but enough has now been said to show the ease with which different schemes can be studied by the present method. It is particularly to be observed that the basis of the method is that the deformations which control compatibility are assumed at the outset and that force patterns, and the corresponding cross-sectional areas, are then examined. The solution is, of course, valid only for a single system of applied loads; the effect of alternative loading systems can only be found by analysis.

In conclusion mention should be made of two other design methods that rely on the same principle of specified deformations. Bleich[14] gives a method for designing tall guyed masts which starts from the requirement that under wind the mast sways a specified amount into an inclined position while remaining straight. In this case the problem is complicated by the non-linear load-deformation characteristic of the guys. A rather different problem was studied by A. van der Neut[15] who described the design of the reinforcing frames of a compressed air wind tunnel on the basis of preliminary estimates of the distortion of the structure. In his discussion van der Neut mentions four limitations which restrict the utility of this method of design.

(1) The arbitrary nature of the external loads, the choice of redundant quantities and the distortions chosen as the starting point will in general result in a design having members of varying cross-section. The complication of manufacturing such members can only be justified if the resulting structure is particularly sound and economical.

(2) Among all the load systems to be supported by the structure one must be of predominating importance.

(3) For each condition of equilibrium at least one dimension must be at choice and this may not always be the case.

(4) The result of the design process may be that a structural member should have a negative stiffness. Since this is impossible a fresh start is required.

These considerations led van der Neut to conclude that design on the basis of assumed deformations may prove to have a rather narrow field of application. It should be recalled, however, that in writing thus he was no doubt thinking of the braced shells on which his investigation had been carried out.

---

† However this is a distinctly academic discovery since we are then dealing with a different frame.

## REFERENCES

[1] GRIOT, G. *Four place influence line tables* . . . Ungar, 1952
[2] KLEINLOGEL, A. *Rahmenformeln.* Ernst. Berlin
[3] KLEINLOGEL, A. *Mehrstielige Rahmen.* Ernst. Berlin
[4] GRAY, et al. *Steel Designers' Manual.* Crosby, Lockwood. 1955
[5] *The structural use of normal reinforced concrete in buildings.* British Standards Institution. CP 114 (1948)
[6] *The use of structural steel in building.* British Standards Institution. BS 449:1948
[7] CHARLTON, T. M. ' The analysis of structures with particular reference to the prediction of deflexions '. *Trans. N.E. Coast Inst. of Engrs and Shipbuilders.* Vol. 74. 1958. p. 163
[8] GRINTER, L. E. *Automatic design of continuous frames.* Macmillan 1939
[9] HAERTLEIN, A. ' Indeterminate truss design methods simplified '. *Engineering News Record.* Vol. 116. 1936. p. 816
[10] PIPPARD, A. J. S. *Strain Energy methods of stress analysis.* Longmans. 1928
[11] FRANCIS, A. J. ' Direct design of elastic statically indeterminate triangulated frameworks for single systems of loads '. *Aust. J. App. Sci.* Vol. 4. 1953. p. 175
[12] FRANCIS, A. J. ' Direct design of non-linear redundant triangulated frameworks '. *Aust. J. App. Sci.* Vol. 6. 1955. p. 13
[13] LIVESLEY, R. K. *The application of an electronic computer to problems of structural analysis and design.* Ph.D. Thesis. Univ. of Manchester. 1954
[14] BLEICH, F. *Stahlhochbauten.* Springer, 1932

## ADDITIONAL REFERENCES

SVED, G. ' The minimum weight of certain redundant structures '. *Aust. J. App. Sci.* Vol. 5. 1954. p. 1

LIVESLEY, R. K. ' The automatic design of structural frames '. *Quart. J. Mech. and App. Maths.* Vol. IV, Pt. 3. 1956

## Appendix C
# ENERGY THEOREMS BY THE ARGYRIS AND KELSEY APPROACH†

The subjects of structural analysis, stress analysis, stability analysis, plate and shell theory, beam theory, the theory of elasticity and so on are based upon four requirements. They are as follows.

(a) When a body is loaded each part or element of the body pushes (or pulls) against those to which it is joined. Since each element is in a state of rest the forces acting upon it must be in equilibrium. Each particle has 6 degrees of freedom (3 of translation along 3 non-coincident axes and 3 of rotation about the same 3 axes). In the case of plane problems each particle has 3 degrees of freedom (2 of translation and 1 of rotation). Corresponding to each of these degrees of freedom an equation of equilibrium can be written down. If each element of a body is in a state of equilibrium the whole of the body and indeed any part of the body is in equilibrium.

(b) When a body is loaded its elements deform usually by a small amount. It is observed that each element deforms in such a way that it fits with its neighbours who in their turn are deformed from their original shape. In other words the structure deforms so that its elements fit together without gaps (and without interference between adjacent elements). It is said that the condition of compatibility is thereby satisfied.

(c) The material has certain mechanical properties which are unique. There is a unique relationship between the stress applied to each element and the resulting strain. This relationship may or may not be linear: for steel at low stresses it is, for plastics it is not. Thus the third requirement of any theory of stress analysis is that the stress-strain relationship of the material must be satisfied. This relationship is usually determined by simple tests, e.g., tension, compression, bending or torsion tests. Here only elastic materials are considered, and by this it is meant that the stress-strain curve is unique and does not depend upon such phenomena as viscosity and hysteresis. In other words on a stress-strain diagram there is only one possible curve representing the states of the material.

(d) The body as a whole satisfies the known boundary conditions. Over the outer surface of the body at every point some information about the applied stress, the applied load, the deflexion in some direction (or directions) or the strain must be specified. What is really being said here is that across the boundary of the body the condition of equilibrium [condition (a) above] and the condition of compatibility [condition (b) above] must be satisfied. For example, if an external

---

† This appendix was contributed by Prof. N. W. Murray of Monash University, Australia.

stress is applied to the surface of a body, the stress just inside the surface acting in a direction which is normal to the surface is the same as the external stress.

If these four conditions are satisfied by a solution at every point in a body, then that solution gives the exact stresses, strains and displacements. Sometimes it is not possible to satisfy all four conditions, and such solutions are only approximate.

There are many energy principles used by engineers but each is based upon these four requirements. From the point of view of analysis it is expedient to look upon the four requirements as fundamental and thereby acknowledge that the energy principles are not fundamental. There are several ways of developing the theory of energy principles but here we first concentrate on the concepts which unify them. All energy principles can be derived easily from either the Principle of Virtual Displacements or from the Principle of Virtual Loads. These will be developed later, and will lead to the idea of the duality of all structural theory. Duality simply means that there are available to us two ways of solving a structural problem. In a later section it will be seen that for every principle or theorem there is a duplicate.

## C.1. LOADING A STRUCTURE

When a structure is loaded it deforms, because no material has an infinite elastic modulus. Before examining the mathematics of deformation it is necessary to have a clear understanding of the physics of the problem. The argument will be developed in general terms so that the conclusions will apply to any structure whose material is elastic and has a non-linear stress-strain curve and which may be prestressed mechanically or thermally. These kinds of cases occur in practice as a result of differential cooling of a weld or through differential settlement in a large building frame. Another example may be the lack of precise fit in the members of a truss. However, it is convenient in this section to consider only a simple linear cantilever shown in FIGURE C.1. Let us study what happens to the energy of the system (i.e., load $P$ and cantilever) when the load $P$ is pushed on to its end.

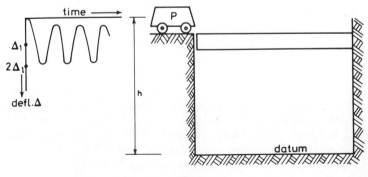

*Figure C.1.*

The graph of deflexion against time shows what would be recorded. Immediately after $P$ is applied it has potential energy $Ph$ where $h$ is the height above datum. Half way through its first downswing it has given up some of its potential energy and has acquired some kinetic energy. In this position the cantilever itself has bent and has thereby acquired some strain energy. At the bottom of the first downswing the weight $P$ has given up potential energy $2P\Delta_1$; it has no kinetic energy and the cantilever has gained some more strain energy which it gives up on the next upswing. If there were no losses from the system this process would go on indefinitely. However, eventually the cantilever does come to rest and the weight has given up potential energy equal to $P\Delta_1$. If the cantilever has a linear load-deflexion characteristic it absorbs strain energy equal to $\frac{1}{2}P\Delta_1$. Where has the other $\frac{1}{2}P\Delta_1$ gone? It has been lost from the system in the form of dissipated energy. (The reader should also consider what happens when he slowly places a load $P$ on the end of the cantilever.) Thus when we look at what happens externally it is seen that three energy quantities must be considered, viz.,

(a) the loss of potential energy of the applied loads.

These may also include the self weight of the structure.

(b) the strain energy gained by the structure in the process of deformation.

(c) the energy which is dissipated.

When the same kind of study is made of a structure whose load-deflexion characteristic is non-linear the same concepts are applicable. These three energies can be represented graphically as areas on a load-deflexion diagram as in FIGURE C.2.

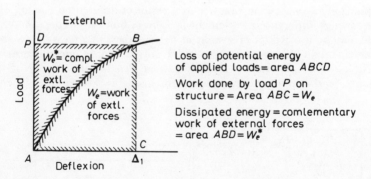

Figure C.2.

Now let us examine what happens internally. A small rectangular parallelopiped of volume $dV$ ($=dx\,dy\,dz$) will be deformed by the internal stresses and as a result will undergo energy changes. At the start of loading it may be in a state of prestress represented by point $A$ in the stress-strain diagram FIGURE C.3. Strictly speaking we can only represent one component of the stress at $A$ (i.e., one of the components shown in FIGURE 3.1(b) of Chapter 3) and the corresponding strain at $A$.

# C.1. LOADING A STRUCTURE

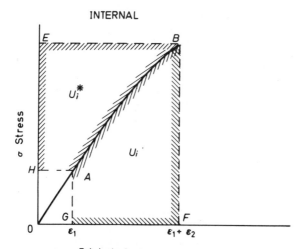

$U_i$ = Area $ABFG$ = change of S.E. of element
$U_i^*$ = Area $ABEH$ = change of compl. S.E. of element

*Figure C.3.*†

Here we use the symbols $\sigma$ and $\epsilon$ to denote corresponding components of stress and strain respectively. However, we could draw six such diagrams for the point $A$ and sum the energy changes. In this way we could calculate the strain energy absorbed by the element $dV$. By integrating this element of strain energy over the whole body the strain energy absorbed by the structure could then be calculated.

When the element changes from state $A$ to state $B$ its increase in strain energy is represented by area $ABFG$. The area $ABEH$ is called the change of the complementary strain energy of the element. The diagram also shows that the prestrain of the element is $\epsilon_1$, the change of strain due to loading is $\epsilon_2$ and the total strain $\epsilon$ is therefore $\epsilon_1 + \epsilon_2$ which is represented by $OF$. Similarly when the element is in state $B$ the total stress is represented by $OE$.

To calculate the change of strain energy of the element in going from state $A$ to state $B$ we must integrate the term $\sigma d\epsilon dV$ and sum for all of the components of stress and strain acting on the element.

i.e., the change of strain energy of the element in going from state $A$ to state $B$

$$= \left[ \sum \int_A^B \sigma d\epsilon \right] dV$$

For the whole structure therefore,

---

† In the text of Chapter 2 strain energy was given the symbol $U$. Here it is given the symbol $U_i$ to emphasise its *internal* nature. The suffix $e$ also emphasises the *external* nature of the work $W_e$. Over the whole structure it will be seen that $U = U_i = W_e$. Similar considerations apply to $C$, $U_i^*$ and $W_e^*$.

$U_i$ = internal strain energy absorbed by the structure

$$= \int_{\text{vol}} \left[ \sum \int_A^B \sigma \, d\epsilon \right] dV$$

Similarly,

$U_i^*$ = internal complementary strain energy of the structure

$$= \int_{\text{vol}} \left[ \sum \int_A^B \epsilon \, d\sigma \right] dV$$

Finally we recall the following quantities
- $\epsilon_1$ = strain of element when the body is unloaded but is prestressed
- $\epsilon_2$ = strain of element due to loading apart from prestress
- $\epsilon$ = $\epsilon_1 + \epsilon_2$ = total strain
- $\sigma$ = total stress in the element, i.e., prestress plus loading stress
- $W_e$ = work of external forces (see FIGURE C.2)
- $W_e^*$ = complementary work of external forces

## C.2. THE EFFECT OF SMALL CHANGES IN A LOADED STRUCTURE

It is not very often that we are interested in the absolute values of quantities such as work $W_e$ of the external forces and the internal strain energy of the structure. Usually what is of greater interest in the loaded structure are the effects of small changes in applied loads, stresses, deflexions and strains. In FIGURES C.2 and C.3 the point $B$ represents the condition of the loaded structure. We shall study the effect of changing the conditions at $B$ by small amounts.

Suppose the structure is prestressed, heated and loaded so that externally and internally conditions are represented by state $B$ (FIGURE C.4). At $B$ the structure is at rest and satisfies the conditions of equilibrium, compatibility, stress-strain relationship, and the boundary conditions. If, however, the external loads are changed or in some other way the state of the loaded structure is varied from that at $B$ to that at $J$, the structure will still have to satisfy these four conditions at $J$. Thus the *variations* of stress, strain, deflexion load, etc., must satisfy equilibrium, compatibility, etc.

In going from $B$ to $J$ the variation in internal strain energy is:
$$\begin{aligned}\delta U_i &= \text{Area } BJQR \\ &= \text{Area } BMQR + \text{Area } BMJ \\ &\approx \text{Area } BMQR \\ &= \sigma_B d\epsilon\end{aligned}$$

where it is understood that the summation over all stress components and integration over the whole volume of the structure should be carried out as discussed earlier. This equation is important because it shows that to calculate changes of $U_i$ we do not have to consider changes of stress but only

## C.2. EFFECT OF SMALL CHANGES IN A LOADED STRUCTURE

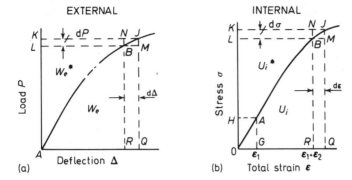

*Figure C.4.*

changes of strain. In other words, when considering small changes in internal strain energy, equilibrium is automatically satisfied but it is necessary to ensure that the variations of strain satisfy compatibility. Similarly by considering external effects the variations in deflexion must form a compatible set but there is no need to consider the condition of equilibrium during these variations. It is also important to see that when $dU_i$ is calculated that the *total stress* $\sigma$ is used, i.e., that we do not fall into the error of neglecting prestressing.

By similar reasoning applied to the areas above the curve $ABJ$ it is seen that when considering small changes in complementary energies compatibility is automatically satisfied but we must ensure that the variations of stresses and forces satisfy equilibrium. Also FIGURE C.4 shows that to calculate $dW_e^*$ and $dU_i^*$ any set of forces and stresses can be used (so long as they satisfy equilibrium) but the *total strain* $\epsilon$ must be used in that calculation and the strain $\epsilon_1$ due to prestressing must not be neglected.

The disturbance which is introduced to the loaded structure can be of any form. Thus in FIGURE C.4(a) although $ABJ$ is drawn as a smooth curve this need not be the case.

For example consider the cantilever shown in FIGURE C.5(a). Its non-linear load deflexion characteristic is shown in FIGURE C.5(b). A large load $P$ is applied at its end $a$. The small disturbance can be in the form of a

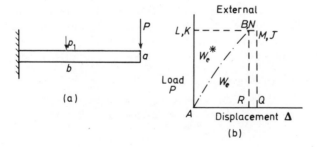

*Figure C.5.*

very small load $P_1$ applied at an arbitrary point $b$ in the cantilever. The load-deflexion curve is then $ABJ$ as shown ($M$ and $N$ coincide with $J$ and $B$ respectively in this case). The previous argument is equally applicable. The small increase in vertical displacement due to the disturbance is a function of $x$ and will satisfy the condition of compatibility. Such a displacement is called a virtual displacement or a set of virtual displacements. The corresponding strains are called virtual strains or a set of virtual strains. The essential ingredient of such sets is that they are compatible. Similarly if a small load $P_1$ is applied this will result in small increases in the internal stresses and the bending moments and shear forces. These increases are in equilibrium with the small increment in applied load. They are called virtual loads, virtual stresses or sets of virtual loads or a set of virtual stresses. Their essential ingredient is that as a set they satisfy equilibrium conditions.

As stated at the beginning of this section it is rare that the energy expressions themselves are of interest. It is more usual that we want to know their variations arising from virtual displacements or virtual loads. Virtual displacements are used to calculate variations in internal strain energy and the external work of the applied loads, i.e., $\delta U_i$ and $\delta W_e$. Virtual loads are used to calculate variations in internal complementary energy and the complementary energy of external forces. These two approaches lead to the idea of duality in the energy principles. In the next section it will be seen that each energy principle (and there are several of them) is based upon either the Principle of Virtual Displacements or the Principle of Virtual Loads.

## C.3. ENERGY THEOREMS AND PRINCIPLES

In this section the principle which arises from a variation of strain or displacement is dealt with first and its counterpart arising from a variation of stress or force will follow with label B. In this way duality in the energy principles will be seen. It should also be noted that the first two principles form the basis for the remainder. It is assumed that the load-deflexion and stress-strain curves are non-linear but that the structure is elastic.

### 1.A. Principle of Virtual Displacements

"An elastic body is in equilibrium if for any virtual displacements from a compatible state $\delta W_e = \delta U_i$."

This follows by applying the law of conservation of energy to our previous reasoning. Thus the strain energy absorbed by the structure during a virtual displacement can be calculated either by calculating the work of the external loads, i.e., $\delta W_e$, or by calculating the change in the internal strain energy, i.e., $\delta U_i$.

### 1.B. Principle of Virtual Loads

"An elastic body is in an elastically compatible state under a given system of forces, prestress and temperature distribution if for any system of virtual forces and stresses $\delta W_e^* = \delta U_i^*$."

The proof follows in the same way as for 1.A above.

## 2.A. Principle of Stationary Potential Energy

"Any small variation of displacements from a state of equilibrium does not give rise to any first order variation of total potential energy."

This follows immediately from 1.A because it is simply another way of looking at the principle of virtual displacements. If we think of $\delta W_e$ as a change of potential energy, i.e., $\delta U_e$, of the external loads then the total energy of the loaded structure is $U = U_i + U_e$. Thus, from 1.A a virtual displacement implies $\delta U = 0$. This principle is illustrated diagrammatically in Figure C.6. The strain energy $U_i$ of a structure increases with deflexion as shown (for a linear structure the curve is a parabola) and the potential energy $U_e$ of the loads decreases with deflexion as shown. The principle of stationary potential energy states that the equilibrium position of a loaded structure is at $B$ where the slopes of the $U_i$ and $U_e$ curves are equal and opposite. Any attempt to deflect the structure away from the equilibrium position $B$ will result in an increase in the total energy of the system but the increase is only of second order. It is seen from this diagram that such a structure requires the addition of energy to disturb it from $B$. It is said that the structure is stable at $B$.

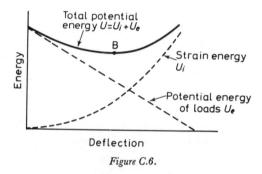

Figure C.6.

## 2.B. Principle of Stationary Complementary Energy

"Any small variation of internal forces from an elastically compatible state does not give rise to any first order variation of total complementary energy."

This principle is proved in the same way as 2.A. The quantity $\delta W_e^*$ is replaced by $-\delta U_e^*$. It should be noted that this principle cannot be applied to statically determinate structures because the internal force system is fixed by equilibrium conditions. For a statically indeterminate structure the arrangement of redundant forces can be changed while at the same time satisfying the equilibrium conditions. The diagram comparable to Figure C.6 is that illustrated in Figure C.7. It shows for example what happens in the case of a continuous beam over three supports when the central support is assumed to take the different values shown on the x-axis. $U_i^*$ is the complementary strain energy of the structure (see Figure C.3) and the quantity $U_e^*$ is the complementary energy of the external loads. $U_e^*$ is in fact equal to the potential energy $U_e$ of the external loads. The difference only becomes apparent when we calculate their differentials.

$\delta U_e$ = change of potential energy of external loads $P_n$ due to a virtual displacement $\delta\Delta_n$

$$= \sum_{\substack{\text{all} \\ \text{loads}}} P_j \delta\Delta_j$$

$\delta U_e^*$ = change of complementary energy of external loads $P_n$ due to virtual loads $\delta P_n$

$$= \sum_{\substack{\text{all} \\ \text{loads}}} \Delta_j \delta P_j$$

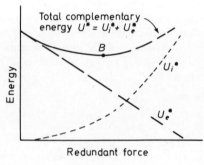

Figure C.7.

## 3.A. Castigliano's Theorem Part 1 (1879)

"In a structure carrying loads $P_j$ the displacements at the load points *and measured in the directions of the applied loads* are

$$\Delta_j, \frac{\partial U_i}{\partial \Delta_j} = P_j"$$

The proof of Castigliano's theorem part 1 follows 1.A. Since $\delta U_i = \delta W_e$ and during a virtual displacement which is zero under all of the loads except $P_j$ we have $\delta W_e = P_j \delta \Delta_j$. Hence, in the limit,

$$\frac{\partial U_i}{\partial \Delta_j} = P_j.$$

## 3.B. First Theorem of Complementary Energy

$$"\frac{\partial U_i^*}{\partial P_j} = u_j"$$

The proof follows from 1.B. $\delta U_i^* = \delta W_e^* = \Delta_j \delta P_j$ and hence the theorem. This theorem is also called the first theorem of complementary energy.

Another theorem which is a direct corollary of the first theorem of complementary energy is called *Engesser's theorem of compatibility*. It is used to find redundant forces in statically indeterminate structures. If a statically indeterminate structure has one member which is too short by an amount $\lambda$ and the force in that member is $X$ then

$$\frac{\partial U_i^*}{\partial X} = \lambda.$$

This theorem can be extended to structures with many degrees of statical indeterminacy, in which case

$$\frac{\partial U_i^*}{\partial X_1} = \lambda_1, \quad \frac{\partial U_i^*}{\partial X_2} = \lambda_2, \quad \frac{\partial U_i^*}{\partial X_3} = \lambda_3, \text{ etc.}$$

In the event of a statically indeterminate structure having no initial lack of fit we have

$$\frac{\partial U_i^*}{\partial X_j} = 0 \text{ for all } j.$$

(This was the form originally presented by Engesser in 1889.) In other words for a structure with no lack of fit the complementary energy expressed as a function of the redundant forces $X$ has a stationary value when compatibility is satisfied.

### 4.A. First Theorem of Minimum Internal Strain Energy

"If $\Delta_j$ is a set of virtual displacements which is zero at the applied loads, at equilibrium the internal strain energy is a minimum with respect to $\Delta_j$, i.e.,

$$\frac{\partial U_i}{\partial \Delta_j} = 0 \quad \text{and} \quad \frac{\partial^2 U_i}{\partial \Delta_j^2} > 0\text{"}$$

If $\delta u_r$ is thought of as a set of virtual displacements and since they are zero at the applied loads then the work $\delta W_e$ of the external loads during this virtual displacement is zero. Hence

$$\delta U_i = \frac{\partial U_i}{\partial \Delta_j} \delta \Delta_j$$

is zero also and it follows that $U_i$ has a stationary value when equilibrium is satisfied. It will be recalled in Figure C.4 that we neglected the small area $BMJ = \frac{1}{2} d\sigma d\epsilon$. Provided that this second order area is positive, any changes will result in a slight increase in $U_i$. The area is positive if the slope of the stress-strain curve is positive which is the case for all structural materials. Hence $U_i$ has a minimum value at equilibrium.

### 4.B. Theorem of Minimum Internal Complementary Strain Energy

"If $X$ is a set of internal virtual stresses (i.e., they are self-equilibrating) or internal forces in a framework the internal complementary strain energy $U_i^*$ is a minimum with respect to $X$," i.e.,

$$\frac{\partial U_i^*}{\partial X} = 0$$

and

$$\frac{\partial^2 U_i^*}{\partial X^2} > 0$$

If the body forces $\omega$ and the surface forces $\phi$ are kept constant but the self equilibrating set of internal stresses $X$ is changed slightly then $\delta W_e^* = 0$ and $\delta U_i^* = (\partial U_i/\partial X)\,\mathrm{d}X$ must also be zero. The small area $BNJ$ in FIGURE C.4 is positive for structural materials so the remainder of the argument is similar to that for 4.A.

## 5.A. APPROXIMATE METHODS

These methods which rely upon the principle of virtual displacements can be lumped together into what is known as the Rayleigh-Ritz method. In this method the form of the deflexion of the structure is guessed and expressed as an equation. For example;

$$\Delta = \Delta_0 + \sum_{n=1}^{N} a_n \Delta_n,$$

where $\Delta_0$ satisfies the kinematic boundary conditions (i.e., conditions which refer to displacements and rotations at the boundary) where these are specified and $\Delta_n$ is zero there but it does not necessarily satisfy the equilibrium conditions at the boundary. The next step in the argument is to express the internal strain energy $U_i$ and the work of the external forces $W_e$ as functions of

$$\Delta = \Delta_0 + \sum_{1}^{N} a_n \Delta_n.$$

A set of virtual displacements can be set up by changing one and only one of the coefficients $a_n$. Thus $\delta\Delta = \Delta_n \delta a_n$. As a result of this virtual displacement the principle of virtual displacements can be applied. The net result is an equation for the coefficient $a_n$ and the process is repeated for all other coefficients.

An alternative way of looking at the Rayleigh-Ritz method is to use the principle of stationary potential energy. The total potential energy $U$ is expressed as a function of $a_1$, $a_2$, $a_3$, etc. For a stationary value of $U$

$$\frac{\partial U}{\partial a_1} = \frac{\partial U}{\partial a_2} = \frac{\partial U}{\partial a_3} = \ldots 0$$

## 5.B. APPROXIMATE METHODS OF STRESS ANALYSIS †

An Airy stress function $F$ is a function which is related in a simple manner to the internal stresses of a loaded body. Its most important property, however, is that it enables the equations of equilibrium to be satisfied. If we assume an Airy stress function of the form $F = F_0 + \Sigma b_n F_n$ where $F_0$ satisfies the stress boundary conditions and $F_n$ produces zero stress there we

---

† This technique is included here for the sake of completeness. It may be ignored by those not familiar with Airy stress functions.

can use $dF = F_n db_n$ to obtain variations of stresses (i.e., virtual stresses). With these virtual stresses we can apply the principle of virtual loads or use the principle of stationary complementary energy.

To summarise this section, it is seen that:

(a) there is duality between the theorems.

(b) one set are derived from the principle of virtual displacements, the other from the principle of virtual loads.

(c) the methods apply to non-linear and linear elastic structures.

There are many methods available but it is not necessary for the reader to be able to use every one of them. It is essential to understand the first two methods and it is a matter of convenience which others are used. Engineers tend to use Castigliano's and Engesser's theorems and the Rayleigh-Ritz method more frequently than the others.

It is fortunate for engineers that most structures behave linearly because there are several other theorems and principles which apply only to linear structures. Thus in these cases we do not have to be careful about which theorem is used. However, when the behaviour of a structure is non-linear none of the theorems and principles listed in the next section should be used.

## C.4. SPECIAL THEOREMS FOR LINEAR STRUCTURES

So far the energy principles have been developed for structures and materials which are non-linear. It has been seen that a change of strain energy is associated with a virtual deformation or a virtual strain whereas a change in complementary energy is associated with a virtual force or a virtual stress. In most structures the load-deflexion and stress-strain curves are straight lines through the origin and for these cases $\delta U_i^* = \delta U_i$ and $\delta W_e^* = \delta W_e$. This simple relationship gives rise to at least four further theorems and methods which are, of course, only valid for linear structures.

6. CASTIGLIANO'S THEOREM PART II

$$"\frac{\partial U_i}{\partial P_j} = u_j,"$$

i.e., the deflexion of a linear structure under a load is the partial derivative of the internal strain energy with respect to that load."

This follows by replacing $\delta U_i^*$ in the first theorem of complementary energy with $\delta U_i$.

7. UNIT DUMMY LOAD METHOD

"The deflexion at a point $A$ in a linear structure can be found by applying a unit load in the required direction. The work of the internal forces which arise from this unit load acting through the corresponding internal displacements which arise from the external loads is equal to the displacement at $A$."

The proof of this method is obtained by using Castigliano's theorem part II or perhaps more easily by looking upon the unit load and the internal forces arising from it as an applied load and the displacements arising from the external loads as a set of virtual displacements. A simple application of the principle of virtual displacements proves the method. The basis for this method will become clearer when a few examples are studied in detail.

## 8. Castigliano's Theorem of Compatibility

"In a linear statically indeterminate structure

$$\frac{\partial U_i}{\partial X_r} = \lambda_r$$

where $\lambda_r$ is the initial lack of fit of the member (measured in the same direction as $X_r$)."

The proof of this follows directly from Engesser's theorem if we substitute $\delta U_i$ for $\delta U_i^*$, a step which is only valid for linear structures.

## 9. The Theorem of Least Work or the Theorem of Minimum Strain Energy

"In a linear structure which is statically indeterminate and has no initial lack of fit

$$\frac{\partial U_i}{\partial X_r} = 0."$$

This theorem follows directly from Castigliano's theorem of compatibility applied to a linear structure in which the members fit exactly. It should be pointed out that because of these quite severe restrictions (which fortunately do not often arise in practice except in prestressed structures) this theorem is far from being a universal law.

## C.5. DISCUSSION OF THE THEOREMS AND PRINCIPLES

Many of the ideas contained in the last two sections are clarified when specific examples are considered. In tackling a problem it is first necessary to decide whether the structure is linear or non-linear. If the latter, theorems 6 to 9 must be avoided. The general theorems and principles follow a pattern and it is not necessary to remember which are general and which are not. In calculating $\delta U_i$ or $\delta W_e$ we use variations of strain or displacement which lie below the curves shown in Figure C.4. Similarly for $\delta U_i^*$, $\delta W_e^*$, stress and load all lie above these curves. Thus by inspection, an equation such as

$$\frac{\partial U_i}{\partial u_j} = P_j$$

is a general theorem whereas

$$\frac{\partial U_i}{\partial P_j} = u_j$$

is only applicable to linear structures.

The second point to decide is whether the structure is prestressed due to initial lack of fit, intentional prestressing or residual thermal stresses. This prestressing must be taken into account when selecting the energy theorem used. It will be noted that the theory has been developed on the assumption that prestressing or thermal effects took place before the external loads were applied. Thus the deflexions $u$ were measured from the position assumed by the structure after prestressing, but before applying the external loads. In other words $u$ is the deflexion due to the external loads only. By developing the argument in this way the initial points $A$ are arranged as shown in FIGURE C.4. In some cases we may be interested in temperature effects on a structure which is already carrying its external load. This changes the interpretation of the term "load" on the diagram. If we are only interested in the total effect of prestressing and external loading on an elastic structure, the order of loading is not important and it is convenient to imagine that prestressing has been carried out first.

Further points which will become clearer when examples are dealt with are:

(a) the principle of virtual displacements or any of the principles based upon it will only give equations of equilibrium

(b) the principle of virtual loads or any of the principles based upon it will only give compatibility conditions.

Therefore, when attempting to analyse a statically indeterminate structure we cannot expect to obtain the solution by using only the principle of virtual displacements. It is equally pointless to try to find the forces in a statically determinate structure by making use of the principle of virtual loads. However, this latter principle can be used to find the deflexions in a statically determinate structure.

Finally we can ask what is the significance of duality and where does it lead us? When a statically indeterminate structure is analysed it has two types of redundancies, viz.,

(a) *statical redundancies* which is a system of internal bending moments and forces $(F)$ in the members of the structure

(b) *kinematical redundancies* which are the deflexions and rotations $(\Delta)$ at the joints of the structure.

It is possible to choose either the set $F$ or the set $\Delta$ as unknowns in the analysis of the structure. The first of these duals is the basis of the flexibility method of analysis and the second the stiffness method.

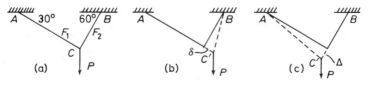

Figure C.8.

## C.6. APPLICATIONS OF THE ENERGY METHODS

*Example 1*: Use the principle of virtual displacements to find the forces $F_1$ and $F_2$ in the bars of the pinjointed frame [FIGURE C.8(a)].

A virtual displacement is one which satisfies compatibility. Allow $AC$ to expand a small amount $\delta$ as shown in FIGURE C.8(b) but at the same time ensure that $BC$ does not expand. Hence $P$ moves vertically down an amount of $\delta/2$. From the principle of virtual displacements, assuming $F_1$ and $F_2$ are tensile,

$$\delta W_e = \delta U_i$$

i.e., $P \cdot (\delta/2) = F_1 \delta$

whence $F_1 = \dfrac{P}{2}$

Similarly from the virtual displacement shown in FIGURE C.8(c)

$$P\left[\frac{\Delta\sqrt{3}}{2}\right] = F_2 \Delta$$

whence $F_2 = \dfrac{P\sqrt{3}}{2}$

Notice the following points.

(a) The principle of virtual displacements results in equilibrium conditions, *not* compatibility equations.

(b) The virtual displacement $\Delta$ appears on each side of the work equation and therefore is cancelled. This *always* happens.

(c) It was convenient to select a pair of virtual displacements in the directions of the members $AC$ and $BC$. Any other pair, e.g., a horizontal and a vertical displacement, would arrive at the same answer.

*Example 2*: Use the principle of virtual displacements to calculate the reaction at $A$ for the simple beam shown in FIGURE C.9(a) and the cantilever bridge which has a hinge at $B$ [FIGURE C.9(b)]. Also sketch the influence lines for the reactions at $A$.

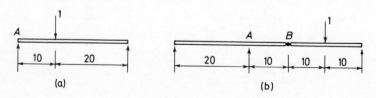

*Figure C.9.*

(a) The beam can be treated as a rigid body and given a virtual displacement $\varDelta$ of unity upwards at $A$ [FIGURE C.10(a)]. From the principle of virtual displacements $\delta W_e = \delta U_i$, i.e., $R_A \varDelta = 1 \cdot \frac{2}{3}\varDelta$.
Hence $R_A = \frac{2}{3}$.

When the unit load is applied at any other point on the beam the reaction at $A$ is found in the same manner. Thus, the dotted line in FIGURE C.10(a) (representing the position of the beam after a virtual displacement) is the influence line for the reaction at $A$.

(b) Again introduce a virtual displacement $\varDelta$ of unity at $A$ [FIGURE C.10(b)]. From purely geometric considerations the displacement under the load is $0.75\varDelta$ and hence ($\delta W_e = \delta U_i$)

$$R_A \varDelta = 1 \times 0.75\varDelta \qquad \text{i.e.} \quad R_A = 0.75$$

The influence line is the dotted line shown in FIGURE C.10(b).

*Figure C.10.*

*Example 3*: Use the principle of virtual displacements to obtain the bending moment and shear force at $B$ [FIGURE C.11(a)].

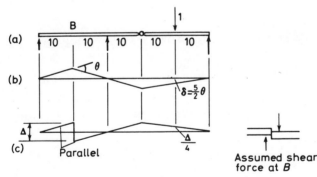

*Figure C.11.*

An imaginary cut is made in the beam at $B$. This changes the beam into a mechanism which can be displaced as shown in FIGURE C.11(b). The angle through which the moment at $B$ acts is $\theta$ as shown. Hence $\delta U_i = M_B \theta$ during the virtual displacement. The work done by the load is $\delta W_e = 1 \cdot \delta = 5\theta/2$. Thus from the principle of virtual displacements $M_B = 5/2$. The sign of $M_B$ is obtained by equating the work done by the external load to the work done *on* the moment $M_B$ during the displacement.

The shear force is obtained by making a cut and introducing the virtual displacement shown in FIGURE C.11(c). Unless the segments of the beam on either side of the cut are parallel after the displacement the moment at $B$ will contribute some energy to the system

$$V_B . \varDelta = -1 . \frac{\varDelta}{4}$$

i.e., $V_B = -1/4$ meaning the shear force is opposite in sign to that shown in FIGURE C.11(c).

*Example 4*: Use the principle of virtual displacements to find the forces $F_1$, $F_2$, $F_3$ in the members of the pin-jointed frame [FIGURE C.12(a)]. All members have the same length but the areas of cross section are $2a$, $a$, $4a$ respectively.

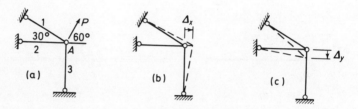

Figure C.12.

Introduce the virtual displacements indicated in FIGURE C.12(b) and (c), and use the principle of virtual displacements, i.e. $\delta W_e = \delta U_i$ assuming the forces are tensile

for $\varDelta_x$:  $P\varDelta_x \cos 60° = F_1 \varDelta_x \cos 30° + P_2 \varDelta_x + P_3 . 0$

$$\text{Hence} \quad \frac{P}{2} = F_1 \frac{\sqrt{3}}{2} + F_2 \tag{1}$$

for $\varDelta_y$:  $P(-\varDelta_y) \sin 60° = F_1 \varDelta_y \sin 30° + F_2 . 0 + F_3(-\varDelta_y)$

$$\frac{P\sqrt{3}}{2} = -\frac{F_1}{2} + F_3 \tag{2}$$

These are two equilibrium equations but since the structure is statically indeterminate it is necessary to introduce a condition of compatibility. One way is to allow the load point $A$ to take up its position when the structure is fully loaded. If this deflexion of $A$ is $\delta_x$ to the right and $\delta_y$ vertically downwards, compatibility is satisfied by ensuring that the three members change their length such that they still join at the new position of $A$. Mathematically we do this by saying that the extensions of the three members are

$$e_1 = \delta_x \cos 30° + \delta_y \sin 30°, \; e_2 = \delta_x, \; e_3 = -\delta_y$$

## C.6. APPLICATIONS OF THE ENERGY METHODS

The stress-strain relationship is introduced at this point of the argument to calculate the stresses

$$\sigma_1 = \frac{Ee_1}{l} = \frac{E}{l}[\delta_x \cos 30° + \delta_y \sin 30°],$$

$$\sigma_2 = \frac{Ee_2}{l} = \frac{E\delta_x}{l}, \quad \sigma_3 = \frac{Ee_3}{l} = -\frac{E\delta_y}{l}$$

Hence the loads in the members can be related to the actual displacement components (viz., $\delta_x$ and $\delta_y$) at $A$ as follows.

$$F_1 = \frac{2aE}{l}[\delta_x \cos 30° + \delta_y \sin 30°], \quad F_2 = \frac{aE\delta_x}{l}, \quad F_3 = -\frac{4aE\delta_y}{l}$$

On substituting into equations (1) and (2) and solving it is found that

$$\delta_x = 0.286\frac{Pl}{aE}, \quad \delta_y = -0.247\frac{Pl}{aE}$$

Hence $F_1 = 0.248P$, $F_2 = 0.286P$ and $F_3 = 0.988P$.

An alternative way of establishing an equation of compatibility needed to complement the equilibrium equations (1) and (2) is to use the principle of virtual loads. This equation can be set up in the following manner. With the loads applied as in FIGURE C.13(a), introduce a disturbance in the form of a small horizontal load $\delta P$ acting to the right at $A$. This load may be equilibrated by an internal force or forces so as to form a system of virtual loads. This system does not have to satisfy compatibility—only the condition of equilibrium. Therefore we first choose to equilibrate $\delta P$ by the internal force $\delta F_2$ as shown in FIGURE C.13(a).

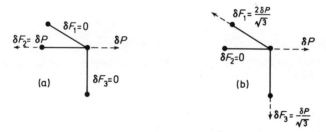

Figure C.13.

From the principle of virtual loads $\delta W_e^* = \delta U_i^*$

i.e., $\delta P . \delta_x = \int_{\text{vol. of structure}} \text{strain} \times \text{change of stress due to } \delta P$

$$= \left[\frac{F_1}{2aE}\right][0][2al] + \left[\frac{F_2}{aE}\right]\left[\frac{\delta P}{a}\right][al] + \left[\frac{F_3}{4aE}\right][0][4al]$$

i.e., $\delta_x = \frac{F_2 l}{aE}$

This is an obvious result and is a compatibility condition. A second compatibility condition can be obtained by equilibrating $\delta P$ by the system of forces shown in FIGURE C.13(b). As before

$$\delta P \cdot \delta_x = \left[\frac{F_1}{2aE}\right]\left[\frac{2\delta P}{\sqrt{3} \cdot 2a}\right]\left[2al\right]$$

$$+ \left[\frac{F_2}{aE}\right]\left[0\right]\left[al\right] + \left[\frac{F_3}{4aE}\right]\left[\frac{\delta P}{\sqrt{3} \cdot 4a}\right]\left[4al\right]$$

$$\delta_x = \frac{F_1 l}{2aE} \cdot \frac{2}{\sqrt{3}} + \frac{F_3 l}{4aE} \cdot \frac{1}{\sqrt{3}}$$

Equating these expressions for $\delta_x$ gives the third equation for which we have been looking.

$$\frac{F_1}{\sqrt{3}} - F_2 + \frac{F_3}{4\sqrt{3}} = 0 \qquad (3)$$

By solving equations (1), (2) and (3) the same result as before is obtained. This last equation can be derived in a number of ways. As an exercise the reader should apply a small load $\delta P$ vertically at $A$ and equilibrate it in two alternative ways. As another exercise the reader should obtain equation (3) above by introducing a small load $\delta P$ into one of the members (say member 2) and equilibrate it by small loads in the other two members. In this latter case $\delta W_e^*$ is zero during the application of the virtual loads.

*Example 5*: Show that for a uniform beam ($EI$ is constant) a set of virtual loads which produce a bending moment $m$ will change the internal complementary strain energy by an amount

$$\delta U_i^* = \int_{\text{length}} \frac{Mm}{EI} \, ds$$

Use this expression to solve the propped cantilever problem illustrated in FIGURE C.14(a).

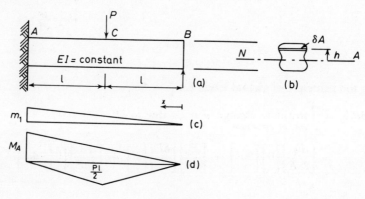

Figure C.14.

## C.6. APPLICATIONS OF THE ENERGY METHODS

The expression for $U_i^*$ in this case simplified to

$$\delta U_i^* = \int_{\text{vol}} \left[\int \epsilon \, d\sigma\right] dV$$

where $\epsilon =$ strain due to external loads $= \dfrac{Mh}{EI}$

and $d\sigma =$ stress due to a system of virtual loads $= \dfrac{mh}{I}$

Hence $\quad \delta U_i^* = \displaystyle\int_{\text{length}} \int_{\text{cross section}} \left[\left(\dfrac{Mh}{EI}\right)\left(\dfrac{mh}{I}\right)\right] dA \, ds$

$$= \int_{\text{length}} \dfrac{Mm}{EI} \, ds$$

since $I = \int h^2 \, dA$

The propped cantilever problem is solved by using this expression in conjunction with the principle of virtual loads. $m$ can be any bending moment system which satisfies equilibrium but obviously it is convenient to choose the simplest. A small decrease in the reaction at $B$ will change the bending moment diagram of FIGURE C.14(d) by that shown in FIGURE C.14(c).

$$m = m_1 \dfrac{x}{2l}, \quad M = M_A \dfrac{x}{2l} - \dfrac{Px}{2} \qquad \text{(B to C)}$$

$$= M_A \dfrac{x}{2l} - \dfrac{P}{2}[2l - x] \qquad \text{(C to A)}$$

$$\delta U_i^* = \dfrac{1}{EI} \int_0^l \left[M_A \dfrac{x}{2l} - \dfrac{Px}{2}\right] \dfrac{m_1 x}{2l} \, dx + \int_l^{2l} \left[M_A \dfrac{x}{2l} - \dfrac{P}{2}(2l - x)\right] \dfrac{m_1 x}{l} \, dx$$

$$= \dfrac{m_1 l}{3EI}\left[2M_A - \dfrac{3Pl}{4}\right]$$

Since the virtual loads are simply a rearrangement of the redundant forces $\delta W_e^* = 0$ and from the principle of virtual loads $\delta U_i^* = 0$. Thus

$$M_A = \dfrac{3Pl}{8}$$

Integrals such as those shown above are sometimes tedious to calculate. It is convenient to evaluate them by means of a table of volume integrals. Readers are urged to check the value of $\delta U_i^*$ by using the following table.

Table of volume integrals $\int_0^L F_1(x)F_2(x)dx$

| $F_2(x) \diagdown F_1(x)$ | $a$ rectangle $L$ | $a$ triangle (right) $L$ | $a$ trapezoid $b$ $L$ |
|---|---|---|---|
| $c$ rectangle $L$ | $Lac$ | $\dfrac{Lac}{2}$ | $\dfrac{Lc(a+b)}{2}$ |
| $c$ triangle (right) $L$ | $\dfrac{Lac}{2}$ | $\dfrac{Lac}{3}$ | $\dfrac{Lc(2a+b)}{6}$ |
| triangle $c$ $L$ | $\dfrac{Lac}{2}$ | $\dfrac{Lac}{6}$ | $\dfrac{Lc(a+2b)}{6}$ |
| $c$ trapezoid $d$ $L$ | $\dfrac{La(c+d)}{2}$ | $\dfrac{La(2c+d)}{6}$ | $\dfrac{La(2c+d)+Lb(c+2d)}{6}$ |
| Parabolic $c$ $d$ $e$ $L/2$ $L/2$ | $\dfrac{La(c+4d+e)}{6}$ | $\dfrac{La(c+2d)}{6}$ | $\dfrac{La(c+2d)+Lb(2d+e)}{6}$ |

*Example 6*: The pin-jointed steel truss shown in FIGURE C.15 has all of its tension members loaded to 12 units of stress and all of the compression members to 10 units of stress. Find the deflexion at $D$.

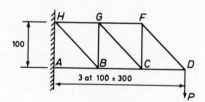

Figure C.15.

The first theorem of complementary energy states;

$$\frac{\partial U_i^*}{\partial P_j} = \Delta_j$$

It may be assumed that the steel structure behaves in a linear manner and therefore Castigliano's theorem part II may be used, i.e.,

$$\frac{\partial U_i}{\partial P_j} = \Delta_j$$

488

## C.6. APPLICATIONS OF THE ENERGY METHODS

$$U_t = \sum_{\text{all members}} \frac{F_n^2 l_n}{2A_n E}$$

i.e.,
$$\Delta_D = \frac{\partial U_t}{\partial P} = \sum_{\text{all members}} \frac{\partial}{\partial F_n}\left[\frac{F_n^2 l_n}{2A_n E}\right]\frac{dF_n}{dP}$$

$$= \sum_{\text{all members}} \frac{F_n l_n}{A_n E}\left[\begin{array}{l}\text{force in member due}\\ \text{to unit load at } D\end{array}\right] \quad (1)$$

| Member | Length | $F_n$ | $dF_n/dP$ | $F_n/A$ | $F_n l_n / A_n$ $\times\ dF_n/dP$ |
|---|---|---|---|---|---|
| DC | 100 | $-P$ | $-1$ | $-10$ | 1,000 |
| CB | 100 | $-2P$ | $-2$ | $-10$ | 2,000 |
| BA | 100 | $-3P$ | $-3$ | $-10$ | 3,000 |
| GF | 100 | $P$ | 1 | 12 | 1,200 |
| GH | 100 | $2P$ | 2 | 12 | 2,400 |
| DF | $100\sqrt{2}$ | $P\sqrt{2}$ | $\sqrt{2}$ | 12 | 2,400 |
| CG | $100\sqrt{2}$ | $P\sqrt{2}$ | $\sqrt{2}$ | 12 | 2,400 |
| BH | $100\sqrt{2}$ | $P\sqrt{2}$ | $\sqrt{2}$ | 12 | 2,400 |
| CF | 100 | $-P$ | $-1$ | $-10$ | 1,000 |
| BG | 100 | $-P$ | $-1$ | $-10$ | 1,000 |

Thus
$$\Delta_D = \frac{18,800}{E}$$

This framework is statically determinate and equation (1) has been derived from Castigliano's theorem part II. It is interesting to note that equation (1) can be given another interpretation. The quantity $\frac{F_n l_n}{A_n E}$ is the extension of the $n$th member due to the load $P$. These extensions (and compressions) together with the displacement of the frame at $D$ form a compatible set of displacements. They can therefore be used as a set of virtual displacements. Thus by removing $P$ and replacing it with a unit load the principle of virtual displacements can be applied ($\delta W_e = \delta U_i$).

1. $\Delta_D = \sum \left[\frac{F_n l_n}{A_n E}\right] \times$ (force in member due to unit load at $D$)

and equation (1) is obtained directly.

*Example 7*: Given that the strain energy of a linear elastic beam $AB$ is

$$U_i = \int_A^B \frac{M^2 \, dx}{2EI}$$

(see equation 3.7), find the deflexion at the end of a uniform cantilever by using the Rayleigh-Ritz method.

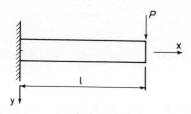

Figure C.16.

Using the arrangement of axes shown in FIGURE C.16 the boundary conditions are

$$\text{at } x = 0 \quad y = \dot{y} = 0 \qquad \left(\dot{y} = \frac{dy}{dx}, \quad \ddot{y} = \frac{d^2y}{dx^2}\right)$$

$$x = l \quad \ddot{y} = 0 \qquad \text{(since the bending moment = 0)}$$

These conditions can be satisfied by assuming the shape of the cantilever is

$$y = a_1 \left[1 - \cos\frac{\pi x}{2l}\right] + a_3 \left[1 - \cos\frac{3\pi x}{2l}\right] + \ldots = \sum_{1,3,5} a_n \left[1 - \cos\frac{n\pi x}{2l}\right]$$

The strain energy of the cantilever is (since $M = EI\ddot{y}$)

$$U_i = \int_0^l \frac{M^2}{2EI} dx = \frac{EI}{2} \int_0^l (\ddot{y})^2 dx$$

i.e.,

$$U_i = \frac{EI}{2} \sum_{1,3,5} \int_0^l a_n^2 \left[\frac{n\pi}{2l}\right]^4 \cos^2\left[\frac{n\pi x}{2l}\right] dx$$

because $\int_0^l \cos\frac{r\pi x}{2l} \cos\frac{s\pi x}{2l} dx = 0 \ (r \neq s)$, and $= \frac{l}{2}$ when $r = s$.

Hence

$$U_i = \sum_{1,3,5} a_n^2 \frac{n^4 \pi^4 EI}{64 l^3}$$

A virtual displacement can be applied to the cantilever from its deflected shape by changing the coefficients $a_n$ in turn. Hence the virtual displacement introduced by changing $a_n$ only is

$$\delta a_n \left[1 - \cos\frac{n\pi x}{2l}\right].$$

## C.6. APPLICATIONS OF THE ENERGY METHODS

At $x = l$ this amounts to a downwards displacement of $\delta a_n$. Hence, during the virtual displacement $\delta W_e = P\delta a_n$. The change of internal strain energy is

$$\delta U_i = a_n \cdot \frac{n^4 \pi^4 EI}{32 l^3} \delta a_n.$$

From the principle of virtual displacements ($\delta W_e = \delta U_i$) we obtain

$$a_n = \frac{32}{\pi^4 n^4} \cdot \frac{Pl^3}{EI}$$

i.e.,
$$y = \frac{Pl^3}{EI} \cdot \frac{32}{\pi^4} \sum \frac{1}{n^4} \left[ 1 - \cos \frac{n\pi x}{2l} \right]$$

At $x = l$,
$$y = \frac{32 Pl^3}{\pi^4 EI} \left[ 1 + \frac{1}{3^4} + \frac{1}{5^4} + \cdots \right]$$

If we take only the first three terms

At $x = l$
$$y = \frac{Pl^3}{3 \cdot 001 EI}$$

*Example 8*: Given a standard pipe bracket (FIGURE C.17) having a vertical load of 1,000 at $A$, a plane second moment of area 7·23 and a polar second moment of area 14·46 find the deflexion at $A$. ($N = 0·4E$)

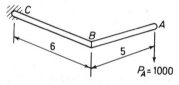

Figure C.17.

Since the structure is linear Castigliano's theorem part II, viz.,

$$\frac{\partial U_i}{\partial P_A} = \Delta_A$$

can be used.

In AB
$$U_i = \int_0^5 \frac{(P_A x)^2}{2EI} dx$$

$$\frac{\partial U_i}{\partial P_A} = \frac{P_A}{EI} \int_0^5 x^2 dx = 5 \cdot 76 \frac{P_A}{E}$$

In BC
$$U_i = \int_0^6 \frac{(P_A y)^2}{2EI} dy + \int_0^6 \frac{(5 P_A)^2}{2 N I_p} dy$$

$$\frac{\partial U_i}{\partial P_A} = \frac{P_A}{EI} \int_0^6 y^2 dy + \frac{25 P_A}{N I_p} \int_0^6 dy = 35 \cdot 87 \frac{P_A}{E}$$

Thus the total
$$\frac{\partial U_i}{\partial P_A} = 41\cdot 63 \frac{P_A}{E}$$

when $P_A = 1{,}000$ therefore the deflexion at $A$ is $41{,}630/E$.
It is found that 62% of this deflexion comes from the torsion of $BC$. Therefore, a designer wishing to reduce the deflexion at $A$ should increase the torsional rigidity of $BC$.

*Example 9*: A uniform beam rests across three supports as shown. What is the effect of a subsidence of 0·1 in the central support FIGURE C.18(a)?

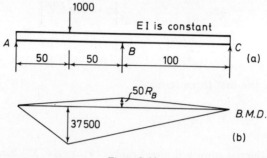

*Figure C.18.*

The beam behaves linearly and therefore Castigliano's theorem of compatibility can be used, i.e.,

$$\frac{\partial U_i}{\partial R_B} = -0\cdot 1;$$

the minus sign arises because $R_B$ and the settlement are measured in opposite directions. The beam is first made statically determinate by removing $R_B$. On applying an upwards force $R_B$ at $B$ the bending moment diagram is as shown in FIGURE C.18(b).

$$U_i = \int \frac{M^2}{2EI} \, ds$$

i.e. $\quad -0\cdot 1 = \dfrac{\partial U_i}{\partial R_B}$

$$= \frac{1}{EI} \int M \frac{\partial M}{\partial R_B} \, ds$$

$$= \frac{1}{EI}\left[ \int_0^{50} \left(-750x + R_B\frac{x}{2}\right)\frac{x}{2}\,dx \right.$$

$$+ \int_{50}^{100} \left(250x - 50{,}000 + \frac{R_B x}{2}\right)\frac{x}{2}\,dx$$

$$+ \left. \int_{100}^{200} \left(250x - 50{,}000 - R_B\frac{x}{2} + 100 R_B\right)\left(-\frac{x}{2} + 100\right)dx \right]$$

## C.6. APPLICATIONS OF THE ENERGY METHODS

$$= \frac{1}{EI}\left[-\frac{750x^3}{6} + \frac{R_B x^3}{12}\right]_0^{50}$$

$$+ \frac{1}{EI}\left[\frac{250x^3}{6} - \frac{50,000x^2}{4} + \frac{R_B x^3}{12}\right]_{50}^{100}$$

$$+ \frac{1}{EI}\left[-\frac{250x^3}{6} + \frac{50,000x^2}{2} + \frac{R_B x^3}{12} - \frac{R_B 100 x^3}{2}\right.$$

$$\left. + 10,000 R_B x - 5,000,000 x\right]_{100}^{200}$$

$$= \frac{1}{EI}[-113,900,000 + 166,700 R_B]$$

i.e. $\quad R_B = \dfrac{-0 \cdot 1 EI + 113,900,000}{166,700} = -0 \cdot 6 EI \times 10^{-6} + 687$

# NAME INDEX

Abeles, P., 330
Alexander, G. W., 38, 84
Allen, D. N. de G., 188, 329
Allen, H. G., 407, 420
Amerikian, A., 331, 375, 381
Anger, G., 331
Antono, A., 331
Argyris, J. Hadji., 85, 447, 448

Baker, J. F., 84, 221, 230, 232, 319, 327, 330, 332, 380
Banerjee, S. P., 331
Bateman, E. H., 265, 329
Batho, C., 309, 330
Beaufoy, L. A., 331, 375, 381
Beggs, G. E., 223, 232
Benscoter, S. U., 432, 447
Berry, A., 389, 420, 442
Bettess, F., 331
Bickley, W. G., 15, 32
Bleich, F., 331, 466, 467
Bolton, A., 2, 32, 293, 299, 329, 406, 420, 442, 448
Bowles, R. E., 305, 330
Brisby, M. D., 331
Brown, E. H., 33, 60, 70, 71, 72, 76, 80, 84
Bryan, C. W., 232, 352, 381
Bull, M. G., 232

Calisev, K. A., 264
Case, J., 124
Castigliano, A., 69, 75, 84
Chandler, D. B., 392, 410, 420
Charlton, T. M., 61, 71, 72, 84, 85, 138, 222, 232, 330, 467
Chen, P. P., 439, 447
Chitty, L., 380
Chu, K-H., 115, 137
Clapeyron, B. P. E., 238, 239, 329
Clough, R. W., 448
Clyde, D. H., 261, 286, 329
Cornish, R. J., 198, 232, 247, 305, 329, 330, 331
Cox, Roxbee, 15, 32
Crandall, S. H., 428, 447
Cross, Hardy, 85, 153, 233, 264, 265, 329, 359, 367, 381
Csonka, P., 331

de la Hire, P., 332
Diwan, A. F. S., 331, 375, 381

Engesser, F., 71, 84
English, J. M., 327, 330
Ennis, L., 318, 330

Fidler, Claxton, 302, 329
Fife, W. M., 8, 32
Fowler, K. T., 375, 381
Francis, A. J., 331, 359, 381, 462, 467
Freeman, R., 318, 330

Gerstner, F. J., 332
Godfrey, G. B., 331
Goldstein, A., 163, 188
Gray, C. S., 331, 467
Griffith, A. A., 375, 381
Grinter, L. E., 286, 329, 375, 381, 462, 467
Griot, G., 331, 467

Haertlein, A., 462, 467
Hartree, D. R., 428, 447
Henderson, J. C. de C., 15, 32
Hendry, A. W., 331
Heyman, J., 327, 330
Hoadley, A., 133, 137, 346
Horne, M. R., 327, 330, 420

James, B. W., 405, 420
Jenkins, Sefton, 188
Johannes, H., 331
Johnson, J. B., 232, 352
Jones, E., 247, 329

Kazinczy, G., 319
Kelsey, S., 85, 468
Kent, L. E., 331
King, J. W. H., 64, 84
Kist, N. C., 319
Kleinlogel, A., 331, 462, 467
Kloucek, C. V., 331
Kron, G., 446, 448

Lamb, H., 50, 84
Landdeck, N. E., 232
Lightfoot, E., 329, 331, 375, 381
Livesley, R. K., 392, 420, 421, 441, 442, 443, 447, 448, 467
Lundquist, E., 405, 420

Macaulay, W. H., 120, 137
Magnel, G., 188
Maier-Leibnitz, H., 319
Makowski, Z. S., 232
Manderla, H., 249

495

# NAME INDEX

Maney, G. A., 249, 329
Manning, G. P., 331
Markland, E., 460
Martin, H. C., 448
Matheson, J. A. L., 18, 32, 79, 85, 264, 285, 329, 331
Maugh, L. C., 331, 375, 381
Maxwell, J. C., 153
Merchant, W., 2, 32, 409, 420, 442, 448
Michalos, J. P., 331
Mitchell, W. A., 331
Mohr, O., 154, 249, 264
Moorman, R. B. B., 359, 375, 381
Morgan, N. B., 381
Morice, P. B., 85, 448
Moseley, H., 332
Müller-Breslau, H., 154, 156, 189, 197
Murray, N. W., 382, 420

Navier, L. M. H., 3
Naylor, N., 290, 292, 299, 329
Neal, B. G., 55, 84, 261, 327, 330
Niles, A. S., 85, 153
Norris, C. H., 2, 11, 32

Palmer, P. J., 232
Parcel, J. I., 359, 375, 381
Pippard, A. J. S., 221, 222, 223, 230, 232, 332, 333, 380, 462, 467
Pollock, P. J., 38, 84
Portland Cement Association, 460

Rankine, J. M., 153
Rashid, C. A., 420
Redshaw, S. C., 232
Rocha, M., 232
Roderick, J. W., 420
Rowan, H. C., 329
Ruge, A. C., 230, 232
Ruppel, W., 460

Salmon, E. H., 302, 330
Samuely, F. J., 188

Schmidt, E. O., 230, 232
Scruton, C., 232
Shaw, F. S., 188
Shepley, E., 306, 330
Skayannis, A. P., 331
Sourochnikoff, B., 330
Southwell, R. V., 1, 32, 39, 69, 70, 84, 143, 166, 265, 309, 420
Steinman, D. B., 304, 305, 306, 330
Stevens, L. K., 133, 137
Stewart, R. W., 331
Straub, H., 332, 380
Sved, G., 467
Symonds, P. S., 55, 84, 327, 330

Taylor, G. I., 375, 381
Timoshenko, S., 2, 11, 15, 24, 32, 52, 84, 137, 264, 303, 329, 332, 342, 350, 380, 381
Topp, L. J., 448
Tranter, E., 380
Tse, K. F., 220, 228, 232
Turneaure, F. E., 232, 352, 381
Turner, J. M., 442, 447, 448

van den Broek, J. A., 84, 319, 330
van der Neut, A., 466
Villarceau, Y., 332

Waddell, J. A. L., 264
Webb, H. A., 124
Westergaard, H. M., 43, 71, 84
Wilbur, J. B., 2, 11, 32
Williams, D., 33, 70, 80, 84, 85, 447, 448
Wilson, G., 156, 188
Wilson, E. H., 331
Wilson, W. M., 3, 249, 329
Winkler, E., 189, 249, 332
Wood, R. H., 308, 329, 420
Wright, J., 232

Young, D. H., 2, 11, 15, 32, 52, 84, 303, 329, 350, 381

# SUBJECT INDEX

Aircraft structures, 442, 446, 461
Alloy, light, 23
Arch, 332
  bowstring, 21
  circular, 347
  effect of axial thrust on, 351
  fixed, 332, 336, 353, 361
  linear, 334, 337, 338
  parabolic, 348
  polygonal, 345, 360
  spandrel-braced, 207
  three-hinged, 8, 50, 335, 339, 343, 359
  tied, 350
  two-hinged, 207, 336, 345
  types of, 335
Arches, interconnected, 375
Automatic design, 462

Beam,
  continuous, 238, 252, 270, 303
  continuous, matrix analysis of, 432
  continuous, on elastic supports, 245
  with fixed ends, 6, 235, 243, 323, 333, 370, 450
Beam-line method, 309
Beams, hyperstatic, 233
Bending moment diagrams, geometry of, 300
Berry functions, 389, 442
Betti's Reciprocal Theorem, 102, 227, 364
Bow's notation, 1, 111

Camber, 104, 108
Cantilever, 36
  effect of axial load on the behaviour of, 386
  propped, 3, 27, 233, 326
Carry-over factor, 266, 304, 308
Castigliano's Theorem of Compatibility
                  ('Least Work'), 79, 80, 83, 84, 138, 176, 177, 183
                  Theorem of Equilibrium, 69
                  Theorem, Part I, 69, 70, 83
                  Theorem, Part II, 75, 83, 98, 176, 184, 345, 354
Centre,
  elastic, 217, 336, 361, 377
  instantaneous, 52, 217, 263
  of moments, 193
  of twist, 102
c-function, 386, 387, 388, 390
Characteristic Point, 301
Cleats, 310, 312, 313, 315

Collapse loads,
  analysis of, 318, 325
  determination by combination of elementary mechanisms, 327
  determination by trial, 326
Column analogy, 217, 359, 367
  comments on, 374
Compatibility, 21, 33, 43, 70, 138, 252, 354, 358, 433, 437
Compatibility method, 138, 153, 421, 447
Composite action, 320
Concrete, 24
Conjugate point, 301, 305
Contraflexure, point of, 27
Convention of signs, 56, 105, 116, 127, 158, 164, 173, 190, 249, 251, 253, 310, 354, 372, 373, 377, 422, 432, 436
Coordinate transformations, 436
Core theory, 342
Corresponding displacements, 56
Creep, 222, 231
Critical configuration, 7, 11
Critical load, 386, 399, 408, 419
Curved member, 377
$\Delta_{ik}$ method, 170, 177


Deflexion,
  by complementary energy, 73
  by integration of beam equation, 117
  by strain energy, 75
  by virtual work, 60
  general expression for, 97
  of beams, 115
  of beams, standard cases, 118
  of hyperstatic structures, 61, 184
  of non-uniform beams, 450
  of non-linear structures, 73, 76
  of pin-jointed frameworks, 103
  point of maximum, 121
Degree of fixity, 306
Design,
  collapse or plastic, 54, 318, 462
  of hyperstatic structures, 461
Determinant, 8, 426
Diagram, load-extension, 23
Digital computer, 387, 421, 443, 444, 447
Direction cosines, 442
Displacements, large, 46
Displacement system, 56
Distribution factor, 268

# SUBJECT INDEX

Duality, 469
Dummy unit load, 99, 104, 115, 124, 202
Dynamic problems, 446

Elastic centre, 355, 356, 359, 364, 371, 372, 374, 375
    instability, 319, 382, 442, 443
    weight, 355
Energy, 33
    complementary, 34, 71
    complementary, first theorem of, 71, 73, 74, 83
    conservation of, 34
    kinetic, 33
    potential, 33, 41, 70, 71
    strain, 33, 34, 35, 69, 75, 92
    strain, due to axial forces, 93
    strain, due to bending and shearing, 94
    strain, under combined stress, 94
Energy theorems, 33, 39
    relations between, 83, 84
Energy, theorem of minimum potential, 40, 57, 82, 83, 84, 138, 142, 163
Energy, theorem of minimum strain, 82, 83
Engesser's Theorem of Compatibility, 76, 83, 138
Equilibrium method, 138, 421, 447
Euler load, 383, 410, 442
Extrados, 341

Fixed point, 301, 303, 308
Flexibility coefficient equations, 153, 177, 226, 245
Flexibility of beams and cleats, 312, 314
Flexural centre, 101
Focal point, 301, 305
Force, 33
Force system, 56, 57, 58
Forces, slowly and suddenly applied, 37
Four moments, theorem of, 240, 330, 331
Fourier series, 412
Frame,
    combination of pin- and rigidly jointed, 18
    multi-storey, 30, 292
    portal, 17, 29, 178, 253, 258, 272, 328
    rigidly jointed, 15, 233, 440, 441
Frameworks, with large axial forces in members,
    actual behaviour of, 411
    combined elastic and plastic behaviour of, 416
    elastic analysis of, 412, 416
    plastic analysis of, 414, 418
Freedom, degree of, 50, 52, 277
Froude Number, 220

Gauss-Seidel method, 429
Graphical integration, 459

Hinge, plastic, 54, 321
Hoadley's method, 133, 345, 347
Hooke's Law, 23, 86, 96
Hyperstatic primary structure, 224
Hyperstatic structures, definition of,

Indirect loading, 192
Influence coefficients, 156
Influence lines, 189
    for arch, 343, 364
    for beams, 189, 199, 218
    for continuous beams, 247, 462
    for hyperstatic structures, 199
    for structures with two redundant reactions, 212
    for trusses, 193, 201, 209, 218
Intrados, 342
Intuitive methods, 27

Jack, 35, 143
Joint equations, 252
Joint,
    pinned, 18
    rigid, 18
    sliding, 18
Joints, method of, 11

Kern point, 342
K-value, 268, 303

Lack of fit of member, 77, 163
Lateral buckling, 409
Lateral loads, 30
Least work, theorem of, 79, 82, 84
Linear algebra, 422
Linked rigidity, 304
Load divide, 195
Load factor, 319, 325

Macaulay's method, 120, 122, 212
Matrix, 423
    inverse of, 426, 430
    methods, 421
    stiffness, 433, 439, 445
    transformation, 428, 437, 440, 442
    unit, 426
Maximum principle, 326, 327
Maxwell's reciprocal theorem, 100, 140, 152, 199
Maxwell-Mohr equations, 138, 154, 158, 166, 176, 178
Mechanism,
    beam, 327, 328
    joint, 328
    sway, 327, 328
Member equations, 252
Mid-ordinate rule, 457

## SUBJECT INDEX

Minimum principle, 326, 327, 328
Mode, collapse, 55
Models of structures, 217
  in which axial stress predominates, 222
  in which bending stress predominates, 222
Model scale factors, 229
Models, structural, 217, 375
  for determining critical loads, 410
  of hyperstatic structures, 219, 221, 224, 229
  of statically determinate structures, 218
Mohr's equation of virtual work, 56, 57, 59, 70, 74, 92, 97, 104, 109
  in terms of axial force, bending moment and shear, 90
Moment-area method, 126, 133
Moment distribution analysis,
  of frames with sway, 277
  of frames without sway, 269
Moment distribution method, 264, 305, 385, 387, 390, 453
Moment,
  fixing, 235, 252, 450
  indicator, 230
  of resistance, plastic, 321, 415
  support, 243
Moving loads, 189
Müller-Breslau's Principle, 197
Müller-Breslau's technique for the special selection of redundants, vii, 214, 216

Naylor's method, 290, 292
Newton's laws of motion, 33
Non-linear structures, 26, 43, 442
Non-uniform beams, 132, 265, 449
No-shear pattern, 296, 298, 299

Oscillation, 38

Parallel, members in, 312
Partitioning techniques in matrix analysis, 444
Plastic design, 318
Plastic hinge, 54, 321, 322, 328, 414
Pratt truss, 201
Pre-stressing, 163, 465
Pre-stressing test, 11
Principal axis, 356, 364, 367, 368

Quasi-external force, 77, 78, 158, 186

Rayleigh-Ritz, 478
Reciprocal relationships, 100, 435
Redundancy, external, 3
  of space structures, 15
Redundancy, internal, 8
  of space structures, 15
Redundancy, internal and external combined, 13, 17
  in space structures, 15

Redundant members, 2
  restraints, 2
Relaxation, 143, 265, 429
  direct, of framework, 150
  operations, 145
  operator, group, 148
  operator, unit block, 148
  solution of linear simultaneous equations, 145
Residuals, 145
Restraint line, 310
Restraints, reactive, 2, 5
Reynolds Number, 220
Ropes, winding, 38

Scalar, 423
Secondary moment, 249, 396
Sections, method of, 11, 193
Series, members in, 312
s-function, 386
Shape factor, 321
Similarity,
  elastic, 218
  kinematic, 218
  of bending deformation, 220
  of extensional deformation, 219
Simpson's rule, 349, 456, 457
Simultaneous equations, 143, 264, 354, 390, 428
Slope-deflexion equations, 140, 249, 310, 317, 375, 385, 390, 432, 451
  application to frames with sway, 255
  application to frames without sway, 251
Southwell plot, 410
Space structures, 15, 166, 441
Spring, 35
Stability of struts and frameworks, 382
Statics, equations of, 1, 5, 33
Steel, structural, 24, 52, 54, 319
Stiffness coefficient equations, 138
Stiffness,
  beam, 266
  factor, 375, 453
Stress components, 1, 17, 18
Stress,
  primary, 16
  secondary, 16
Structure,
  primary, 21, 207, 212, 240, 336, 353, 359, 371
  primary, auxiliary, 356, 363
Subsidence, 240, 246, 275
Substitute frame, 274, 309
Suffix notation, 423
Superposition,
  of deflexions, 39
  of forces, 39, 74
  principle of, 23, 26, 38, 62, 384, 389
Support, fixed, hinged, roller, 5, 18

## SUBJECT INDEX

Sway, 251, 391, 453
    effects, correction by proportion, 277
    effects, correction by succession approximation, 286
    equation, 255, 257
Sydney Harbour Bridge, 318

Temperature effects, 108, 165, 351
Tension coefficients, 165
Theorem complementary to Castigliano's first theorem, 69
Theorem, first, of minimum strain energy, 70
Theorem, general, in mechanics, 70
Theorems governing plastic collapse, 325
Three moments, equation of, 238, 303
Trapezoidal rule, 456

Truss, complex, 11
    compound, 10
    roof, 52
    Vierendeel, 17, 21

Vector, 423
    column, 432
Velocities, virtual, 50
Viaduct, 375
Voussoir, 333

Williot-Mohr diagram, 109, 133, 198, 202, 210, 214, 217
Work, 33, 35, 69
    virtual, vii, 49, 70, 325, 328
    virtual, application to statically determinate frameworks, 50, 197
    virtual, in terms of stresses, 86

Yield, 319, 321, 322